Engineering Thermodynamics
Work and Heat Transfer

By the same authors

Thermodynamic and Transport Properties
of Fluids: SI (Blackwell)

Thermodynamic Properties of Fluids
and other Data: Btu (Blackwell)

By H. Cohen, G. F. C. Rogers and H. I. H. Saravanamuttoo

Gas Turbine Theory: 2nd Edition, SI (Longman)

Engineering Thermodynamics Work and Heat Transfer

SI Units

G. F. C. Rogers   **Y. R. Mayhew**

Professor of Engineering   *Senior Lecturer*
Thermodynamics

Department of Mechanical Engineering, University of Bristol

Longman

London and New York

Longman Group Limited London

*Associated companies, branches and representatives
throughout the world*

*Published in the United States of America
by Longman Inc., New York*

First published (British thermal units) 1957 (Eight impressions)
Second edition (SI Units) 1967 (Nine impressions)
Third edition 1980

British Library Cataloguing in Publication Data

Rogers, Gordon Frederick Crichton
    Engineering thermodynamics. — 3rd ed.
    1. Heat engineering 2. Thermodynamics
    I. Title   II. Mayhew, Yon Richard
    621.4'021   TJ265   79-40493

    ISBN 0-582-30500-4

Printed in Great Britain by
Richard Clay (The Chaucer Press) Ltd,
Bungay, Suffolk

Preface to Third Edition

Although minor additions and improvements have been made to the text and references to take account of recent work and current applications of thermodynamics, no substantial changes have been made in this Third Edition. Its prime purpose is to convey to students the outcome of much international effort to standardise nomenclature, units and methods of presenting technical data. The importance of this should not be underestimated. If the ever-increasing output of technical literature is to be assimilated, unremitting attention must be paid to ways of making it easily digestible. Readers can no longer afford the time to battle with ambiguities and idiosyncracies. It is vital that students—the authors of tomorrow—should acquire good habits during their undergraduate studies.

When preparing the Second Edition in 1966 it was necessary to anticipate certain decisions relating to the introduction of SI (Système Internationale d'Unités). The authors failed to anticipate the choice of K instead of °K as a symbol for the kelvin unit of temperature, and failed to foresee the new definition of the mole. (The formal decisions were taken in 1968 and 1971 respectively.) This Third Edition has afforded the authors the opportunity to bring the book up to date in these respects, and in others associated with a more rigorous attitude to the concept of a physical quantity and its units which has repercussions in the way equations should be written and data presented in graphs and tables. The reader is referred particularly to the rewritten Appendix A on Dimensions and Units.

G.F.C.R.
Y.R.M.

Bristol 1980

Preface to Second Edition

This edition of the book in SI units has been produced in the hope that it will play its part in furthering the change from British to metric units which is to take place during the next few years. Examples and problems have been worked out using data contained in the Second Edition of *Thermodynamic and Transport Properties of Fluids* (*SI Units*) (Basil Blackwell, 1967). The first edition of the book, and the tables in British units, will continue to be available as long as required.

While preparing this edition, the opportunity has been taken to bring the material up-to-date in some major and many minor respects. A large part of Chapter 8 has been rewritten to accommodate changes in the approach to the definition of temperature units and the adoption of the triple point as the datum state for steam tables. The treatment of regenerative cycles has been revised. Material has been added to Chapters 7 and 15 to include an introduction to the concept of availability and the closely associated use of the Gibbs function and other chemical thermodynamic data: the section on calorific values has been brought into line with the new British Standard. A chapter has been added on 'Direct Conversion'; although inevitably little more than descriptive, it is hoped that it will encourage some students to pursue the subject further as a postgraduate study. Owing to the widespread use of computers, numerical methods of dealing with unsteady heat flow problems have been substituted for the less flexible graphical methods described in the previous edition. Additional material is included on liquid metals as heat transfer media, and on boiling. Various other topics which have assumed importance since the book was first written, such as cycles for nuclear power plant and the Wankel engine, have also been included, and the references have been revised to bring greater benefit from 'further reading'.

The general plan of the book is unchanged: the authors still believe that classical macroscopic thermodynamics provides the best introduction to the subject for mechanical engineers, and that any real attempt to deal with statistical thermodynamics and irreversible thermodynamics is best left to postgraduate courses.

We wish to acknowledge the invaluable help given by Dr M. Hollingsworth, who reworked the Problems of Appendix B in SI units to provide the reader with answers. And once again we thank Mr C. J. Spittal, our librarian, for his work in checking the references.

G. F. C. R.
Y. R. M.

Bristol 1967

Preface to First Edition

This book is intended for engineering students and covers the fundamentals of applied thermodynamics courses up to Honours Degree standard. The following presentation has been adopted.

An exposition of the principles of thermodynamics is given in Part I without reference to the behaviour of any particular substance. Part II opens with a discussion of the properties of fluids, and the principles of Part I are then applied to closed and open systems assuming that the properties are best related (a) in tables and (b) by simple equations. This is followed by an analysis of power and refrigeration cycles. Part II ends with a study of gas and vapour mixtures and combustion processes.

Parts I and II are chiefly concerned with relations between fluid properties and the quantities of work and heat which accompany changes of state. The way in which the work and heat transfers are effected in practice is described in Parts III and IV respectively. Part III is divided broadly into work transfers associated with (a) positive-displacement machines and (b) nonpositive-displacement machines. For the latter, a more detailed analysis of flow processes is required than is given in Parts I and II, and therefore a chapter on one-dimensional flow is included. This chapter is designed as a bridge between the sciences of thermodynamics and fluid dynamics and, to avoid obscuring the fundamental links, concepts such as total temperature and Mach number have not been used. The usual classification of modes of heat transfer is adopted in Part IV, with the emphasis placed on methods of approach rather than on the presentation of established empirical data.

We do not suggest that our order of presentation is suitable as a direct introduction to the subject, and the book is not designed as a substitute for a course of lectures. Most students are best introduced to a subject by a careful juxtaposition of principle and application, and the book leaves it open to the lecturer to choose any preferred order of presentation. But, although a mixture of principle and application is necessary in a lecture course, it can leave the student with serious misconceptions, and it by no means follows that he will gain a sufficient grasp of the principles to be able to tackle unfamiliar problems. Our object has been to produce a book which will enable the student to distinguish principle from application: a reading of the first six chapters, as revision late in his first year, should enable him to make this distinction. It should also help him to appreciate the elegance and power, as well as the limitations, of the science he is using.

Our second concern has been to provide a book of reference to which the student can turn for the elucidation of difficult points. A rigorous approach cannot often be used in a lecture course, partly for lack of time and partly because it would be inappropriate when introducing the student to a new idea. In this book, however, a rigorous proof or explanation is given wherever practicable for the benefit of the better student who is not satisfied with simplifications.

A few words are necessary about two of the appendixes. Many of the problems provided in Appendix B are intended to amplify points only touched on in the text. To this extent they should be regarded as an integral part of the book. The answers to the problems, and to the worked examples in the text, have been obtained using the abridged tables of properties arranged by us and published separately—*Thermodynamic Properties of Fluids and Other Data* (Basil Blackwell, 1957). Most of the references, given in Appendix C, are to be regarded as suggestions for further reading and also as acknowledgement of sources. A few classical papers of academic interest are also included.

Finally we wish to acknowledge our particular indebtedness to two books which we have found a source of inspiration: J. H. Keenan's *Thermodynamics* (Wiley, 1941), and E. R. G. Eckert's *Introduction to the Transfer of Heat and Mass* (McGraw-Hill, 1950). We also wish to thank our colleagues at Bristol—Dr. B. Crossland, Mr. T. V. Lawson, and Dr. P. Woodward—for their useful suggestions after reading parts of the manuscript. Mr. C. J. Spittal, of the University Library, has been most helpful in obtaining papers and checking the references.

G. F. C. R.

Y. R. M.

Bristol, 1957

Prologue—Why Study Thermodynamics?

Since the Industrial Revolution there has been an exponential growth of population and of material wealth. This growth has been associated with the ever-increasing use of energy in machines. The revolution began with coal, and has progressed through the use of petroleum, natural gas and uranium. Hydroelectric and geothermal power have made, and can make, only a small contribution on a world scale, although highly significant to certain countries with no indigenous resources of fossil fuel. Easily exploited reserves of both fossil fuel and uranium are limited, and many will approach exhaustion in the lifetime of present students. There are two obvious consequences: firstly, ways have to be found of using our energy resources more efficiently, and secondly, in the long term, other sources of power must be developed.

It is the science of thermodynamics which enables us to deal quantitatively with the analysis of machines which are used to convert chemical or nuclear energy into useful work. It is therefore an essential study for those hoping to improve the effectiveness with which we use our existing energy resources. The optimisation of total energy schemes involving combined electricity production and process or district heating, or the assessment of combined gas and steam turbine plant ('Cogas' plant), to take but two examples, inevitably involve thermodynamic analysis. One of the important consequences of the Second Law of Thermodynamics is that unnecessarily large temperature differences used for the transfer of heat incur a waste of energy. A knowledge of the associated subject of heat transfer, dealt with in Part IV of this book, is also essential in an energy-conscious world.

Thermodynamics is likely to fill a vital role in the solution of the long-term problem also. It may not help in itself to throw up suggestions for new sources of power, but it is an essential tool for evaluating correctly the potential of new ideas. For example, it will be used to assess proposals for more effective ways of using solar energy, whether in devices which convert it into electricity, or less directly by photosynthesis in plant life from which alcohol or methane fuels can be obtained. It will also be used if ways can be found of harnessing the fusion process. Like fission, fusion will provide heat which can be converted to electricity in more or less conventional power plant. Such solutions would provide mankind with safe and non-polluting sources of power for an indefinitely long period.

Are these massive sources of power really needed? Certainly it is possible for a few small groups to live in self-supporting communes. However, if we care for

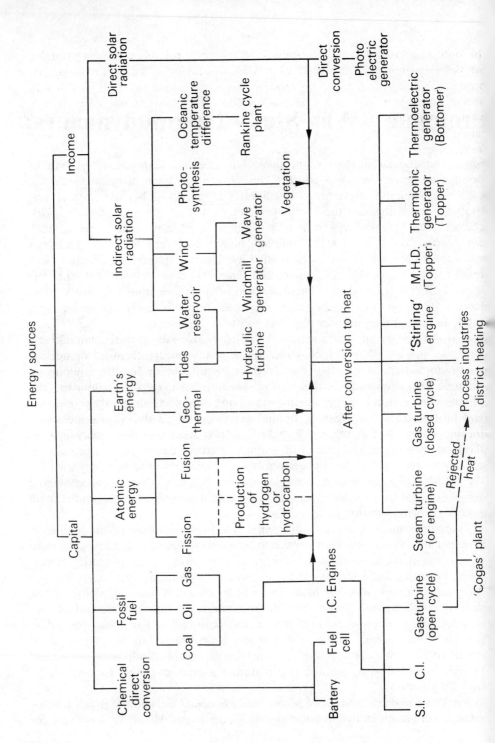

x

he bulk of mankind (approximately $3\cdot7\times10^9$ and probably doubling in under ;0 years), continued expansion of power production is essential if only to produce the energy required for modern methods of food production. But it is needed not merely for this: experience in the developed nations suggests that world population will be stabilised only when the standard of living has been raised substantially in the developing countries. Failure to stabilise it will lead inevitably to famine and conflict over living space. Finally, some of the other reasons why more energy is likely to be needed in the future are:

(a) To enable lower-grade metallic ores to be used, and to operate waste reclamation plant to recover scarce materials.
(b) To produce secondary fuels for land, sea and air transport, and the necessary hydrocarbon feedstock for the chemical industry, when natural oil and gas become scarce.
(c) To reduce environmental pollution from power plant and industrial processes, which will become increasingly important as the population increases.

The increase in demand will be offset to some extent by energy conservation measures, and by new inventions which will enable society's needs to be met with a smaller expenditure of scarce resources. The development of microelectronics, and of silicon light fibres as a substitute for copper wire, are two recent examples of such inventiveness. There is little doubt, however, that there will be a large net increase in demand from the world as a whole because, at present, two-thirds of the world use energy at about one-tenth of the rate used in the developed countries. In studying engineering in general, and thermodynamics in particular, the student is laying a foundation which will enable him to make a substantial contribution to the welfare of mankind.

To whet the appetite, the accompanying diagram shows the range of energy resources at man's disposal and the ways in which they might be used. Not all of the devices mentioned have been developed to date, and some may never be economic, but most of them receive some attention in this book. Other devices dealt with, which do not appear in this schematic diagram because they do not involve the production of power, are refrigerators, heat pumps, air-conditioning plant and heat exchangers.

Contents

Part I Principles of Thermodynamics

Part III Work Transfer

Part I Principles of Thermodynamics

Introduction

In this book the student will be introduced to a wide range of engineering plant: steam turbines, reciprocating engines, turbo-jets and rockets, combustion systems, refrigerators and heat pumps, direct conversion devices such as fuel cells, compressors, boilers, condensers, cooling towers and heat exchangers. What determines their place in the book is that they are concerned with the transfer of heat and work, in many cases with the ultimate object of converting one form of energy into another. In every case they make use of the way a fluid behaves as it is compressed or expanded, heated or cooled. These features imply that the design of the equipment must involve the science of *thermodynamics*, because this is the science dealing with the relations between the properties of a substance and the quantities 'work' and 'heat' which cause a change of state. Before a satisfactory analysis of such changes can be made, it is necessary to introduce some carefully defined concepts.

The idea of a *system* plays an important part in thermodynamics; it may be defined as a region in space containing a quantity of matter whose behaviour is being investigated. This quantity of matter is separated from its surroundings by a *boundary*, which may be a physical boundary, such as the walls of a vessel, or some imaginary surface enveloping the region. Before any thermodynamic analysis is attempted it is essential to define the boundary of the system, because it is across the boundary that work and heat are said to be transferred. The term *surroundings* is restricted to those portions of matter external to the system which are affected by changes occurring within the system. When the same matter remains within the region throughout the process under investigation it is called a *closed system*, and only work and heat cross the boundary. An *open system*, on the other hand, is a region in space defined by a boundary across which matter may flow in addition to work and heat.

Part I is primarily concerned with the behaviour of a particular form of matter—a fluid. The fluid may be a gas (e.g. air), a vapour (e.g. steam), a liquid, or a mixture of these substances provided they do not react chemically with one another. Closed and open systems containing such fluids are the main concern of engineering thermodynamics. Chemical and physical thermodynamics deal with a wider variety of systems and phenomena, e.g. solids, surface films, electric cells, mixtures undergoing chemical reactions, thermoelectric effects, and magnetic substances [see Refs. (5) and (8)]. These are not the main concern of this book, although some consideration is given in Part II to the chemical reactions involved in the combustion of fuel.

Here, therefore, a system may be defined less generally as a region in space containing a quantity of fluid. If the fluid is flowing into, out of, or through the region, it is an open system. Some examples are (1) a gas expanding from a container through a nozzle, (2) steam flowing through a turbine, and (3) water entering a boiler and leaving as steam. Examples of a closed system are (1) a mixture of water and steam in a closed vessel, and (2) a gas expanding in a cylinder by displacing a piston. From the last

1

example it will be seen that the boundary separating a closed system from its surroundings need not be fixed; it may expand or contract to accommodate any change of volume undergone by the fixed quantity of fluid. The processes undergone by the fluid in a closed system are described as *non-flow processes*, whereas those undergone by the fluid in an open system are referred to as *flow processes*.

Since chemical, surface tension, electrical and magnetic effects are absent, an closed system considered here can only change its condition or 'state' for one or both of the following reasons: (1) there will be a change of state if a part of the boundary is displaced by a force so that some mechanical work is done, e.g. when a force is applied to a piston to compress a gas; (2) there will be a change of state if the system is brought into contact with a portion of the surroundings which is at a different temperature i.e. if heat is allowed to flow across the boundary. Thus the engineer is concerned to know the way in which the state of a system changes when work and heat cross the boundary.

The *state** of a system at any given instant is determined by the values of its *properties* at that instant. The sort of characteristic which can be used to describe the condition of any object under investigation must always be found by experience, and will depend upon the nature of the investigation. From the standpoint of thermodynamics, many observable characteristics will be irrelevant, e.g. shape and colour are irrelevant characteristics of the systems considered in this book. On the other hand shape would be an important characteristic of any thermodynamic system involving surface tension effects. Those characteristics of a system which can be used to describe its condition or state, and which are relevant to the investigation, are termed the properties of the system. These properties are not all independent of one another, e.g. for a perfect gas it is known that the temperature must have a definite value if the pressure and volume are fixed, in accordance with the well-known equation of state $pv = RT$. Classical thermodynamics has been built up from a study of closed systems, and it has been found by experiment† that *only two properties are necessary and sufficient to determine the state of a closed system containing a simple fluid*. This state may be termed the *thermodynamic state* of the fluid, and the properties of the fluid from which these two may be chosen are called the *thermodynamic properties*. When the thermodynamic state has been determined by a knowledge of two thermodynamic properties, any third thermodynamic property can be found from a relation expressing it is a function of the two known properties. It will be assumed in Part I that these relations have been determined by experiment for all the fluids likely to be used in engineering systems; the equation of state of a perfect gas is merely one such relation.

The study of open systems is complicated by the fact that the fluid is in motion. If attention is concentrated on a portion of the fluid entering the system, it will be realised that it possesses a relevant *mechanical state* in addition to a thermodynamic state. The *mechanical properties* necessary to determine its mechanical state are

---

*The term 'state' is sometimes used in textbooks on heat to refer to the solid liquid or vapour form of a substance, but these forms are referred to as 'phases' in thermodynamics.

†Advanced students will find an interesting logical analysis of this procedure in Ref. (2), Chapter 1.

(1) velocity relative to the fixed boundary of the system and (2) position in the gravitational field, i.e. height above some datum level. Both these characteristics may change during the passage of the fluid through the system, and they may affect events which occur within the system and the work and heat which cross the boundary; they cannot therefore be omitted from an analysis of open systems. The two mechanical properties are independent of one another and, moreover, are independent of the thermodynamic properties. For example, the fact that a perfect gas is flowing through a pipe with a certain velocity in no way changes the fact that at any point the pressure, volume and temperature of an element of the gas are related by $pv = RT$. All the relations between the thermodynamic properties which will be developed in the following pages are similarly independent of whether the fluid is in motion or changing its position in the gravitational field. Thus the mechanical state is something which can be superimposed upon the thermodynamic state without disturbing the relations between the thermodynamic properties. It will appear later that this makes it possible for the body of knowledge built up from a study of closed systems to be applied to the analysis of open systems.

We are now in a position to summarise the contents of Part I, which present the fundamentals *without reference to any particular fluid*. Firstly, we note that experiment has shown that the state of a closed system is determined when two thermodynamic properties are known, and that all the other thermodynamic properties are then determined. We shall assume that sufficient experimental information is available to enable any third thermodynamic property to be found if two are known. Secondly, the state of a system changes when work or heat cross the boundary. Thermodynamics provides a means of relating the quantities of heat and work with the change of state, i.e. with the variation of the properties of the fluid. The structure of thermodynamics rests upon two important principles called the First and Second Laws of Thermodynamics. These cannot be proved and are treated as axioms; their validity rests upon the fact that neither they, nor any deductions made from them, have ever been disproved by experience.

The First Law is initially stated with reference to closed systems, and it is applied to various processes which such systems may undergo. It is then applied to open systems. The First Law is an expression of the principle of the conservation of energy. With the aid of this law it is possible to calculate the quantities of heat and work which cross the boundary of a system when given changes in properties occur, e.g. the work done by steam expanding through a given range of pressure in a turbine, the work required to produce a given pressure rise in an air compressor, or the heat required to generate steam at a given pressure in a boiler.

A second class of problems is concerned with how much of the heat, normally supplied by burning fuel, can be converted into work by a heat engine. Or, again, how much work is required to extract a given amount of heat from a body in a refrigerator. These are problems concerning the efficiency of power plant, and they can be solved with the aid of the Second Law of Thermodynamics. The Second Law is an expression of the fact that it is impossible to convert all the heat supplied to an engine into work;

some heat must always be rejected, representing a waste of energy. As with the First Law, the Second is enunciated with reference to closed systems, and then its consequences for open systems are deduced.

In the final chapter, *which may be omitted on a first reading*, some important general relations between the thermodynamic properties are deduced from the First and Second Laws.

This Introduction opened with a list of engineering plant, in the design of which thermodynamics plays a vital role. It may not be out of place to end it with a reminder that the complete design of any of the plant mentioned must involve other engineering sciences: such as mechanics of machines and properties of materials. The engineer will be concerned not only with the behaviour of the working fluid, but also with the stresses and strains and vibration problems that are inevitable consequences of trying to meet the thermodynamic requirements as economically as possible. No student of engineering—however science-based his courses may be—should forget that his ultimate aim is the design of new equipment which is both more efficient in operation, and cheaper to produce and maintain, than its predecessor. The natural resources of our planet are not unlimited.

1 Fundamental Concepts

Before the basic structure of thermodynamics can be outlined, it is necessary to define some of the concepts more thoroughly than has been attempted in the Introduction, paying particular attention to temperature, work and heat. Accordingly, this chapter is devoted to a discussion of such fundamental concepts.

1.1 Thermodynamic properties

The idea of a system has been presented in the Introduction. A closed system is fully defined when the following details are known:

(a) the fluid, e.g. whether air, hydrogen, water, etc.,
(b) the boundary between the fluid under consideration and its surroundings,
(c) the mass of fluid within the boundary.

The state of the system has yet to be determined; for this a knowledge of its properties is required.

To define the state of a closed system, only the thermodynamic properties of the fluid are relevant. Any such property must be some characteristic that can be measured, and it must have a unique numerical value when the fluid is in any particular state. Pressure, volume and temperature are obvious examples. The value of a property must be independent of the process through which the fluid has passed in reaching that state. It follows that a change in the value of a property depends only on the initial and final states of the system, and is independent of the process undergone by the system during the change of state.

Using pressure as an example, a gas at state 1 may have a pressure of 5 atm, and after a compression process to state 2 a pressure of 8 atm. The change in pressure is an increase of 3 atm however the compression is carried out. If dP is any infinitesimal change in a property during a process, the total change between states 1 and 2 can be written as

$$\int_1^2 dP = P_2 - P_1$$

Mathematically speaking, dP is an *exact differential.**

Consider the following mechanical analogy. A cyclist travels from point 1, z_1 m above sea level, to point 2 on a hill z_2 m above sea level. Treating the cyclist as the system, his height above sea level can be regarded as a property of the system because the change in height $(z_2 - z_1)$ is independent of the path, i.e. the process, by which the cyclist travels from 1 to 2. The work done by the cyclist, however, is not independent of the process; it clearly depends upon whether he has chosen a short or long route between 1 and 2 and whether he has a following or head wind. Consequently the work done, although a measurable quantity, is not a property. If, during any infinitesimal part of a process, a measurable quantity which depends upon the process is denoted by dX, dX is not an exact differential and

$$\int_1^2 dX \neq X_2 - X_1$$

When the quantity is not a property, the integration between state 1 and state 2 must be written as

$$\int_1^2 dX = X_{12} \quad \text{(or simply as } X\text{)}$$

Further discussion of quantities which are not properties will be deferred to section 1.3, which deals with the concepts of work and heat.

We have already noted the experimental fact that when the values of two properties are known, the state of a closed system is completely determined. An important assumption is implicit in this statement, i.e. that the system is in *thermodynamic equilibrium.* A system is said to be in thermodynamic equilibrium if no further changes occur within it when it is isolated from the surroundings in such a way that no heat and work can cross the boundary. The properties, such as pressure and temperature, must be uniform throughout the system when it is in equilibrium. If the pressure is not uniform, due to turbulence for example, spontaneous internal changes occur in the isolated system until the turbulence has died down and the pressure has become uniform. Similarly, if there are temperature gradients in the isolated system, a spontaneous redistribution of temperature occurs until all parts of the system are at the same temperature. Only under conditions of equilibrium can single values of pressure and temperature be ascribed to the system, and thus be used to determine its state. It will always be assumed in this book that the system is in equilibrium at the beginning and end of any process, and therefore the end states are determined when the values of two properties are fixed.

It follows that the initial and final states of any closed system can be located as points on a diagram using two properties as coordinates. If the system is imagined to pass through a continuous series of equilibrium states during the process, the intermediate states could also be located on the diagram, and a line representing the *path*

*The concept of an exact differential is defined in Chapter 7.

of the process could be drawn through all the points. Such a process is called a *reversible process*; it will be found to be an ideal process analogous to the frictionless process so often referred to in mechanics. In all real processes, however, the system is not in equilibrium in any of the intermediate states. These states cannot be located on the coordinate diagram because the properties do not have single unique values throughout the system. Such processes are called *irreversible processes*. They are best represented on a diagram by a dotted line joining the initial and final state points, to indicate that the intermediate points are indeterminate. The full significance of the ideas of reversibility and irreversibility will appear later.

Pressure, specific volume and temperature are three properties observable by the senses. Pressure p is defined as the force exerted by the fluid on unit area of the boundary, and the unit* we normally employ is the *bar* (1 bar $= 10^5$ N/m² $= 1/1·01325$ atm). The specific volume v is the volume of unit mass of fluid, usually quoted in m³/kg. It is the reciprocal of the density, defined as mass per unit volume, but it is conventional to use specific volume rather than density as one of the properties in thermodynamics. Temperature is a property deserving special attention and it will be introduced in section 1.2. Three more thermodynamic properties—internal energy, enthalpy and entropy—will emerge as consequences of the First and Second Laws of Thermodynamics. Although they are not directly observable by the senses they are essential to the structure of thermodynamics.

From these six properties, two may be selected to determine the state of a closed system in thermodynamic equilibrium, and the values of the remaining four are then fixed. Care must be taken to see that the two chosen properties are independent of one another, i.e. it must be possible to vary one of these properties without changing the other. Obviously density and specific volume are not independent properties because one is simply the reciprocal of the other; this fact is recognised in that density is not included in the six properties to which reference was made. But, less obviously, pressure and temperature are not always independent properties: when a liquid is in contact with its vapour in a closed vessel it is found that the temperature at which the liquid and vapour are in equilibrium is always associated with a particular pressure and one cannot be changed without the other. Pressure and temperature cannot be used to determine the state of such systems, although other pairs, such as pressure and specific volume, may be so used.

As stated in the Introduction, it will be assumed throughout Part I that if any two independent properties are known, the others can be found from experimental data relating to the particular fluid under consideration. If, for example, p and v are known, any third property x must be a function of these two, i.e. $x = \phi(p, v)$. For some fluids the functions are simple, and algebraic equations are used to relate the properties; for others the function is complicated and the experimental data are most conveniently presented in tabular form. Fig. 1.1 shows such a table, the dots denoting values of temperature corresponding to the various combinations of p and v. Similar tables can

---

*The N/m² is called the pascal (Pa), so 1 bar $= 10^5$ Pa. Dimensions and units are discussed in Appendix A.

be prepared with internal energy, enthalpy or entropy as the third property. It is assumed in Part I that either a set of equations or a set of tables is available for each fluid likely to be encountered in engineering systems.

	v_a	v_b	— — — — — — —	v_y	v_z
p_a	•	•		•	•
p_b	•	•		•	•
			Values of temperature (or internal energy or enthalpy or entropy)		
p_v	•	•		•	•
p_s	•	•		•	•

Fig. 1.1   Table of Properties

1.2 Temperature

Two systems are said to have *equal temperatures* if there is no change in any of their observable characteristics when they are brought into contact with one another. Consider a system A consisting of a gas enclosed in a vessel of fixed volume fitted with a pressure gauge. If there is no change of pressure when this system is brought into contact with a second system B, say a block of metal, then the two systems are said to have the same temperature. Experiment shows that if the system A is brought into contact with a third, C, again with no change of pressure, then B and C when brought into contact will also show no change in their characteristics. That is, if two bodies are each equal in temperature to a third body they are equal in temperature to each other. This principle of *thermal equilibrium* is often called the *zeroth law of thermodynamics*; the possibility of devising a means of measuring temperature rests upon this principle.

The temperatures of any group of systems may be compared by bringing a particular system, known as a thermometer, into contact with each in turn. System A performs this function in the foregoing example. Such a system must possess an easily observable characteristic termed a thermometric property, e.g. the pressure of a gas in a closed vessel, the length of a column of mercury in a capillary tube, or the resistance of a platinum wire. A scale of temperature could be defined by assigning numbers to divisions on the pressure gauge, to points spaced at intervals on the capillary tube, or to divisions on the meter registering the resistance of the wire: units so formed are called degrees of temperature. These temperature scales would be quite arbitrary, and each instrument would have its own scale. The necessary conditions for a temperature

cale to be reproducible when any given thermometric property is used will be discussed in Chapter 8, as well as the means adopted for comparing scales resulting from
the use of different thermometric properties.

The reason for postponing the definition of temperature scales is that the practical
procedure of constructing an empirical scale of temperature is irrelevant to the
structure of thermodynamics. It will be shown that a temperature scale can be defined
as a consequence of the Second Law of Thermodynamics without reference to any
particular thermometric property, and consequently this has been adopted as the
ultimate standard. In Part I, where the analysis is carried out only in symbolic form
no numerical scale need be assigned to the property temperature. The qualitative
terms 'high' and 'low' will be used, however, thereby adopting the conventional
terminology which originates from our sense of hotness and coldness.

1.3 Work and heat

The limitations placed upon the closed systems considered here,* imply that only the
application of work or heat can cause a change of state, or conversely, that a change of
state must be accompanied by the appearance of work or heat at the boundary.

As in mechanics, work is said to be done when a force moves through a distance.
If part of the boundary of a system undergoes a displacement under the action of a

Fig. 1.2

pressure, the work done, W, is the product of the force (pressure × area), and the
distance it moves in the direction of the force. Fig. 1.2a illustrates this with the conventional piston and cylinder arrangement, the heavy line defining the boundary of
the system. The basic unit of work in the SI system is the newton metre (N m), also
called the joule (J): we shall use the more convenient kilojoule (kJ) which is 10^3 N m.
Fig. 1.2b illustrates another way in which work might be applied to a system. Here the
boundary really consists of the inner surface of the vessel and the surfaces of the

---

*See Chapter 20 for a brief discussion of a broader concept of work.

paddle. A force is exerted by the paddle as it changes the momentum of the fluid, and since this force moves during the rotation of the paddle some work is done.

The analogy of the cyclist was used in section 1.1 to indicate that work is a quantity which is *not* a property of a system. Work is a transient quantity which only appears at the boundary while a change of state is taking place within a system. *Work is 'something' which appears at the boundary when a system changes its state due to the movement of a part of the boundary under the action of a force.* Although there cannot be said to be any work in a system either before or after the change has taken place, work may be said to 'flow' or 'be transferred' across the boundary. This is a somewhat loose way of speaking; it will be shown in the next chapter that it is 'energy' which is transferred, and that work is the particular form of energy transfer associated with the movement of a force at the boundary.

It is necessary to adopt a sign convention to indicate the direction in which the work 'flows'. If work is done *by* the system *on* the surroundings, e.g. when a fluid expands pushing a piston outwards, the work is said to be *positive.* Conversely, if work is done *on* the system *by* the surroundings, e.g. when a force is applied to a rotating paddle, or to a piston to compress a fluid, the work is said to be *negative.*

Heat, denoted by the symbol Q, may be defined in an analogous way to work as follows. *Heat is 'something' which appears at the boundary when a system changes its state due to a difference in temperature between the system and its surroundings.* Again anticipating the next chapter, the 'something' is a form of energy transfer; this time the form associated with a temperature difference. Heat, like work, is a transient quantity which only appears at the boundary while a change is taking place within the system. It is apparent that neither dW nor dQ are exact differentials (section 1.1), and therefore any integration of the elemental quantities of work or heat which appear during a change from state 1 to state 2 must be written as

$$\int_1^2 dW = W_{12} \quad \text{(or } W\text{)}, \quad \text{and} \quad \int_1^2 dQ = Q_{12} \quad \text{(or } Q\text{)}$$

Although in thermodynamics there cannot be said to be any heat *in* a system either before or after a change of state, loosely speaking heat may be said to 'flow' or 'be transferred' across the boundary. Strictly speaking it is energy which is transferred, but to say 'the heat transferred' is a shorthand way of saying 'the energy transferred by virtue of a temperature difference'. If heat flows *into* a system *from* the surroundings, the quantity is said to be *positive* and, conversely, if heat flows *from* the system *to* the surroundings it is said to be *negative.*

The exact definition of the *unit* of heat must be deferred until the concept of energy has been introduced formally. The old calorimetric definition may be mentioned however; it is so called because it originates from the early experiments on heat carried out with calorimeters. The unit of heat was defined as that quantity which would raise the temperature of unit mass of water, at standard atmospheric pressure, through one degree on some temperature scale. Not only does the unit depend upon the temperature scale used, but careful experiments showed that a slightly different quantity of heat was necessary at different points on the chosen temperature scale, i.e. the unit depended on the temperature of the water at the beginning of the one degree rise. Thus a unit called the *kilocalorie* (kcal) was defined originally as the heat required to raise one kilogram of water one degree on the Celsius scale of temperature. (Similarly the *British Thermal Unit* (Btu) and *Celsius Heat Unit* (Chu) were defined as the quantities of heat required to raise one pound of water one degree Fahrenheit and one degree Celsius respectively.) Later, more carefully defined units were used, such as the 15 °C kcal for which the temperature rise must be from 14·5 to 15·5 °C. So called 'calorimetric' heat units are still in use, which for most practical purposes are identical with calorimetric units defined in terms of the properties of water, but these 'calorimetric' units are now defined in terms of work units. This is discussed further in section 2.3.

Fig. 1.3

Summarising: neither heat nor work are properties of a system, but both are transient quantities, only appearing at the boundary while a change of state occurs within the system. Although they cannot, therefore, be used to describe the state of a system, heat and work can be used to describe the process undergone by the system during a change of state. For example, using the cyclist analogy, the work done by the cyclist might indicate whether he had chosen a short or long route. Or again, consider a vessel of constant volume containing a gas. Let the gas undergo a change of state in such a way that the pressure and temperature are increased. This change could be obtained, as in Fig. 1.3, by placing the vessel in contact with a body at a higher temperature, or as in Fig. 1.2b by churning the gas with a rotating paddle. Thus the same final state could be reached by the provision of either heat or work. Moreover, both heat and work could be added simultaneously in an infinite number of different proportions to produce the same final state. From this example it is clear that the quantities of heat and work are not only functions of the initial and final states of the system, but also of the process undergone by the system.

2 The First Law of Thermodynamics

The structure of thermodynamics rests upon two fundamental laws. The First Law, introduced in this chapter, is concerned with the principle of conservation of energy as applied to closed systems which undergo changes of state due to transfers of work and heat across the boundary. The First Law cannot be proved; its validity rests upon the fact that neither it, nor any of its consequences, have ever been contradicted by experience.

2.1 The cycle

A closed system is said to undergo a cyclic process, or *cycle*, when it passes through a series of states in such a way that its final state is equal in all respects to its initial state.

Fig. 2.1

This implies that all its properties have regained their initial values. The system is then in a position to be put through the same cycle of events again, and the procedure may be repeated indefinitely. Work can be transferred continuously by devising machines which undergo cyclic processes.

Consider the simple, idealised, mechanical system consisting of a frictionless pulley and a weightless rope shown in Fig. 2.1a. Let a force be applied to raise the weight

12

through a certain distance, a force only infinitesimally greater than the weight. Work is done on the system, and hence the quantity is negative. Now let the weight descend slowly to its original position by reducing the force until it is infinitesimally smaller than the weight (Fig. 2.1b). Since the pulley is frictionless, an infinitesimal difference between the weight and applied force is sufficient to set the system in motion. The system does work on the surroundings in overcoming the applied force (it could, in fact, be used to raise another weight), so that this time the quantity is positive. The system has now returned to its original state and, in the limit, both quantities of work are equal in magnitude. For the whole cyclic process, therefore,

Fig. 2.2

$$\sum \delta W = 0$$

A circle on the summation sign is used to signify that the algebraic summation of all the increments of work δW is carried out over a complete cycle.

Now consider what happens if, for the second part of the cycle, the force is removed from the rope and a brake shoe is applied to the pulley to restrain the downward acceleration of the weight (Fig. 2.2). When the weight has been restored to its original position, it will be noticed that the system has not reached its initial state in *all respects*, for the temperature of the pulley and brake shoe is higher than it was originally. To restore the system to its original state and so complete the cycle, heat may be permitted to flow across the boundary of the system. If desired, the quantity of heat may be measured by allowing it to raise the temperature of water in a calorimeter. This process is illustrated diagrammatically in Fig. 2.2.

As no force moves in the surroundings of the system during the last part of the cycle, no work is done, and

$$\sum \delta W \neq 0$$

The net work done during this cycle is a negative quantity equal to the work required to raise the weight initially. On the other hand, a quantity of heat has crossed the boundary into the surroundings (i.e. the calorimeter). Since no heat crosses the boundary in the first part of the cycle and a negative quantity appears in the second,

$$\sum \delta Q \neq 0$$

and is negative.

From this experiment it appears that when a system is taken through a complete cycle, the algebraic summation of the quantities of either work or heat which cross the boundary may have some value other than zero. Since the change in the value of any property over a complete cycle must be zero, because the initial and final states are identical, this is a further illustration of the fact that work and heat are not properties of a system.

2.2 The First Law of Thermodynamics

Experiments similar to those described in section 2.1 were first carried out by Joule (1843). He found that the net quantity of heat delivered by the system was directly proportional to the net quantity of work done on the system during the cycle. The First Law of Thermodynamics may be expressed as a generalisation of this result covering all cyclic processes, including those during which a net amount of heat is supplied *to* the system and a net amount of work is done *by* the system. In fact, since this is the more important type of cycle, the First Law is usually stated in the following way. *When any closed system is taken through a cycle, the net work delivered to the surroundings is proportional to the net heat taken from the surroundings.* This may be expressed by

$$\sum \delta Q \propto \sum \delta W \qquad (2.1)$$

An important type of cyclic process which a closed system may undergo may be described with the aid of Fig. 2.3. The system is a gas enclosed by a cylinder and piston. If a source of heat is applied to the system the gas expands, pushing the piston before it and doing work on the surroundings; the gas may then be returned to its original state by applying a sink of heat, i.e. a body at a lower temperature than the gas. In the second part of the cycle, the gas contracts and some work is done by the surroundings as the piston moves inwards. This is a straightforward closed system undergoing a series of non-flow processes during which the fluid does not move bodily but merely expands and contracts.

Fig. 2.3

The First Law is adopted as an axiom and provides one of the two foundation stones upon which the structure of classical thermodynamics rests. Joule's experiments, and others of the same nature, are not *proofs* of the First Law; they merely lend support for its adoption as an axiom by providing instances which do not contradict the law. A proof could only be formulated if the First Law could be shown to be a consequence of some even broader and more general proposition about nature.* Certain propositions about the behaviour of thermodynamic systems can be deduced from the laws of thermodynamics; they may be termed corollaries. These corollaries are often most conveniently proved by 'reductio ad absurdum', i.e. the contrary of a proposition is assumed to be true, and the consequences are then shown to lead to a contradiction of the fundamental axiom.

Before deducing the corollaries of the First Law, which help to bring out all its implications, the constant of proportionality implicit in the expression (2.1) will be discussed.

---

*Indeed this has now been attempted by Keenan and Hatsopoulos who show that both the First and Second Laws are consequences of a more basic generalisation about the existence of stable states (i.e. about the fact that there are states which cannot undergo a finite change without a permanent finite change in the environment). This work is described in Ref. (11).

2.3 Mechanical equivalent of heat

If the constant of proportionality is denoted by $\mathcal{J}$, expression (2.1) can be rewritten as

$$\mathcal{J} \sum \delta Q - \sum \delta W = 0 \tag{2.2}$$

$\mathcal{J}$ is called the *mechanical equivalent of heat* or *Joule's equivalent*; it expresses the number of work units which are equivalent to one heat unit. Once a work unit and a heat unit have been adopted, experiments such as those carried out by Joule can be used to find the value of $\mathcal{J}$. For such experiments it has been customary to use a calorimetric definition of the unit of heat. If the kJ and the kcal are chosen as the units of work and heat respectively, the value of $\mathcal{J}$ is found to be approximately 4·19 kJ/kcal; similarly in the other common systems of units

$$\mathcal{J} \approx 427 \frac{\text{kgf m}}{\text{kcal}} \approx 778 \frac{\text{ft lbf}}{\text{Btu}} \approx 1400 \frac{\text{ft lbf}}{\text{Chu}}$$

These values are sufficiently accurate for most engineering calculations where both calorimetric and mechanical units are employed.

Strictly speaking, once the First Law has been established and heat and work are interpreted as forms of energy transfer, the need for separate units vanishes. Certainly there is no justification for two sets of absolute standards, e.g. the primary standards of mass, length and time to which the unit of work may be referred, and a standard thermometer and mass of water of a certain purity to which the unit of heat may be referred. This was recognised in 1929 when the First International Steam Table Conference decided to *define* a heat unit in terms of a work unit. The definition accepted today was laid down at the Fifth International Conference on Properties of Steam (1956) in terms of the joule (J) as the standard unit of work [see Appendix A]. A heat unit defined in terms of the joule depends only on the primary standards of mass, length and time. This *International Table kilocalorie*, unit symbol kcal_{IT} although often written without the suffix, is defined as

$$1 \text{ kcal}_{\text{IT}} = 4·1868 \text{ kJ}$$

The number was chosen so that calorimetric methods of measuring quantities of heat can still be used for ordinary purposes.* Another kilocalorie defined in terms of the Joule, and still used in physical chemistry, is the *thermochemical kilocalorie* defined as

$$1 \text{ kcal}_{\text{th}} = 4·184 \text{ kJ}$$

---

*Bearing in mind that 1 kgf = 9·806 65 N, the kcal is related to the metric technical unit of work by

$$1 \text{ kcal} = \frac{4·1868 \times 10^3}{9·806\ 65} = 426·935 \text{ kgf m}$$

The British units are related to the kJ by the definitions

$$1·8 \frac{\text{Btu}}{\text{lb}} = 1 \frac{\text{Chu}}{\text{lb}} = 4·1868 \frac{\text{kJ}}{\text{kg}} \quad \text{which lead to} \quad 1 \text{ ft lbf} = \frac{1}{778·169} \text{ Btu} = \frac{1}{1400·70} \text{ Chu}$$

These definitions ensure that specific heats in calorimetric units are numerically identical in all systems of units—see section 8.2(ii).

Since no fundamental distinction can now be made between the unit of heat and the unit of work, and the 'mechanical equivalent of heat' $\mathcal{J}$ is recognised merely to be a conversion factor between different units, $\mathcal{J}$ can be dropped from equation (2.2) which can now be written as

$$\sum \delta Q - \sum \delta W = 0 \tag{2.3}$$

It is understood that the same unit, for us the kJ, is used consistently for both heat and work. Equation (2.3), in conjunction with our sign convention, can now be taken to express the First Law.

Some of the corollaries of the First Law may now be considered.

2.4 Corollaries of the First Law

COROLLARY 1. *There exists a property of a closed system such that a change in its value is equal to the difference between the heat supplied and the work done during any change of state.*

Property X

Property Y

Fig. 2.4

PROOF. Assume that the converse of the proposition be true; i.e. that $\sum_1^2 (\delta Q - \delta W)$ depends upon the process, as well as upon the end states 1 and 2, and so cannot be equal to the change in a property. The procedure is to show that this contradicts the First Law.

Consider any two processes A and B by which a system can change from state 1 to state 2 (Fig. 2.4). The assumption is that in general

$$\sum (\delta Q - \delta W)_{\text{A}} \neq \sum (\delta Q - \delta W)_{\text{B}} \tag{2.4}$$

The suffixes A and B are used to indicate the process for which the summation is carried out. In each case let the system return to its original state by a third process C. For each of the complete cycles AC and BC so formed, we have

$$\sum (\delta Q - \delta W)_{\text{AC}} = \sum_1^2 (\delta Q - \delta W)_{\text{A}} + \sum_2^1 (\delta Q - \delta W)_{\text{C}}$$

and

$$\sum (\delta Q - \delta W)_{\text{BC}} = \sum_1^2 (\delta Q - \delta W)_{\text{B}} + \sum_2^1 (\delta Q - \delta W)_{\text{C}}$$

If the assumption described by the inequality (2.4) is true, it follows that

$$\sum (\delta Q - \delta W)_{\text{AC}} \neq \sum (\delta Q - \delta W)_{\text{BC}}$$

But this contradicts the First Law expressed by equation (2.3), which implies that these quantities must be equal since they are both zero. Therefore the original proposition, that $\sum_1^2 (\delta Q - \delta W)$ is independent of the process, must be true.

16

If the property so discovered is denoted by U, the corollary can be expressed mathematically as

$$\sum_1^2 (\delta Q - \delta W) = U_2 - U_1$$

Or writing Q and W for the net quantities of heat and work crossing the boundary during the change of state,

$$Q - W = (U_2 - U_1) \tag{2.5}$$

The property U is called the *internal energy* of the system and equation (2.5) is called the *non-flow energy equation*. Expressed verbally: any quantity of heat supplied to a closed system must equal the increase of internal energy plus the work done by the system. The adjective 'internal' has been adopted to distinguish this form of energy from mechanical forms which might be superimposed on the system. Thus if the system is moving as a whole and changing its position in the gravitational field, we know from mechanics that the system will possess kinetic and potential energy in addition to internal energy.

It should be noted that the First Law is more than a mere definition of the internal energy by a statement to the effect that $\delta U = \delta Q - \delta W$. It states that the internal energy is a quantity *different in kind* from the quantities of work and heat. Internal energy, being a property, can be said to reside in the system and it can be increased or decreased by a change of state. It is only energy which can truly be said to flow across the boundary, and the terms work and heat simply refer to two different causes of the flow of energy. As already stated, if the flow occurs as the result of the movement of a force at the boundary it is termed work, and if it occurs because of a temperature difference between the system and surroundings it is called heat.

Suppose a system consists of two bodies at different temperatures, the boundary being imagined to be an insulating wall isolating the two bodies from their surroundings (Fig. 2.5a). This would represent a system in a non-equilibrium state, since at least one property, temperature, is not uniform throughout the system. Energy will flow from one body to another until the temperatures are identical, when the system will be in equilibrium and any property will have the same value in all parts of the system. The term heat has no meaning in this situation as no energy crosses the boundary. On the other hand, if one body is considered to be the system and the other to be part of its surroundings as in Fig. 2.5b, and energy flows from one to the other due to the temperature difference, then heat is said to cross the boundary. It follows that the first step in the enunciation of any thermodynamic problem must be to define the system.

The terms in any physical equation must have the same dimensions and be expressed in the same units. Since Q and W are to be expressed in kJ, U must be in kJ; but if Q and W were in kcal then U would also be in kcal. It is important to notice that the energy equation is concerned with *changes* in internal energy. The First Law suggests no means for assigning an absolute value to the internal energy of a system at any

17

given state. For practical purposes, however, it is possible to adopt some state, say state 0, at which the energy is arbitrarily taken to be zero, and then find the changes of energy when the system proceeds from this reference state to a series of states 1, 2, 3, etc. In this way values of U_1, U_2, U_3, etc., can be found and tabulated as in Fig. 1.1, although they are strictly values of $(U_1 - U_0)$, $(U_2 - U_0)$, $(U_3 - U_0)$, etc. In all engineering applications of thermodynamics only changes in energy are of interest and consequently what state is chosen as the reference state is unimportant. For example, between states 2 and 3, defined say by the values (p_2, v_2) and (p_3, v_3), the change of internal energy can be found from the table of properties as $(U_3 - U_2)$. The quantity $(U_3 - U_2)$ is strictly $(U_3 - U_0) - (U_2 - U_0)$ but, since U_0 cancels, the reference state is irrelevant.

Although latent forms of energy due to chemical and nuclear bonds are present in the fluid, these forms remain constant in any change of state during which nuclear

Fig. 2.5

and chemical reactions do not take place. The only forms of energy which can change are those associated, on the kinetic theory of fluids, with the kinetic energy of the molecules and the potential energy due to intermolecular forces; 'random molecular energy' might be used as a collective term when thinking of the mechanical model. It is this random molecular energy which is envisaged when the term internal energy is applied to a closed system in equilibrium. When the system is not in equilibrium, the internal energy can include macroscopic mechanical forms, both kinetic and potential, associated with the movement of relatively large scale elements of fluid; under these circumstances the symbol E will be used in place of U. The internal energy quoted in tables of properties always refers to the fluid when in a state of equilibrium.

COROLLARY 2. *The internal energy of a closed system remains unchanged if the system is isolated from its surroundings.*

No formal proof of this proposition is required once the first corollary has been established; it follows directly from equation (2.5). If the system is isolated from its surroundings, Q and W are both zero and hence ΔU must be zero. The system represented by Fig. 2.5a is an example of an isolated system. All that happens in this case is a spontaneous redistribution of energy between parts of the system which continues until a state of equilibrium is reached; there is no change in the total quantity of energy within the system during the process.

Corollary 2 is often called the Law of Conservation of Energy. It is sometimes loosely applied to the whole universe to suggest that the energy of the universe is constant. Such an extrapolation may be admissible only if the universe can be regarded as finite, because the proposition has been inferred from our experience of finite systems.

COROLLARY 3. *A perpetual motion machine of the first kind is impossible.*

The perpetual motion machine was originally conceived as a purely mechanical contrivance which, when once set in motion, would continue to run for ever. Such a machine would be merely a curiosity of no practical value, and we know that the presence of friction makes it impossible. What would be of immense value is a machine producing a continuous supply of work without absorbing energy from the surroundings; such a machine is called a *perpetual motion machine of the first kind.*

It is always possible to devise a machine to deliver a limited quantity of work without requiring a source of energy in the surroundings. For example, a gas compressed behind a piston will expand and do work at the expense of the internal energy of the gas. Such a device cannot produce work continuously, however; for this to be achieved the machine must be capable of undergoing a succession of cyclic processes. But equation (2.3) states that if a net amount of heat is not supplied by the surroundings during a cycle, no net amount of work can be delivered by the system. Thus the First Law implies that a perpetual motion machine of the first kind is impossible.

3 Non-flow Processes

In this chapter the non-flow energy equation expressing the First Law will be given particular forms relating to various special processes which occur in closed systems. It will be shown that these processes may be classified further into reversible and irreversible processes. For the former class the energy equation may be expressed in a particularly useful form because the quantities of heat and work can each be stated in terms of changes in the properties of the system. The equations so derived are quite general in that they apply to any fluid.

3.1 The energy equation and reversibility

In section 2.4 the non-flow energy equation was presented in the form

$$Q - W = (U_2 - U_1)$$

It has already been stated that we shall only consider systems which are in equilibrium in their initial and final states. When the system is in equilibrium all the properties are uniform throughout the system and, in particular, each unit mass of the system will have the same internal energy.* Thus, if u represents the internal energy of unit mass, and m is the mass of the system, it is possible to write $U = mu$. In the same circumstances the volume V is equal to mv. u is termed the *specific internal energy* just as v is the specific volume, although the adjective 'specific' is often omitted when referring to internal energy. It is often convenient to work with specific quantities throughout an analysis and then multiply the final result by the mass of the system. Thus the non-flow energy equation is usually written as

$$Q - W = (u_2 - u_1) \tag{3.1}$$

implying that Q and W are the quantities of heat and work per unit mass of fluid.

For the particular class of processes known as reversible processes, the system is imagined to pass through a continuous series of equilibrium states. In such cases the equation may be applied to any infinitesimal part of the process between the end states, and the energy equation may be written in the differential form

$$dQ - dW = du \tag{3.2}$$

*This statement only applies to systems consisting of a fluid in a single phase or, when more than one phase is present, if the phases are homogeneously mixed.

The analysis of a reversible process may therefore be carried out by considering an infinitesimal part of the process undergone by unit mass, and overall changes which occur between the end states can be found by an integration performed at the conclusion of the analysis followed by multiplication by the mass of the system.

During irreversible processes, on the other hand, the system does not pass through a series of equilibrium states. It is not possible to write mu for the internal energy of the system in any intermediate state because (a) a single value of u cannot be used to define the random molecular energy of the fluid, and (b) this energy is not the only relevant form if the fluid is turbulent. Each elemental mass δm may have a different specific internal energy e which must include the kinetic and potential energy of the mass δm. It is possible to consider an infinitesimal part of an irreversible process, but only by writing

$$dQ - dW = dE$$

where $E = \Sigma e\,\delta m$. Consequently, to evaluate dE it would be necessary to find the change in the specific internal energy of each of the elemental masses of which the system is composed—an impossible task. When considering irreversible processes, therefore, the energy equation can only be used in the integrated form, equation (3.1), and it is then sufficient to be able to assign values of u to the end states. This can be done since these are always assumed to be states of equilibrium.

For the system to pass through a series of equilibrium states, and therefore to undergo a reversible process, it may be shown that the work and heat must cross the boundary under certain particular conditions. These conditions will be discussed for work and heat in turn.

(i) *Work and reversibility*

Let us consider a closed system where a part of the boundary is allowed to move under such conditions that the external restraining force is only infinitesimally smaller than that produced by the pressure of the system. Fig. 3.1 illustrates such a system. The area of the piston is A, the pressure of the fluid at any instant is p, and the restraining force exerted by the surroundings is infinitesimally smaller than pA. If p is assumed constant during an infinitesimal movement of the piston over a distance dl, the work done by the fluid in moving the external force pA through this distance must be $pA\,dl$. But $A\,dl = dV$, the infinitesimal change of volume, and therefore

Fig. 3.1

$$dW = p\,dV$$

If the expansion occurs from a pressure p_1 to a pressure p_2, in such a way that the restraining force is changed continuously so that it is never materially different from

pA, the total work done can be found by summing up all the increments of work $p\,\mathrm{d}V$, i.e.

$$W = \int_1^2 p\,\mathrm{d}V \tag{3.3}$$

The conditions under which this expansion has been imagined to proceed are just those necessary for the expansion to be reversible. No pressure gradients or eddies are

set up in the fluid, and the properties are uniform throughout the system at all times. The system passes through a series of equilibrium states, and the process may be represented by a full line on the p-v diagram, Fig. 3.2. The area under this curve is given by $\int_1^2 p\,\mathrm{d}v$. Comparison with equation (3.3) shows that the area represents the work done per unit mass of fluid.

Fig. 3.2   Reversible process

The reason for the use of the term 'reversible' is now apparent; the expansion may be stopped at any point and reversed by only an infinitesimal change in the external force. The system will then return through the same series of states, and exactly the same quantity of work will be done by the surroundings on the system as has been done by the system on the surroundings in the original process. $\mathrm{d}V$ would be negative for the compression, and hence $p\,\mathrm{d}V$ would appear as a negative quantity in agreement with the sign convention adopted for work.

It is easy to show that $\int p\,\mathrm{d}v$ is not always equal to the work done per unit mass, and that the equality only holds when the system passes through a series of equilibrium states. Fig. 3.3a shows a container divided into two compartments by a sliding partition. One part contains a mass of gas at a pressure p_1 and specific volume v_1, and the other is evacuated. When the partition is withdrawn, the gas undergoes an expansion which is not restrained in any way by an opposing force at the moving boundary; the process is called a *free expansion*. When the system has settled down to a new state of equilibrium it may have a pressure p_2 and specific volume v_2. The end states could be located on the p-v diagram, but the fact that a process had occurred could only be indicated by a dotted line. The inter-

(a)

(b)

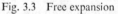

Fig. 3.3   Free expansion

mediate states are indeterminate because no single values of p and v can be ascribed to the system as a whole. Some intermediate points could be found, however, if the process were carried out in a number of steps as in Fig. 3.3b. After each partition is withdrawn, and the gas has settled down to an equilibrium state, the pressure and volume could be measured and the corresponding state point plotted on the *p-v* diagram. If a dotted curve is drawn through all such points, as in Fig. 3.4, the area under the curve could be found and this would be equal to $\int_1^2 p \, dv$. Yet no work has been done by the system on the surroundings at all, because no external force has been moved through a distance.

This is an extreme case; in general there is some restraining force at a moving boundary, and some work will be done by the fluid during the expansion. But, unless the process is reversible, the work done per unit mass will always be less than $\int p \, dv$. This point will be elaborated in section 5.4, where it is also noted that for a compression the work done on the system is always greater than $\int p \, dv$ unless the process is reversible.

Fig. 3.4   Irreversible process

The condition for reversibility, that the external force in the surroundings should differ only infinitesimally from the internal force due to the pressure of the fluid, means that there is no excess force available for accelerating the fluid or boundary. Thus it implies that the reversible expansion or compression is an infinitely slow process. Since any real process must occur within a finite time, it is clear that the reversible process is an ideal process, i.e. one which may be approached but never actually achieved in practice.

(ii) *Heat and reversibility*

The part of the surroundings which exchanges energy with a system by virtue of the fact that it is at a different temperature is called a *heat reservoir*; it may be a source or sink of heat. Usually the term reservoir is restricted to mean a source or sink of such capacity that any quantity of heat which crosses the boundary during a process is insufficient to change its temperature. A practical source of large capacity might be the gases produced by the continuous combustion of a fuel, or a mass of condensing vapour; a practical sink might be a river, the earth's atmosphere, or a mass of melting solid.

Consider two methods of raising the temperature of a fluid contained in a closed vessel from 100 to 200 degrees of temperature. An infinite heat source of, say, 300 degrees, might be used for the purpose. Under these circumstances heat will flow at a finite rate across the boundary. There will be a temperature gradient set up in the fluid because the fluid nearest to the source will always be hotter than the fluid at the opposite end of the vessel. The system cannot, therefore, be in a state of equilibrium during the transfer of energy. On the other hand, if this source is replaced by a series of heat sources ranging in temperature from $(100 + d\theta)$ to $(200 + d\theta)$, so that the active source is never more than infinitesimally different in temperature from the system, temperature gradients will not be produced in the fluid and the system may be presumed to pass through a series of equilibrium states. Moreover, at any point in the process only an infinitesimal change in the temperature of the source is necessary for it to become a sink and for the direction of the process to be reversed.

Thus a transfer of energy by virtue of a temperature difference can only be carried out reversibly if the temperature difference is infinitesimally small. This implies that the rate at which the transfer can proceed is infinitely slow, and once again it appears that the reversible process cannot be completely achieved in practice. Practical methods by which a reversible heat transfer process can be approached are available, and one is described in section 12.8.

A number of particular non-flow processes may now be considered, each of which may or may not be performed reversibly. When they are performed irreversibly, the non-flow energy equation may be applied in the integrated form

$$Q - W = (u_2 - u_1)$$

because the end states are always assumed to be states of equilibrium. When they are performed reversibly, the energy equation may also be applied in the differential form

$$dQ - dW = du$$

And, in addition, the work done can be expressed in terms of the properties of the system, i.e. $dW = p\,dv$, so that

$$dQ - p\,dv = du$$

It might be supposed, by analogy, that when heat is transferred during a reversible process it too can be expressed in terms of properties. As temperature is clearly associated with heat, it might be one of the properties. For example, we might be able to write $dQ = \theta\,dx$, where θ is the temperature and x some other property. This is in fact the case. As a consequence of the Second Law of Thermodynamics it will be shown in section 6.4 that two suitable properties can be defined, and that $dQ = T\,ds$ where T is the *thermodynamic temperature*, and s is the property *entropy*. We shall not, at this stage, presuppose the relationship $dQ = T\,ds$, but will point out the type of problem in which it might prove useful.

3.2 Constant volume process

The discussion based on Figs. 1.2b and 1.3, in section 1.3, has already shown that there are an infinite number of processes which can produce a given change of state and during which the volume remains constant. Any combination of quantities of heat and work can be used which will satisfy the equation

$$Q - W = u_2 - u_1$$

remembering that u_1 and u_2 are fixed when the initial and final states are given. The only restriction is that W must be zero or negative, because during a constant volume process work can only be done in practice by some method of churning the fluid.

The process of increasing the internal energy of a system by using the viscous or frictional effects inherent in any movement of a fluid is one to be avoided. Such a process involves a degradation of energy because, as will be shown later, no practical method exists of making all the internal energy available again as work. Consequently, whenever reference is made to a constant volume process, it will always be to one in which the work done is zero unless otherwise stated. Thus the energy equation for a constant volume process is usually written as

$$Q = u_2 - u_1 \qquad (3.4)$$

That is, the quantity of heat is equal to the change of internal energy.

If, in addition to the work being zero, it is stipulated that the heat is transferred by virtue of an infinitesimally small temperature difference, then the process is reversible and the energy equation can be written in differential form as

$$dQ = du \qquad (3.5)$$

There is one, and only one, constant volume process which can join two given end states when the process is reversible.

It is worth noting that the integrated form of (3.5) is equation (3.4), so that the overall quantity of heat transferred during a constant volume process from state 1 to state 2 does not depend upon the provision of a reversible method of transferring the heat, but merely on the assumption that the work done is zero.

3.3 Constant pressure process

A closed system undergoing a process at constant pressure is illustrated in Fig. 3.5. The fluid is enclosed in a cylinder by a piston on which rests a constant weight. If heat is supplied, the fluid expands and work is done by the system in overcoming the constant force; if heat is extracted, the fluid contracts and work is done on the system by the constant force. In the general case

Fig. 3.5

$$Q - W = (u_2 - u_1)$$

Work might also be done on the system simultaneously by churning the fluid with a paddle, and this negative quantity of work must then be included in the term W. If no paddle work is done on the system, and the process is reversible,

$$dQ - p\,dv = du \qquad (3.6)$$

Since p is constant, this can be integrated to give

$$Q - p(v_2 - v_1) = (u_2 - u_1) \qquad (3.7)$$

If the initial and final states are known, e.g. if p, v_1 and v_2 are given, a definite quantity of heat Q is transferred during the process which can be calculated from (3.7). There is

25

therefore only one reversible constant pressure process which can join the two given states.

A further simplification of the energy equation is possible if a new property is introduced. Since p is constant, $p \, dv$ is identical with $d(pv)$. Thus the energy equation can be written as

$$dQ - d(pv) = du$$

or

$$dQ = d(u + pv)$$

The quantity $(u + pv)$ occurs so frequently in thermodynamics, particularly in equations relating to open systems, that it has been given a special name and symbol, the *enthalpy h*. Because enthalpy is defined merely as a combination of the properties u, p and v, it is itself a property. That is, a change in enthalpy between two states depends only upon the end states, and is independent of the process. Values of the enthalpy are included in tables of properties of fluids and, as with internal energy, the values given are always differences between the enthalpy at various states and the enthalpy at some reference state where it is arbitrarily assumed to be zero. Since only changes of enthalpy are of interest, this information is adequate. In terms of specific quantities,

$$h = u + pv \tag{3.8}$$

and for any mass of fluid m, in a state of equilibrium,

$$H = U + pV \tag{3.9}$$

where $H = mh$, $U = mu$, and $V = mv$. The units of enthalpy must be those of energy; therefore we shall express enthalpy in kJ and the specific enthalpy in kJ/kg.

Using this derived property, the energy equation for a reversible constant pressure process becomes

$$dQ = dh \tag{3.10}$$

or in the integrated form

$$Q = (h_2 - h_1) \tag{3.11}$$

Thus the heat added in a reversible constant pressure process is equal to the increase of enthalpy,* whereas it was equal to the increase of internal energy in the reversible constant volume process. Equation (3.11) does not apply if the process is irreversible; it is an integrated form of the special equation (3.6), and has not been deduced directly from the general equation.

---

*It is this particular relation which led to the adoption of the term 'total heat' as an alternative to enthalpy. Such a name is misleading because the enthalpy is related to the heat supplied only in the special case of a reversible constant pressure process, whereas the property is used in the analysis of many other processes where it bears no direct relation to a quantity of heat.

3.4 Polytropic process

The constant volume and constant pressure processes can be regarded as limiting cases of a more general type of process in which both the volume and pressure change, but in a certain specified manner. In many real processes it is found that the states during an expansion or compression can be described approximately by a relation of the form $pv^n =$ constant, where n is a constant called the *index of expansion or compression*, and p and v are average values of pressure and specific volume for the system. Compressions and expansions of the form $pv^n =$ constant are called *polytropic* processes. When $n = 0$ the relation reduces to $p =$ constant, and when $n = \infty$ it can be seen to reduce to $v =$ constant by writing it in the form $p^{1/n}v =$ constant.

For the *reversible* polytropic process, single values of p and v can truly define the state of a system, and $dW = p\, dv$. The work done per unit mass during a change from state 1 to state 2 may then be found by integration as follows. For the initial, final and any intermediate state,

$$p_1 v_1^n = p_2 v_2^n = pv^n$$

Therefore

$$W = \int_1^2 p\, dv = p_1 v_1^n \int_1^2 \frac{dv}{v^n} = \frac{p_1 v_1^n (v_2^{1-n} - v_1^{1-n})}{1-n}$$

$$= \frac{p_2 v_2^n v_2^{1-n} - p_1 v_1^n v_1^{1-n}}{1-n} = \frac{p_2 v_2 - p_1 v_1}{1-n} \tag{3.12}$$

Fig. 3.6   Polytropic processes

The integrated form of the energy equation for a reversible polytropic process may therefore be written as

$$Q - \frac{(p_2 v_2 - p_1 v_1)}{1-n} = (u_2 - u_1) \tag{3.13}$$

For any given reversible polytropic process, i.e. one joining two given end states, definite quantities of heat and work cross the boundary. For example, if the end states are defined by (p_1, v_1) and (p_2, v_2), the index n can be found, and W and Q calculated from (3.12) and (3.13).

Some reversible polytropic processes are represented in Fig. 3.6 on the p-v diagram. The lower right-hand quadrant contains expansion processes starting from an initial state represented by point 1, and the upper left-hand quadrant contains compression processes starting from the same initial state. The area under the curve or line represents the work done in each case, i.e. under (a) it is zero, under (b) it is $p(v_2 - v_1)$, and under (c) it is $(p_2 v_2 - p_1 v_1)/(1 - n)$.

The larger the value of n, the more nearly does the polytropic curve approach the vertical line representing the constant volume process. This fact could also be found by differentiating $pv^n = $ constant to give

$$v^n \, dp + pnv^{n-1} \, dv = 0$$

or

$$\frac{dp}{dv} = -\frac{np}{v} \tag{3.14}$$

Thus the slope of the curve increases in the negative direction with increase of n.

Although it is physically possible for processes to fall into the other two quadrants, they do not often arise in practical systems. These would be processes during which the pressure and volume increase or decrease simultaneously, and would imply a negative value of n if they were polytropic processes. An example of a system in which

Fig. 3.7

this occurs is illustrated in Fig. 3.7, wherein a piston is controlled by a spring. When heat is supplied to the fluid the restraining force increases as the expansion of fluid compresses the spring. The pressure of the fluid is always equal to the spring force divided by the piston area, and consequently the pressure and volume increase simultaneously. The process would be represented by a straight line in the upper right-hand quadrant of Fig. 3.6; it would of course only be a polytropic process if the line passed through the origin (i.e. when $n = -1$).

3.5 Adiabatic process

The term *adiabatic* is used to describe any process during which heat is prevented from crossing the boundary of the system. That is, an adiabatic process is one undergone by a system which is thermally insulated from its surroundings. For adiabatic non-flow processes, therefore, the energy equation reduces to

$$-W = (u_2 - u_1) \tag{3.15}$$

In an adiabatic expansion work is done by the system at the expense of its internal energy; in an adiabatic compression the internal energy is increased by an amount equal to the work done on the system. If the initial and final states are known, the values u_1 and u_2 are determined and the work done can be calculated. We shall proceed to show that in the case of a *reversible* adiabatic process there is only one process which can join the end states, i.e. that there is a definite relation between p and v. In this case, therefore, the work done can be predicted when the initial state and only one property of the system in its final state are known.

We shall suppose that p_1 and v_1 are known, and that it is required to find the work done by the system during a reversible adiabatic expansion to p_2. For this process the energy equation may be written in differential form as

$$p \, dv = -du \tag{3.16}$$

If a known algebraic equation $u = \phi(p, v)$ relates p, v and u, (3.16) may be rewritten as

$$p \, dv = -d\phi(p, v)$$

and integrated. The result will be a definite relation between p and v, e.g. $v = f(p)$. v_2 can be found by putting $p = p_2$ in f(p). Consequently u_2 and u_1 can be found from $\phi(p, v)$, and the work can be calculated using equation (3.15).

Fig. 3.8   Reversible adiabatic process

If no simple algebraic equation relates p, v and u, the following graphical solution may be used.

(a) First, a series of constant pressure lines are drawn on a diagram using u and v as coordinates, from the information contained in the table of properties (Fig. 1.1). For each pressure in the table a series of values of u and v are given, and these pairs of values can be represented by points on the diagram. All points relating to one pressure, p_a say, are joined up to form a constant pressure line. This is repeated for a series of pressures $p_b, \ldots, p_m, p_n$, covering the range p_1 to p_2 as shown in Fig. 3.8.

(b) Secondly, the path of the reversible process is traced out on the diagram starting from the initial state 1. Recasting equation (3.16) in the form $du/dv = -p$, it is seen that at any point in the process the slope of the curve is negative and equal in magnitude to the pressure at that point. Through each constant pressure line a series of short, parallel lines can be drawn with a slope equal to $-p$. The path of the process will be represented by a curve drawn from point 1 so that its slope is always the same as that of the short parallel line where it intersects each constant pressure line.

(c) Lastly, since the final pressure p_2 is known, the final state can be located on the curve, and the value of u_2 can be read off the diagram. The work done can then be calculated from (3.15).

It is clear that such graphical procedures are extremely lengthy, and it is easy to show that the calculation can be considerably simplified if we can write $dQ = T\,ds$ for a reversible process (see section 3.1). Since $dQ = 0$ for an adiabatic process, $ds = 0$ when it is reversible and the entropy must remain constant. The foregoing problem may be solved by finding s_1 corresponding to p_1 and v_1 from the tables of properties, and then, since $s_2 = s_1$, finding v_2 corresponding to p_2 and s_2. Corresponding values of u_1 and u_2 can also be found from the tables and the problem is solved. This illustrates one possible use of the property 'entropy'.

3.6 Isothermal process

When the quantities of heat and work are so proportioned during an expansion or compression that the temperature of the fluid remains constant, the process is said to be *isothermal*. Since temperature gradients are excluded by the definition, the reversibility of an isothermal process is implied. If it were not reversible a single value of the temperature could not be allotted to the system as a whole in all the intermediate states. It should be noted, however, that the term isothermal is often ascribed to processes which are not reversible but during which the average temperature of the bulk of the fluid remains constant.

It is possible to show that for a reversible isothermal process a certain definite relation must exist between p and v, and consequently the work done has a definite value which can be predicted. For any reversible process the energy equation is

$$dQ - p\,dv = du$$

or

$$Q - \int_1^2 p\,dv = (u_2 - u_1)$$

A relation between p, v and temperature θ must exist, either as some equation of state $\theta = \phi(p, v)$, or in the form of a table of properties. Consequently, if the temperature remains constant, a definite relation exists between p and v which may be determined analytically or graphically. If p_1, v_1 and p_2 are given, v_2 can be found from this particular p-v relation. The work done may be found either by integrating $p \, dv$ between these limits, or by finding the area under the curve on a p-v diagram. Since the end states are known, u_1 and u_2 can be found, and consequently Q can be calculated from the energy equation if it is required.

There is no need to outline the graphical procedure required if no simple function relates p, v and the temperature θ, because once again the problem is much more easily solved by the use of the entropy s and the thermodynamic temperature T. For a reversible process $dQ = T \, ds$, and since T is constant in the case of an isothermal process it is possible to integrate this expression directly. Thus for a reversible isothermal process, $Q = T(s_2 - s_1)$, where $T = T_1 = T_2$. If p_1 and v_1 are known, s_1, T_1 and u_1 can be found from tables of properties. Since p_2 and $T_2 (= T_1)$ are known, s_2 and u_2 can also be found from the tables. Thus Q can be calculated from $T(s_2 - s_1)$, and the work done can then be found from

$$W = Q - (u_2 - u_1)$$

4 Flow Processes

This chapter is devoted to a discussion of various flow processes which occur in ope
systems. Since the energy equation expressing the First Law has been deduced wit
reference to closed systems, the first step is to deduce an equivalent equation for ope
systems. Flow processes may be classified into steady-flow and nonsteady-flo
processes, and since the former are of greater inter
est they are considered first. Finally, the idea of
cycle consisting of steady-flow processes is pre
sented. The application of the concept of reversibilit
to flow processes will be deferred until Chapters
and 6.

4.1 The steady-flow energy equation

Fig. 4.1 Open system with heating
chamber and turbine

A diagrammatic sketch of a typical open system
is shown in Fig. 4.1. Fluid flows through the system
at a uniform rate, while heat and work are trans
ferred between the fluid and surroundings also at a
uniform rate. The work is done in this instance by
allowing the fluid to impinge on a row of turbine
blades. The boundary has been drawn outside the
container for clarity, but it should strictly be considered to encompass only the fluid. The open parts
of the boundary* are chosen to lie across the entrance and exit pipes where the properties of the
fluid are known.

At any given point in the flow, an element of the fluid has a relevant mechanical
state in addition to a thermodynamic state. Any of the mechanical and thermodynamic properties may change as the element passes through the system, so that both
types of property are relevant to the analysis. For the flow to be regarded as *steady*,
the mass flow (kg/s say) must be constant and the same at inlet and outlet, and the
properties of the fluid at any point in the open system must not vary with time. The
latter condition means that all elements of fluid flowing past a given point will

*In fluid mechanics the boundary is often referred to as a *control surface* and the system as a *control volume*.

ways have the same mechanical and thermodynamic states. The properties will, in
general, be different at different points in the system, and therefore any given element
of fluid passes through a series of states on its way through the system.

It is possible to regard the continuous-flow process as a series of non-flow processes
undergone by an imaginary closed system, and Fig. 4.2 illustrates this.* The boundary
of the closed system includes the fluid contained in the open system at any instant,
plus a portion of the fluid about to enter the system. The imaginary closed system
must then be supposed to undergo a deformation as indicated by the succession of
Figs. 4.2a, b and c. Continuous flow may be regarded as a succession of these non-flow
processes, carried out for every element of fluid entering the open system. It now becomes possible to apply the energy equation which has been developed from a study
of closed systems.

First consider the state of the imaginary closed system in the configuration shown
in Fig. 4.2a. Fig. 4.3 shows an enlarged view of the part of the system containing the

Fig. 4.2   Imaginary non-flow process

element of fluid which is about to enter the open system. The thermodynamic and
mechanical states of the element of mass δm may be assumed to be known. The random
molecular energy of the element is U_1 say, and if the element is moving with a velocity
C_1 at a height z_1 above some reference level, it also possesses kinetic energy $\delta m \, C_1^2/2$
and potential energy $\delta m \, gz_1$. The total energy of the mass δm is therefore

$$U_1 + \tfrac{1}{2}\delta m \, C_1^2 + \delta m \, gz_1$$

If we assume that the element is small enough for the thermodynamic properties to be
uniform throughout its extent, this becomes

$$\delta m(u_1 + \tfrac{1}{2}C_1^2 + gz_1)$$

At the instant when mass δm reaches the boundary of the open system, let the fluid
within the open system possess a quantity of internal energy E'. E' includes the kinetic
and potential energies of the fluid elements within the boundary of the open system;
the possibility of using the term 'internal energy' in this way has already been noted in
section 2.4. For reasons which will appear later, however, there is no need to specify

*This approach was suggested in a paper by Gillespie and Coe, Ref. (9), and used by Keenan in his
Thermodynamics, Ref. (6).

the different forms of energy of which E' consists, nor does it matter that this quantit
of fluid is not in a state of equilibrium and that the mass of fluid is unknown. The tot
internal energy of the imaginary closed system in its initial state (Fig. 4.2a) is

$$E' + \delta m(u_1 + \tfrac{1}{2}C_1{}^2 + gz_1)$$

Now consider the state of the closed system in the configuration shown in Fig. 4.2
Since, for steady flow, the properties of the fluid at any point in the open system do nc
change with time, the quantity of energy contained within the original boundary mu:
still be E'. Therefore, if a mass δm leaves the open system at some part of the boundar

Fig. 4.3

where the properties are denoted by suffix 2, the final internal energy of the imaginar
closed system (Fig. 4.2c) will be

$$E' + \delta m(u_2 + \tfrac{1}{2}C_2{}^2 + gz_2)$$

Lastly, we must consider how the imaginary closed system changes its state fror
that in Fig. 4.2a to that in Fig. 4.2c. In this period of time we may assume that δQ
units of energy are transferred to the system as heat, and that δW units of energy ar
transferred to the surroundings as work via a turbine shaft. δW is not the only wor
done as far as the imaginary closed system is concerned because parts of its boundar
move at sections 1 and 2. For the element δm to enter the open system, the imaginar
closed system must be compressed, its volume decreasing by $\delta m\, v_1$. This is accompanie
by a force $p_1 A$ moving a distance $l = \delta m\, v_1/A$, where A is the cross-sectional area o
the element. The work done by the surroundings on the system is therefore $\delta m\, p_1 v$.
Similarly, it can be shown that the work done by the closed system on the surrounding
as δm is expelled from the open system, is $\delta m\, p_2 v_2$.* The net work done *by* the syster
during the change represented in Fig. 4.2 is therefore

$$\delta W + \delta m(p_2 v_2 - p_1 v_1)$$

*The product pv has been called the flow work, flow energy and pressure energy by various authors. Suc
names may be misleading, and there is no need to give this quantity any special name at all. The term arise
out of the need to view the flow process as a series of non-flow processes, and so far as the open system
concerned it is merely the product of two properties, pressure and specific volume; it is not a part of th
energy of the system nor does it represent any work done on or by the *open* system.

We may now apply the energy equation to the imaginary closed system and write

$$\delta Q - \{\delta W + \delta m(p_2 v_2 - p_1 v_1)\}$$
$$= \{E' + \delta m(u_2 + \tfrac{1}{2}C_2{}^2 + gz_2)\} - \{E' + \delta m(u_1 + \tfrac{1}{2}C_1{}^2 + gz_1)\} \qquad (4.1)$$

The quantity E' cancels, and by writing the enthalpy h for $(u + pv)$, the equation reduces to

$$\delta Q - \delta W = \delta m(h_2 + \tfrac{1}{2}C_2{}^2 + gz_2) - \delta m(h_1 + \tfrac{1}{2}C_1{}^2 + gz_1)$$

The continuous steady-flow process consists of the sum total of all the elemental mass transfers across sections 1 and 2: it may therefore be represented by the equation

$$\Sigma \delta Q - \Sigma \delta W = \Sigma \delta m(h_2 + \tfrac{1}{2}C_2{}^2 + gz_2) - \Sigma \delta m(h_1 + \tfrac{1}{2}C_1{}^2 + gz_1)$$

This equation can be simplified considerably if the properties are uniform over the cross-section of flow at inlet and outlet. Thus if every δm enters the system with the properties h_1, C_1, z_1 and leaves with the properties h_2, C_2, z_2, the equation can be written as

$$\Sigma \delta Q - \Sigma \delta W = (h_2 + \tfrac{1}{2}C_2{}^2 + gz_2)\Sigma \delta m - (h_1 + \tfrac{1}{2}C_1{}^2 + gz_1)\,\Sigma \delta m$$

We may now write m for $\Sigma \delta m$, where m denotes any mass of fluid, usually the mass entering and leaving per unit time. $\Sigma \delta Q$ and $\Sigma \delta W$ are the heat and work transfers associated with the flow of mass m through the system. If we divide throughout by m, we can write the equation finally as

$$Q - W = (h_2 - h_1) + \tfrac{1}{2}(C_2{}^2 - C_1{}^2) + g(z_2 - z_1) \qquad (4.2)$$

where Q and W are the heat and work transfers per unit mass flowing through the system. Equation (4.2) expresses the relation between heat and work and the properties of the fluid at the inlet and outlet of an open system. It is called the *steady-flow energy equation*, and provides the basic means for studying most open systems of importance in engineering.

The assumptions upon which equation (4.2) is based may be summarised as follows:

(a) The mass flow at inlet is constant with respect to time, and equal to the mass flow at outlet.
(b) The properties at any point within the open system do not vary with time.
(c) The properties are constant over the cross-section of the flow at inlet and outlet.
(d) Any heat or work crossing the boundary does so at a uniform rate.

Note that no assumption has been made about the nature or configuration of the steady flow *within* the open system, and the steady-flow energy equation applies whether there is viscous friction or not. It corresponds to the general non-flow energy equation (3.1).

It is not possible for all these conditions to be satisfied in any practical process. For example, when a fluid flows along a pipe, it is known that viscous friction causes the velocity to fall rapidly to zero in the layer adjacent to the wall. Friction between the layers of fluid will also slightly affect the distribution of thermodynamic properties across the pipe. Thus assumption (c) is not completely justified. Nevertheless, experiments have shown that equation (4.2) can be applied in these circumstances with sufficient accuracy for practical purposes if mean values of the properties are used. To take another example, it is not necessary for assumption (b) to be satisfied rigorously. The properties at any point within the open system may be allowed to fluctuate.

Fig. 4.4   Simple water tube boiler

Provided that they return to the same values periodically, the equation can still be applied. Under these circumstances, the quantities Q and W represent the average heat transferred and work done per unit mass, over a period of time appreciably longer than that required for one cyclic variation of the internal energy E'.

Finally we must note an important equation which follows directly from assumptions (a) and (c). It is known as the *continuity equation* and expresses the principle of conservation of mass in steady flow. If m is the rate of mass flow, we have

$$m = \frac{A_1 C_1}{v_1} = \frac{A_2 C_2}{v_2} \qquad (4.3)$$

The steady-flow energy equation will now be applied to various open systems of particular importance. In all these examples the potential energy term $g(z_2 - z_1)$ is ither zero or negligible compared with the other terms. It has been included in quation (4.2) for the sake of completeness, but will henceforth be omitted. It may ppear, for example, in the analysis of hydraulic machinery where the fluid is a liquid nd the enthalpy and heat terms are negligible; but such systems are not considered n this book.

4.2 Open systems with steady flow

i) *Boiler and condenser*

Fig. 4.4 illustrates a boiler, the fluid entering as a liquid and leaving as a vapour at a onstant rate. In this case no work is done on or by the fluid as it passes through the ystem, and hence $W = 0$. The velocities of flow are usually quite low, so that the

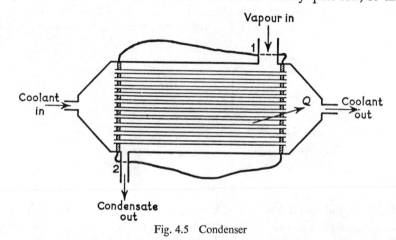

Fig. 4.5   Condenser

difference between the kinetic energies at inlet and outlet is negligible compared with the other terms of the equation. This is verified by a numerical example in section 10.1. The steady-flow energy equation for this open system reduces to

$$Q = (h_2 - h_1) \tag{4.4}$$

where Q is the heat required to produce unit mass of vapour. Or, if m kg of vapour are produced per second, and we write $Q = m(h_2 - h_1)$, Q is the heat required per second. All that is necessary for the quantity of heat to be determined is a knowledge of the thermodynamic state of the fluid as it enters and leaves the boiler.

Fig. 4.5 illustrates a simple surface condenser. The vapour passes over a bank of tubes, and is condensed as it comes into contact with the surface of the tubes. The tubes are maintained at a lower temperature than the vapour by a flow of cooling water. The cooling water is not part of the fluid of this open system, but acts as a sink of heat in the surroundings. Equation (4.4) also applies to this system, although in this case h_1 is greater than h_2 and Q is negative.

(ii) *Nozzle and diffuser*

A nozzle is a duct of varying cross-sectional area so designed that a drop in pressure from inlet to outlet accelerates the flow. A simple convergent nozzle is shown in Fig. 4.6a. The flow through a nozzle usually occurs at very high speed, and there is little time for the fluid to gain or lose energy by a flow of heat through the walls of the nozzle as the fluid passes through it. The process is therefore always assumed to be adiabatic. Also, no work crosses the boundary during the process. The steady-flow energy equation becomes

$$\tfrac{1}{2}(C_2{}^2 - C_1{}^2) = (h_1 - h_2) \qquad (4.5)$$

i.e. the gain in kinetic energy is equal to the decrease of enthalpy.

If the thermodynamic state at inlet and outlet and the inlet velocity are known, the final velocity can be calculated. Normally only one property in the final thermodynamic state, namely the pressure, is known, and some additional assumption must

(a) Nozzle   (b) Diffuser

Fig. 4.6

be made about the flow before h_2, and hence the final velocity, can be predicted. This additional assumption is that the flow is reversible. The way in which the concept of reversibility can be applied to flow processes will be described after the Second Law has been introduced (sections 5.4 and 6.6).

The function of a diffuser is the reverse of that of a nozzle; the diffuser is a duct so shaped that the fluid flowing through it decelerates, the pressure increasing from inlet to outlet. Fig. 4.6b shows a simple divergent diffuser. Equation (4.5) applies to this system also, although in this case both sides of the equation will be negative.

(iii) *Turbine and compressor*

A turbine is a means of extracting work from a flow of fluid expanding from a high pressure to a low pressure. Fig. 4.7 is a diagrammatic sketch of one type of turbine. Briefly, the fluid is accelerated in a set of fixed nozzles, and the resulting high-speed jets of fluid then change their direction as they pass over a row of curved blades attached to a rotor. A force is exerted on the blades equal to the rate of change of momentum of the fluid, and this produces a torque at the rotor shaft. At the same time the velocity of the fluid is reduced to somewhere near the value it possessed before entering the nozzles. As a first approximation the velocity of flow at the inlet and outlet of the turbine can be assumed equal. Since the average velocity of flow through a turbine is very high, the process can be assumed to be adiabatic. Hence the energy equation becomes,

$$W = (h_1 - h_2) \tag{4.6}$$

If the mass flow is m kg/s, and we write $W = m(h_1 - h_2)$, then W is the work done per second.

The rotary compressor can be regarded as a reversed turbine, work being done on the fluid to raise its pressure. In this case work is done on the fluid by a bladed rotor driven from an external source. This increases the velocity of the fluid. The velocity

Fig. 4.7   Turbine

is then reduced in a set of fixed diffusers to some value approximating to that at inlet to the compressor, and the pressure is increased. Equation (4.6) is applicable, but in this case W is negative because work is done on the system, and $h_2 > h_1$.

Again, normally the final thermodynamic state is not known. The outlet pressure p_2 may be known, but a second property must be determined by making some additional assumption about the flow before h_2 can be found and the work of expansion or compression predicted.

(iv) *Throttling*

A flow of fluid is said to be *throttled* when some restriction is placed in the flow. A partially closed valve in a pipe line is a simple example, Fig. 4.8. The term 'throttling' is usually applied to relatively low-speed flow, i.e. low enough for any difference between the kinetic energy at inlet and outlet to be negligible. A numerical example in section 10.3 shows the order of magnitude. Any heat transfer across the boundary

39

can be neglected; this time for the reason that throttling takes place in such a short length of pipe that the surface area across which heat can flow is very small. Also, no work crosses the boundary. Thus the energy equation reduces to

$$h_1 = h_2 \qquad (4.7)$$

i.e. the throttling process is an adiabatic steady-flow process such that the enthalpy is the same at inlet and outlet.

(v) *Reciprocating compressor*

Fig. 4.9 illustrates a reciprocating compressor. The remarks made about assumption (b) in the last part of section 4.1 apply to this system. If a receiver is inserted in the pipe line at the outlet of the compressor, of sufficient capacity to damp out the pulsations, the properties of the fluid at the outlet can be assumed to be constant. Atmospheric conditions usually determine the properties at the inlet. The steady-flow energy equation can then be applied to relate the average heat transfer and work done per unit mass to the properties of the fluid at inlet and outlet. Some heat transfer does occur in a reciprocating compressor because the flow occurs at a relatively low speed, and the surface area of the cylinder walls is appreciable. The difference between the kinetic energies at inlet and outlet, however, can be neglected. Thus the equation becomes

Fig. 4.8   Throttling process

$$Q - W = (h_2 - h_1) \qquad (4.8)$$

4.3 Nonsteady-flow processes

A fluid expanding through an orifice in the side of a vessel provides a simple example of an open system undergoing a nonsteady-flow process. Neither the mass of fluid within the open system, nor the quantity of internal energy possessed by this mass, remains constant with respect to time. That is, the quantity denoted by E' varies, and it cannot be eliminated from the energy equation. Moreover, in the general case, the properties of the fluid at inlet and outlet of the open system also vary during the process. If E' is to be capable of being specified at any instant, the mechanical and thermodynamic state of each element of fluid within the system must be known. The total change $\sum \delta E = (E'' - E')$ between the initial and final states of the system denoted by single and double primes respectively, can be calculated from

$$(E'' - E') = \sum_{V''} \delta m \left(u'' + \frac{C''^2}{2} + gz'' \right) - \sum_{V'} \delta m \left(u' + \frac{C'^2}{2} + gz' \right)$$

where V', V'' denote summations carried out over the initial and final volumes of the system.

40

We shall now simplify the problem by assuming that the initial and final states of the system are (a) states of thermodynamic equilibrium implying that u is uniform throughout the volume of the system initially and finally, and (b) states of rest implying that the kinetic energy terms are zero. For most practical unsteady-flow processes it is possible to choose a boundary and boundary conditions which fulfil these assumptions. This is fortunate because otherwise the determination of E' and E'' would be very difficult. The above expression then reduces to

$$(E'' - E') = m''(u'' + gz'') - m'(u' + gz')$$

Inlet valve
shut on
upstroke

Delivery valve
shut on
downstroke

1

Receiver

Q

W

2

Fig. 4.9   Reciprocating compressor

where z' and z'' must be interpreted as the heights of the centre of gravity of the initial and final masses m' and m''. Even if the position of the centre of gravity does not change, the potential energy would change if the mass of the system changes.

We shall consider the general case depicted in Fig. 4.1, starting as before with the elemental process undergone by an imaginary closed system as illustrated in Fig. 4.2. The difference in this case is that as the imaginary closed boundary is deformed, a mass δm_1 enters the system and a different mass δm_2 leaves it. During this elemental process, increments of heat δQ and work δW cross the boundary of the open system, and the total internal energy of the fluid within the original open boundary changes by an amount $\delta E'$. The primed quantities refer to the whole mass contained in the open system; a single prime for the initial equilibrium state and a double prime for the final equilibrium state. Suffixes 1 and 2 are used to denote the properties of the fluid as it enters and leaves the system at any instant. The energy equation for the non-flow process becomes,

$$\delta Q - \{\delta W + (\delta m_2 p_2 v_2 - \delta m_1 p_1 v_1)\}$$
$$= \delta E' + \delta m_2(u_2 + \tfrac{1}{2}C_2{}^2 + gz_2) - \delta m_1(u_1 + \tfrac{1}{2}C_1{}^2 + gz_1)$$

Although the properties at inlet and outlet may again be assumed constant during this elemental process, they may have different values for each successive elemental pro-

41

cess, i.e. the properties vary with time. The quantities δQ, δW and $\delta E'$ may also vary We may write down similar equations for each elemental process between the initia and final states of the open system. Adding these equations, we have

$$\Sigma \, \delta Q - \{\Sigma \, \delta W + (\Sigma \, \delta m_2 p_2 v_2 - \Sigma \, \delta m_1 p_1 v_1)\}$$
$$= \Sigma \, \delta E' + \Sigma \, \delta m_2 (u_2 + \tfrac{1}{2} C_2{}^2 + g z_2) - \Sigma \, \delta m_1 (u_1 + \tfrac{1}{2} C_1{}^2 + g z_1$$

If we write Q and W for the total quantities of heat and work crossing the boundar during the whole nonsteady-flow process, substitute $m''(u'' + gz'') - m'(u' + gz')$ fo $\Sigma \, \delta E'$, and introduce the property enthalpy, this reduces to

$$Q - W = \{m''(u'' + gz'') - m'(u' + gz')\} + \Sigma \, \delta m_2 (h_2 + \tfrac{1}{2} C_2{}^2 + g z_2) -$$
$$\Sigma \, \delta m_1 (h_1 + \tfrac{1}{2} C_1{}^2 + g z_1) \tag{4.9}$$

Although this equation is still very cumbersome in its general form, it can b simplified further when it is applied to particular systems. One example will be con sidered in general terms, and other numerical examples are given in section 10.6.

Consider the system shown in Fig. 4.10. A fluid under pressure in a container i allowed to expand through a turbine to atmospheric pressure. We require an expres

Fig. 4.10 Nonsteady-flow process

sion for the work done by the turbine as the fluid in the container changes from on state of equilibrium to another, i.e. during the time the valve is allowed to remain open The state of the fluid in the container is assumed to be known before the valve i opened and after it is closed. For this system the potential energy terms can b neglected.

A major simplification in equation (4.9) is achieved because no mass enters th system. Also, since the process will be rapid, it may be assumed adiabatic. Finally as a first approximation the kinetic energy of the fluid leaving the turbine will b ignored; a turbine is usually designed for this to be small as it represents a waste o energy. Thus equation (4.9) reduces to

$$W = (m'u' - m''u'') - \Sigma \, \delta m_2 h_2 \tag{4.10}$$

In other words, the work done by the turbine is equal to the decrease of internal energy in the system minus the enthalpy of the fluid that has left the system. A description of the way in which this equation can be applied in a practical case must be left to Part II.

4.4 Cycles consisting of steady-flow processes

During a steady-flow process the open system itself does not, strictly speaking, undergo any process, because the properties remain constant at every point within it: nevertheless individual elements of fluid undergo a process as they flow through the system. It is therefore possible to imagine a series of open systems through which the

Fig. 4.11   Closed-cycle plant

fluid circulates continuously, each element of fluid passing through a cycle of mechanical and thermodynamic states.

Fig. 4.11 illustrates a simple steam power plant in which the fluid undergoes a continuous series of cyclic processes. Water at a certain pressure and temperature enters the boiler and is converted to steam; the steam expands to a low pressure in the turbine and is condensed at that low pressure; finally the water is returned to the boiler by the feed pump at its original pressure and temperature. The individual components—boiler, turbine, condenser and feed pump—are open systems, and the energy equation for each may be written as:

Boiler: $Q_{12} = m(h_2 - h_1)$

Turbine: $-W_{23} = m(h_3 - h_2)$

Condenser: $Q_{34} = m(h_4 - h_3)$

Feed pump: $-W_{41} = m(h_1 - h_4)$

Adding these equations we have

$$(Q_{12} + Q_{34}) - (W_{23} + W_{41}) = 0$$

or

$$\Sigma\, \delta Q - \Sigma\, \delta W = 0$$

43

Thus the net rate of heat addition to the whole plant equals the net rate of work done by the plant.

This equation is analogous to the original expression of the First Law made for a closed system undergoing a cycle of non-flow processes. It could have been deduced by regarding the whole plant as a closed system. The total internal energy of this system does not vary because the properties are constant with respect to time at every point within it. Consequently if heat and work cross the boundary at various points at a uniform rate, the net rate of heat entering the closed system must be equal to the

Fig. 4.12   Open-cycle plant

net rate of work done by it. If this were not so, the total internal energy would increase or decrease.

Fig. 4.11 is an example of a plant operating on a *closed cycle*, i.e. one in which the same fluid passes round the circuit continuously. Fig. 4.12 illustrates an *open-cycle* plant, where the fluid enters at 'a' and leaves at 'b' with the same properties.* This can be regarded as a single open system, so that the steady-flow energy equation can be applied directly to give

$$(Q_{12} + Q_{34}) - (W_{23} + W_{41}) = (h_b - h_a) + \tfrac{1}{2}(C_b{}^2 - C_a{}^2)$$

Since the fluid enters and leaves the system in the same state, the right-hand side of the equation is zero, and again

$$\sum \delta Q - \sum \delta W = 0$$

All the properties revert to their initial values, and the fact that fresh quantities of fluid act as a vehicle for these properties is irrelevant.

*The term 'open-cycle plant' is often used to describe a power plant which uses a continuous fresh supply of working fluid, but where not all the properties are identical at inlet and outlet: see section 12.1.

5   The Second Law of Thermodynamics and Reversibility

In this chapter the idea of a cycle efficiency is introduced, and the Second Law is then stated and distinguished from the First. One of the corollaries of the Second Law is deduced, but the development of the other corollaries requires a fuller treatment of the concept of reversibility than has so far been given. These corollaries are therefore dealt with in the next chapter, while the remainder of this chapter is devoted to the formal definition of a reversible process and to a discussion of its implications both for non-flow and steady-flow processes.

5.1 Cycle efficiency

When a closed system undergoes a series of non-flow processes, during which the fluid passes through a cycle of thermodynamic states, it has been seen that the First Law is expressed by

$$\Sigma \, \delta Q = \Sigma \, \delta W$$

Up to now the symbol Q has been used for a positive or negative quantity of heat. *In what follows it will be more convenient to use separate symbols for positive and negative quantities.* Furthermore, the symbols will be used to stand for the positive numerical values, and the accompanying sign will indicate the direction of flow in accordance with our convention. Thus, if Q_1 is the heat supplied, Q_2 the heat rejected, and W the net work done by the system during the cycle, the First Law becomes

$$Q_1 - Q_2 = W \tag{5.1}$$

A system operating in a cycle and producing a net quantity of work from a supply of heat is called a *heat engine*. The greater the proportion of the heat supply converted into work, the better is the engine. Consequently the *cycle efficiency* of a heat engine is defined as

$$\eta = \frac{\text{work done}}{\text{heat supplied}} = \frac{W}{Q_1} \tag{5.2}$$

Hence, from equations (5.1) and (5.2), we have

$$\eta = \frac{Q_1 - Q_2}{Q_1} \tag{5.3}$$

The quantities Q_1 and Q_2 may be expressed per unit mass of the system, or they may be regarded as the rate of heat supplied and rejected. The latter alternative is possible because the numerator and denominator of equation (5.2) can be multiplied by the number of cycles per unit time without affecting their ratio.

The efficiency of any heat engine operating on either a closed cycle or an open cycle, consisting of steady-flow processes, can also be expressed by equation (5.3). In this case Q_1 and Q_2 must be interpreted either as the heat supplied and rejected per unit mass flowing round the circuit, or as the rates of heat supplied and rejected.

5.2 The Second Law of Thermodynamics

There is nothing implicit in the First Law to say that some proportion of the heat supplied to an engine must be rejected, and therefore that the cycle efficiency cannot

$$(a)$$
Heat engine

$$(b)$$
Heat pump or refrigerator

Fig. 5.1

be unity. All that the First Law states is that net work cannot be produced during a cycle without *some* supply of heat, i.e. that a perpetual motion machine of the first kind is impossible. The *Second Law* is an expression of the fact that some heat must always be rejected during the cycle, and therefore that the cycle efficiency is always less than unity. The law may be stated in the following form.* *It is impossible to construct a system which will operate in a cycle, extract heat from a reservoir, and do an equivalent amount of work on the surroundings.* Thus the First Law says that the net work can never be greater than the heat supplied, while the Second Law goes further and states that it must always be less. If $W < Q_1$, then it follows from equation (5.1) that Q_2 must have some finite value.

If energy is to be supplied to a system in the form of heat, the system must be in contact with a reservoir at a temperature higher than that of the fluid at some point in the cycle. Similarly, if heat is to be rejected, the system must at some time be in contact with a reservoir of lower temperature than the fluid. Thus the Second Law implies that if a system is to undergo a cycle and produce work, it must operate between at least two reservoirs of different temperature, however small this difference may be. Fig. 5.1a will be used to represent a heat engine. A machine which will

*This is essentially the statement given by Planck in Ref. (7).

produce work continuously, while exchanging heat with only a single reservoir, is known as a perpetual motion machine of the second kind; such a machine would contradict the Second Law.

It is now possible to see why a ship could not be driven by an engine using the ocean as a source of heat, or why a power station could not be run using the atmosphere as a source of heat. There is nothing in the First Law to say that these desirable projects are not feasible. Neither project would contravene the principle of conservation of energy; their impossibility is a consequence of the Second Law. They are impossible because there is no natural infinite sink of heat at a lower temperature than the atmosphere or ocean, and they would therefore be perpetual motion machines of the second kind.

The cycles described so far have all been those of heat engines. However, we can devise cycles during which the net work is done *on* the system while a net amount of heat is *rejected*. In this case the quantities in equation (5.1) are of opposite sign, i.e. Q_1 is the heat rejected, Q_2 is the heat supplied, and W is the net work done on the system. This would be the cycle of a refrigerator or heat pump, and it can be regarded as a reversed heat engine cycle. The refrigerator or heat pump is represented symbolically by Fig. 5.1b. It should be noted that the Second Law does not imply that work cannot be continuously and completely converted into heat. Indeed any process involving friction achieves this without the need for the system to operate in a cycle. For example, a fluid in a closed vessel may have work done on it by a paddle at the same time as heat is allowed to cross the boundary into the surroundings. The rates of flow of work and heat may be made equal, the internal energy of the system remaining constant. But this is an extremely wasteful method of producing heat because, by using the work to drive a heat pump, a larger quantity of heat may be delivered to the surroundings for the same expenditure of work. The quantity of heat delivered is then greater than the net work done on the system by an amount equal to the heat drawn from the colder reservoir. An important consequence of the Second Law is, therefore, that *work is a more valuable form of energy transfer than heat*: heat can never be transformed continuously and completely into work, whereas work can always be transformed continuously and completely into heat, and, if properly used, can even result in a supply of heat which is greater than the quantity of work expended.

This section has been devoted primarily to distinguishing between the First and Second Laws. It should be clear that the Second Law cannot be deduced from the First; it is a separate axiom. Like the First Law, its validity rests upon the fact that neither the law, nor any of its consequences, have ever been disproved by experience. The following statements summarise the more obvious consequences of the Second Law:

(a) If a system is taken through a cycle and produces work, it must be exchanging heat with at least two reservoirs at different temperatures.

(b) If a system is taken through a cycle while exchanging heat with only one reservoir, the work done must be either zero or negative.

(c) Since heat can never be converted continuously and completely into work, whereas work can always be converted continuously and completely into heat, work is a more valuable form of energy transfer than heat.

Other consequences of the Second Law will be developed in the form of corollaries, and the first is deduced in the next section.

5.3 The Clausius statement of the Second Law

The first corollary of the Second Law has often been used as a statement of the law; it was first used in this way by Clausius. When so used, the statement of the law already given can easily be shown to be a corollary of this proposition.

COROLLARY 1. *It is impossible to construct a system which will operate in a cycle and transfer heat from a cooler to a hotter body without work being done on the system by the surroundings.*

PROOF. Suppose the converse of the proposition be true. The system could be represented by a heat pump for which $W = 0$, as in Fig. 5.2. If it takes Q units of heat from the cold reservoir, it must deliver Q units to the hot

Fig. 5.2

reservoir to satisfy the First Law. A heat engine could also be operated between the two reservoirs; let it be of such a size that it delivers Q units of heat to the cold reservoir while performing W units of work. Then the First Law states that the engine must be supplied with $(W + Q)$ units of heat from the hot reservoir. In the combined plant the cold reservoir becomes superfluous because the heat engine could reject its heat directly into the heat pump. The combined plant represents a heat engine extracting $(W + Q) - Q = W$ units of heat from a reservoir, and delivering an equivalent amount of work. This is impossible according to the Second Law. Consequently the converse of the proposition cannot be true and the original proposition must be true.

5.4 Reversibility and irreversibility

The idea of a reversible process has so far only been developed with respect to non-flow processes. It was presented as a process between two states during which the system passes through a series of equilibrium states. For this to happen, it was shown that the heat and work must be transferred across the boundary in a particular way, i.e.

(a) heat must flow by virtue of an infinitesimally small temperature difference, and
(b) work must only be done when the force exerted by the surroundings on the moving boundary is infinitesimally different from the force due to the pressure of the fluid in the system.

When these conditions are fulfilled, only an infinitesimal change is necessary in the temperature of the reservoir, or in the force applied to the boundary by the surroundings, to reverse the process and make the system return through the same series of equilibrium states to the initial state. If this is done, any work previously delivered to the surroundings is returned to the system, and any heat supplied to the system is returned to the reservoir. No effects of either the original process or the reversed process are left in the system or the surroundings, because any changes required to reverse the process are infinitesimal.

So that the idea of reversibility can be applied to non-flow and steady-flow processes, the following criterion of reversibility will be adopted: *When a fluid undergoes a reversible process, both the fluid and its surroundings can always be restored to their original states.* Steady-flow processes undergone by a fluid in an open system can easily be imagined to be reversible according to this criterion. For example, imagine that a fluid, after expanding adiabatically in a turbine, is compressed adiabatically in a rotary compressor. If this can be carried out so that the fluid is returned to its initial thermodynamic and mechanical states, and so that the work obtained from the turbine is just sufficient to drive the compressor, then the expansion in the turbine satisfies the criterion for a reversible process. Similarly, it is theoretically possible for the fluid leaving a nozzle to be returned to its initial state, by allowing it to flow through a diffuser, without leaving any effects of the combined process in the surroundings. On the other hand the original conception of a reversible process, as one during which a system passes through a series of equilibrium states, is too narrow to enable it to be applied to a steady-flow process. We have already pointed out that a steady-flow system, as such, cannot be said to undergo a process at all, because the properties of the fluid within it are everywhere constant with respect to time; moreover, it is clearly never in a state of equilibrium.

We shall proceed to show that if either of the following phenomena is present during a process it cannot be reversible:

(i) friction, and
(ii) heat transfer across a finite temperature difference.

Since at least one of these is always present in some degree, no real process can be reversible. The reversible process must be regarded as a limiting case—an ideal which can never quite be achieved. Any particular process involving (i) or (ii) can be shown to be irreversible, although no general proof can be given. The procedure will be to find processes by which the fluid can be returned to its initial state, and then show that the surroundings cannot also be returned to their original state without contradicting the Second Law.

(i) *Friction*

Friction is always present when one part of a fluid moves relatively to another, or to a solid boundary, with finite velocity. This is inherent in the viscous nature of fluids. The effect of viscous friction is always to produce a conversion of some of the kinetic energy of the fluid into random molecular energy. Non-flow and steady-flow processes will be considered in turn.

(a) *Friction in non-flow processes.* Consider the simple case of a fixed volume of fluid being stirred by a paddle. Work is done on the system and the First Law states that

$$-W = (u_2 - u_1)$$

If the process is reversible it must be possible to return the quantity W to the surroundings, and restore the internal energy of the system to u_1 without change of volume. The internal energy can be restored to its original value by allowing heat to flow across the boundary into the surroundings equal in amount to the former increase in internal energy. We then have to think of a device which will convert all this heat into work without it itself introducing further changes of state in the surroundings.

Fig. 5.3

Only a cyclic device, i.e. a heat engine, could do this, but such an engine is impossible according to the Second Law. Therefore the stirring process does not satisfy the criterion of reversibility.

In practical processes, any deliberate stirring of the fluid is avoided whenever possible. Nevertheless it might be supposed that friction would make its presence felt during the ordinary non-flow expansion or compression of a fluid in the cylinder of a reciprocating machine operating at a reasonable speed. With any such process there will always be a finite difference between the force displaced in the surroundings, and the opposing force due to the pressure of the fluid in the system. Thus the process must be accompanied by an acceleration of the moving boundary and working fluid, and there will be some conversion of kinetic energy into random molecular energy due to friction. As an example, consider the behaviour of the adiabatic system containing unit mass of fluid shown in Fig. 5.3. For simplicity the piston may be assumed weightless and frictionless, so that the surroundings exert a constant force on the fluid due to the atmospheric pressure p_a. Let the initial pressure of the fluid, p_1, be greater than this, a stop holding the piston in position. When the stop is removed, the fluid expands and the piston accelerates under the action of the unbalanced force $(p_1 - p_a)A$, where A is the area of the piston.

Before the system can settle down to a new state of equilibrium, where the pressure of the fluid has fallen to p_a, the kinetic energy of the fluid must be dissipated. This is accomplished by an oscillation which is damped out by the viscous forces in the fluid. Since the system is thermally insulated from the surroundings, the whole of the kinetic energy is reconverted to random molecular energy u by fluid friction.* The decrease of internal energy from one equilibrium state to the other is therefore smaller than it would have been had the expansion from p_1 to p_a proceeded infinitely slowly, i.e. less internal energy has left the system.

*A column of air in the surroundings will also have been accelerated and viscous forces in the surroundings will contribute to the damping; for simplicity we may neglect this effect, e.g. by assuming the surroundings to have negligible viscosity.

The process can be represented approximately on the p-v diagram as the dotted curve in Fig. 5.4. The end states are defined by (p_1, v_1) and (p_a, v_2). Although the pressure is not uniform throughout the system during the process, an average pressure might be measured to give the intermediate points some meaning. From the First Law for an adiabatic non-flow process,

$$W = (u_1 - u_2)$$

When carried out infinitely slowly, this might be written

$$W' = (u_1 - u_2')$$

where $u_2' < u_2$ and $W' > W$. When the pressure is constant, a decrease of inter energy of any practical fluid is always accompanied by a decrease of volume. Thus v_2' is less than v_2, and the perfectly restrained process might be shown by the full line 1–2'.

Fig. 5.4

The work done during the actual process, in moving the constant force p_aA a distance $(v_2 - v_1)/A$, is equal to $p_a(v_2 - v_1)$. This is represented by the area of the shaded rectangle. It is clearly less than the area under the dotted curve, which is given by $\int_1^2 p \, dv$ where p is interpreted as the average pressure at any point in the expansion.

By a similar argument it can be shown that a compression, carried out under similar conditions, would require an amount of work equal to $p_2(v_1 - v_2)$ to be done on the system by a constant force p_2A in the surroundings. This work is equal to the area of the shaded rectangle in Fig. 5.5. In this case the work done is greater than $\int_1^2 p \, dv$.

The work of expansion can be increased, and the work of compression reduced, by applying the external force in a series of steps decreasing successively from p_1A to p_aA in the former case, and increasing successively from p_aA to p_2A in the latter. The difference between the positive work of expansion and the negative work of compression is then reduced. In the limit, when the external force never departs by more than an infinitesimal amount from that due to the pressure of the fluid, the work done will be numerically equal to $\int_1^{2'} p \, dv$ in each case and the processes become reversible.

In an actual reciprocating engine the mass of the piston and other moving parts is considerable, but most of the kinetic energy acquired by these is stored in the flywheel and recovered. The piston is prevented from undergoing the oscillation which would occur if the piston were free, and the kinetic energy acquired by the fluid exists in the form of eddies which are damped out by viscous friction. The creation and dissipation of kinetic energy is therefore a continuous process throughout the expansion or compression. With normal piston speeds the quantity of energy dissipated in this way is very small, and the effect of fluid friction can usually be ignored. Thus, when a calculation of the work done during a practical non-flow expansion or compression is made on the assumption that it is a reversible process, it provides not only a criterion of perfection but also a close approximation to the actual value.* It will not be such a good approximation when special devices are used to introduce turbulence, as in some compression-ignition engines.

(*b*) *Friction in steady-flow processes.* When considering the effect of friction during flow processes, it is usually found to be by no means negligible. In nozzles, turbines and rotary compressors, the speed of flow is high and the friction is considerable.

Fig. 5.5

Consider first the effect of friction in a nozzle. Let the thermodynamic state of the fluid at the inlet be defined by (p_1, h_1), and let the inlet velocity be C_1. Assume that the outlet pressure is fixed at some value p_2, this being chosen as the independent property of the final state. The appropriate steady-flow energy equation (4.5) states that

$$\tfrac{1}{2}(C_2^2 - C_1^2) = (h_1 - h_2)$$

Friction always produces a force opposing the motion, and after the expansion to pressure p_2 the stream of fluid moves more slowly than it would have done had there been no friction. If the values which the dependent properties would have after frictionless flow are denoted by primes, this means that $C_2 < C_2'$. Consequently it follows that $h_2 > h_2'$, since the First Law must be satisfied, i.e.

$$\tfrac{1}{2}(C_2'^2 - C_1^2) = (h_1 - h_2')$$

During the process with friction, some of the directional kinetic energy $(C^2/2)$ is being continually transformed into random molecular energy (u). But at any given pressure an increase of u must involve an increase in the specific volume v, and consequently an increase in enthalpy $(u + pv)$. In effect, some of the initial enthalpy which could have produced an increase in kinetic energy has been prevented from doing so by friction.

It is often helpful to represent a steady-flow process on a coordinate diagram, although it cannot be completely defined in this way. The thermodynamic state of the fluid at the inlet and outlet may be represented by points on a diagram using two thermodynamic properties as coordinates, and a line joining the points can be used

---

*We are here referring to the work done on or by the fluid: some of this work will be dissipated externally by piston friction, but this is expressed separately by the mechanical efficiency of the machine—see section 16.2.

to represent the process.* A full line will always be used to denote a frictionless process, and a dotted line to denote a process where friction is present. Such diagrams only partially represent the process because the changes in mechanical properties are not shown. Fig. 5.6a shows the expansion process in a nozzle on a p-h diagram.

Fig. 5.6a can equally well represent the expansion in a turbine between pressures p_1 and p_2. With the usual assumptions, i.e. adiabatic flow and equal inlet and outlet velocities, the energy equation is

$$W = (h_1 - h_2)$$

If the flow is frictionless it may be written

$$W' = (h_1 - h'_2)$$

where $W' > W$ and $h'_2 < h_2$. Similarly, Fig. 5.6b can represent the processes in either a diffuser or a rotary compressor. As always, friction results in the final enthalpy

Fig. 5.6

being greater than it would have been after a frictionless process to the same final pressure. Considering the case of a rotary compressor, this means that the actual work of compression $(h_2 - h_1)$ is greater than the frictionless work $(h'_2 - h_1)$.

As an example, we may show that friction makes an expansion in a turbine irreversible. The dotted curve 1–2 in Fig. 5.7 illustrates this process. As before, we shall find processes by which the fluid can be returned to its initial state, and then show that the surroundings cannot be returned to their initial state without contradicting the Second Law. Let the fluid flow through a rotary compressor on leaving the turbine, so that it is compressed without friction to the original pressure (curve 2–3). The

---

*Strictly speaking, the intermediate points only have meaning if every element of fluid flowing through the system passes through the same series of thermodynamic states. This implies that the properties are uniform at right angles to the flow throughout the system—not merely at the inlet and outlet, which was all we had to assume when deriving the steady-flow energy equation. When this additional assumption can be made, the flow is called 'one-dimensional' (see Chapter 18).

enthalpy drop in the turbine must have been less than the enthalpy rise in the compressor, since friction was present in the expansion. Therefore some net work, given by

$$(h_3 - h_2) - (h_1 - h_2) = (h_3 - h_1)$$

must be supplied by the surroundings. The fluid can be returned to its final state by cooling it at constant pressure without change of velocity, the quantity of heat entering the surroundings being equal to $(h_3 - h_1)$ from equation (4.4). To return the surroundings to their initial state, a heat engine would be required to convert $(h_3 - h_1)$ units of heat into $(h_3 - h_1)$ units of work. This is impossible according to the Second Law, and therefore the process in the turbine cannot have been reversible.

(ii) *Transfer of heat across a finite temperature difference*

Consider the process of allowing heat to flow from a source to a fluid at a lower temperature, when the difference between the temperature of the source and the fluid is finite. If the fluid is a closed system having fixed boundaries, the quantity of heat supplied is equal to the increase of internal energy. Alternatively, if the heat is supplied to a fluid flowing in a duct under conditions such that the velocity remains unchanged, the quantity of heat is equal to the increase of enthalpy, equation (4.4). In either case the fluid may be restored to its initial state by allowing the same quantity of heat to flow across the boundary into a sink. Even if this second process is carried out so that the temperature of the sink is never more than infinitesimally lower than that of the fluid, the difference in temperature between the source and the sink will be finite. If the original process is to be reversible, it must be possible to transfer the quantity of heat from the sink to the source without incurring any further changes in the surroundings. This is impossible because according to Corollary 1 of the Second Law some work would have to be supplied by the surroundings. Thus the process of heat transfer across a finite temperature difference is irreversible. We have already noted that a temperature difference is essential if any heat transfer is to take place in a finite time, so that all real transfers of heat must be irreversible.

Fig. 5.7

5.5 Reversible and irreversible cycles

A cycle is reversible if it consists only of reversible processes. When such a cycle is reversed, all the quantities of work and heat crossing the boundary are reversed in direction without their magnitude being affected. The cycle may consist of reversible non-flow processes, or it may be an open or closed cycle consisting of reversible steady-flow processes.

The original conception of a reversible cycle is due to Sadi Carnot [Ref. (4)]. He conceived of a particular cycle, called the *Carnot cycle*, which will be examined in detail in Part II. It is sufficient to note here that it is one of a special class of reversible cycles during which all the heat enters the system from a source at constant temperature, and all the heat leaving it is rejected to a sink at constant temperature. It follows that the processes during which heat is exchanged with the surroundings must be isothermal, the temperature of the fluid never differing by more than an infinitesimal amount from the fixed temperature of the source or sink.

A cycle may be thought of as reversible without this limitation. If the temperature of the fluid changes during any heat transfer process, all that is necessary is that the source or sink be a series of reservoirs covering the required range of temperature. At any instant during the process, heat must be exchanged between the system and the particular source or sink which differs only infinitesimally in temperature from the fluid in the system. This idea has already been introduced in section 3.1 (ii). Engines operating on this type of reversible cycle will be shown symbolically by Fig. 5.8. The quantity Q_1, for example, can be regarded as the sum of the infinitesimal quantities of heat received from the hot reservoirs.

Series of hot reservoirs

Series of cold reservoirs

Fig. 5.8

It is the limited class of reversible cycles, of which the Carnot cycle is a member, that plays an important part in the deduction of the corollaries of the Second Law. If the term 'reservoir' is restricted in the manner suggested in section 3.1 (ii), to a source or sink of invariable temperature, then this class of reversible cycles will comprise those of engines or heat pumps operating between only two reservoirs.

If any of the processes in a cycle are irreversible, the whole cycle is irreversible. The efficiency of an engine in which irreversible processes occur must always be less than that of the hypothetical reversible engine. For example, if friction occurs during the expansion in the turbine of the steam power plant shown in Fig. 4.11, the work done is reduced. Since the enthalpy at exit from the turbine is increased, more heat must be rejected in the condenser to bring the fluid back to the same state before it passes through the feed pump. The First Law is still satisfied, i.e. $Q_1 = W + Q_2$, but the cycle efficiency, W/Q_1, is reduced. If the efficiency of a cycle is calculated on the assumption that all the processes are reversible, this provides a figure for the maximum possible efficiency of power plant operating on that cycle.

6 Corollaries of the Second Law

The Clausius statement of the Second Law has been presented as the first corollary. Other corollaries are concerned with: (a) the characteristics of reversible engines operating between only two reservoirs, (b) the thermodynamic scale of temperature, (c) the characteristics of engines operating between a series of reservoirs, and (d) the property 'entropy'.

The way in which entropy can be used to simplify calculations concerning non-flow processes has already been indicated in Chapter 3. This chapter ends with a description of the way entropy may be used in the analysis of steady-flow processes, followed by a few remarks on the limitations of the laws of thermodynamics and a summary of Chapters 1 to 6.

6.1 Reversible engines operating between only two reservoirs

COROLLARY 2. *It is impossible to construct an engine operating between only two heat reservoirs, which will have a higher efficiency than a reversible engine operating between the same two reservoirs.*

PROOF. Assume the converse of this proposition to be true; let X be such an engine, having an efficiency η_X. Let it receive heat Q_1 from the source, do work W_X, and reject

Fig. 6.1

heat $(Q_1 - W_X)$ to the sink. Then it is assumed that $\eta_X > \eta_R$, where η_R is the efficiency of a reversible engine R operating between the same two reservoirs (Fig. 6.1a). If the reversible engine also receives heat Q_1 from the source, it will do work W_R such that $W_R < W_X$, and the heat rejected will be $(Q_1 - W_R)$ which is greater than $(Q_1 - W_X)$.

56

Let the reversible engine be reversed and act as a heat pump (Fig. 6.1b). It now receives heat $(Q_1 - W_R)$ from the low-temperature reservoir, receives work W_R from the surroundings, and rejects heat Q_1 to the high temperature reservoir. If the engine X is coupled to the heat pump, and the latter feeds heat Q_1 directly into the former, the combined plant represents a heat engine receiving heat $(Q_1 - W_R) - (Q_1 - W_X)$ $= (W_X - W_R)$ from the surroundings, and delivering an equivalent amount of work. According to the Second Law this is impossible, and the assumption that $\eta_X > \eta_R$ cannot be true. Consequently the original proposition must be true.

COROLLARY 3. *All reversible engines operating between the same two reservoirs have the same efficiency.*

Once the second corollary has been proved, no special proof is required for the third. If two reversible engines differ in efficiency when operating between the same two reservoirs, then one must have a higher efficiency than the other. This is impossible if Corollary 2 be true. Since all reversible engines operating between the same two reservoirs have the same efficiency, this efficiency must depend upon the only feature which is common to them all, viz. the temperatures of the reservoirs: it is called the *Carnot efficiency.*

6.2 Thermodynamic temperature scale

COROLLARY 4. *A scale of temperature can be defined which is independent of any particular thermometric substance, and which provides an absolute zero of temperature.*

Fig. 6.2

Let us consider the reversible engine operating between only two reservoirs represented symbolically in Fig. 6.2a. Corollary 3 implies that its efficiency depends only on the temperature of the reservoirs, and therefore that it is independent of the particular properties of the working fluid and of the quantity of heat Q_0 supplied to the engine. But

$$\eta = \frac{Q_0 - Q}{Q_0} = 1 - \frac{Q}{Q_0}$$

and hence it follows that Q/Q_0 is a function only of the temperatures of the reservoirs.

Let us now allot a positive number T_0 to some reservoir in a convenient reference state which is easily reproducible (e.g. a pure substance melting at a definite pressure); further, let us define the temperature T of any other reservoir by the equation

$$T = T_0 \frac{Q}{Q_0} \qquad (6.1)$$

The value T is then fully and uniquely determined by (a) the arbitrary choice of T_0 and (b) the ratio Q/Q_0 which, as a consequence of the Second Law, is a fixed and definite quantity for the two given reservoirs. If Q is measured for several sinks at different temperatures and plotted against T obtained from (6.1), we obtain by definition a scale of temperature which is linear in Q. The slope of the line in Fig. 6.2b is Q_0/T_0. The *unit of temperature* now follows as the fraction $1/T_0$ of the interval from $T = 0$ to T_0. The temperature scale defined in this way is called the *thermodynamic scale*, because it is dependent solely on the laws of thermodynamics and not upon the properties of any particular substance. It is an *absolute* scale because it presents us with the idea of an absolute zero, i.e. $T = 0$ when $Q = 0$. The Second Law implies that Q can in fact never be zero, and we may therefore deduce that absolute zero is a conceptual limit and not a temperature which can ever be reached in practice.*

It may be helpful if we visualise the scale as being constructed by a series of reversible engines, Fig. 6.3a, each operating between only two reservoirs and each producing the same quantity of work. Each sink is a source for the following engine, the heat entering a reservoir being equal to the heat leaving it. If the temperatures of the reservoirs are defined in the way suggested by equation (6.1), then

$$\frac{Q_0}{Q_1} = \frac{T_0}{T_1}, \quad \frac{Q_1}{Q_2} = \frac{T_1}{T_2}, \quad \frac{Q_2}{Q_3} = \frac{T_2}{T_3}, \ldots \qquad (6.2)$$

And the efficiencies of the engines therefore become

$$\eta_0 = \frac{T_0 - T_1}{T_0}, \quad \eta_1 = \frac{T_1 - T_2}{T_1}, \quad \eta_2 = \frac{T_2 - T_3}{T_2}, \ldots \qquad (6.3)$$

But since

$$W = \eta_0 Q_0 = \eta_1 Q_1 = \eta_2 Q_2 = \ldots$$

it follows that

$$\frac{Q_0}{T_0}(T_0 - T_1) = \frac{Q_1}{T_1}(T_1 - T_2) = \frac{Q_2}{T_2}(T_2 - T_3) = \ldots$$

Using equations (6.2), this reduces to

$$(T_0 - T_1) = (T_1 - T_2) = (T_2 - T_3) = \ldots \qquad (6.4)$$

Thus the differences between the temperatures of successive reservoirs are the same and can provide intervals or units of temperature. T_0 may be as high as we wish

---

* Strictly speaking, it is necessary to invoke the Third Law (cf. p. 114n.) to deduce the unattainability of absolute zero: this is discussed in Ref. (8).

to make it, and the intervals may be made as small as desired by increasing the number of engines in the series. The series of reservoirs could theoretically be used as a standard set of temperatures with which to calibrate any practical thermometer.

Figs. 6.3b and c illustrate the effect of increasing the number of engines. The last possible engine in the series is one which rejects an amount of heat equal to the work W; according to the Second Law the next engine in the series is an impossibility. It is evident that the temperature of the sink of the last possible engine is reduced as the number of engines is increased, but as long as W is finite the scale must stop short of

Fig. 6.3   Thermodynamic temperature scale

absolute zero. Note that the choice of 100 units for Q_0 in the scales of Figs. 6.3b and c is quite arbitrary; a discussion of the accepted way in which numbers are assigned to the thermodynamic scale is left to section 8.1.

Although it has been internationally agreed to define the thermodynamic temperature by writing $T = T_0(Q/Q_0)$, it could have been defined in an infinite variety of ways without losing its character of being independent of the properties of any particular substance. For example, we could define a scale, on which the temperature is denoted by ξ say, by writing $\xi = \xi_0 \ln (Q/Q_0)$; on this scale the temperature ξ would range from $-\infty$ to $+\infty$. This was in fact the first scale proposed by Kelvin. The simplest expression has been adopted, however, not because the scale is linear in Q when formed by reversible engines producing equal work, but because it simplifies the relations between temperature and the other thermodynamic properties. (This point will be appreciated after a study of Chapter 8.) It is the existence of these relations which makes the measurement of thermodynamic temperature a practical possibility, since they enable the temperature to be deduced from measurements of other properties. The practical details of measurement need not concern us here; they are referred to in section 8.1, Part II. It will be assumed for the remainder of Part I that all temperatures are measured on the thermodynamic scale, and they will be denoted by T.

As a result of Corollaries 3 and 4, the most important characteristics of engines operating between only two reservoirs may now be stated as follows:

(a) All reversible engines, operating between a source at temperature T_1 and a sink at temperature T_2, have an efficiency equal to $(T_1 - T_2)/T_1$: see equation (6.3).

(b) For a given value of T_2, the efficiency increases with increase of T_1. Since the lowest possible temperature of a practical infinite sink is fixed within close limits, i.e. the temperature of the atmosphere or the sea, it may be said that a given quantity of heat is more useful for producing work the higher the temperature of the source from which it is received.

(c) For all reversible engines operating between only two reservoirs, $Q_1/T_1 = Q_2/T_2$. If an engine is irreversible, its efficiency is less than $(T_1 - T_2)/T_1$. For the same heat received, the heat rejected must be greater than that rejected by a reversible engine operating between the same two reservoirs. Therefore, for irreversible engines operating between only two reservoirs, $Q_1/T_1 < Q_2/T_2$.

It is important to note that when applying statements (a), (b) and (c) to reversible engines, the temperature T refers both to the temperature of the reservoir and to the temperature of the fluid exchanging heat with it. On the other hand, when considering irreversible engines, T refers to the temperature of the reservoir only: the temperature of the fluid may then not be the same as that of the reservoir and different parts of the fluid may have different temperatures.

The three results (a), (b) and (c) will now be used in the deduction of the characteristics of reversible engines operating between more than two reservoirs.

6.3 Engines operating between more than two reservoirs

In many practical cycles the heat is received and rejected during processes which involve a continuous change in the temperature of the fluid. It has been explained that these cycles can still be considered reversible if the source and sink are each assumed to consist of an infinite number of reservoirs differing infinitesimally from one another in temperature. At any instant during a heating and cooling process, heat must be exchanged between the system and a source or sink which differs only infinitesimally in temperature from the fluid in the system.

We have seen that the efficiency with which a given quantity of heat can be converted into work by a reversible engine operating between reservoirs at T_1 and T_2 is given by $(T_1 - T_2)/T_1$ or $(1 - T_2/T_1)$. Now let T_1' and T_2' be the maximum and minimum temperatures of the working fluid in a reversible engine operating between more than two reservoirs. Only a fraction of the heat supplied can be received at T_1' and only a fraction of the heat rejected can be rejected at T_2'. The remaining part of the heat supplied must be converted into work with an efficiency less than $(1 - T_2'/T_1')$, because the temperatures of the remaining sources are less than T_1' and the temperatures of the remaining sinks are greater than T_2'. Thus the efficiency with which the total heat received is converted into work must be less than $(T_1' - T_2')/T_1'$. This result may be summarised as:

COROLLARY 5. *The efficiency of any reversible engine operating between more than two reservoirs must be less than that of a reversible engine operating between two reservoirs which have temperatures equal to the highest and lowest temperatures of the fluid in the original engine.*

The most important characteristic of engines operating between more than two reservoirs is presented as the sixth corollary of the Second Law; it is known as the *Clausius Inequality*.

COROLLARY 6. *Whenever a system undergoes a cycle, $\oint(dQ/T)$ is zero if the cycle is reversible and negative if irreversible, i.e. in general $\oint(dQ/T) \leqslant 0$.*

PROOF. First consider an engine operating between a finite number of reservoirs. The engine is represented in Fig. 6.4a; it may be reversible *or* irreversible. Let quantities of heat Q_{1a}, Q_{1b}, . . ., Q_{1n} be supplied from sources at temperatures T_{1a},

(a)

(b)

Fig. 6.4

T_{1b}, . . ., T_{1n}, and quantities of heat Q_{2p}, Q_{2q}, . . ., Q_{2z} be rejected to sinks at T_{2p}, T_{2q}, . . ., T_{2z}. Now let the sources be supplied with their quantities of heat by a series of *reversible* heat pumps receiving quantities of heat Q_{0a}, Q_{0b}, . . ., Q_{0n} from a single reservoir at T_0. Similarly, let the sinks reject their quantities of heat to a series of *reversible* heat engines which in turn reject quantities of heat Q_{0p}, Q_{0q}, . . ., Q_{0z} to the single reservoir at T_0. T_0 may be any temperature, providing it is less than the temperature of any of the reservoirs of the original engine. The whole plant is represented in Fig. 6.4b.

The work delivered by the original engine is given by

$$\sum_a^n Q_1 - \sum_p^z Q_2$$

the work delivered by the auxiliary engines is

$$\sum_p^z (Q_2 - Q_0)$$

and the work supplied to the heat pumps must be

$$\sum_a^n (Q_1 - Q_0)$$

The series of sources and sinks plays no effective part in the transformations of energy, and the complete plant is a system undergoing a cyclic process while exchanging heat with a single reservoir (of temperature T_0). It follows from the Second Law that the net work done by the system cannot be positive; it must be either zero or negative. Thus the net work done by the composite system can be written as

$$\sum_a^n Q_1 - \sum_p^z Q_2 + \sum_p^z (Q_2 - Q_0) - \sum_a^n (Q_1 - Q_0) \leqslant 0$$

which reduces to

$$\sum_a^n Q_0 - \sum_p^z Q_0 \leqslant 0$$

But for each of the reversible heat engines and heat pumps which operate between only two reservoirs, we can write equations of the form:

$$\frac{Q_1}{T_1} = \frac{Q_0}{T_0} \quad \text{between } a \text{ and } n$$

and

$$\frac{Q_2}{T_2} = \frac{Q_0}{T_0} \quad \text{between } p \text{ and } z$$

Therefore, substituting for the Q_0 terms in the previous equation, we have

$$\sum_a^n \frac{Q_1}{T_1} T_0 - \sum_p^z \frac{Q_2}{T_2} T_0 \leqslant 0$$

Dividing throughout by the constant temperature T_0, we get finally

$$\sum_a^n \frac{Q_1}{T_1} - \sum_p^z \frac{Q_2}{T_2} \leqslant 0 \tag{6.5}$$

We may now revert to the *original convention of writing Q for any quantity of heat whether positive or negative* (see beginning of section 5.1), so that the algebraic summation of all such quantities as $(\delta Q/T)$ for the original cycle becomes

$$\sum \frac{\delta Q}{T} \leqslant 0$$

In the limiting case, where the temperature changes continuously during the heat exchanges, the number of reservoirs becomes infinite and the individual quantities of heat become infinitesimal. The equation can then be written as

$$\oint \frac{dQ}{T} \leqslant 0 \tag{6.6}$$

We have now to show that the equality holds in the special case of a reversible engine. The foregoing argument can be repeated for a heat pump, all the quantities of heat and work flowing in directions opposite to those shown in Fig. 6.4b. If the quantities of heat are denoted by symbols with primes, we should find that equations (6.5) and (6.6) are unchanged in other respects, e.g. equation (6.6) is simply

$$\oint \frac{dQ'}{T} \leqslant 0 \tag{6.7}$$

Now if an engine is reversible, and it is reversed to form a heat pump, all the quantities of heat will be unchanged in magnitude although reversed in sign. It follows that $dQ' = -dQ$. Equation (6.7) can therefore be written as

$$-\oint \frac{dQ}{T} \leqslant 0 \quad \text{or} \quad \oint \frac{dQ}{T} \geqslant 0$$

The only way in which this result can be satisfied simultaneously with equation (6.6) is for the equality to hold. Therefore, if the engine is reversible,

$$\oint \frac{dQ}{T} = 0$$

The engine operating between only two reservoirs is easily seen to provide a special example of the Clausius Inequality. In this case

$$\oint \frac{dQ}{T} = \frac{Q_1}{T_1} - \frac{Q_2}{T_2}$$

and we have already noted that the right-hand side is zero if the engine is reversible and negative if irreversible (section 6.2).

It must be emphasised that the Clausius Inequality is a relation between the quantities of heat received or rejected by a series of reservoirs and the temperatures of those reservoirs, when a system is taken through a cycle. Only in the special case of a reversible cycle can the temperature be associated with the temperature of the fluid undergoing the cycle. This is a most important case however, and the following section will show how it may be used to deduce a new property of a system.

So far, the Second Law has been stated with reference to cycles, and various characteristics of these cycles have been deduced. All the conclusions refer equally to cycles composed of either non-flow or steady-flow processes. The implication of the Second Law for the individual processes of which the cycles may be composed must now be considered. It will be necessary to deal with non-flow and steady-flow processes separately, and the former class will be considered first.

6.4 Consequences of the Second Law for non-flow processes

COROLLARY 7. *There exists a property of a closed system such that a change in its value is equal to $\int_1^2 \frac{dQ}{T}$ for any reversible process undergone by the system between state 1 and state 2.*

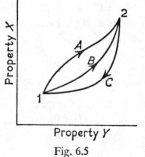

Fig. 6.5

PROOF. Assume that the converse is true, i.e. that $\int_1^2 \frac{dQ}{T}$ depends upon the reversible process as well as upon the end states. The procedure is to show that this contradicts Corollary 6. Here the temperature T under consideration is the temperature of the fluid undergoing the process. However, since the process is to be reversible, T must also be equal to the temperature of the reservoir with which the system is exchanging the quantity of heat dQ. This is why we can base the proof upon Corollary 6.

Consider any two *reversible* processes A and B by which a system can change from state 1 to state 2 (Fig. 6.5). The assumption is that

$$\int_1^2 \left(\frac{dQ}{T}\right)_A \neq \int_1^2 \left(\frac{dQ}{T}\right)_B$$

In each case let the system return to its original state by a third reversible process C. For each of the complete reversible cycles AC and BC we have

$$\oint \left(\frac{dQ}{T}\right)_{AC} = \int_1^2 \left(\frac{dQ}{T}\right)_A + \int_2^1 \left(\frac{dQ}{T}\right)_C$$

and

$$\oint \left(\frac{dQ}{T}\right)_{BC} = \int_1^2 \left(\frac{dQ}{T}\right)_B + \int_2^1 \left(\frac{dQ}{T}\right)_C$$

f the assumption described by the inequality is true, it follows that

$$\oint \left(\frac{\mathrm{d}Q}{T}\right)_{\mathrm{AC}} \neq \oint \left(\frac{\mathrm{d}Q}{T}\right)_{\mathrm{BC}}$$

But these cyclic integrals must be equal because, according to Corollary 6, they are ero for any reversible cycle. Therefore the original proposition, that $\int_1^2 \frac{\mathrm{d}Q}{T}$ is inde-pendent of the path of the reversible process and represents the change in a property of the system, must be true.

If the property is denoted by S, this corollary can be expressed mathematically as

$$\int_1^2 \left(\frac{\mathrm{d}Q}{T}\right)_{\mathrm{rev}} = S_2 - S_1 \tag{6.8}$$

or in differential form as

$$\mathrm{d}S = \left(\frac{\mathrm{d}Q}{T}\right)_{\mathrm{rev}} \tag{6.9}$$

The suffix 'rev' is added as a reminder of the fact that the relation only holds for a reversible process. The property S is called the *entropy* of the system.

Before describing the property entropy in more detail, one final corollary will be deduced.

COROLLARY 8. *The entropy of any closed system which is thermally isolated from the surroundings either increases or, if the process undergone by the system is reversible, remains constant.*

PROOF. There are two cases to be considered; that of a closed system undergoing a reversible process presents no difficulty. For any thermally insulated system $\mathrm{d}Q = 0$. But for any reversible process $\mathrm{d}S = (\mathrm{d}Q/T)$. Therefore, for any reversible adiabatic process the change of entropy must be zero. For this reason the reversible adiabatic process is called an *isentropic process*. That the entropy of a system undergoing an irreversible adiabatic process must increase may be proved as follows.

Let any closed system change from state 1 to state 2 by an irreversible adiabatic process A. There may be a change of entropy of the fluid in this case because equation (6.9) is not applicable. But since the process is adiabatic, $\mathrm{d}Q = 0$ and therefore

$$\int_1^2 \left(\frac{\mathrm{d}Q}{T}\right)_{\mathrm{A}} = 0$$

Now let the system be returned to its original state by some reversible process B, in this case not necessarily adiabatically. For this process it is possible to write

$$\int_2^1 \left(\frac{\mathrm{d}Q}{T}\right)_{\mathrm{B}} = S_1 - S_2$$

The cycle as a whole is irreversible, and hence from Corollary 6

$$\oint \frac{dQ}{T} = \int_2^1 \left(\frac{dQ}{T}\right)_B < 0$$

Therefore,

$$(S_1 - S_2) < 0, \quad \text{or } S_2 > S_1$$

Before proceeding to deduce the consequences of the Second Law for steady-flow processes, we shall discuss some of the more important characteristics of the property 'entropy'.

6.5 Characteristics of entropy

(i) *Determination of values of entropy*

The property 'entropy' arises as a consequence of the Second Law, in much the same way as the property internal energy arises from the First Law. There is, however, an important difference. The change in internal energy can be found directly from knowledge of the heat and work crossing the boundary during *any* non-flow process undergone by a closed system. The change in entropy, on the other hand, can be found from a knowledge of the quantity of heat transferred only during a *reversible* non-flow process.

Since equations (6.8) and (6.9) apply to reversible processes, during which the system passes through a series of equilibrium states, they may be applied to any unit mass of the system. The entropy per unit mass, or specific entropy, will be denoted by s, and $S = ms$ when the system is in equilibrium. It is possible, therefore, to write the equations as

$$\int_1^2 \left(\frac{dQ}{T}\right)_{rev} = s_2 - s_1 \tag{6.10}$$

$$ds = \left(\frac{dQ}{T}\right)_{rev} \tag{6.11}$$

implying that dQ is the increment of heat transferred per unit mass of fluid in the system.

As with internal energy, only changes of entropy are normally of interest. The entropy at any arbitrary reference state can be made zero, and the entropy at any other state can then be found by evaluating (dQ/T) for any reversible process by which the system can change from the reference state to this other state. Since no real process is reversible, values of entropy cannot be found from measurements of Q and T in a direct experiment. The entropy is a thermodynamic property, however, and it can be expressed as a function of other thermodynamic properties which can be measured in experiments involving real processes. Two important relations of this kind can be obtained by combining the equations expressing the First and Second Laws. Thus, the First Law yields the equation

$$(dQ)_{rev} = du + p \, dv$$

and combining this with equation (6.11), we have

$$T \, ds = du + p \, dv \tag{6.12}$$

Alternatively, substituting $du = dh - d(pv)$, we get

$$T \, ds = dh - v \, dp \tag{6.13}$$

Equations (6.12) and (6.13) are general relations between properties which apply to any fluid. Moreover, when integrated they give the difference in entropy between any two equilibrium states, regardless of whether any particular process joining them is carried out reversibly or not.* We may show how equation (6.12) can be used for compiling tables of values of entropy as follows.

The entropy at state 1, defined by (p_1, v_1) say, relative to the entropy at some reference state 0, defined by (p_0, v_0), is given by

$$s_1 - s_0 = \int_0^1 \frac{du}{T} + \int_0^1 \frac{p}{T} \, dv \tag{6.14}$$

The integrals can be evaluated graphically by the following procedure. The end states can be plotted on a p-v diagram and joined by any arbitrary curve as in Fig. 6.6a. The curve can be chosen arbitrarily because the change of entropy depends only on

Fig. 6.6   Graphical determination of entropy

the end states.) For several pairs of values of p and v along the curve, equilibrium values of u and T can be found from tables of properties. Finally, curves of $1/T$ versus u, and p/T versus v can be plotted as in Figs. 6.6b and c. The shaded areas under these curves will be equal to the integrals on the right-hand side of equation (6.14), and hence s_1 can be found assuming s_0 is arbitrarily chosen to be zero. Values of s for any

*The only proviso is that the states must be capable of being joined *in principle* by a reversible process, because this is assumed in deriving (6.12) and (6.13). Normally this is so: an example to the contrary is when a vapour in a metastable state (supersaturated) changes to a wet vapour in a stable state [see sections 8.2 (i) and 18.2]. Tables of entropy do not refer to metastable states and this special case can be ignored here.

other state, defined by different values of p and v, may be obtained in a similar way and the results can be presented in a table of the form shown in Fig. 1.1.

When the properties of a fluid can be related by simple algebraic equations, the integrals of equation (6.14) can be evaluated analytically, and a simple equation for s in terms of p and v, or any other two independent properties, can be determined. Once the functions relating the entropy and other properties have been established, whether expressed algebraically or in the form of a table, there is no need to find the change of entropy by evaluating $\int (dQ/T)_{rev}$. It can be found directly from a knowledge of other properties in the initial and final states.

(ii) *Entropy as a criterion of reversibility*

One use of the property 'entropy' has already been indicated in sections 3.5 and 3.6, where it was shown to simplify calculations concerned with reversible adiabatic and isothermal non-flow processes. Another use is suggested by Corollary 8, namely that the change of entropy can be used as a criterion of reversibility. If a non-flow process is adiabatic, this corollary can be applied directly; it states that the entropy will remain unchanged if the process is reversible. When the process is not adiabatic the criterion may be applied in the following way. Any system together with its surroundings can be regarded as one single system. If the term 'surroundings' is used in a special sense to mean only that portion of matter outside a system which is affected by the changes under consideration, the system plus surroundings may be called the 'universe'. The inverted commas will be retained as a reminder that the word is used in this restricted sense. The 'universe' is an isolated system, and its entropy must either increase or remain constant in accordance with Corollary 8. If its entropy is determined before and after a change, an increase indicates that the process must have been irreversible. The way in which the entropy changes can therefore still be used as a criterion of reversibility, although when the process is not adiabatic it is necessary to refer to the entropy of the 'universe' and not merely to that of the system.

Before presenting an example of this use of the property entropy, we must see how the change in the entropy of a reservoir may be evaluated. It will be remembered that a reservoir is assumed to be a source or sink of such capacity that the quantity of heat leaving or entering it does not cause its temperature to change by more than an infinitesimal amount. Any change in the specific entropy (or any other specific property) is similarly infinitesimal, but since the mass of the reservoir is large, the total entropy change is finite. Because the elements of the reservoir undergo only infinitesimal changes, the process undergone by the reservoir as a whole when the quantity of heat Q enters or leaves it must be considered reversible. If the constant temperature of the reservoir is T, it follows that the entropy change of a reservoir is simply Q/T.

Let us now consider the process of transferring a quantity of heat Q from a reservoir at T_1 to a system at a constant lower temperature T_2, via an intervening medium which is a poor conductor, Fig. 6.7. The source and conductor are the surroundings of the system. There is a temperature gradient along the conductor during the passage of energy through it. The heat Q enters the system by virtue of an infinitesimal temperture difference at the boundary. The system therefore passes through a series of quilibrium states and, *as far as the system is concerned*, it undergoes a reversible hange. Since its temperature remains constant at T_2, there is an increase in the entropy of the system equal to Q/T_2. The entropy change in the surroundings must now be considered. The change in the entropy of the reservoir is a decrease of Q/T_1. No change of entropy occurs in the conductor since it is in xactly the same state after the quantity of energy has assed through it as it was initially. Thus the entropy of the universe' has increased because $Q/T_2 > Q/T_1$. The com- plete process must therefore be irreversible.

Fig. 6.7

The form of irreversibility just discussed is usually des- cribed as *external* because no dissipation of energy takes place within the system; the irreversibility occurs in the urroundings. Energy can be said to be 'wasted' or 'degraded' because the quantity Q has been transferred to a lower temperature level without producing any of the work of which it is capable. If T_0 is the lowest temperature of any natural infinite sink, such as the atmosphere or ocean, the maximum efficiency with which a quantity of heat at temperature T can be transformed into work is $(T - T_0)/T$. Therefore the maximum quantity of work which Q at T_1 could theoretically have produced is $Q(T_1 - T_0)/T_1$, while after its conduction to T_2 it is only capable of producing $Q(T_2 - T_0)/T_2$ units of work. No means exist for returning this quantity of heat to the original temperature level without some expenditure of work (Corollary 1), and so some energy has been irretrievably degraded in the sense that its capacity for pro- ducing work has been permanently decreased.

An external irreversibility is inevitably present when heat is conducted from one fluid to another through a dividing wall.* For example, when heat is transferred from

Fig. 6.8

the furnace gases to the water in a boiler, there is always a con- siderable temperature difference between the main body of hot gas and the main body of water. Briefly, this is because there are relatively stationary boundary layers of fluid on both sides of the boiler tube which are poor conductors of heat (Fig. 6.8). The tem- perature drop across them is therefore considerable. The quantity of heat transferred to the working fluid in an engine can be cal- culated from the change in the average values of the properties of the fluid during the heat transfer process, and this quantity is the same whether the source is at a much higher, or only infini- tesimally higher, temperature. Thus although an external irreversi- bility results in the loss of some capacity for producing work, this loss is *not* expressed directly in the cycle efficiency of a heat engine. The cycle efficiency of the engine illustrated in Fig. 6.9 is still given by $W/Q_1 = (Q_1 - Q_2)/Q_1$, and the values of Q_1 and Q_2 are independent of the temperature differences $(T_1' - T_1)$ and $(T_2 - T_2')$. The temperature differences adopted in practice

*A detailed analysis of this process is given in Part IV.

69

merely govern the area of the heat transfer surfaces required. Of course, if the cycle were to be modified to operate over the full range of temperature available $(T_1' \to T_2')$, a higher efficiency would be obtained, so that it is a question of balancing capital and running costs. Fluid friction, on the other hand, does cause a dissipation of energy within the system and it is said to produce an *internal* irreversibility. The changes in the fluid properties are affected by friction, as explained in section 5.4 (i) and the cycle efficiency is reduced by this form of irreversibility.

6.6 Consequences of the Second Law for steady-flow processes

It must be emphasised that entropy is a thermodynamic property. The entropy of any element of fluid is a function only of the thermodynamic state and, as already pointed out, such functions are independent of whether the fluid is flowing or not. Each element of fluid of mass δm in an open system may have a different value of the specific entropy s'. The total entropy S' of the fluid in the system at any given moment is given by $S' = \Sigma\, \delta m\, s'$ During any steady-flow process S' remains constant because the properties at all points within the system do not change with time. We may imagine the steady-flow process to consist of a succession of elementary non flow processes as in section 4.1.

Fig. 6.9

First let us consider a steady-flow process during which no heat is allowed to cross the boundary, although work may do so. The entropy of the imaginary closed system in its initial configuration (Fig. 4.2a) is given by

$$\delta m\, s_1 + S'$$

where s_1 denotes the specific entropy of the element of fluid which is about to enter the open system. Similarly, the entropy of the closed system in its final configuration is

$$\delta m\, s_2 + S'$$

Since the imaginary non-flow process is adiabatic, the final entropy of the closed system must either be greater than or be equal to the initial entropy (Corollary 8). Therefore,

$$\delta m\, s_2 + S' \geqslant \delta m\, s_1 + S', \quad \text{or} \quad \delta m\, s_2 \geqslant \delta m\, s_1$$

For the steady-flow process we therefore have

$$\Sigma\, \delta m\, s_2 \geqslant \Sigma\, \delta m\, s_1$$

Since each element of fluid enters the open system with the same properties, denoted by suffix 1, and leaves with the same properties, denoted by suffix 2, this expression reduces to

$$s_2 \geqslant s_1 \tag{6.15}$$

Thus for an adiabatic steady-flow process, the specific entropy of the fluid leaving the system must be greater than, or equal to, the specific entropy at entry. If the process is reversible, i.e. frictionless, $s_2 = s_1$; otherwise $s_2 > s_1$. The reversible adiabatic steady-flow process may therefore be called an isentropic steady-flow process.

Since a large number of practical steady-flow processes may be assumed adiabatic, the relation (6.15) is most useful. Firstly, the principle of increasing entropy informs us of the direction of any real adiabatic flow process; it must occur so that the fluid changes from a state of lower entropy to a state of higher entropy. Sometimes this principle can be used effectively to show that certain flow processes are impossible. For example, in the study of gas dynamics it is used to show that a flow of gas through a shock wave can only be accompanied by a rise in pressure because a fall in pressure implies a decrease of entropy of the gas. Secondly, the equality of entropy at the inlet and outlet of an open system, when the adiabatic process is reversible, can simplify the analysis of many steady-flow processes.

Consider for example an adiabatic expansion in a turbine, where the fluid enters with an initial thermodynamic state defined by (p_1, v_1), and leaves with a pressure p_2. The values of h_1 and s_1 can be found from tables of properties. If the expansion is assumed to be reversible, s_2 will be known because it is equal to s_1. The final state is now completely determined by the properties p_2 and s_2, and the value of h_2 can also be found from the appropriate tables. The work done can then be found from the steady-flow energy equation

$$W = (h_1 - h_2)$$

If the adiabatic process is irreversible, the final state cannot be predicted; all that is known is that $s_2 > s_1$. For the same value of p_2 this can be found from the tables of properties to imply that h_2 is greater than it would have been after a reversible expansion. This result is in agreement with the earlier discussion of the effect of irreversibility on the expansion in a turbine (section 5.4). But, as already pointed out, the effect of friction cannot be neglected in most steady-flow processes. In practice it is usual to calculate the required quantity on the assumption that the process is reversible, and then multiply the result by a *process efficiency* to give a more realistic estimate. The process efficiency in the case of a turbine is defined by

$$\text{Turbine efficiency} = \frac{\text{Work done during the real expansion}}{\text{Work done during an isentropic expansion}}$$

Process efficiencies must be found empirically, i.e. by carrying out tests on actual components. Their definition is somewhat arbitrary, and no fixed form can be adopted which will apply to all processes.

71

So far we have only considered adiabatic steady-flow processes. The analysis leading up to the relation (6.15) must now be repeated for the case of an open system which is exchanging heat with the surroundings. Consider the case of the reversible steady-flow process. When it is reversible, the individual non-flow processes of which it is assumed to consist must also be reversible. In general the heat crosses the boundary along a finite length of the system as the temperature of the fluid changes in the direction of flow. Let increments of heat $\delta Q_a, \delta Q_b, \ldots, \delta Q_n$ cross the boundary at points where the fluid temperature is $T_a, T_b, \ldots, T_n$, during the change of configuration of the imaginary closed system (Fig. 4.2). Then equation (6.8) applied to this process becomes

$$(\delta m \, s_2 + S') - (\delta m \, s_1 + S') = \sum_a^n \frac{\delta Q}{T}$$

and hence

$$\delta m \, s_2 - \delta m \, s_1 = \sum_a^n \frac{\delta Q}{T}$$

Similar equations can be written for every elemental mass passing through the open system. Adding these equations we have, for the whole steady-flow process,

$$\sum \delta m \, s_2 - \sum \delta m \, s_1 = \sum \left(\sum_a^n \frac{\delta Q}{T} \right)$$

To simplify this equation we must assume not only that each elemental mass δm enters with the specific entropy s_1 and leaves with s_2, but also that each element exchanges $\delta Q_a, \delta Q_b, \ldots$ with the surroundings when the temperature is $T_a, T_b, \ldots$. If we make this additional assumption the equation reduces to

$$s_2 - s_1 = \sum_a^n \frac{\delta Q}{T}$$

where δQ is the quantity of heat per unit mass, transferred across the boundary at the point where the fluid temperature is T. Also, if the temperature increases continuously from T_1 to T_2 during the process, the equation may be written in the form

$$s_2 - s_1 = \int_1^2 \frac{dQ}{T} \tag{6.16}$$

In the special case where T is constant during the process, e.g. when the fluid is a liquid which is being vaporised at constant pressure, equation (6.16) becomes

$$s_2 - s_1 = \frac{1}{T} \int_1^2 dQ$$

and, therefore,

$$Q = T(s_2 - s_1)$$

where Q is the total rate of heat transfer per unit mass between inlet and outlet. Or, f m is the rate of mass flow and Q the rate of heat transfer per unit time,

$$Q = mT(s_2 - s_1) \qquad (6.17)$$

It must be emphasised that equations (6.16) and (6.17) only apply to an internally reversible process, i.e. a frictionless process,* because they are derived from equation (6.8). Nevertheless, in many practical cases where heat is received or rejected during a steady-flow process these equations can be applied with sufficient accuracy. For example, the velocity of flow in a boiler or condenser is low, and the effect of friction is negligible.

6.7 The validity and limitations of the laws of thermodynamics

With its emphasis on logical structure, and its reference to axioms and corollaries, this account of thermodynamics may have obscured the important fact that thermo-dynamics is essentially an experimental science. It is not a branch of mathematics, like Euclidean geometry, although it makes use of mathematical tools as do all the physical sciences. The two 'axioms' of thermodynamics are hypotheses put forward to explain the results of a vast number and variety of experiments. Every experiment which confirms a deduction from a hypothesis confirms and strengthens the validity of the hypothesis itself. The mathematical tools are merely used to draw every possible observable and verifiable consequence from the hypothesis.

When a hypothesis is sufficiently well established, it achieves the status of a law. The First and Second Laws of Thermodynamics are so well established that we may confidently assume they will remain unchanged in the face of new developments. They may, of course, be shown to be of more limited application than was first sup-posed, and to be special cases of a more general hypothesis—in much the same way that Newtonian mechanics has been subsumed in the theory of relativity. Indeed this has already happened to some extent with the discovery that mass can be transformed into energy. But this has in no way invalidated the application of classical thermo-dynamic theory to systems in which mass is effectively conserved.

It is worth noting some possible restrictions upon the generality of the laws of thermodynamics.† We have already observed that it may not be permissible to apply the First Law to the whole universe to suggest that its total energy is constant. A similar extrapolation of the Second Law is sometimes made which is equally question-able. Since all real processes are irreversible, the entropy of all 'universes' must

---

*Since friction results in an increase of entropy, it is possible to imagine an irreversible steady-flow process in which the amount of heat rejected results in a decrease of entropy which is just equal to the increase caused by friction. In such circumstances s_2 will equal s_1. Although strictly speaking this would be an isentropic process, it is an unimportant kind and less ambiguity arises if the term isentropic is restricted to mean a reversible adiabatic process as will be done throughout this book.

†Advanced students will find a stimulating discussion of this in Ref. (2).

increase whenever a change occurs within them. This has led to the broad generalisation that the entropy of the universe as a whole is increasing. But the Second Law, like the First, is an expression of the observed behaviour of finite systems and it is not certain that the universe can be regarded as finite. Moreover, the significance of the Second Law for systems consisting of living organisms is not yet clear.

There is a definite limit to the applicability of the laws of thermodynamics at the other end of the scale of magnitude; the laws cannot be applied to microscopic systems which contain only a small number of molecules. The reasons for making this statement are as follows.

The thermodynamic state of any system can be completely described by 'macroscopic' properties, i.e. properties which are directly observable by the senses, such as pressure, volume and temperature. Other properties defined in terms of these or deduced as consequences of the laws of thermodynamics are also used but, though more abstract, they are none the less macroscopic in the sense that they are quite independent of any theory as to the molecular or atomic structure of matter. By constructing hypotheses about the 'microscopic' nature of matter, it is possible to regard these macroscopic properties as averages of microscopic quantities. Thus pressure can be regarded, on the kinetic theory of fluids, as the average value of all the rates of change of momentum due to molecular impacts on a unit area of the boundary. These hypotheses must yield results in agreement with the observed macroscopic behaviour of matter, and changes in such hypotheses do not require any modification to be made to the structure of classical thermodynamics. An assumption as to the nature of matter can, however, place a restriction on the field of application of thermodynamics. The kinetic theory indicates that a thermodynamic system must be large enough to contain a great number of molecules. If a system consists of only a few molecules the concept of pressure, for example, is meaningless. The force exerted by the system on its boundaries would be finite at the points of impact of the molecules and zero at all other points: thus the 'pressure' would not be uniform over the surface nor constant with respect to time, and it would be useless as a characteristic for describing the state of the system. Since, for example, one cubic centimetre of air at atmospheric pressure and temperature contains about 25×10^{18} molecules, this restriction is of little practical importance to engineers.

The way in which the laws of thermodynamics lose their meaning when applied to microscopic systems is illustrated by the following discussion of a microscopic engine which appears to contradict the Second Law. On the kinetic theory, the molecules of a gas are rushing about at random with various velocities. If a plate is suspended in the gas it will be bombarded on both faces by the molecules and, on the average, the number and force of the impacts on each face will be equal. If the plate is made so small that only a few molecules are bombarding each face at any given instant, then sometimes the force exerted on one side will be larger than that on the other. The plate could be made to operate a ratchet mechanism, the ratchet wheel only moving when the force on one particular side of the plate predominates. The wheel might turn a winch and raise a weight, so doing work on the surroundings. This microscopic engine does work continuously by merely extracting energy from the gas, e.g. the atmosphere, and no second reservoir is necessary. Such an engine contradicts the Second Law of Thermodynamics. This argument is open to refutation, on the grounds that the microscopic engine is impossible in principle because it would itself be subject to the random motion of the molecules.

We have shown how the Second Law loses its meaning when applied to microscopic systems or to an infinite universe. When applied to macroscopic finite systems it has been found to be valid, but it is worth emphasising that its usefulness is limited. The limitation can be seen more clearly from some of the corollaries, such as the Clausius Inequality, rather than from the statement of the law itself. The limitation results from the fact that the corollaries of the Second Law only provide definite quantitative statements about *reversible* processes. Only for such processes is it possible to predict the work and heat transfers crossing the boundary of a system. For irreversible processes the law merely provides statements of trend, e.g. that the entropy of an isolated system must increase, and quantitative predictions of energy transfers cannot be made.

In engineering calculations empirical 'correction factors' (e.g. process efficiencies) are often used to supplement the Second Law. In effect they express the degree of reversibility of real processes, and they will be introduced formally in Part II. Sometimes a process efficiency is insufficiently illuminating and recourse is made to additional empirical relations which are found to be applicable to a limited range of phenomena. Relations of this kind are used in Part IV, i.e. *Fourier's law of heat conduction* and *Newton's law of viscosity*. They all take the form of proportionalities between a quantity transferred in an irreversible process and a property gradient. The constants of proportionality in the examples mentioned are the 'thermal conductivity' and 'viscosity' respectively. These 'laws' are approximate empirical relations, whose accuracy can be determined by direct experiment—their inaccuracy is manifested by the fact that the 'constants of proportionality' are not in fact constant but vary with the conditions of the experiment (temperature, pressure, or their gradients, etc.). The laws of thermodynamics, like Newton's laws of *motion*, are quite different in kind, being fundamental hypotheses upon which the respective sciences as a whole are based, and which can only be true, false, or shown to be part of an even broader generalisation about nature.

We began this section by trying to dispel a misconception that might have resulted from the systematic presentation of the theory, and we may end on the same note. It should not be supposed from this exposition that the science of thermodynamics developed historically in neat logical steps. In common with every science, the concepts which led to a systematic ordering and satisfactory understanding of experience, and which now seem so obvious, required for their birth centuries of experiment and the flashes of insight of some remarkable men. Although a definitive history of thermodynamics has yet to be written, some brief but stimulating accounts can be found. The student will be well advised to supplement his systematic study with a reading of the short historical survey given in Ref. (10).

6.8 Summary of chapters 1 to 6

With the thoughts of section 6.7 in mind, we may now return to our logical account, and survey briefly the ground we have covered. We opened by defining terms such as 'system', 'property', 'temperature', 'work' and 'heat', although they did not all permit of simple definition with equal rigour. For example, we were able to note a few of the characteristics of the concepts of work and heat, such as the fact that they are not properties, but we were unable to define them completely until the First Law was enunciated. This is always the case with any really fruitful concept in an experimental science. Such concepts cannot be described completely in everyday language, and are only definable in terms of the way in which they play their part in the logical framework of the theory. The more fruitful they are in enabling observable consequences to be deduced from the general law, the more they must rely on the logical structure for their definition*; energy and entropy are excellent examples of such concepts.

The notion of energy arose when we applied the First Law to a non-cyclic process. We saw that when heat and work cross the boundary of a closed system, there is a property of the system which increases by an amount equal to the net quantity of heat and work entering the system. We called this property the 'internal energy', and we explained the phenomenon by suggesting that heat and work are merely different ways in which 'energy' can be transferred between a system and its surroundings. Energy is the first outstandingly fruitful concept in thermodynamics, and it should be noted that it cannot be defined in terms we have already used without circularity. For example, it is pointless to say that energy is the capacity for doing work, and then define work as a form of transfer of energy. It is nearer the truth to say that energy is a concept which is necessary for the deduction of a variety of observable consequences from the First Law.

The First Law refers to any cycle, whether that of a heat engine or heat pump— and there is nothing in it to say that heat and work are not equivalent in all respects. We might have a cycle (a) in which all the heat supplied is converted into work, or (b) in which all the work supplied is converted into heat. The necessity for the Second Law arises from the fact that heat and work are *not* equivalent in all respects. The Second Law is not reciprocal like the First, in the sense that it states (a) to be impossible without denying the possibility of (b). It is this lack of reciprocity in the behaviour of cycles which is the fundamental reason why the concept of reversibility has to be introduced. When the consequences of the Second Law for a non-cyclic process were deduced, we found that only in the absence of friction and heat transfer across a temperature difference can a change take place without a degradation of energy. When either of these causes of irreversibility is present during a process, some inherent capacity for producing work, in the system or its surroundings, is permanently lost. When the process is reversible, on the other hand, any changes brought about in the system and its surroundings can be undone—nothing irretrievable has occurred.

*For the advanced student who is philosophically inclined, an interesting discussion of why this is so can be found in Ref. (1).

The implications of the Second Law can only be properly expressed in terms of measurable consequences, by using our mathematical tools to deduce a new property of a system—namely entropy. This property could not be deduced in quite such a direct way as internal energy. We had first to show that all reversible engines operating between the same two reservoirs must have the same efficiency. This, in turn, led us to infer that the efficiency must be a function only of the temperature of the reservoirs— since in every other respect the engines may differ from one another. When the temperatures of the two reservoirs are defined in such a way that their ratio is equal to the ratio of the quantities of heat leaving and entering the reservoirs, we found that we could devise a temperature scale which was independent of the behaviour of any particular substance and which was an absolute scale in the sense that a meaning could be given to the idea of zero temperature.

Once we had decided to use temperatures measured on this thermodynamic scale, we were able to widen the scope of our enquiry and investigate the behaviour of engines operating between more than two reservoirs. We found that when the cycle is reversible, the integration of all such quantities as (dQ/T) round the complete cycle must be equal to zero. If the cycle is reversible, there can be no more than an infinitesimal difference in temperature between the reservoirs and the system undergoing the cycle at the points where each quantity of heat dQ is transferred across the boundary. Therefore, T can refer to the system as well as the reservoir. It becomes probable that (dQ/T) represents a change in a property of the system since the total change for the complete reversible cycle is zero. We proved that this is in fact the case, by showing that the integral of (dQ/T) between the end states of any reversible process is independent of the process. The property whose change is given by $\int(dQ/T)_{rev}$ is called the entropy of the system. It differs from the internal energy, whose change is given by $\int(dQ - dW)$, because the change in entropy can be evaluated from $\int(dQ/T)$ only when the process is reversible. This is no great drawback since we can usually find a reversible process which can be imagined to connect any two given end states and for which we can evaluate $\int(dQ/T)$ in terms of other properties.

The lack of reciprocity in the Second Law can now be expressed in terms of the property entropy. We showed that for any isolated system the entropy must either remain constant or increase; it increases if any process occurring in the isolated system is irreversible. The degradation of energy consequent upon an irreversibility is thereby manifested as an increase in the entropy of the (system + surroundings) or 'universe'. All real processes must occur in such a direction that the state of the 'universe' changes from one of lower to one of higher entropy.

Although the reasoning required to deduce the existence of entropy from the Second Law is more complex than that required to deduce the internal energy from the First, the two concepts are of the same type. They are both consequences of empirical laws and are to this extent grounded in experience. If entropy appears the more mysterious, it is only because it is less familiar—the student meets the concept of energy in his early study of mechanics. It is because each is the consequence of an empirical law of great generality that these properties have such significance—much greater significance, for example, than the property enthalpy which was arbitrarily defined as a combination of other properties purely for convenience.

The utility of the concepts of internal energy and entropy only becomes apparent when the principles of thermodynamics are applied to particular systems. For the closed systems we are considering, it is an experimental fact that any property can be expressed as a function of two other independent properties when the system is in equilibrium. This fact, together with the laws of thermodynamics, enables a great many useful relations between the properties to be deduced which are universally valid for all such systems. These general relations are deduced in the next chapter, although they need concern only the advanced student. They are necessary because values of the internal energy and entropy are most conveniently obtained indirectly from measurements of other properties—indeed the latter can be obtained in no other way since no system can be put through a reversible process in practice. The next chapter indicates how these general relations may be used when tables of properties are to be compiled. So far we have assumed, and we shall do so again in Part II, that tables of properties are available for any substance we need consider, e.g. that given the values of p and v at any state we shall know the corresponding values of the other properties. We have already shown that this information enables the behaviour of both closed and open systems to be analysed and predicted when the processes are reversible, and also that some of the calculations would be extremely tedious, or even impossible, without the use of the concept of entropy. All this part of the exposition in these first six chapters will be further illustrated in Part II by numerical calculations relating to systems consisting of particular fluids.

7   General Thermodynamic Relations*

With the aid of the two laws of thermodynamics and the differential calculus, it is possible to deduce general relations between the thermodynamic properties which hold for any of the limited class of systems considered in this book. Since relations between thermodynamic properties are independent of whether the substance is in motion or not, we need refer only to closed systems.

The distinguishing characteristics of the systems considered here are that, when in equilibrium, (a) they only change their state when heat and mechanical work cross the boundary, and (b) only two independent properties are necessary to determine their thermodynamic state. These conditions are satisfied, in the absence of surface, gravitational, electric, and magnetic effects, by any system consisting of a single substance, even if more than one phase is present. They are also satisfied by systems consisting of a mixture of substances, provided that no chemical reaction is taking place: for example, air is normally treated as a single substance. When mixtures which react chemically are considered, it must be emphasised that only for *stable* states of chemical equilibrium are two properties sufficient to define the system completely. A mixture of H_2 and O_2 at atmospheric temperature is in an *apparently* stable state only because the reaction $H_2 + \frac{1}{2}O_2 \rightarrow H_2O$ leading to the most stable state is very slow at that temperature. When such quasi-stable states of chemically reacting systems are to be considered, more than two properties are required to define the state —in our foregoing example, the proportions of H_2, O_2 and H_2O in the mixture must be specified as a third property. For a treatment of systems whose composition is variable, the reader is referred to texts on chemical thermodynamics.

Some of the more important thermodynamic relations are deduced in this chapter; principally those which are useful when tables of properties are to be compiled from limited experimental data, and those which may be used when calculating the work and heat transfers associated with processes undergone by a liquid or solid. It should be noted that the relations only apply to a substance in the solid phase when the stress, i.e. the pressure, is uniform in all directions; if it is not, a single value for the pressure cannot be allotted to the system as a whole. The chapter ends with an introduction to the concept of availability and its relation to the Gibbs function.

*This chapter should be omitted on the first reading.

7.1 Properties to be related

We are already acquainted with six properties which may be used to describe the thermodynamic state of a system: pressure, volume, temperature, internal energy, enthalpy and entropy. The first three are directly measurable characteristics. The absolute measurement of pressure and volume presents no difficulty, and we have seen that the introduction of the thermodynamic scale as a consequence of the Second Law makes an absolute measurement of temperature possible. The internal energy u was shown to be a property as a consequence of the First Law and, since the enthalpy h is defined as a function of u, p and v, h is itself a property. Lastly, the entropy s was shown to be a property as a consequence of the Second Law.

Two other properties will be added to the list, although they are of minor significance to engineers. These are the *Helmholtz function* f and the *Gibbs function* g, arbitrarily defined in a similar way to enthalpy by the expressions:

$$f = u - Ts$$

$$g = h - Ts$$

h, f and g are sometimes referred to as *thermodynamic potentials*, and g is often referred to as the *free energy* for reasons which will be apparent after a reading of section 7.9. Both f and g are useful when considering chemical reactions, and the former is of fundamental importance in statistical thermodynamics. The Gibbs function is also useful when considering processes involving a change of phase.

Of the eight properties p, v, T, u, h, s, f and g, only the first three are directly measurable. We shall find it convenient to introduce other combinations of properties which are relatively easy to measure and which, together with measurements of p, v and T, enable the values of the remaining properties to be determined. These combinations of properties might be called 'thermodynamic gradients'; they are all defined as the rate of change of one property with another while a third is kept constant. Before proceeding to define them, and to deduce general relations between the properties, it is necessary to review certain theorems concerning the mathematical properties of differentials.

7.2 Exact differentials

When the state of a system can be defined by two independent properties, it follows that any third property must be a function of these two. In general terms, any property z can be expressed as $z = \phi(x, y)$ where x and y are two independent properties. This implies that if x and y change to new values x' and y', z will have a new definite value z' irrespective of the process. The function can therefore be represented by a three-dimensional surface, Fig. 7.1a. Now consider an infinitesimal change from a state defined by the point x, y to that defined by $(x + dx)$, $(y + dy)$. It should be clear from Fig. 7.1b that the change in z is given by

$$dz = \left(\frac{\partial z}{\partial x}\right)_y dx + \left(\frac{\partial z}{\partial y}\right)_x dy \tag{7.1}$$

In mathematics it is usual to omit the suffixes of the partial differential coefficients because it is obvious which variable is being maintained constant. However, in thermodynamics it is often necessary to refer to these coefficients out of context of such equations as (7.1). Since any one of the many thermodynamic properties might be kept constant during the change, it is essential to retain the suffix so that we know which third property is being considered.

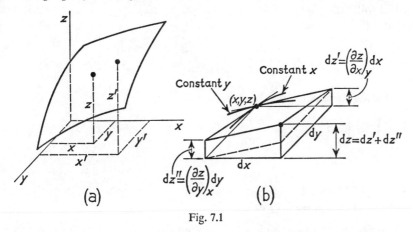

Fig. 7.1

Equation (7.1) is quite general and applies to any infinitesimal change of state. Suppose an infinitesimal change occurs in such a way that z is constant. Then dz is zero, and

$$\left(\frac{\partial z}{\partial x}\right)_y dx = -\left(\frac{\partial z}{\partial y}\right)_x dy$$

But since z is constant,

$$dx = \left(\frac{\partial x}{\partial y}\right)_z dy$$

and substituting for dx in the previous equation we have,

$$\left(\frac{\partial z}{\partial x}\right)_y \left(\frac{\partial y}{\partial z}\right)_x \left(\frac{\partial x}{\partial y}\right)_z = -1 \tag{7.2}$$

This is a relation between the three gradients at any point on the three-dimensional surface of Fig. 7.1a, i.e it is a general relation between the three properties x, y and z at any state and does not refer to a process.

Equations frequently arise in thermodynamics in the form

$$dz = M \, dx + N \, dy \tag{7.3}$$

where, although x and y are known to be properties, z is not necessarily a property. The coefficients M and N may be constants or functions of x and y. An example is provided by the energy equation for a reversible non-flow process: $dQ = du + p \, dv$. If an equation such as (7.3) can be expressed in the form of (7.1), it may be inferred that z is a property and that $z = \phi(x, y)$. If this cannot be done, the equation cannot be directly integrated to give $z = \phi(x, y)$; z is not a property, and the change in z depends upon the path of the process as well as on the values of x and y at the end states. A simple test by which equations (7.1) and (7.3) can be compared for identity may be devised as follows.

(7.1) and (7.3) are identical if

$$M = \left(\frac{\partial z}{\partial x}\right)_y \quad \text{and} \quad N = \left(\frac{\partial z}{\partial y}\right)_x$$

Differentiating these equations we have,

$$\left(\frac{\partial M}{\partial y}\right)_x = \left\{\frac{\partial}{\partial y}\left(\frac{\partial z}{\partial x}\right)_y\right\}_x = \frac{\partial^2 z}{\partial x \, \partial y}$$

and

$$\left(\frac{\partial N}{\partial x}\right)_y = \left\{\frac{\partial}{\partial x}\left(\frac{\partial z}{\partial y}\right)_x\right\}_y = \frac{\partial^2 z}{\partial y \, \partial x}$$

But if $z = \phi(x, y)$, the magnitude of the second derivative of z with respect to x and y is independent of the order of differentiation, and hence

$$\left(\frac{\partial M}{\partial y}\right)_x = \left(\frac{\partial N}{\partial x}\right)_y \tag{7.4}$$

It follows that if (7.4) is satisfied, equations (7.1) and (7.3) are identical, and the quantity z in (7.3) must be capable of being expressed as a function of x and y. dz is then referred to as an *exact* or *perfect differential*. Conversely, if the quantity z in (7.3) is known to be a property, (7.4) must be satisfied.

Reverting to the example provided by the energy equation for a closed system, $M = 1$ and $N = p$ in this case. Therefore

$$\left(\frac{\partial M}{\partial y}\right)_x = \left(\frac{\partial 1}{\partial v}\right)_u = 0$$

and

$$\left(\frac{\partial N}{\partial x}\right)_y = \left(\frac{\partial p}{\partial u}\right)_v$$

Reflection on the process of heating a gas at constant volume indicates that in general the rate of change of pressure with internal energy at constant volume cannot be zero. Thus the two partial differential coefficients are not equal, so that dQ cannot be an exact differential and Q is not a property. In order to integrate such an equation the path of the process must be specified in addition to the end states. In this case the path is fixed when the variation of p with v is known.

When $(M\,dx + N\,dy)$ is not an exact differential, it can always be made exact by multiplying throughout by an *integrating factor*. For our example, the integrating factor is $1/T$, because the expression then becomes equal to (dQ/T), the differential of the property entropy. ds can be substituted for (dQ/T) in this case because the equation $dQ = du + p\,dv$ applies to a reversible process. It is important to realise that we have not introduced entropy in the same way as h, f, or g—as an arbitrary combination of other properties. The fact that $(du + p\,dv)/T$ is an exact differential is a consequence of the Second Law; it is neither self-evident nor is it a purely mathematical deduction.* This accounts for the considerable significance which attaches to entropy. As the following sections will show, it is possible to deduce a large number of useful relations between the properties of a substance which do not contain entropy at all, but which could not have been deduced without its help.

7.3 Some general thermodynamic relations

The First Law applied to a closed system undergoing a reversible process states that

$$dQ = du + p\,dv$$

From the Second Law

$$ds = \left(\frac{dQ}{T}\right)_{\text{rev}}$$

Combining these equations we have

$$T\,ds = du + p\,dv$$

or

$$du = T\,ds - p\,dv \tag{7.5}$$

The properties h, f and g may also be put in terms of T, s, p and v as follows:

$$dh = du + p\,dv + v\,dp$$
$$= T\,ds + v\,dp \tag{7.6}$$

*In this book we are dealing mainly with systems whose state is determined by two independent properties. The Second Law is not required to show that an integrating factor *exists* for a total differential which is a function of two variables—this is an established fact in the theory of functions. But the same does not necessarily apply to functions of more than two variables, and the Second Law tells us that even for such cases (e.g. systems involving electric, magnetic or surface tension effects) there always exists an integrating factor for dQ, *and* that this factor is always $1/T$.

$$df = du - T\,ds - s\,dT$$

$$= -p\,dv - s\,dT \tag{7.7}$$

$$dg = dh - T\,ds - s\,dT$$

$$= v\,dp - s\,dT \tag{7.8}$$

Each of these equations is a result of the two laws of thermodynamics.
Since du, dh, df and dg are exact differentials we can express them as

$$du = \left(\frac{\partial u}{\partial s}\right)_v ds + \left(\frac{\partial u}{\partial v}\right)_s dv$$

$$dh = \left(\frac{\partial h}{\partial s}\right)_p ds + \left(\frac{\partial h}{\partial p}\right)_s dp$$

$$df = \left(\frac{\partial f}{\partial v}\right)_T dv + \left(\frac{\partial f}{\partial T}\right)_v dT$$

$$dg = \left(\frac{\partial g}{\partial p}\right)_T dp + \left(\frac{\partial g}{\partial T}\right)_p dT$$

Comparing these equations with (7.5) to (7.8) we may equate the corresponding
coefficients. For example, from the two equations for du we have

$$\left(\frac{\partial u}{\partial s}\right)_v = T \quad \text{and} \quad \left(\frac{\partial u}{\partial v}\right)_s = -p$$

The complete group of such relations may be summarised as follows.*

$$\left(\frac{\partial u}{\partial s}\right)_v = T = \left(\frac{\partial h}{\partial s}\right)_p \tag{7.9}$$

$$\left(\frac{\partial u}{\partial v}\right)_s = -p = \left(\frac{\partial f}{\partial v}\right)_T \tag{7.10}$$

$$\left(\frac{\partial h}{\partial p}\right)_s = v = \left(\frac{\partial g}{\partial p}\right)_T \tag{7.11}$$

$$\left(\frac{\partial f}{\partial T}\right)_v = -s = \left(\frac{\partial g}{\partial T}\right)_p \tag{7.12}$$

---

*These relations may be obtained by an alternative argument. Consider equation (7.5) for example; it is
quite general and applies to any infinitesimal process. Applying it first to an infinitesimal process at constant
volume, we have

$$(du)_v = T(ds)_v, \quad \text{or} \quad (\partial u/\partial s)_v = T$$

Applying it to an infinitesimal process at constant entropy we have

$$(du)_s = -p(dv)_s, \quad \text{or} \quad (\partial u/\partial v)_s = -p$$

Finally, making use of the fact that equation (7.4) must hold for equations (7.5) to (7.8), we have

$$\left(\frac{\partial T}{\partial v}\right)_s = -\left(\frac{\partial p}{\partial s}\right)_v \tag{7.13}$$

$$\left(\frac{\partial T}{\partial p}\right)_s = \left(\frac{\partial v}{\partial s}\right)_p \tag{7.14}$$

$$\left(\frac{\partial p}{\partial T}\right)_v = \left(\frac{\partial s}{\partial v}\right)_T \tag{7.15}$$

$$\left(\frac{\partial v}{\partial T}\right)_p = -\left(\frac{\partial s}{\partial p}\right)_T \tag{7.16}$$

These last four equations are known as the *Maxwell relations*.

It must be emphasised that equations (7.9) to (7.16) do not refer to a process, but simply express relations between properties which must be satisfied when any system is in a state of equilibrium. Each partial differential coefficient can itself be regarded as a property of state. The state may be defined by a point on a three-dimensional surface, the surface representing all possible states of stable equilibrium. Suppose, for example, that the coordinates of Fig. 7.1a are u, v and s. If u and v are specified, s is determined, and any gradient of the surface at the point (u, v, s) is also determined. The gradient of a curve formed by the intersection of a horizontal plane with the surface is $(\partial u/\partial v)_s$, and equation (7.10) informs us that the gradient at any point on this curve is always negative and numerically equal to the pressure at that point. All the relations (7.9) to (7.16) can be interpreted graphically in this way.

To deduce equations (7.9) to (7.12) we took four forms of the equation expressing the laws of thermodynamics, and compared each with an equation expressing one property in terms of two others. The two independent properties were so chosen that a direct comparison could be made of the differential coefficients, e.g. we took $u = \phi(s, v)$ for comparison with equation (7.5). However, we need not limit the analysis in this way. For example, writing $u = \phi(T, v)$ we have

$$du = \left(\frac{\partial u}{\partial T}\right)_v dT + \left(\frac{\partial u}{\partial v}\right)_T dv$$

Equating this to (7.5) we get

$$\left(\frac{\partial u}{\partial T}\right)_v dT + \left(\frac{\partial u}{\partial v}\right)_T dv = T\, ds - p\, dv$$

or

$$T\, ds = \left(\frac{\partial u}{\partial T}\right)_v dT + \left\{p + \left(\frac{\partial u}{\partial v}\right)_T\right\} dv$$

In this case we may take $s = \phi(T, v)$ to provide an equation for direct comparison of coefficients, i.e.

$$\mathrm{d}s = \left(\frac{\partial s}{\partial T}\right)_v \mathrm{d}T + \left(\frac{\partial s}{\partial v}\right)_T \mathrm{d}v$$

Then

$$\left(\frac{\partial s}{\partial T}\right)_v = \frac{1}{T}\left(\frac{\partial u}{\partial T}\right)_v \tag{7.17}$$

and

$$\left(\frac{\partial s}{\partial v}\right)_T = \frac{1}{T}\left\{p + \left(\frac{\partial u}{\partial v}\right)_T\right\} \tag{7.18}$$

Since there are eight properties, and in general each can be written as a function of any other two, it is clear that a very large number of relations can be derived. No attempt will be made here to expand the number of relations systematically,* although a few others will be deduced in the course of illustrating the use of these general relations.

It must be remembered that relations obtained by using expressions of the form $z = \phi(x, y)$ are only valid if x and y are independent. Thus relations based on the assumption that p and T are independent cannot be applied to a mixture of phases. For example, the right-hand equations of (7.11) and (7.12) were obtained by writing $g = \phi(p, T)$. It is shown in section (7.7) that when the temperature or pressure of a mixture of phases is maintained constant, g is constant also. If equations (7.11) and (7.12) were valid they would imply that v and s are zero under these conditions, which is obviously untrue.

7.4 Measurable quantities

Of the eight thermodynamic properties listed in section (7.1), only p, v and T are directly measurable. We must now examine the information that can be obtained from measurements of these primary properties, and then see what other easily measurable quantities can be introduced.

(i) *Equation of state*

Imagine a series of experiments in which the volume of a substance is measured over a range of temperature while the pressure is maintained constant, this being repeated for various pressures. The results might be represented graphically by a three-dimensional surface, or by a family of constant pressure lines on a v-T diagram. It is useful if an equation can be found to express the relation between p, v and T, and this can always be done over a limited range of states. No single equation will hold for all

---

* A comprehensive collection can be found in Ref. (3).

phases of a substance, and usually more than one equation is required even in one phase if the accuracy of the equation is to match that of the experimental results. Equations relating p, v and T are called *equations of state* or *characteristic equations*. Accurate equations of state are usually complicated, a typical form being

$$pv = A + \frac{B}{v} + \frac{C}{v^2} + \ldots$$

where A, B, C, . . . are functions of temperature which differ for different substances.

An equation of state of a particular substance is an empirical result, *and it cannot be deduced from the laws of thermodynamics*. Nevertheless the general form of the equation may be predicted from hypotheses about the microscopic structure of matter. This type of prediction has been developed to a high degree of precision for gases, and to a lesser extent for liquids and solids. The simplest postulates about the

Fig. 7.2   Determination of coefficient of expansion from p-v-T data

molecular structure of gases lead to the concept of the perfect gas which has the equation of state $pv = RT$. Experiments have shown that the behaviour of real gases at low pressure and high temperature agrees well with this equation. Additional refinements to the assumptions lead to the more complicated equations of state. A brief statement of the main assumptions is given in Chapter 8, but for a description of the theory itself the reader is referred to books on the kinetic theory of gases and statistical mechanics.

(ii) *Coefficient of expansion and compressibility*

An equation of state is not the only useful information that can be obtained from p-v-T measurements. When the experimental results are plotted as a series of constant pressure lines on a v-T diagram, as in Fig. 7.2a, the slope of a constant pressure line at any given state is $(\partial v/\partial T)_p$. If the gradient is divided by the volume at that state, we have the value of a property of the substance called its *coefficient of cubical expansion* β. That is,

$$\beta = \frac{1}{v} \left(\frac{\partial v}{\partial T} \right)_p \qquad\qquad (7.19)$$

Values of β can be tabulated for a range of pressures and temperatures, or plotted graphically as in Fig. 7.2b. For solids and liquids over the normal working range of pressure and temperature, the variation of β is small and can often be neglected. In tables of physical properties β is usually quoted as an average value over a small range of temperature, the pressure being atmospheric. This average coefficient may be symbolised by $\bar\beta$ and it is defined by

$$\bar\beta = \frac{v_2 - v_1}{v_1(T_2 - T_1)}$$

Fig. 7.2a can be replotted to show the variation of volume with pressure for various constant values of temperature, Fig. 7.3a. In this case the gradient of a curve at any state is $(\partial v/\partial p)_T$. When this gradient is divided by the volume at that state, we have a property known as the *compressibility* κ of the substance. Since this gradient is always

Fig. 7.3   Determination of compressibility from *p-v-T* data

negative, i.e. the volume of a substance always decreases with increase of pressure when the temperature is constant, the compressibility is usually made a positive quantity by defining it as

$$\kappa = -\frac{1}{v}\left(\frac{\partial v}{\partial p}\right)_T \tag{7.20}$$

Again, for solids and liquids κ can be regarded as a constant for many purposes. In tables of properties it is often quoted as an average value over a small range of pressure at atmospheric temperature, i.e.

$$\bar\kappa = -\frac{v_2 - v_1}{v_1(p_2 - p_1)}$$

When β and κ are known, equation (7.2) may be used to find the third partial differential coefficient involving p, v and T. Thus (7.2) becomes

$$\left(\frac{\partial p}{\partial T}\right)_v \left(\frac{\partial T}{\partial v}\right)_p \left(\frac{\partial v}{\partial p}\right)_T = -1$$

Since

$$\left(\frac{\partial v}{\partial T}\right)_p = \beta v \quad \text{and} \quad \left(\frac{\partial v}{\partial p}\right)_T = -\kappa v$$

$$\left(\frac{\partial p}{\partial T}\right)_v = \frac{\beta}{\kappa} \tag{7.21}$$

When the equation of state is known, the coefficients of cubical expansion and compressibility can be found by differentiation. For a perfect gas, for example, we have

$$\left(\frac{\partial v}{\partial T}\right)_p = \frac{R}{p} \quad \text{and} \quad \left(\frac{\partial v}{\partial p}\right)_T = -\frac{RT}{p^2}$$

Hence

$$\beta = \frac{1}{v}\left(\frac{\partial v}{\partial T}\right)_p = \frac{R}{pv} = \frac{1}{T}$$

and

$$\kappa = -\frac{1}{v}\left(\frac{\partial v}{\partial p}\right)_T = \frac{RT}{p^2 v} = \frac{1}{p}$$

It will be found in section (7.5) that information about the relation between p, v and T is not sufficient to enable the other properties to be determined, and other measurable quantities must be introduced. These will be discussed in the remainder of this section.

(iii) *Specific heats*

There are three more partial differential coefficients which are relatively easily determined experimentally. Consider first the quantity $(\partial u/\partial T)_v$. During a process at constant volume, the First Law informs us that an increase of internal energy is equal to the heat supplied. If a calorimetric experiment is conducted with a known mass of substance at constant volume, the quantity of heat Q required to raise the temperature of unit mass by ΔT may be measured. We can then write

$$\left(\frac{\Delta u}{\Delta T}\right)_v = \left(\frac{Q}{\Delta T}\right)_v$$

The quantity obtained in this way is known as the mean specific heat at constant volume over the temperature range ΔT. It is found to vary with the conditions of the experiment, i.e. with the temperature range and the specific volume of the substance. As the temperature range is reduced the value approaches that of $(\partial u/\partial T)_v$, and the true *specific heat at constant volume* is defined by

$$c_v = \left(\frac{\partial u}{\partial T}\right)_v$$

This is a property of the substance, and in general its value varies with the state of the substance, e.g. with temperature and pressure.

The First Law also implies that the heat supplied is equal to the increase of enthalpy during a reversible constant pressure process. Therefore a calorimetric experiment carried out with a substance at constant pressure gives us

$$\left(\frac{\Delta h}{\Delta T}\right)_p = \left(\frac{Q}{\Delta T}\right)_p$$

which is the mean specific heat at constant pressure. As the range of temperature is made infinitesimally small, this becomes the rate of change of enthalpy with temperature at a particular state defined by T and p, and this is the true *specific heat at constant pressure* defined by

$$c_p = \left(\frac{\partial h}{\partial T}\right)_p$$

c_p also varies with the state, e.g. with pressure and temperature.

Descriptions of experimental methods of determining c_p and c_v can be found in texts on physics. When solids and liquids are considered, it is not easy to measure c_v owing to the stresses set up when such a substance is prevented from expanding. Fortunately a relation between c_p, c_v, β and κ can be found as follows, from which c_v may be obtained if the remaining three properties have been measured.

From equations (7.5) and (7.6) we can write

$$T \, ds = du + p \, dv = dh - v \, dp$$

Putting $u = \phi(T, v)$ and $h = \phi(T, p)$, we can also write respectively

$$du = \left(\frac{\partial u}{\partial T}\right)_v dT + \left(\frac{\partial u}{\partial v}\right)_T dv = c_v \, dT + \left(\frac{\partial u}{\partial v}\right)_T dv$$

and

$$dh = \left(\frac{\partial h}{\partial T}\right)_p dT + \left(\frac{\partial h}{\partial p}\right)_T dp = c_p \, dT + \left(\frac{\partial h}{\partial p}\right)_T dp$$

Substituting for du and dh in the above equation for $T \, ds$, we arrive at

$$c_v \, dT + \left\{ p + \left(\frac{\partial u}{\partial v}\right)_T \right\} dv = c_p \, dT - \left\{ v - \left(\frac{\partial h}{\partial p}\right)_T \right\} dp$$

As this equation is true for any process, it will also be true for the case when $dp = 0$ and hence

$$(c_p - c_v)(dT)_p = \left\{ p + \left(\frac{\partial u}{\partial v}\right)_T \right\} (dv)_p$$

or

$$c_p - c_v = \left\{ p + \left(\frac{\partial u}{\partial v}\right)_T \right\} \left(\frac{\partial v}{\partial T}\right)_p$$

Now from equations (7.15) and (7.18) we have

$$\left(\frac{\partial p}{\partial T}\right)_v = \left(\frac{\partial s}{\partial v}\right)_T = \frac{1}{T} \left\{ p + \left(\frac{\partial u}{\partial v}\right)_T \right\}$$

and therefore

$$c_p - c_v = T \left(\frac{\partial p}{\partial T}\right)_v \left(\frac{\partial v}{\partial T}\right)_p$$

Finally, from equations (7.19) and (7.21),

$$c_p - c_v = \frac{\beta^2 T v}{\kappa} \tag{7.22}$$

Thus at any state defined by T and v, c_v can be found if c_p, β and κ are known for the substance at that state.

The values of T, v and κ are always positive and, although β may sometimes be negative (e.g. between 0 and 4 °C water contracts on heating at constant pressure), β^2 is always positive. It follows that c_p is always greater than c_v.

By making use of equation (7.9) other expressions for c_p and c_v can be obtained. Thus since

$$c_v = \left(\frac{\partial u}{\partial T}\right)_v = \left(\frac{\partial u}{\partial s}\right)_v \left(\frac{\partial s}{\partial T}\right)_v$$

we have

$$c_v = T \left(\frac{\partial s}{\partial T}\right)_v \tag{7.23}$$

Similarly,

$$c_p = \left(\frac{\partial h}{\partial T}\right)_p = \left(\frac{\partial h}{\partial s}\right)_p \left(\frac{\partial s}{\partial T}\right)_p$$

and consequently

$$c_p = T \left(\frac{\partial s}{\partial T}\right)_p \tag{7.24}$$

These two relations will be found useful when expressing entropy in terms of measurable quantities (section 7.5).

(iv) *Joule-Thomson coefficient*

Consider the partial differential coefficient $(\partial T/\partial p)_h$. We have already noted in section 4.2 (iv) that if a fluid is flowing through a pipe, and the pressure is reduced by a throttling process, the enthalpies on either side of the restriction may be equal. Fig. 7.4a illustrates the process: the velocity increases at the restriction, with a consequent decrease of enthalpy, but this increase of kinetic energy is dissipated by friction as the eddies die down after the restriction. The steady-flow energy equation implies that the enthalpy of the fluid is restored to its initial value if the flow is adiabatic and if the velocity before the restriction is equal to that downstream of it. These conditions are very nearly satisfied in the following experiment; it was first carried out by Joule and Thomson (Lord Kelvin), and is usually referred to as the Joule-Thomson experiment.

A fluid is allowed to flow steadily from a high pressure to a low pressure through a porous plug inserted in a pipe. The pipe is well lagged so that any heat flow to or from

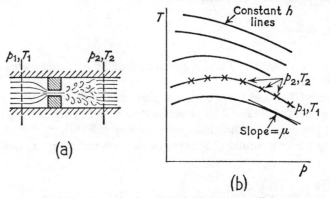

Fig. 7.4   Determination of Joule-Thomson coefficient

the fluid is negligible when steady conditions have been reached. Furthermore, the velocity of flow is kept low, and any difference between the kinetic energy upstream and downstream of the plug is negligible. A porous plug is used because the local increase of directional kinetic energy, caused by the restriction, is rapidly reconverted to random molecular energy by viscous friction in the fine passages of the plug. Irregularities in the flow die out a very short distance downstream of the plug, and temperature and pressure measurements taken there will be values for the fluid in a state of thermodynamic equilibrium.

With the upstream pressure and temperature maintained constant at p_1 and T_1, the downstream pressure p_2 is reduced in steps and the corresponding temperature T_2 is measured. The fluid in the successive states defined by the values of p_2 and T_2 must always have the same value of the enthalpy, namely the value of the enthalpy corresponding to the state defined by p_1 and T_1. From these results, points representing equilibrium states of the same enthalpy can be plotted on a T-p diagram, and joined up to form a curve of constant enthalpy. The curve does *not* represent the throttling process itself, which is irreversible. During the actual process the fluid undergoes first a decrease and then an increase of enthalpy, and no single value of the specific enthalpy can be ascribed to all elements of the fluid. If the experiment is repeated with different values of p_1 and T_1, a family of curves may be obtained covering a range of values of enthalpy, Fig. 7.4b. The slope of a curve at any point in the field is a function only of the state of the fluid; it is the *Joule-Thomson coefficient* μ, defined by

$$\mu = \left(\frac{\partial T}{\partial p}\right)_h$$

The change of temperature due to a throttling process is small and, if the fluid is a gas, it may be an increase or decrease. At any particular pressure there is a temperature, the *temperature of inversion*, above which a gas can never be cooled by a throttling process. If a gas is to be liquefied by the Linde process it must first be cooled below the temperature of inversion.

It may be seen that both c_p and μ are defined in terms of p, T and h. The third partial differential coefficient based on these three properties can therefore be found from equation (7.2), i.e.

$$\left(\frac{\partial h}{\partial p}\right)_T \left(\frac{\partial p}{\partial T}\right)_h \left(\frac{\partial T}{\partial h}\right)_p = -1$$

Hence,

$$\left(\frac{\partial h}{\partial p}\right)_T = -\mu c_p \qquad\qquad (7.25)$$

Fig. 7.5

Summarising: experimental methods are available for determining the p-v-T relation (and hence β and κ), and also values of c_p, c_v and μ at any state. We must now see how the values of the remaining properties (s, h, u, f and g) can be determined from these experimental data.

7.5 Graphical determination of entropy and enthalpy

Let us assume that v is expressed as a function of p and T in the form of a family of curves, Fig. 7.5. For practical purposes, and in particular for compiling tables of properties, it is sufficient if we can determine the change in entropy and enthalpy between any state (p, T) and an arbitrary reference state (p_0, T_0). The values of these properties at the reference state may be denoted by s_0 and h_0; they are usually put equal to zero. Since the change in a property is independent of the process, any convenient path may be considered. We shall proceed from the reference state to any other state by means of a reversible constant pressure process at p_0 followed by a reversible isothermal process at T, Fig. 7.5.

93

The change in entropy along any constant pressure line can be found by integrating $(\partial s/\partial T)_p$ with respect to T. It is not possible to derive an expression for $(\partial s/\partial T)_p$ in terms of p, v and T, but we may make use of equation (7.24) from

$$\left(\frac{\partial s}{\partial T}\right)_p = \frac{c_p}{T}$$

The change of entropy between state (p_0, T_0) and state $(p_0; T)$ is then given by

$$s_{0T} - s_0 = \left[\int_{T_0}^{T} \frac{c_p}{T}\,dT\right]_{p_0}$$

where s_{0T} is used to denote the entropy at a state defined by p_0 and any temperature T. Experimental data for c_p at pressure p_0 may be expressed by a single curve of c_p versus T, from which Fig. 7.6a can easily be constructed. The integral is then given

Fig. 7.6

by the area under the curve between T_0 and any required temperature T.

The change of entropy along any isothermal can be found by integrating $(\partial s/\partial p)_T$ with respect to p. This coefficient can be expressed in terms of primary properties, since the Maxwell relation (7.16) states that

$$\left(\frac{\partial s}{\partial p}\right)_T = -\left(\frac{\partial v}{\partial T}\right)_p$$

Hence, on integration between the state (p_0, T) and the state (p, T), we have

$$s - s_{0T} = -\left[\int_{p_0}^{p} \left(\frac{\partial v}{\partial T}\right)_p\,dp\right]_T$$

The integration may be performed graphically if values of $(\partial v/\partial T)_p$ are plotted against pressure as in Fig. 7.6b. This can be done by measuring the slopes of the constant pressure lines in Fig. 7.5 at the required temperature T. The integral is then given by the area under the curve between p_0 and p.

The complete expression for the entropy at any state defined by (p, T), in terms of the entropy at the reference state and the measurable properties p, v, T and c_p, becomes

$$s = s_0 + \left[\int_{T_0}^{T} \frac{c_p}{T} \, dT \right]_{p_0} - \left[\int_{p_0}^{p} \left(\frac{\partial v}{\partial T} \right)_p dp \right]_T \qquad (7.26)$$

An expression for the enthalpy of a substance can be found in a similar way. The change in enthalpy along a constant pressure line can be found by integrating $(\partial h / \partial T)_p$. Since this partial differential coefficient is equal to c_p, the change of enthalpy between the reference state (p_0, T_0), and any other state where the pressure is also p_0, is given by

$$h_{0T} - h_0 = \left[\int_{T_0}^{T} c_p \, dT \right]_{p_0}$$

In this case the integral can be evaluated directly from the curve of c_p versus T. Again, only values of c_p at the pressure p_0 need be known.

To find the change of enthalpy along the isothermal path, we must integrate the partial differential coefficient $(\partial h / \partial p)_T$. This quantity may be expressed in terms of p, v and T, although the expression is not given directly by a Maxwell relation as in the case of $(\partial s / \partial p)_T$. The required expression can be obtained by repeating the analysis used to determine equation (7.18), but in this case starting with the equation expressing h as a function of p and T. Alternatively, the argument contained in the footnote in section (7.3) may be used as follows. Applying (7.6) to an infinitesimal isothermal process, we have

$$(dh)_T = (T \, ds)_T + (v \, dp)_T$$

Dividing throughout by the change of pressure this becomes

$$\left(\frac{\partial h}{\partial p} \right)_T = T \left(\frac{\partial s}{\partial p} \right)_T + v$$

The Maxwell relation (7.16) may be used to eliminate the entropy term to give

$$\left(\frac{\partial h}{\partial p} \right)_T = v - T \left(\frac{\partial v}{\partial T} \right)_p \qquad (7.27)$$

Equation (7.27) could now be integrated as it stands, although it would involve the evaluation of two integrals. By substituting τ for $1/T$, one of the integrals is eliminated. Thus (7.27) can be written as

$$\left(\frac{\partial h}{\partial p} \right)_T = v - \frac{1}{\tau} \left(\frac{\partial v}{\partial \tau} \right)_p \left(\frac{\partial \tau}{\partial 1/\tau} \right)_p$$

$$= v + \tau \left(\frac{\partial v}{\partial \tau} \right)_p$$

and hence

$$\left(\frac{\partial h}{\partial p}\right)_T = \left(\frac{\partial v_T}{\partial \tau}\right)_p \tag{7.28}$$

Integrating between state (p_0, T) and state (p, T) we have

$$h - h_{0T} = \left[\int_{p_0}^{p} \left(\frac{\partial v_T}{\partial \tau}\right)_p \mathrm{d}p\right]_T$$

The integration can be performed graphically if Figs. 7.7a and b are constructed. Fig. 7.7a can be deduced directly from the p-v-T data in Fig. 7.5. Then, by measuring the slopes of the constant pressure lines at the required temperature T (and therefore

Fig. 7.7

a particular value of τ), the appropriate curve of Fig. 7.7b is obtained. The integral is equal to the area under this curve between the limits p_0 and p.

The complete expression for the enthalpy may be written as

$$h = h_0 + \left[\int_{T_0}^{T} c_p \, \mathrm{d}T\right]_{p_0} + \left[\int_{p_0}^{p} \left(\frac{\partial v_T}{\partial \tau}\right)_p \mathrm{d}p\right]_T \tag{7.29}$$

As already stated, s_0 and h_0 are usually assigned the arbitrary value of zero.

It is worth emphasising that the p-v-T data alone are insufficient for the determination of the entropy and enthalpy. Some additional information about the substance along a line of reference states is evidently necessary. It is not essential for the additional data to consist of experimental values of c_p; values of the Joule-Thomson coefficient can be used instead. Thus, from equation (7.25) we have

$$c_p = -\frac{1}{\mu}\left(\frac{\partial h}{\partial p}\right)_T$$

and hence from (7.28),

$$c_p = -\frac{1}{\mu}\left(\frac{\partial v_T}{\partial \tau}\right)_p$$

The partial differential coefficient $(\partial v_T/\partial_T)_p$ is already used in equation (7.29), and values of this quantity at the reference pressure p_0 and any temperature T can be read from Fig. 7.7b. The required values of c_p can then be determined if μ is known for various temperatures at pressure p_0.

Once the entropy and enthalpy have been obtained, the remaining properties u, and g are easily found. For example, $u = (h - pv)$ and both h and v are known for any pair of values of p and T. Similarly, f is found from $(u - Ts)$ and g from $(h - Ts)$. Note that the value of u at the reference state cannot be zero if h_0 is put equal to zero. At the reference state (p_0, T_0) the internal energy is

$$u_0 = h_0 - p_0 v_0 = - p_0 v_0$$

It should also be noted that it has been assumed that no discontinuity occurs in the experimental curves during the change from the reference state to the state (p, T). This assumption is invalid if a change of phase occurs, and in such cases it is necessary to determine the changes of entropy and enthalpy in more than two steps. This matter will be referred to again in section (7.7).

The following section shows how expressions for the entropy and enthalpy may be deduced when the p-v-T data are given in the form of an equation of state. After a general example, we shall consider two important hypothetical substances which are characterised by two particular equations of state. When the equation of state is known, it is usually easier to work from first principles rather than use equations (7.26) and (7.29).

7.6 Analytical determination of entropy and enthalpy

A typical equation of state giving v as an explicit function of p and T is provided by

$$v = \frac{RT}{p} + A - \frac{B}{T^2}$$

where R, A and B are constants. We have seen that in addition to this information we require a knowledge of c_p or μ along a line of reference states. Let us assume that, over the required range of temperature, experiment shows c_p to vary linearly with T at zero pressure. That is, when $p = 0$,

$$c_p = C + DT$$

where C and D are constants. The problem is to find expressions for s and h at any state defined by pressure p and temperature T, and we may assume that the values of s and h are put equal to zero at an arbitrary reference state (p_0, T_0). We shall work from first principles, and check the result by applying (7.26) and (7.29).

Considering s as a function of p and T,

$$ds = \left(\frac{\partial s}{\partial T}\right)_p dT + \left(\frac{\partial s}{\partial p}\right)_T dp$$

97

Using equation (7.24) and the Maxwell relation (7.16), this becomes

$$ds = \frac{c_p}{T}\,dT - \left(\frac{\partial v}{\partial T}\right)_p dp \qquad (7.30)$$

The next step is to find an expression for c_p in terms of p and T. This can be done because the variation of c_p along a line of reference states is given. Using (7.24) again and differentiating,

$$c_p = T\left(\frac{\partial s}{\partial T}\right)_p$$

and

$$\left(\frac{\partial c_p}{\partial p}\right)_T = T\left(\frac{\partial^2 s}{\partial T\,\partial p}\right)$$

On substitution of the Maxwell relation (7.16) this becomes

$$\left(\frac{\partial c_p}{\partial p}\right)_T = -T\left(\frac{\partial^2 v}{\partial T^2}\right)_p$$

Differentiating the characteristic equation, treating p as constant

$$\left(\frac{\partial v}{\partial T}\right)_p = \frac{R}{p} + \frac{2B}{T^3}$$

and

$$\left(\frac{\partial^2 v}{\partial T^2}\right)_p = -\frac{6B}{T^4}$$

Hence

$$\left(\frac{\partial c_p}{\partial p}\right)_T = \frac{6B}{T^3}$$

Integrating, and remembering that when $p = 0$, $c_p = C + DT$, we have

$$c_p = \left(\int \frac{6B}{T^3}\,dp\right)_T = C + DT + \frac{6Bp}{T^3}$$

Substituting for $(\partial v/\partial T)_p$ and c_p in equation (7.30),

$$ds = \left(C + DT + \frac{6Bp}{T^3}\right)\frac{dT}{T} - \left(\frac{R}{p} + \frac{2B}{T^3}\right) dp \qquad (7.31)$$

$$= \left(\frac{C}{T} + D\right)dT - \frac{R}{p}\,dp + 2B\left(\frac{3p}{T^4}\,dT - \frac{1}{T^3}\,dp\right)$$

$$= \left(\frac{C}{T} + D\right)dT - \frac{R}{p}\,dp - 2B\,d\left(\frac{p}{T^3}\right)$$

By integration between the reference state (p_0, T_0) and any other state (p, T), we obtain

$$s = C \ln \left(\frac{T}{T_0} \right) + D(T - T_0) - R \ln \left(\frac{p}{p_0} \right) - 2B \left(\frac{p}{T^3} - \frac{p_0}{T_0^3} \right)$$

The enthalpy can be obtained from equation (7.6),

$$dh = T \, ds + v \, dp$$

Substituting for $T \, ds$ from (7.31), and for v from the equation of state, we have

$$dh = \left(C + DT + \frac{6Bp}{T^3} \right) dT - \left(\frac{RT}{p} + \frac{2B}{T^2} \right) dp + \left(\frac{RT}{p} + A - \frac{B}{T^2} \right) dp$$

$$= (C + DT) \, dT + A \, dp + 3B \left(\frac{2p}{T^3} \, dT - \frac{1}{T^2} \, dp \right)$$

$$= (C + DT) \, dT + A \, dp - 3B \, d \left(\frac{p}{T^2} \right)$$

Therefore,

$$h = C(T - T_0) + \frac{D}{2} (T^2 - T_0^2) + A(p - p_0) - 3B \left(\frac{p}{T^2} - \frac{p_0}{T_0^2} \right)$$

As a check, we may deduce these expressions for s and h from equations (7.26) and (7.29). We have already obtained expressions for c_p and $(\partial v / \partial T)_p$, so equation (7.26) becomes

$$s = s_0 + \left[\int_{T_0}^{T} \left(\frac{C}{T} + D + \frac{6Bp}{T^4} \right) dT \right]_{p_0} - \left[\int_{p_0}^{p} \left(\frac{R}{p} + \frac{2B}{T^3} \right) dp \right]_T$$

$$= s_0 + C \ln \frac{T}{T_0} + D(T - T_0) - 2Bp_0 \left(\frac{1}{T^3} - \frac{1}{T_0^3} \right) - R \ln \frac{p}{p_0} - \frac{2B}{T^3} (p - p_0)$$

$$= s_0 + C \ln \frac{T}{T_0} + D(T - T_0) - R \ln \frac{p}{p_0} - 2B \left(\frac{p}{T^3} - \frac{p_0}{T_0^3} \right)$$

Since $s = 0$ when $p = p_0$ and $T = T_0$, $s_0 = 0$, and this becomes identical with the previous result.

In order to use (7.29) we must first find $(\partial v_T / \partial_T)_p$. From the equation of state,

$$v_T = \frac{R}{p} + A_T - B_T^3$$

and

$$\left(\frac{\partial v_T}{\partial_T} \right)_p = A - 3B_T^2$$

Hence (7.29) becomes

$$h = h_0 + \left[\int_{T_0}^{T}\left(C + DT + \frac{6Bp}{T^3}\right)dT\right]_{p_0} + \left[\int_{p_0}^{p}(A - 3B\tau^2)\,dp\right]_{T}$$

$$= h_0 + C(T - T_0) + \frac{D}{2}(T^2 - T_0^2) - 3Bp_0\left(\frac{1}{T^2} - \frac{1}{T_0^2}\right) + A(p - p_0) - \frac{3B}{T^2}(p - p_0)$$

$$= h_0 + C(T - T_0) + \frac{D}{2}(T^2 - T_0^2) + A(p - p_0) - 3B\left(\frac{p}{T^2} - \frac{p_0}{T_0^2}\right)$$

This again equals the former result when h_0 is put equal to zero.

We may now consider two simple characteristic equations which are of particular significance. All gases at low pressures and high temperatures have an equation of state approximating to the perfect gas equation

$$pv = RT$$

A better approximation is provided by van der Waals' equation

$$\left(p + \frac{a}{v^2}\right)(v - b) = RT$$

R, a and b are constants for any particular gas. The significance of these equations is discussed more fully in Chapter 8, but a few of their consequences will appear here in the course of deducing the relevant expressions for entropy and enthalpy.

(i) Perfect gas

Certain information about the internal energy, enthalpy and specific heats of a perfect gas is implicit in the equation of state. Transforming equation (7.18), we have

$$\left(\frac{\partial u}{\partial v}\right)_T = T\left(\frac{\partial s}{\partial v}\right)_T - p$$

which, on substituting the Maxwell relation (7.15), can be written

$$\left(\frac{\partial u}{\partial v}\right)_T = T\left(\frac{\partial p}{\partial T}\right)_v - p \qquad (7.32)$$

But from the equation of state $pv = RT$,

$$\left(\frac{\partial p}{\partial T}\right)_v = \frac{R}{v} = \frac{p}{T}$$

Therefore,

$$\left(\frac{\partial u}{\partial v}\right)_T = 0$$

This means that when T is constant, u does not vary with v. But the existence of a characteristic equation of any kind implies that when T is constant p and v are not independent properties. Consequently it also follows that u does not vary with p when T is constant. The final conclusion must be that *for a perfect gas u is a function only of T*.

Since the enthalpy is defined as $(u + pv)$, and $pv = RT$, it follows immediately that h also is a function only of T. It is of interest to note that this implies that the temperatures before and after a throttling process must be equal, and therefore that the Joule-Thomson coefficient of a perfect gas is zero.

The specific heats are defined by

$$c_v = \left(\frac{\partial u}{\partial T}\right)_v \quad \text{and} \quad c_p = \left(\frac{\partial h}{\partial T}\right)_p$$

But for a perfect gas u and h are not functions of v and p, and the partial notation may be dropped. Thus

$$c_v = \frac{du}{dT} \quad \text{and} \quad c_p = \frac{dh}{dT}$$

and the specific heats are either constants or functions of temperature only. A relation between c_p, c_v and R follows immediately. Since $dh = du + d(pv)$, we have in this case

$$dh = du + R\,dT$$

and hence

$$c_p - c_v = R \tag{7.33}$$

Thus, although c_p and c_v may vary with temperature, their difference is constant. This important result can also be obtained in another way. From (7.22) we have the general relation

$$c_p - c_v = \frac{\beta^2 Tv}{\kappa}$$

When $pv = RT$ it has already been shown in section 7.4 (ii) that β and κ are equal to $1/T$ and $1/p$ respectively. Therefore

$$c_p - c_v = \frac{pv}{T} = R$$

Expressions for enthalpy and internal energy can now be written as

$$h = h_0 + \int_{T_0}^{T} c_p\,dT$$

and

$$u = u_0 + \int_{T_0}^{T} c_v \, dT$$

In this case h and u may be put equal to zero along a line of reference states at T_0, since they are independent of pressure.

We have seen that the perfect gas equation implies that c_p and c_v are independent of p and v; they must therefore be either constant or some function of temperature. A perfect gas is sometimes defined as one which both satisfies the equation $pv = RT$ and has constant specific heats. The enthalpy and internal energy of a perfect gas defined in this way are therefore given by the expressions

$$h = c_p(T - T_0), \quad \text{and} \quad u = c_v(T - T_0)$$

when u_0 and h_0 are put equal to zero. And the changes in these properties between any two states at T_1 and T_2 are

$$h_2 - h_1 = c_p(T_2 - T_1) \tag{7.34}$$

and

$$u_2 - u_1 = c_v(T_2 - T_1) \tag{7.35}$$

It now remains to find an expression for the entropy. The simplicity of the equation of state enables equation (7.6) to be used. Writing $c_p \, dT$ for dh, and R/p for v/T, (7.6) becomes

$$ds = \frac{c_p}{T} \, dT - \frac{R}{p} \, dp$$

Therefore, when s is put equal to zero at the reference state (p_0, T_0),

$$s = \int_{T_0}^{T} \frac{c_p}{T} \, dT - R \ln \frac{p}{p_0}$$

If c_p is constant, this reduces to

$$s = c_p \ln \frac{T}{T_0} - R \ln \frac{p}{p_0}$$

And the change of entropy between states defined by (p_1, T_1) and (p_2, T_2) is

$$s_2 - s_1 = c_p \ln \frac{T_2}{T_1} - R \ln \frac{p_2}{p_1} \tag{7.36}$$

In this section, and in Chapter 8, we have introduced the perfect gas by defining it as a substance which has the equation of state $pv = RT$. It then follows that u and h are functions only of T. When approaching the subject in this way, we are assuming that the absolute thermodynamic temperature T can be measured independently of the concept of a perfect gas. Now it is true that this can be done in principle (by using a Carnot engine), and even in practice by making use of the Clapeyron equation (see section 7.7). Or we can adopt the argument that all this is a matter of history and that the International Practical Scale on which our instruments are calibrated is essentially equivalent to the thermodynamic scale. But it is a fact that, over a wide range of temperature, thermodynamic temperatures used for fixed points on the International Practical Scale have been measured by gas thermometry. It is therefore instructive to see how the procedure can be justified without involving a circular argument. What we have to do is to define a perfect gas independently of the concept of thermodynamic temperature, and to show that an empirical temperature measured by a thermometer using such a gas is identical with the thermodynamic temperature. This we can do as follows.

Real gases are known to approximate asymptotically to two laws as their pressures are lowered. Firstly they obey *Boyle's law*

$$pv = \text{constant}]_{\theta=\text{const.}} \tag{7.37}$$

Secondly they follow the *Joule-Thomson law* that

$$h = f(\theta) \tag{7.38}$$

which can be established from the Joule-Thomson porous plug experiment described in section 7.4 (iv). Here θ denotes empirical temperature measured on any arbitrary scale, and the function $f(\theta)$ depends on the scale used as well as on the particular gas considered.

Boyle's law enables us to define a gas temperature θ_G which is a linear function of pv, i.e.

$$pv = A\theta_G \tag{7.39}$$

We now have to show that θ_G is identical with the absolute thermodynamic temperature T.

From a combination of the First and Second Laws we can write

$$ds = \frac{dh}{T} - \frac{v\,dp}{T} = \frac{1}{T}\left[\frac{\partial f(\theta)}{\partial T}\right]dT - \left(\frac{v}{T}\right)dp$$

But because s is a property and therefore ds is a perfect differential, it follows that

$$\frac{\partial}{\partial p}\left[\frac{1}{T}\frac{df(\theta)}{dT}\right]_{T} = -\left[\frac{\partial}{\partial T}\left(\frac{v}{T}\right)\right]_{p} \tag{7.40}$$

the derivative within brackets on the LHS being total in view of equation (7.38). The LHS of equation (7.40) must be zero, as the term inside the square brackets is merely some function of temperature, and hence from the RHS it follows that

$$\left[\frac{\partial}{\partial T}\left(\frac{v}{T}\right)\right]_{p} = 0$$

and therefore

$$\frac{v}{T} = \phi(p) \tag{7.41}$$

From equations (7.39) and (7.41) we can write

$$v = T\phi(p) = \frac{A\theta_G}{p}$$

or

$$p\phi(p) = \frac{A\theta_G}{T} \tag{7.42}$$

Equation (7.42) can only be true for all values of the independent variables p and T i[...] each side is equal to a constant, say K, and hence

$$T = \frac{A}{K}\theta_G \tag{7.43}$$

From equation (7.39) it is clear that A can be chosen arbitrarily, and thus A/K can b[...] made equal to unity. Thus we have established that the perfect gas temperature θ_G[...] and the absolute thermodynamic temperature T are identical.

(ii) Van der Waals gas

The characteristic equation first suggested by van der Waals is one of a class of such[...] equations in which v is not expressed as an explicit function of p and T. It can be[...] recast, however, with p expressed explicitly as a function of v and T, i.e.

$$p = \frac{RT}{v - b} - \frac{a}{v^2}$$

In this case it will obviously be more suitable to consider v and T as the independent[...] properties defining the state of the gas.

An expression for entropy in terms of v and T can be found as follows.

$$ds = \left(\frac{\partial s}{\partial T}\right)_v dT + \left(\frac{\partial s}{\partial v}\right)_T dv$$

Using (7.23) and the Maxwell relation (7.15), this becomes

$$ds = \frac{c_v}{T} dT + \left(\frac{\partial p}{\partial T}\right)_v dv \tag{7.44}$$

This is the equivalent of equation (7.30). For the van der Waals gas,

$$\left(\frac{\partial p}{\partial T}\right)_v = \frac{R}{v - b}$$

So that (7.44) becomes

$$ds = \frac{c_v}{T}\, dT + \frac{R}{v-b}\, dv$$

and on integration,

$$s = s_0 + \int_{T_0}^{T} \frac{c_v}{T}\, dT + R \ln \frac{v-b}{v_0-b} \tag{7.45}$$

And if c_v can be assumed constant,

$$s = s_0 + c_v \ln \frac{T}{T_0} + R \ln \frac{v-b}{v_0-b}$$

When an equation of state has the form $p = \phi(v, T)$ it is easier to find the internal energy first and then deduce the enthalpy. Treating u as a function of v and T,

$$du = \left(\frac{\partial u}{\partial T}\right)_v dT + \left(\frac{\partial u}{\partial v}\right)_T dv$$

But $(\partial u/\partial T)_v = c_v$, and from (7.32) we have

$$\left(\frac{\partial u}{\partial v}\right)_T = T\left(\frac{\partial p}{\partial T}\right)_v - p = \frac{RT}{v-b} - p = \frac{a}{v^2}$$

therefore

$$du = c_v\, dT + \frac{a}{v^2}\, dv$$

Integrating, we get

$$u = u_0 + \int_{T_0}^{T} c_v\, dT + a\left(\frac{1}{v_0} - \frac{1}{v}\right) \tag{7.46}$$

For any state defined by (v, T), the value of u is given by (7.46), p is given by the equation of state, and consequently the enthalpy can be determined from $(u + pv)$. It is apparent that van der Waals' equation, unlike the perfect gas equation, implies that the internal energy and enthalpy vary with both v and T.

To find whether c_v also varies with v, we may derive a suitable expression for $(\partial c_v/\partial v)_T$. Differentiating (7.23) we get

$$\left(\frac{\partial c_v}{\partial v}\right)_T = T\left(\frac{\partial^2 s}{\partial T\, \partial v}\right)$$

Using the Maxwell relation (7.15) this becomes

$$\left(\frac{\partial c_v}{\partial v}\right)_T = T\left(\frac{\partial^2 p}{\partial T^2}\right)_v$$

From van der Waals' equation,

$$\left(\frac{\partial p}{\partial T}\right)_v = \frac{R}{v - b}$$

and

$$\left(\frac{\partial^2 p}{\partial T^2}\right)_v = 0$$

Therefore,

$$\left(\frac{\partial c_v}{\partial v}\right)_T = 0$$

and hence c_v is independent of v. We may again infer that c_v is also independent of p and that it can therefore only be a function of T. It is evident that even this more accurate equation of state implies that c_v is independent of p and v, and this is not true for any real gas.

7.7 Change of phase

While a substance is undergoing a change of phase the temperature and pressure are not independent variables: at any given pressure the change takes place at a definite temperature—the saturation temperature T_{sat}. The field of equilibrium states covering a change of phase from liquid to vapour is represented by the series of constant pressure lines on the T-v diagram of Fig. 7.8. An important expression for the rate of change of p with T_{sat} can be deduced with the aid of the Gibbs function. From equation (7.8) we have

Fig. 7.8

$$\mathrm{d}g = v\,\mathrm{d}p - s\,\mathrm{d}T$$

Since p and T are constant during the change of phase, $\mathrm{d}g$ is zero, i.e. the Gibbs function is constant. Equating the values of g at the two extreme conditions we have

$$g_f = g_g$$

where suffixes f and g refer to the saturated liquid and vapour states. If the pressure is changed by an infinitesimal amount from p to $(p + \mathrm{d}p)$, the saturation temperature will change from T_{sat} to $(T + \mathrm{d}T)_{\text{sat}}$, and the Gibbs function from g to $(g + \mathrm{d}g)$. At this new pressure we must also have

$$(g + \mathrm{d}g)_f = (g + \mathrm{d}g)_g$$

It follows that

$$dg_f = dg_g$$
$$v_f \, dp - s_f \, dT_{sat} = v_g \, dp - s_g \, dT_{sat}$$

or

$$(v_g - v_f) \, dp = (s_g - s_f) \, dT_{sat}$$

Now the change of state from f to g occurs at constant temperature and pressure, and thus

$$s_g - s_f = \frac{h_g - h_f}{T_{sat}}$$

The previous equation therefore becomes

$$\frac{dp}{dT_{sat}} = \frac{h_g - h_f}{T_{sat}(v_g - v_f)} \qquad (7.47)$$

This is known as *Clapeyron's equation*; it applies equally to changes from the solid to the liquid phase (fusion, $i \rightarrow f$) and from the solid to the vapour phase (sublimation, $i \rightarrow g$).

For a small change of pressure Δp, the latent heat $(h_g - h_f)$ and difference of specific volumes $(v_g - v_f)$ can be considered constant, and the average rate of change of pressure with saturation temperature can be written as $(\Delta p / \Delta T_{sat})$. In this form Clapeyron's equation permits the latent heat to be estimated from measurements of v_f, v_g and the saturation temperatures at two nearby pressures.

The Clapeyron equation also points to one possible way of measuring absolute thermodynamic temperature. Equation (7.47) can be rewritten as

$$\frac{dT_{sat}}{T_{sat}} = \frac{v_{fg}}{h_{fg}} \, dp$$

and integrated between (T_0, p_0) defined by some fixed point, and (T, p) at which the temperature is to be measured. (T_0, p_0) may, for example, be the triple point of water discussed in section 8.2. This integration yields

$$\ln \left(\frac{T}{T_0} \right)_{sat} = \int_{p_0}^{p} \frac{v_{fg}}{h_{fg}} \, dp$$

Measurements of h_{fg} and v_{fg} over the range p_0 to p can be made independently of temperature, and T can thus be found. By using solid-vapour, solid-liquid and liquid-vapour changes of phase, as well as different substances, a continuous and wide range of T can be covered. This method of measuring absolute thermodynamic temperature is not used in practice because gas thermometry is more accurate.

At the end of section 7.5 it was pointed out that equations (7.26) and (7.29) cannot be applied directly if a change of phase occurs between the reference state and the state at which the entropy and enthalpy are required. When a change of phase occurs, the p-v-T data in Fig. 7.5 will appear as in Fig. 7.9. When the reference state lies within one phase region, and the second state lies within another, the changes in entropy and enthalpy may be found by adding the changes associated with each of four steps, e.g. (a) from state (p_0, T_0) to state 1, (b) from state 1 to state 2, (c) from state 2 to state (p_0, T), and (d) from state (p_0, T) to state (p, T). The changes of entropy and enthalpy during processes (a), (c) and (d) can be found in the way already described. During the constant pressure process (b), however, the specific heat is infinite since the temperature remains constant. For this step the latent heat at p_0 must be known. The value of this quantity may be found by a calorimetric experiment, or by using Clapeyron's equation as already explained. The change of enthalpy between state 1 and state 2 is then known, since it is equal to the latent heat, and the change of entropy can be found from

$$s_2 - s_1 = \frac{h_2 - h_1}{T_1}$$

The derivation of Clapeyron's equation has been given here as an example of the use of the Gibbs function. Another example of interest to engineers is provided by the following. We have seen that the Gibbs function has a unique value at any given pressure (or temperature) providing two phases are present. This fact can be used to simplify the calculation of the change of enthalpy during an isentropic process when the final state is in the two-phase region. Such calculations are frequently made when analysing the performance of steam turbine plant.

Fig. 7.9

By definition $g = h - Ts$, so that the change of enthalpy during the process can be written as

$$h_2 - h_1 = (g_2 + T_2 s_2) - h_1 \qquad (7.48)$$

Since the process is isentropic, $s_2 = s_1$, and assuming the initial state of the substance is known, the values of h_1 and s_1 can be found from tables of properties. If the final pressure is given, both T_2 and g_2 can also be found from the tables, and consequently (7.48) enables $(h_2 - h_1)$ to be calculated.* Values of h in the two-phase region are not given directly in tables because they depend upon the relative amounts of liquid and vapour in the mixture. The somewhat lengthier method of calculating the enthalpy change when values of g are not available is given in section 10.2.

*The last Callendar Steam Tables (1939) tabulate values of a quantity denoted by G, for water-steam mixtures at low pressures. It should be noted that (a) G refers to unit mass, although a capital letter is used, and (b) it is defined as $(Ts-h)$, i.e. $G = -g$. Equation (7.48) therefore becomes $h_2-h_1 = (T_2 s_2 - G_2) - h_1$.

7.8 Processes undergone by solids and liquids

In a study of heat engines we are primarily concerned with processes undergone by a gas, or a vapour in contact with its liquid, and Part II is devoted to the consideration of such systems. Some processes undergone by solids or liquids will be considered briefly in this section, however, as they provide simple examples of the use of the general thermodynamic relations. Only *reversible* non-flow processes are considered, but the equations deduced apply equally to solids and liquids—with the proviso mentioned in the introduction that for solids the stress must be uniform in all directions.

(i) *Constant volume process*

It was pointed out in section 7.4 that the specific heat at constant volume of a solid or liquid is not easy to measure because of the large pressure required to maintain the substance at constant volume. The rise of pressure consequent upon an increase of temperature at constant volume can be found as follows.

Equation (7.21) states that

$$\left(\frac{\partial p}{\partial T}\right)_v = \frac{\beta}{\kappa}$$

For solids and liquids the variations of β and κ with pressure and temperature are small. These properties may be assumed constant for many purposes and mean values, $\bar{\beta}$ and $\bar{\kappa}$, can be used. With this approximation, the above relation can be integrated for a constant volume process to give

$$(p_2 - p_1) \approx \frac{\bar{\beta}}{\bar{\kappa}}(T_2 - T_1) \tag{7.49}$$

(ii) *Isothermal process*

The change of volume accompanying a change of pressure during an isothermal process can be found from the compressibility of a substance. Thus from equation (7.20)

$$-\frac{1}{v}\left(\frac{\partial v}{\partial p}\right)_T = \kappa$$

Since the volume of a solid or liquid changes only slightly with pressure, a mean value $\bar{v}$ may be chosen. This, together with a mean value of κ, enables the expression to be integrated. Since T is constant for an isothermal process,

$$dv = -v\kappa\, dp$$

and

$$v_2 - v_1 \approx -\bar{v}\bar{\kappa}(p_2 - p_1) \tag{7.50}$$

The work done during an isothermal process is given by

$$W = \left(\int p\, dv\right)_T \approx -\bar{v}\bar{\kappa}\int_{p_1}^{p_2} p\, dp \approx -\frac{\bar{v}\bar{\kappa}}{2}(p_2^2 - p_1^2) \tag{7.51}$$

The heat supplied during the isothermal process can be found by using equation (7.30), viz.

$$ds = \frac{c_p}{T}\, dT - \left(\frac{\partial v}{\partial T}\right)_p dp$$

From (7.19), $(\partial v/\partial T)_p$ is equal to βv, and for an isothermal process dT is zero. Hence

$$ds = -\beta v\, dp$$

or

$$dQ = T\, ds = -T\beta v\, dp$$

Again assuming constant mean values $\bar{\beta}$ and $\bar{v}$, this can be integrated to give

$$Q \approx -T\bar{\beta}\bar{v}(p_2 - p_1) \tag{7.52}$$

The change in internal energy can be found from the energy equation, i.e. by substituting (7.51) and (7.52) in

$$u_2 - u_1 = Q - W$$

(iii) *Isentropic process*

During an isentropic process, the change in pressure, temperature, or volume, accompanying a given change in any one of these properties, can be found as follows. Using equation (7.30) again, and rewriting it as

$$T\, ds = c_p\, dT - \beta v T\, dp$$

we have, when ds is zero,

$$c_p\, dT = \beta v T\, dp$$

Since the temperature rise will be small, a mean value of T can be used in the right-hand side of the equation, so that

$$(T_2 - T_1) \approx \frac{\bar{\beta}\bar{v}\bar{T}}{\bar{c}_p}(p_2 - p_1) \tag{7.53}$$

Similarly, to find T in terms of v, equation (7.44) can be rewritten as

$$T\, ds = c_v\, dT + T\left(\frac{\partial p}{\partial T}\right)_v dv$$

But from (7.21), $(\partial p/\partial T)_v = \beta/\kappa$, and therefore when ds is zero

$$c_v \, dT = - \frac{T\beta}{\kappa} \, dv$$

And hence on integration

$$(T_2 - T_1) \approx - \frac{T\bar{\beta}}{\bar{c}_v\bar{\kappa}} (v_2 - v_1) \tag{7.54}$$

Finally, eliminating $(T_2 - T_1)$ between (7.53) and (7.54),

$$(p_2 - p_1) \approx - \frac{1}{\bar{v}\bar{\kappa}} \frac{\bar{c}_p}{\bar{c}_v} (v_2 - v_1) \tag{7.55}$$

The work done during an isentropic process is

$$W = \left(\int p \, dv \right)_s$$

Using the differential form of equation (7.55), it follows that

$$W \approx - \frac{\bar{\kappa}\bar{v}\bar{c}_v}{\bar{c}_p} \int_{p_1}^{p_2} p \, dp$$

$$\approx - \frac{\bar{\kappa}\bar{v}}{2} \frac{\bar{c}_v}{\bar{c}_p} (p_2^2 - p_1^2) \tag{7.56}$$

From the energy equation it follows that this must be equal to the change in internal energy.

7.9 Availability and the Gibbs function

We close this chapter with an introduction to the concept of availability, and a further discussion of the use of the Gibbs function. *Availability* (sometimes known as *exergy*) is a measure of the maximum useful work that can be done by a system interacting with an environment which is at a constant pressure p_0 and temperature T_0. The simplest case to consider is that of a heat reservoir (i.e. a heat source of infinite capacity and therefore invariable temperature) of temperature T. We have seen that the maximum efficiency with which heat withdrawn from a reservoir may be converted into work is the Carnot efficiency. Thus the availability of a reservoir at T providing a rate of heat transfer Q, in surroundings at T_0, is $Q(T-T_0)/T$.

The case of greatest interest in engineering is that of a system undergoing a non-cyclic process while doing work. Any power plant, even if part of it involves a secondary working fluid being taken through a cycle, when considered as a whole is a system undergoing a non-cyclic process (e.g. see Fig. 11.1). Such a system can be considered to undergo the process while interacting ultimately with a single reservoir, namely the atmosphere at p_0 and T_0, and the fluid will be incapable of producing any further work when it has reached a state of equilibrium with the surroundings. A comparison of the actual work performed with the maximum possible, starting from a given initial state, is an indication of the effectiveness of the plant used. It will be convenient to consider closed systems and steady-flow open systems separately.

In what follows we shall consider systems whose initial states are such that they produce positive work when degenerating to state (p_0, T_0). Similar arguments can be pursued for systems requiring work done on them in order to bring them to state (p_0, T_0), but for simplicity only the former type of system will be considered.

(i) *Availability in closed systems*

Let us consider a cylinder containing a compressed fluid at (p_1, T_1), the cylinder being closed by a piston. We shall calculate the useful work done when the system has reached a state which is in equilibrium with the surroundings and which therefore can be denoted by suffix 0. The initial and final internal energies U_1 and U_0 are fixed by the end states (p_1, T_1) and (p_0, T_0), and the work done by the fluid during the expansion, from the First Law, is

$$W = -(U_0 - U_1) + Q \tag{7.57}$$

The useful work done on the piston is less than this, because work $p_0(V_0 - V_1)$ is done on the surrounding medium which is at a constant pressure p_0.* Thus the useful work W' is

$$W' = -(U_0 - U_1) + Q - p_0(V_0 - V_1) \tag{7.58}$$

The principle of increasing entropy states that the sum of the changes of entropy of the surroundings ΔS_0 and of the system ΔS is greater than or equal to zero. Now the changes of entropy of the surroundings which are at a constant temperature T_0, when Q units of heat are transferred from them to the system, is $-Q/T_0$. We can therefore write

$$\Delta S + \left(-\frac{Q}{T_0}\right) \geqslant 0$$

or

$$Q \leqslant T_0 \Delta S$$

and hence

$$W' \leqslant -(U_0 - U_1) + T_0(S_0 - S_1) - p_0(V_0 - V_1)$$

---

*When calculating the work done on a piston elsewhere in the book, we do not subtract the work $p_0(V_0 - V_1)$ because we always consider the boundary of the system to be the face of the piston, thus ignoring the question of whether this work is entirely transferred to the piston or whether some of it is transferred to the atmosphere behind the piston. When we are concerned with a sequence of processes constituting a *machine* cycle, the net work done on the atmosphere in this way is zero, and the total work and the useful work over the machine cycle are equal.

The maximum value is obtained when the process is reversible and the equality holds, and thus the availability of the system can be written

$$W_{max} = (U_1 + p_0 V_1 - T_0 S_1) - (U_0 + p_0 V_0 - T_0 S_0)$$
$$= A_1 - A_0 \tag{7.59}$$

$A = (U + p_0 V - T_0 S)$ is called the *non-flow availability function*, and it is seen to be a composite property: composite because it is a property not only of the system but also of its surroundings.

In general it will be possible to think of an infinite variety of paths linking states (p_1, T_1) and (p_0, T_0), both irreversible and reversible; equation (7.59) implies that of these the reversible paths will all yield the same useful work, viz. W_{max}. Because the surroundings are at the invariable temperature T_0, the process can be conceived

(a)   (b)

Fig. 7.10

reversible only if either (a) all the heat is transferred when the system is at T_0 [e.g. during $i \to 0$ in Fig. 7.10a], or (b) the heat exchange with the surroundings is via appropriate reversible Carnot engines working between the system and the surroundings [e.g. during $i \to 0$ in Fig. 7.10b].

It is interesting to note that when the system is in state 0, A is identical with the Gibbs function G defined in section 7.1. Thus $G = H - TS$ which, when expanded and referred to state 0, becomes

$$G_0 = H_0 - T_0 S_0 = U_0 + p_0 V_0 - T_0 S_0 = A_0 \tag{7.60}$$

Let us now consider what appears to be a trivial case, viz. when state $(p_1, T_1) \equiv (p_0, T_0)$. This would be so for successive isentropic compression and expansion of a gas yielding no net work. It could equally well represent the case of a fluid changing phase isothermally, say from saturated liquid f to saturated vapour g. But as p and T do not then fix the state, this would not be a process between identical initial and final states and it would be meaningful to ask what useful work is produced. Applying equation (7.59),

$$W_{max} = (U_{f0} + p_0 V_{f0} - T_0 S_{f0}) - (U_{g0} + p_0 V_{g0} - T_0 S_{g0})$$
$$= -(H_{fg0} - T_0 S_{fg0}) \tag{7.61}$$

or using equation (7.60),

$$W_{max} = G_{f0} - G_{g0}$$

Since $H_{fg} = TS_{fg}$, it must be concluded that no useful work can be derived from this process, and all the work done resulting from the absorption of latent heat from the surroundings is expended in doing useless work on the surroundings against the pressure p_0. It follows that $G_f = G_g$, a result already derived in section 7.7.

Another case where the initial and final values of p and T are the same, but where the states are different, is when a chemical reaction occurs and reactants R at (p_0, T_0) change to products P at (p_0, T_0). Then we have for the maximum useful work

$$W_{max} = (U_{R0} + p_0 V_{R0} - T_0 S_{R0}) - (U_{P0} + p_0 V_{P0} - T_0 S_{P0})$$
$$= (H_{R0} - H_{P0}) + T_0(S_{P0} - S_{R0}) = -\Delta H_0 + T_0(S_{P0} - S_{R0}) \tag{7.62}$$

or

$$W_{max} = (G_{R0} - G_{P0}) = -\Delta G_0 \tag{7.63}$$

ΔH_0 is the *enthalpy of reaction* discussed in section 15.8 (and in sections 15.4 to 15.6 under the less general title of enthalpy of combustion). It is thus seen that the maximum useful work that can be derived from a closed system containing reactants at (p_0, T_0) in surroundings at (p_0, T_0) is not in general equal to $-\Delta H_0$, because $T_0(S_{P0} - S_{R0})$ usually differ from zero. In calculating the entropy difference $(S_{P0} - S_{R0})$, the entropies of the reactants and products must be known relative to a common datum. Such information is available as a result of the *Third Law of Thermodynamics** which defines the absolute entropy of substances at absolute zero and thus provides the required common datum. 'Products' in this context must be further qualified as constituents which are in chemical equilibrium taking account of dissociation, and this is discussed more fully in section 7.9 (iii).

(ii) *Availability and steady-flow open systems*

Let us imagine that we have an infinite reservoir feeding fluid at (p_1, T_1) with velocity C_1 at height z_1 to an open system at a steady rate. In order to derive the maximum amount of work from the system when discharging the fluid into surroundings at

---

*The Third Law of Thermodynamics postulates that for any pure substance in thermodynamic equilibrium the entropy approaches zero as the thermodynamic temperature approaches absolute zero. Nernst first stated the Law in 1907 in a different form, and it is often referred to as the Nernst Theorem. A discussion of the Law is outside the scope of this book, the Law's main sphere of application being in chemical thermodynamics and low-temperature physics.

(p_0, T_0), the fluid must leave at (p_0, T_0) with zero velocity and at the lowest possible datum level z_0. Thus from the steady-flow energy equation we get

$$W = -(H_0 + mgz_0 - H_1 - \tfrac{1}{2}mC_1{}^2 - mgz_1) + Q$$

$$= [H_1 + m(\tfrac{1}{2}C_1{}^2 + gz_1)] - [H_0 + mgz_0] + Q \qquad (7.64)$$

In practice, particularly when doing an overall analysis of plant such as boilers, turbines or condensers, the kinetic and potential energy terms can be omitted because they are insignificant. When analysing say a turbine, stage by stage in detail, the kinetic energy terms are significant and have to be carried throughout the analysis. For simplicity we shall omit the kinetic and potential energy terms in what follows.

(a)   (b)

Fig. 7.11

With open systems, the whole of W is useful work, the work done by and on the surroundings as the fluid enters and leaves the system being included in the enthalpy terms. We can use the same argument as before to establish Q in terms of entropy, and we therefore have

$$W \leqslant -(H_0 - H_1) + T_0(S_0 - S_1) \qquad (7.65)$$

Again the maximum work will be done when the process is reversible and the equality holds. Thus the availability of the fluid in state 1 is

$$W_{\max} = (H_1 - T_0S_1) - (H_0 - T_0S_0) = B_1 - B_0 \qquad (7.66)$$

where the quantity $B = (H - T_0S) = (U + pV - T_0S)$ is called the *steady-flow availability function*. B differs from the non-flow availability function $A = (U + p_0V - T_0S)$ in the pressure term. When state 1 $\equiv$ state 0, $A \equiv B \equiv G$.

When a system works between states 1 and 2, 2 being intermediate between 1 and 0, the maximum useful work quantities associated with the end states are

$$(W_{\max})_1 = B_1 - B_0 \quad \text{and} \quad (W_{\max})_2 = B_2 - B_0$$

and thus the maximum work available between states 1 and 2 is

$$(W_{\max})_{1-2} = B_1 - B_2 \tag{7.67}$$

As an illustration of the use of B let us consider a gas expanding in a turbine from state (p_1, T_1) to a pressure p_0. Since state 1 is an arbitrary state, the temperature reached after isentropic expansion will not necessarily be T_0; let us say it will be T_i' as in Fig. 7.11a. The work attainable during the isentropic expansion, per unit mass flow, is

$$W_{\text{isen}} = -(h_i' - h_1) = -c_p(T_i' - T_1) \tag{7.68}$$

whereas during an actual adiabatic expansion the work will be

$$W = -c_p(T_i - T_1) \tag{7.69}$$

The isentropic efficiency of the turbine is then conventionally expressed as

$$\eta_{\text{isen}} = \frac{W}{W_{\text{isen}}} \tag{7.70}$$

Now the maximum useful work per unit mass flow, as derived from the specific availability function $b = B/m$, is

$$W_{\max} = b_1 - b_0 = (h_1 - h_0) - T_0(s_1 - s_0)$$

$$= c_p(T_1 - T_0) - c_p T_0 \ln \frac{T_1}{T_0} + RT_0 \ln \frac{p_1}{p_0}$$

$$= c_p(T_1 - T_i') + c_p(T_i' - T_0) - c_p T_0 \ln \left(\frac{T_1}{T_i'} \frac{T_i'}{T_0} \right) + RT_0 \ln \frac{p_1}{p_0}$$

$$= c_p(T_1 - T_i') + c_p T_0 \left[\left(\frac{T_i'}{T_0} \right) - 1 - \ln \frac{T_i'}{T_0} \right] \tag{7.71}$$

We may call the ratio $W/W_{\max}$ the effectiveness ε of the process. It is clear that $\eta_{\text{isen}} \neq \varepsilon$ except when $c_p T_0[(T_i'/T_0) - 1 - \ln (T_i'/T_0)] = 0$, and this is possible only in the special case when $T_i' = T_0$. Thus the isentropic efficiency is not a measure of the fraction of the maximum useful work actually utilised, but merely the fraction of work available from a particular type of process, i.e. from an adiabatic expansion. Fig. 7.11b illustrates how, in principle, the exhaust gas at T_i' from an isentropic turbine could supply heat to a series of Carnot engines in order to obtain $W_{\max}$ as calculated from the availability function.

(iii) *The Gibbs function and the steady-flow system*

A steady-flow system of particular practical importance is one in which the fluid both enters and leaves at (p_0, T_0), so that $B = G$ at entry and exit. When the fluid is also identical in *state* at entry and exit, this leads to the trivial case $W_{\max} = G_0 - G_0 = 0$. Or where the fluid changes phase, say from saturated liquid f to saturated vapour g, this again leads to $W_{\max} = G_{f0} - G_{g0} = 0$, because the Gibbs function is constant during a change of phase at constant temperature. The case is of particular importance, however, when considering systems in which chemical reactions occur as in external and internal combustion power plant (see Figs. 11.1 and 12.3). For such plant the maximum amount of work would appear as

$$W_{max} = G_{R0} - G_{P0} = -\Delta G_0 \tag{7.72}$$

or in terms of enthalpies and entropies

$$W_{max} = (H_{R0} - T_0 S_{R0}) - (H_{P0} - T_0 S_{P0}) = -\Delta H_0 + T_0(S_{P0} - S_{R0}) \tag{7.73}$$

That the above expressions are identical with the equivalent equations (7.62) and (7.63) applying to a closed system is not surprising: here we are neglecting the kinetic and potential energy terms, and for the closed system we subtracted the $p_0 \Delta V$ work from W. It is of course only under the special conditions of $(p_1, T_1) \equiv (p_0, T_0)$ that the two expressions for W_{max} are identical.

Equations (7.72) and (7.73) have to be qualified in two respects: (a) If the products leaving at (p_0, T_0) can still react with the atmosphere, further work could still be done. In combustion, excess air is normally one of the reactants and such further reaction is then negligible, but with other reactions this point may be of importance. (b) Strictly speaking additional work could be derived from the products by letting each expand isothermally and reversibly to the partial pressure at which that particular product exists in the atmosphere, perhaps by using devices with semi-permeable membranes (see sections 14.3 and 15.7). For example, the partial pressure of CO_2 in the atmosphere is quite low, while its partial pressure in common products is quite high. Because such processes are impracticable at present, they will be ignored in evaluating W_{max}.

If we define the effectiveness of a combustion power plant as W/W_{max}, where W_{max} is given by equation (7.73), it is a matter of indifference whether ΔH_0 is the value with water in the products as a liquid or a vapour, because the difference between using the one or the other is automatically compensated by the change in the value of the entropy term. That this must be so is also obvious from a comparison of equations (7.72) and (7.73) and the fact that the Gibbs function does not vary during an isothermal change of phase. This point is of interest in relation to the discussion about the use of gross and net calorific values in defining thermal efficiencies (see sections 15.6 and 17.5).

When discussing in section 17.5 the possibility of using W/W_{max} in preference to conventional thermal efficiencies for assessing the performance of combustion power plant, we made* use of the results of section 15.7 (ii) for the purpose of evaluating W_{max}. We considered the simple reaction $CO + \frac{1}{2}O_2 \rightarrow CO_2$ as a specific example, and showed that the maximum work attainable from this reaction, per kmol of CO, in surroundings at (p_0, T_0) was

$$W_{max} = R_0 T_0 [\ln K + \ln(p_0)^{1/2}]$$

---

*As stated at the beginning, the authors' intention is that this chapter be left until a preliminary study of the book has been completed, and we underline this here by using the past tense.

where K was the equilibrium constant given by

$$K = \frac{p_{CO_2}}{(p_{CO})(p_{O_2})^{1/2}}$$

The form of the relation for W_{max} is quite general for all chemical reactions between perfect gases, except for the power of the p_0 term which depends on the difference between the number of kilomoles of reactants and products. If p_0 is 1 atm, however, and both p_0 and K are expressed in atmosphere units, the expression for W_{max} reduces to the truly common form

$$W_{max} = R_0 T_0 \ln K \tag{7.74}$$

The remainder of this section is devoted to a discussion of the implications of equating the two expressions for W_{max}, viz. (7.72) and (7.74). It follows immediately that

$$R_0 \ln K = -\frac{\Delta G_0}{T_0} \tag{7.75}$$

where ΔG_0 is the change of Gibbs function $(G_{P0} - G_{R0})$. It should be remembered that K could, for example, refer to the equation $CO + \frac{1}{2}O_2 \rightarrow CO_2$ or to the equation $2CO + O_2 \rightarrow 2CO_2$, giving the numerically different equilibrium constants

$$\frac{p_{CO_2}}{(p_{CO})(p_{O_2})^{1/2}} \quad \text{and} \quad \frac{(p_{CO_2})^2}{(p_{CO})^2(p_{O_2})}$$

and it is important to evaluate K and ΔG_0 in equation (7.75) for consistent molar quantities. It was proved in section 15.7 (ii) that K is independent of the total pressure in the reaction vessel (e.g. pressure p in the van't Hoff box), but it was pointed out there that K could, and was in fact found to be, a function of the second independent property T. Equation (7.75) implies that this is true also for ΔG_0. To investigate whether the form of the relation between K and T can be predicted, let us proceed by finding $(\partial \ln K/\partial T)_p$ at T_0 first, and then vary the reference temperature T_0. The partial notation can be omitted because K has been found to be independent of p.

From equation (7.12) then, we can write

$$S_{P0} = -\frac{dG_{P0}}{dT} \quad \text{and} \quad S_{R0} = -\frac{dG_{R0}}{dT}$$

and substituting in equation (7.73) we obtain

$$-\frac{\Delta G_0}{T_0} = -\frac{\Delta H_0}{T_0} - \left(\frac{dG_{P0}}{dT} - \frac{dG_{R0}}{dT}\right) = -\frac{\Delta H_0}{T_0} - \frac{d\Delta G_0}{dT}$$

or

$$\frac{1}{T_0}\frac{d\Delta G_0}{dT} - \frac{\Delta G_0}{T_0^2} = \frac{d}{dT}\left(\frac{\Delta G_0}{T_0}\right) = -\frac{\Delta H_0}{T_0^2}$$

Since, from equation (7.75),

$$\frac{d}{dT}\left(\frac{\Delta G_0}{T_0}\right) = -R_0\frac{d}{dT}(\ln K)$$

it follows that

$$\frac{d}{dT}(\ln K) = \frac{\Delta H_0}{R_0 T_0^2} \tag{7.76}$$

or

$$\frac{d}{d\left(\frac{1}{T}\right)}(\ln K) = -\frac{\Delta H_0}{R_0} \tag{7.77}$$

Equation (7.76), or its equivalent form (7.77), is known as the *van't Hoff isobar*: it provides a relation between K, temperature, and the enthalpy of reaction which is itself a function of temperature.

If $\ln K$ is measured over a range of temperature and plotted against $1/(R_0 T)$ (i.e. we are now considering the reference temperature T_0 as a variable), the slope of the curve so obtained will be equal to $-\Delta H_0$. Since ΔH_0 is a weak function of temperature, as is shown in Example 15.6, such plots yield nearly straight lines (Fig. 7.12). We have thus established, not that the relation between K and T can be predicted *a priori*, but at least that experimental data for equilibrium constants, enthalpies of reaction, and changes of Gibbs function are interchangeable. Such relations as (7.75) and (7.77) can therefore be used to check the consistency of data from different types of experiment. Furthermore, just as ΔH can be calculated at any temperature T from a single value of ΔH_0 together with specific heat data (see section 15.4),

Fig. 7.12

equation (7.77) makes it possible for the value of K to be calculated at any temperature T from the same data. This is explained in greater detail in section 15.8.

To sum up: in this section we have introduced the concept of availability, related two availability functions to other thermodynamic functions, and have indicated how the concept can provide the basis for less arbitrary definitions of the effectiveness of processes than those in common use. The reader can find a detailed discussion of the rational definition of efficiency of internal-combustion plant in Ref. (12) together with derivations of overall optimum working conditions for several types of external-combustion power plant. Furthermore, it is claimed that an 'availability balance' is one of the best methods for pinpointing the processes which contribute most to the overall loss due to irreversibilities in complex plant (e.g. in regenerative steam power plant). Certainly it can contrast the relative effects of internal and external irreversibilities, which a mere consideration of cycle efficiency cannot do. To distinguish it from the simple energy balance, such an analysis is often called a *Second Law analysis*: for further details the reader may consult Ref. (13). Finally, we have employed the concept of maximum work to derive useful relations between the equilibrium constant K, enthalpy of reaction ΔH_0 and corresponding change of Gibbs function ΔG_0.

Part II Applications to Particular Fluids

Introduction

The principles of thermodynamics have been explained in Part I without reference to any particular fluid. In Part II we shall apply these principles to closed and open systems consisting of either (a) liquids and vapours whose properties are most conveniently related in tabular form, or (b) gases whose thermodynamic properties can be related by simple algebraic equations. We shall still be concerned only to relate the heat and work crossing the boundary with the changes in the properties of the fluid; the detailed analyses of the ways in which the work and heat transfers are effected are dealt with in Parts III and IV respectively. For example, we shall calculate the theoretical work done when a fluid enters a turbine with a certain state and is expanded to a lower pressure, but we shall not stop to consider how the turbine blading should be designed so that the actual work is as near the theoretical value as possible.

Part II opens with a description of the characteristics of typical working fluids used in power plant. This is followed by examples of non-flow, steady-flow and nonsteady-flow processes undergone by water and steam, considered as an example of a liquid and vapour, and air, considered as an example of a gas. The performance of the more important thermodynamic cycles is then analysed, e.g. it is shown how the cycle efficiency and work ratio can be calculated when the working fluid is taken through a given cycle of states. When engines of the same sort are being compared, the work ratio, i.e. the ratio of net work to positive work, provides an indication of the size of plant required for a given total output and also an indication of how far the actual efficiency will fall short of the theoretical value.

Finally, there is a chapter showing how the behaviour of systems consisting of a chemically inert mixture of fluids can be deduced from a knowledge of the properties of the constituents, and a chapter which is devoted to the thermodynamic aspects of combustion.

8   Properties of Fluids

Before describing the way in which the properties of a real fluid are related, with the aid of numerical examples, some system of units must be adopted. Of the three primary properties—pressure, specific volume and temperature—the adoption of units for the first two presents little difficulty. Pressure will always be expressed in bar, and specific volume in cubic metres per kilogram (see Appendix A). A pressure gauge or manometer usually measures the pressure relative to atmospheric pressure. The absolute value, which we shall always imply by the symbol p, can easily be obtained by adding the barometric pressure to the gauge pressure.

The unit of temperature needs more careful attention, and the first section is devoted to this topic. This is followed by a description of the characteristics of a liquid and vapour. The most convenient method of tabulating values of the thermodynamic properties is discussed, and the more important graphical representations are described. Finally, the special case of a gas is dealt with, and it is shown how the assumption of a simple equation of state can lead to an easy analytical treatment of the relations between the properties.

8.1   The unit of temperature and the International Practical Scale

Temperature measurement is now a field of study by itself, and only the briefest outline can be given here.* Let us first consider the procedure for setting up any empirical scale of temperature. As explained in section 1.2, temperature may be defined in terms of the variation of some thermometric property such as the length of a column of liquid in a tube, the pressure of a gas maintained at constant volume, the resistance of a wire, or the e.m.f. produced by a thermocouple. Three items of information must be specified if a scale of temperature is to be reproducible:

(a) the particular thermometric substance and property used,
(b) the numbers to be attached to two standard fixed points,
(c) the relation between the temperature and the thermometric property which is to be used for interpolating between the fixed points and extrapolating beyond them.

We shall discuss items (b) and (c) first.

*A comprehensive survey of the whole field can be found in Ref. (22), and short summaries in Refs. (23) and (32).

Two reference points, called the upper and lower fixed points, may be located on the scale of the thermometer by bringing it into contact with two systems in turn, each at a definite and reproducible temperature. For example, the lower fixed point may be provided by a mixture of ice and water in equilibrium at standard atmospheric pressure, and the upper fixed point by a mixture of water and steam in equilibrium at standard atmospheric pressure. When in a state of equilibrium both these mixtures provide definite and reproducible temperatures. On the Celsius (or centigrade) scale the ice and steam points are assigned the numbers 0 and 100 respectively, while on the Fahrenheit scale the corresponding numbers are 32 and 212. There are therefore 100 units or degrees of temperature between the fixed points on the Celsius scale, and 180 on the Fahrenheit scale.

We have now to decide how to interpolate between the fixed points and extrapolate beyond them. In the early days of thermometry the simplest procedure was adopted, i.e. the scale was divided into a number of equal parts. This implies that the temperature is defined as a linear function of the thermometric property. Thus, if θ and X are the numerical values of the temperature and thermometric property,

$$\theta = aX + b$$

The constants a and b are determined by the numbers allotted to the fixed points. On the Celsius scale, a and b can be found by solving the simultaneous equations

$$0 = aX_i + b, \quad 100 = aX_s + b$$

where X_i and X_s are the values of the thermometric property at the ice and steam points. The solution is

$$a = \frac{100}{X_s - X_i}, \quad b = -\frac{100\, X_i}{X_s - X_i}$$

Substituting these values in the original equation for the temperature, we get

$$\theta_C = 100\, \frac{X - X_i}{X_s - X_i}$$

If this procedure is repeated for the Fahrenheit scale we have

$$\theta_F = 32 + 180\, \frac{X - X_i}{X_s - X_i}$$

It follows that the relation between the temperatures on the two scales is

$$\theta_F = 32 + 1 \cdot 8\, \theta_C$$

Provided the same thermometric substance and property are used, there is no fundamental difference between the two scales we have been considering. A temperature measured on one scale can always be converted into a temperature measured on the other with mathematical precision. For the moment then, let us assume we are using

125

the Celsius system of numbering, and consider what happens when the substance or property is changed. For example, we might construct the following two scales. From a liquid-in-glass thermometer we should have

$$\theta_C = 100\,\frac{l - l_i}{l_s - l_i}$$

where l is the length of the column of liquid. And from a constant-volume gas thermometer, where p is the pressure, we should have

$$\theta_C = 100\,\frac{p - p_i}{p_s - p_i}$$

If linear interpolation is used in this way, we find that when any two thermometers using either different thermometric properties or substances are brought into contact with the same heat reservoir, *the readings will not be identical except at the fixed points.* For example, if the temperature measured by a constant-volume hydrogen thermometer is $40\,^{\circ}\mathrm{C}$, the temperature recorded by a typical mercury-in-glass thermometer will be $40 \cdot 11\,^{\circ}\mathrm{C}$ and by a platinum resistance thermometer $40 \cdot 36\,^{\circ}\mathrm{C}$. If one thermometric property does not vary linearly with the other, the two temperature scales, defined linearly in terms of these properties, cannot be identical except at the fixed points. Unfortunately there is nothing in the foregoing method of setting up a temperature scale to suggest why we should adopt one type of thermometer rather than another as a standard instrument. Moreover the scales are not absolute, because the zero is assigned to some arbitrary point fixed only by convention.

Now we have seen (section 6.2) that an absolute scale of temperature arises as a consequence of the Second Law, and that this thermodynamic scale is independent of any particular thermometric substance or property. Also, the temperature defined in this way enables us to introduce the useful concept of entropy. This theoretical scale is clearly a good candidate for adoption as the ultimate standard. It will be recalled that the thermodynamic temperature T must satisfy the relation

$$T = T_0\,\frac{Q}{Q_0} \tag{8.1}$$

where Q and Q_0 are the quantities of heat absorbed and rejected by a reversible engine operating between reservoirs at T and T_0. But this relation does not completely determine the scale—we have still to fix the size of the interval or unit of temperature. It was pointed out in section 6.2 that this could be varied by altering the number of engines in the series shown in Fig. 6.3, and there are two ways of approaching the problem.

Firstly, we may choose two fixed points, such as the ice point T_i and steam point T_s, and define the number of degrees between them. For example, we can write $T_s - T_i = 100$ units. This is equivalent to placing 100 reversible engines in the series between the fixed points. When used in conjunction with the equation $T_s/T_i =$

Q_s/Q_i, this arbitrary choice fully defines a linear thermodynamic scale. Q_s/Q_i has a unique value because the efficiencies $(=1-Q_i/Q_s)$ of all reversible engines operating between T_s and T_i are the same, and thus the two simultaneous equations yield unique values of T_s and T_i. The main advantage of this approach to the definition of the thermodynamic scale is that it provides a scale which coincides exactly at the fixed points with all empirical scales based on the same fixed points.

The second approach, adopted internationally in 1954, is the more obvious one of assigning an arbitrary number to T_0 at one easily reproducible fixed point. Then any other temperature is uniquely defined by $T = T_0(Q/Q_0)$ for the reason just discussed. This approach is equivalent to fixing the number of engines in Fig. 6.3 between the fixed point and absolute zero. The fixed point chosen for the definition of the thermodynamic scale is the triple point of pure water. A detailed discussion of this state is given in the next section: here it suffices to say that it is the state at which the solid, liquid and vapour phases can coexist in equilibrium and it is easily reproducible. The pressure of the triple point is very low, i.e. 0·006 112 bar, but the temperature is only slightly above that of the ice point at atmospheric pressure. The internationally agreed number for the triple point T_0 is 273·16: thus *the unit of thermodynamic temperature is the fraction* 1/273·16 *of the thermodynamic temperature of the triple point of water*. The unit is called the *kelvin*, and in this book it is denoted by the unit symbol K.

It is pertinent to ask why such an odd number has been chosen. The reason lies in the fact that with this method of definition no *exact* value can now be attached to $T_s - T_i$ and it becomes a matter of experiment. The choice of 273·16 K for T_0, however, makes $T_s - T_i \approx 100$ K within 0·005 K, and thus the new unit is still for many practical purposes the same size as on all previous empirical Celsius scales.

The kelvin is not the only thermodynamic unit in use. The other thermodynamic unit of importance is the *rankine*, which is the fraction $1/(1\cdot8 \times 273\cdot16) = 1/491\cdot688$ of the thermodynamic temperature of the triple point of water. Thus the kelvin is 1·8 times as large as the rankine (unit symbol R).

Non-absolute versions of the thermodynamic scales are obtained by shifting the zero datum of the scales. For example, the *Celsius thermodynamic scale* employs the same size of unit or interval as the Kelvin scale, but has its datum at 273·15 K, i.e. at a temperature 0·01 K below the triple point. It should be noted that, although the temperature unit on the thermodynamic Celsius scale is identical with the kelvin, it is customary to denote temperatures measured relative to the 273·15 K datum by the symbol °C. Thus we can write

   0 °C = 273·15 K exactly

By choosing this datum we find that

   0 °C ≈ ice point, 100 °C ≈ steam point, (both at 1 atm)

and therefore have a scale directly comparable with empirical Celsius scales described at the beginning of this section. Similarly, the *Fahrenheit thermodynamic scale* has the zero datum moved to 459·67 R, i.e. 32·018 R below the triple point, thereby yielding ice and steam points of approximately 32 °F and 212 °F. By shifting the zero we do not alter the fact that the scales are based on the laws of thermodynamics and not on the characteristics of a particular substance, but it does mean that the temperatures are no longer absolute. Temperatures on these non-absolute thermodynamic scales will be symbolised by t. The relations between the numerical values T_K, T_R, t_C and t_F on the four scales are summarised in Fig. 8.1.

	T_K	t_C	T_R	t_F
$T_K =$	T_K	$t_C + 273 \cdot 15$	$\dfrac{1}{1 \cdot 8} T_R$	$\dfrac{1}{1 \cdot 8}(t_F + 459 \cdot 67)$
$t_C =$	$T_K - 273 \cdot 15$	t_C	$\dfrac{1}{1 \cdot 8}(T_R - 491 \cdot 67)$	$\dfrac{1}{1 \cdot 8}(t_F - 32)$

Fig. 8.1   Relations between thermodynamic scales of temperature

The foregoing is a logical account of the *definition* of a thermodynamic unit of temperature: the *determination* of a temperature by using a reversible engine and meauring Q/Q_0 is of course a practical impossibility. Various indirect methods have been devised of which some details can be found in Ref. (20). These methods usually involve the use of an elaborate type of gas thermometer and are the province of the physicist rather than the engineer. The measurement of a single temperature requires months of careful scientific work and the procedures are too intricate for the direct calibration of even precision instruments. For this reason a procedure for establishing a practical scale was adopted, originally at the Seventh General Conference of Weights and Measures in 1927; the scale was most recently revised at the Thirteenth Conference in 1967, and it was named the *International Practical Temperature Scale* 1968 (*IPTS*-68). The procedure can be summarised as follows.

The thermodynamic temperatures of various easily reproducible states are measured—namely triple, freezing and boiling points of very pure substances, covering a wide range of temperature and providing a skeleton of fixed points. Also specified are the methods of interpolation to be used between the fixed points: the specification consists of detailed statements about the thermometric substances and properties to be used, and about the functions to be used for relating temperature to the thermometric property in each range. For example, from the freezing point of antimony (903·89 K) to that of gold (1337·58 K) the temperature T_{68} on the IPTS-68 is defined by the equation

$$E = a + b(T_{68} - 273 \cdot 15 \text{ K}) + c(T_{68} - 273 \cdot 15 \text{ K})^2$$

here E is the electromotive force of a standard thermocouple of rhodium-platinum alloy and platinum, when one junction is at the temperature $T_{68} = 73\cdot15$ K and the other is at the unknown temperature T_{68}. The suffix 68 is added to indicate that it is an IPTS-68 temperature and thus distinguish it from the thermodynamic temperature proper. The constants a, b and c are calculated from the values of E at the two fixed points mentioned and at the freezing point of silver (1235·08 K).

The complete scale is formed by using (i) a platinum resistance thermometer between the triple point of hydrogen (13·81 K) and the freezing point of antimony, (ii) a standard thermocouple between the freezing points of antimony and gold as explained previously, and (iii) above the gold point methods involving Planck's law of radiation (see section 23.1). Further details can be found in Refs. (25) and (32).

The IPTS-68 is used by standards laboratories for calibrating scientific instruments. Direct reference to the thermodynamic scale is unnecessary, because the fixed points and interpolation formulae are chosen in such a way that the temperature measured on IPTS-68 closely approximates the thermodynamic temperature; the difference is within the limits of the present accuracy of measurement. It will be assumed in the following chapters that all temperatures are measured by instruments calibrated to read the IPTS-68: they are thereby virtually thermodynamic temperatures and the symbols T (and t), not T_{68} (and t_{68}), will be used.

8.2 Properties of liquids and vapours

We noted in section 1.1 that a knowledge of two independent properties suffices to determine the thermodynamic state of a fluid when it is in equilibrium, and therefore that any other thermodynamic property is a function of the chosen pair of independent properties. We shall consider first the relation between the primary properties p, v and T, the equation expressing this relation for any particular fluid being called the *equation of state* or *characteristic equation* of the fluid. Except for gases, considered in sections 8.5 and 8.6, no simple algebraic equation can be found which will cover all possible states, and the relation is best expressed in either graphical or tabular form.

It must be emphasised that the equation of state is essentially an *empirical* relation; it cannot be deduced by thermodynamic reasoning. Assumptions as to the microscopic structure of a liquid or vapour can lead to a prediction of the form of the equation, but there still remain constants which must be determined by experiment for each substance. We shall refer to this type of prediction again when we consider the behaviour of gases; for the moment we will consider only the graphical expression of p-v-T relations obtained by direct measurement. Although we are concerned mainly with liquids and vapours it will be necessary to refer briefly to the solid phase also, because of the adoption of the triple point of water as a standard fixed point.

(i) p-v-T data

Since we have three variables to consider, the obvious procedure is to measure the variation of one with another while the third is kept constant, and to repeat this for a range of values of the third variable. Imagine unit mass of ice below freezing point, enclosed in a cylinder by a piston under a constant load such that the pressure is one atmosphere (1 atm = 1·01325 bar). Let us follow the events which occur as heat is transferred to the cylinder while the pressure is kept constant. The temperature rises and the ice expands until a temperature of 273·15 K (0 °C) is reached, as indicated by the full line AB in Fig. 8.2. Further heating does not continue to raise the temperature of the ice, but causes a change to the liquid phase (BC). The change of phase occurs at constant temperature and is accompanied by a reduction in specific volume, the heat required for this process being called the *latent heat of fusion*. With the system of units used in this book, the latent heat will be measured in kJ/kg. In contrast to latent heat, the heat transferred to a substance while the temperature changes is

Fig. 8.2

sometimes referred to as *sensible* heat. Further heating, of what is now liquid, results in a rise of temperature, a contraction in volume until the temperature is about 4 °C (point D), and subsequent expansion until a temperature of 373·15 K (100 °C) is reached (E). At this point a second phase change occurs at constant temperature with a large increase in volume until the liquid has been vaporised (F). The heat required in this case is called the *latent heat of vaporisation*. When vaporisation is complete the temperature rises once more (FG). Water behaves in an anomalous way compared with other substances, in that it contracts on melting and also contracts before it expands when its temperature is raised above the melting point. The dotted curve in Fig. 8.2 shows the behaviour of normal pure substances. Constant pressure lines such as those in Fig. 8.2 are sometimes called *isobars*. It must be emphasised that in this and subsequent property diagrams the change in volume of the solid and liquid has been greatly exaggerated in comparison with the volume changes due to the liquid-vapour phase change or the heating of the vapour: the diagrams are not to scale.

Let us now consider what happens to water when the same sequence of events is followed through at a lower pressure, Fig. 8.3a. Firstly, there is a slight rise in the melting point [as predicted by the Clapeyron equation (7.47)]. Secondly, there is a marked drop in the boiling point and a marked increase in the change of volume which accompanies evaporation. When the pressure is reduced to 0·006 112 bar, the melting and boiling temperatures become equal and the change of phase, ice-water-steam, is represented by a single horizontal line (bcf). As explained in the previous section, the temperature at which this occurs has been accepted internationally as a fixed point

(a) Water   (b) Normal substance

Fig. 8.3

for the absolute temperature scale and is by definition 273·16 K. Only at this temperature, and the pressure of 0.006 112 bar, can ice, water and steam coexist in thermodynamic equilibrium in a closed vessel, and bcf is called the *triple point line*. If the pressure is reduced still further the ice, instead of melting, *sublimates* directly into steam.

Consider now the behaviour at pressures above atmospheric. The shape of the curves is similar to that of the atmospheric isobar, but there is a marked reduction in the change of volume accompanying evaporation. At a sufficiently high pressure this change of volume falls to zero, and the horizontal portion of the curve reduces to a point of inflexion. This is referred to as the *critical point*, and the properties at this state are known as the critical pressure p_c, critical temperature T_c and critical volume v_c. The values for water are

$$p_c = 221·2 \text{ bar}, \quad T_c = 647·3 \text{ K}, \quad v_c = 0·003 \ 17 \text{ m}^3/\text{kg}$$

At pressures above critical, there is no definite transition from liquid to vapour and the two phases cannot be distinguished visually. The latent heat of vaporisation falls to zero at the critical pressure and at higher pressures the term has no meaning.

Fig. 8.3b shows the corresponding behaviour of normal substances. As the only difference is in the behaviour near the triple point line, only the left-hand portion of the diagram has been shown. It is apparent that normally both the melting and boiling points rise with increase in pressure and that liquids expand on heating at all pressures and temperatures.

It is instructive to show graphically how the sublimation, melting and boiling temperatures vary with pressure as in Fig. 8.4, because the origin of the term 'triple point' lies in this method of presenting the information. The boiling point curve is

Fig. 8.4 Triple point of water

obtained by following arrow 1 in Fig. 8.3 from the triple point line to the critical point, and the sublimation point curve by following arrow 2 as the pressure is reduced below the triple point value. Two melting point curves are shown, the full line being for water and the dotted curve for normal substances. These are obtained by following arrow 3 in Figs. 8.3a and b respectively. The numerical values in Fig. 8.4 refer only to water.

In the foregoing we have considered how T varies with v at constant p, and for the purpose of introducing the pattern of behaviour of the three phases this is very convenient. We have seen, however, that a p-v diagram is more useful for depicting practical processes. By following a succession of constant temperature (i.e. horizontal) lines in Fig. 8.3, it may be seen that the corresponding p-v diagram of Fig. 8.5 is obtained (the points E, F, G and c, f, g correspond to those in Fig. 8.3a). Alternatively, each of the isothermals can be imagined to be obtained by allowing a liquid to expand isothermally behind a piston. Since we shall have no need in practice to refer to behaviour in the solid phase, only the portion referring to a liquid and vapour is shown and it is unnecessary to distinguish between water and other substances in what follows. Various terms have been introduced to characterise the different states through which the substance passes during the expansion 1 2 3 4. From 1 to 2 the substance is an *unsaturated liquid*, while at 2 it is a *saturated liquid*. Between 2 and 3 it is a *two-phase mixture* of liquid and vapour or *wet vapour*. At 3 it is a *saturated vapour*, and from 3 to 4 an *unsaturated vapour*. If all the points representing saturated liquid and vapour states are joined to form a curve, i.e. cNf in Fig. 8.5, the line separating the liquid and two-phase regions is called the *saturated liquid line*, and the line separating the vapour and two-phase regions is called the *saturated vapour line*. The two lines meet at the critical point N.

It is evident that when a substance exists in a saturated state, or as a wet vapour, the pressure and temperature are not independent properties. For example, if p and T are given, the state is not determined because it might lie anywhere on the line 2–3. If the substance is known to be a saturated liquid, or a saturated vapour, then the value of *either p or T* is sufficient to determine its state: and if the substance is a wet

Fig. 8.5 *p–v* diagram for water

vapour a knowledge of either p and v, or T and v, may be used to determine its state. The pressure at which a liquid vaporises, or a vapour condenses, is called the *saturation pressure* corresponding to the given temperature; alternatively, the temperature at which these phenomena occur is called the *saturation temperature* corresponding to the given pressure. The choice of phrasing depends on whether the temperature or pressure is being used as one of the independent variables.

Substances referred to as gases are those which lie in the region above the critical isothermal T_c at ordinary pressures and temperatures, i.e. they are substances which have a very low critical temperature. This should not be regarded as a rigid definition, however, and the terms 'gas' and 'vapour' are often used synonymously. A substance in a state lying above T_c cannot be liquefied by isothermal compression; it must first be cooled below the critical temperature. This explains why some substances, e.g. oxygen and hydrogen, were originally called 'permanent gases'. The critical temperatures of these substances are far below the temperature of any freezing mixture that was available in the early days of the science, and of course more sophisticated methods of liquefaction had not then been devised.

A vapour in a state lying along the saturated vapour line fN is sometimes called a *dry saturated vapour*, although the additional adjective 'dry' is superfluous in this instance. The term *dry vapour* may refer to a saturated or unsaturated vapour. When the vapour is unsaturated, i.e. in a state lying to the right of fN, it is more usual to speak of it as a *superheated vapour*. The reason for this last term becomes clear when it is realised that the temperature of an unsaturated vapour is always greater than the saturation temperature corresponding to the pressure of the vapour. For example, an unsaturated vapour in state G on curve 3–4 (Fig. 8.5) has the temperature T_G

Fig. 8.6 *p-v-T* surface for water

which is higher than the saturation temperature corresponding to the pressure p_G. This saturation temperature is the temperature of an isothermal passing through point F. The difference between the actual temperature, and the saturation temperature corresponding to the actual pressure, is called the *degrees of superheat*. Similarly, an alternative term is also used for an unsaturated liquid, namely *compressed liquid*. This is because liquid in a state lying to the left of cNM always has a pressure which is greater than the saturation pressure corresponding to the actual temperature of the liquid.

Figs. 8.3 and 8.5 are two-dimensional representations of the relation between three variables. They illustrate effectively the practical method of obtaining such a relation, i.e. by observing the way one property varies with another while the third is kept constant. A three-dimensional model, however, gives a clearer picture of the fact that only two properties need to be fixed to determine the state of the substance. If p, v and T are plotted along three mutually perpendicular axes as in Fig. 8.6, all possible equilibrium states lie on a surface called the *p-v-T* surface. Any reversible process, i.e. succession of equilibrium states, can be represented by a line on the surface. A minor qualification is necessary—under certain conditions it is possible for a substance to be in a state of equilibrium represented by a point not on the surface. But such a state is not one of *stable* equilibrium; it must be either an *unstable* or *metastable* state. If it is unstable, any infinitesimal disturbance causes the substance to undergo a spontaneous irreversible process which returns it to a state of stable equilibrium lying on the *p-v-T* surface. If the state is metastable a finite disturbance is necessary. An example of this kind of phenomenon is discussed in greater detail in section 18.2.

Having found the empirical relation between the properties *p*, *v* and *T*, the problem that remains is the determination of the internal energy, enthalpy and entropy in terms of the two primary properties chosen to describe the state of the substance. The procedure is somewhat complicated because *u*, *h* and *s* are not directly measurable properties. A proper analysis of the procedure is given in Chapter 7 for advanced students; all that is attempted in the next sub-section is an indication of one possible method in simple terms.

(ii) *u-h-s data and specific heats*

A method of computing the value of the entropy *s* at any state defined by (p, v), relative to the value at some reference state (p_0, v_0) has been described in section 6.5 (i). The equations expressing the First and Second Laws were combined to yield the equation

$$s_1 - s_0 = \int_0^1 \frac{1}{T}\,\mathrm{d}u + \int_0^1 \frac{p}{T}\,\mathrm{d}v \tag{8.2}$$

and it was shown how the integrals could be evaluated graphically.

Now we assumed for this process of integration that, in addition to the *p-v-T* data, we had a table of properties giving *u* for various pairs of values of *p* and *v*. This latter information is just as empirical as the *p-v-T* data. Moreover, the *p-v-u* relation cannot be deduced from the *p-v-T* relation; some additional experimental information is necessary. Since $h = u + pv$, values of enthalpy can be obtained directly from the *p-v-u* relation for ranges of values of *p* and *v*, and no additional experiments are required to determine this property. We have thus arrived at the position that both *s* and *h* can be expressed in terms of *p* and *v* when the *p-v-T* and *p-v-u* relations have been determined experimentally.

We have already described how the *p-v-T* data can be obtained; it is the determination of *p-v-u* data that we have now to discuss. A method of measuring changes of internal energy is suggested by the application of the non-flow energy equation to a constant volume process, i.e. $Q = (u_2 - u_1)$. We can imagine unit mass of the liquid being heated in a closed container of volume v_0. The heat required to raise the pressure from p_0 to p_1, p_2, p_3, etc., can be measured, and these quantities of heat will be equal to the changes of internal energy between state (v_0, p_0) and the various states (v_0, p_1), (v_0, p_2), etc. This procedure can be repeated for other values of *v* and, if the value of *u* at state (p_0, v_0) is put equal to zero, we shall have a series of values of u for a range of states defined by values of p and v. We can then proceed to calculate values of h and s in the manner suggested.

It is not essential that the empirical information required to supplement the p-v-T data be in the form of p-v-u data. In Chapter 7 we show how specific heat data can be used for the purpose. Chapter 7 also describes a method of calculating values of s which only requires sufficient supplementary information to cover a single line of reference states, e.g. a series of states at constant pressure, instead of the whole field of possible states. This method obviously requires much less experimental work to be performed. We shall not pursue this matter here, but since we shall find the idea of specific heat particularly useful when dealing with the properties of gases, we will end this sub-section with a discussion of this concept.

Originally the specific heat was defined as the heat required to raise the temperature of unit mass of substance one degree. Since the quantity of heat depends upon the process employed to raise the temperature, and an infinite number of processes are possible, there are an infinite number of specific heats. However, only two came to be used in practice. The specific heat at constant volume (c_v) was defined as the heat required to raise the temperature of unit mass one degree during a reversible constant volume process, and the specific heat at constant pressure (c_p) as the heat required to raise unit mass one degree during a reversible constant pressure process.* Now the quantities of heat added during these two processes are equal to the changes in internal energy and enthalpy respectively. The foregoing definitions can therefore be written as

$$c_v = \frac{Q_v}{\Delta T} = \frac{\Delta u}{\Delta T}, \quad c_p = \frac{Q_p}{\Delta T} = \frac{\Delta h}{\Delta T} \tag{8.3}$$

The values of c_v and c_p, defined in this way, are found to vary with the value of ΔT, i.e. with the range of temperature used in the experiment, and also with the state of the substance at the beginning of the heating process. To make the definitions yet more precise, it is the practice in thermodynamics to use the following definitions:

$$c_v = \left(\frac{\mathrm{d}u}{\mathrm{d}T}\right)_v, \quad c_p = \left(\frac{\mathrm{d}h}{\mathrm{d}T}\right)_p \tag{8.4}$$

These definitions† are the limit of the original definitions (8.3) as ΔT approaches zero. The specific heats are to be regarded as the rates of change of one property with another while a third is kept constant. They have definite values at any given state and vary only with the state of the substance; they are therefore properties of the substance. This will be made clear if we outline a method for their measurement.

---

*It is only necessary that the constant volume process be internally reversible, i.e. that there should be no paddle work, because then the quantity of heat is quite definite and depends only on the equilibrium end states (see section 3.2). However, the constant pressure process must be carried out slowly to make the effect of friction negligible during the expansion, and this implies a slow rate of heat addition.

†It is more usual to use the partial differential notation as in Chapter 7, but for elementary purposes this can be dispensed with.

c_v can easily be found by carrying out the experiment already described for the purpose of obtaining the *p-v-u* data. If the temperature is measured instead of the pressure, the results can be expressed on a *u-T* diagram, as in Fig. 8.7a. The slopes of the constant volume lines can be measured at various temperatures to give values of c_v corresponding to various values of *v* and *T*. These may be plotted as in Fig. 8.7b.

Fig. 8.7

Note that the arbitrary choice of a reference state, at which *u* is put equal to zero, does not affect the values of c_v. The choice of reference state merely fixes the origin of the coordinates in relation to the family of curves in Fig. 8.7a, and does not affect the slope of the curves at any given state defined by values of *T* and *v*.

Fig. 8.8 Specific heat at constant pressure

A similar plot of c_p versus *T*, this time for various constant values of the pressure, can be obtained by heating the substance at constant pressure, plotting *h* versus *T* for various values of *p*, and measuring the slopes of these curves. The actual variation of c_p with *T* and *p*, for water and steam, is depicted in Figs. 8.8a and b. It will be noticed that the curves only cover a range of states up to the line of saturation states. This is because the concept of specific heat is only meaningful when the substance exists in a single phase. When more than one phase is present, as in the case of a wet vapour, a substance undergoes a change in enthalpy without any change of temperature during a constant pressure process. The specific heat at constant pressure would therefore be infinite.

The difference between c_p and c_v for liquids is very small. Liquids have a low co-efficient of expansion, i.e. the change of volume with temperature during a constant pressure process is negligible. Such a process is therefore effectively also a constant volume process, and we can write

$$\left(\frac{dh}{dT}\right)_p \approx \left(\frac{dh}{dT}\right)_v$$

Also, when p is constant the increase of enthalpy is effectively equal to the increase of internal energy because

$$dh = du + d(pv) = du + p\,dv \approx du$$

Therefore

$$\left(\frac{dh}{dT}\right)_p \approx \left(\frac{dh}{dT}\right)_v \approx \left(\frac{du}{dT}\right)_v$$

It is also worth noting that over normal ranges of temperature and pressure the variation of c_p and c_v is small—this applies to both liquids and vapours. In general, temperature has a greater effect than pressure, but even the variation with temperature can often be neglected for approximate calculations. Thus, when the range of temperature and pressure is not too large in any process considered, mean values of the specific heats can be chosen and assumed constant for that particular process. As we shall show later, this procedure considerably simplifies many calculations.

Finally, a word about units. In the SI the unit of specific heat is kJ/kg K. When calorimetric heat units are used, it is evident from their definition (see footnote in section 2.3) that the value of the specific heat will be the same in any system, e.g.

$$1 \text{ kcal/kg K} = 1 \text{ Chu/lb K} = 1 \text{ Btu/lb R}$$

and that all such units of specific heat are 4·1868 times greater than the kJ/kg K

8.3 Tables of properties

We are now in a position to describe the most convenient way of setting out the relations between p, v, T, u, h and s in tabular form. The simple type of table illustrated in Fig. 1.1 is not used in practice because, by adopting the following method, the necessary information can be presented in a much more condensed form. All the properties for the whole field of Fig. 8.5 can be given in three tables, and for many purposes sufficiently accurately in only two.

138

(i) *Saturation table*

We have noted that when a substance is known or specified to be either a saturated liquid or saturated vapour, its state is completely determined when the value of one property is given. If pressure is chosen to be the independent variable, the corresponding saturation values of all other properties of the liquid and vapour can be set out as in Fig. 8.9. The suffixes f and g refer to saturated liquid and saturated vapour respectively. The suffix s is added to temperature because this is the same for both phases in the saturated state: when temperature is chosen to be the argument of the table, p carries the suffix s and these two columns are interchanged.

$\dfrac{p}{\text{[bar]}}$	$\dfrac{t_s}{\text{[°C]}}$	v_f $\left[\dfrac{m^3}{kg}\right]$	v_g $\left[\dfrac{m^3}{kg}\right]$	u_f $\left[\dfrac{kJ}{kg}\right]$	u_g $\left[\dfrac{kJ}{kg}\right]$	h_f $\left[\dfrac{kJ}{kg}\right]$	h_g $\left[\dfrac{kJ}{kg}\right]$	s_f $\left[\dfrac{kJ}{kg\,K}\right]$	s_g $\left[\dfrac{kJ}{kg\,K}\right]$
0·006112	0·01	0·0010002	206·1	0	2375	0·0006	2501	0	9·155
0·010	7·0	0·0010001	129·2	29	2385	29	2514	0·106	8·974
↓									
1·01325	100·0	0·001044	1·673	419	2507	419	2676	1·307	7·355
↓									
220	373·7	0·00269	0·00368	1949	2097	2008	2178	4·289	4·552
221·2	374·15	0·00317	0·00317	2014	2014	2084	2084	4·430	4·430

Fig. 8.9   Table for saturated water and steam

Sometimes, for brevity, the internal energy is not listed because it can easily be obtained from the values of p, v and h, i.e.

$$u_f = h_f - pv_f, \quad \text{and} \quad u_g = h_g - pv_g$$

The values of v_f may also be omitted, or given in a separate shorter table, because the specific volume of a liquid varies so little with pressure. Thus, for water, v_f varies only from 0·001 00 m³/kg at the triple point to 0·001 29 m³/kg at a pressure of 50 bar, whereas when the liquid is transformed into saturated vapour the respective specific volumes become 206·1 and 0·0394 m³/kg. Indeed it is often sufficiently accurate to assume a constant value for v_f, and for water at moderate pressures the value of 0·001 m³/kg is frequently used.

On the other hand, where brevity is not important, the differences $(h_g - h_f)$ and $(s_g - s_f)$ are allotted columns in the table because they are so frequently required. For convenience they are denoted by the symbols h_{fg} and s_{fg}. Since the heat added in a reversible constant pressure process is equal to the increase of enthalpy, the latent heat of vaporisation must be equal to h_{fg}.

It will be appreciated that the tables only give the difference between the internal energy, enthalpy or entropy at any state and the value of the respective property at a reference state. The internal energy and enthalpy cannot be zero at the *same* reference state, however, because they are related by $h = u + pv$. Thus if u is put equal to zero at a state defined by (p_0, v_0), the value of h at this state is $p_0 v_0$. For steam tables, based on the International Skeleton Tables 1963, the liquid at the triple point is chosen as the datum state and both u_f and s_f are put equal to zero at this state.* The value of the enthalpy at this state is therefore

$$h_f = u_f + pv_f$$

$$= 0 + 0 \cdot 006\ 112[\text{bar}] \times 10^5 \left[\frac{\text{N/m}^2}{\text{bar}}\right] \times 0 \cdot 001\ 000\ 2\left[\frac{\text{m}^3}{\text{kg}}\right] \times 10^{-3} \left[\frac{\text{kN m}}{\text{N m}} \text{ or } \frac{\text{kJ}}{\text{J}}\right]$$

$$= 0 + 0 \cdot 006\ 112 \times 0 \cdot 001\ 000\ 2 \times 10^2 = 0 \cdot 000\ 611\ 2\ \text{kJ/kg} \qquad (8.5)$$

The saturation table for steam extends from the triple point pressure of $0 \cdot 006\ 112$ bar to the critical pressure of $221 \cdot 2$ bar. At the critical pressure the properties with suffix f become equal to the corresponding properties with suffix g, i.e. u_{fg}, h_{fg} and s_{fg} are all zero.

A knowledge of the properties along the saturation line enables the properties at any state in the wet region to be easily computed, and no separate table is required. The reason is that a wet vapour is a mixture of saturated liquid and saturated vapour. When two masses of the same substance, at the same pressure and temperature, are brought together, they will be in equilibrium with one another and there will be no change in pressure or temperature. The same argument can be applied to any *intensive* property, i.e. to any property whose value is independent of the mass of the system. It cannot be applied to *extensive* properties, such as V, U, H and S, whose values depend upon the amount of substance under consideration. An extensive property of a combination of the two masses is equal to the sum of the corresponding extensive properties of the individual masses. For example, if two masses m_1 and m_2 at the same pressure and temperature have specific volumes v_1 and v_2, the total volume on combination is given by

$$V_3 = V_1 + V_2 = m_1 v_1 + m_2 v_2$$

It will be meaningful to speak of the specific volume of the combination if the two masses are homogeneously mixed. Then we have

$$v_3 = \frac{V_3}{m_3} = \frac{m_1 v_1 + m_2 v_2}{m_1 + m_2}$$

All the specific properties, which are derived from extensive properties, can be treated in this way.

---

*In earlier steam tables the datum state used was saturated liquid at 0°C, whereas the lowest temperature at which saturated liquid can exist is $0 \cdot 01$°C (see the boiling curve in Fig. 8.4). Thus the datum state was an unstable state arrived at by extrapolation. Furthermore, it was h_f and not u_f which was put equal to zero at the datum state. The shift to the new datum has little effect on numerical values, amounting only to -0.0416 kJ/kg for enthalpy and $-0 \cdot 000\ 15$ kJ/kg K for entropy.

The foregoing provides the clue to the treatment of a wet vapour. Unit mass of wet vapour can be regarded as a mixture of x kg of saturated vapour and $(1 - x)$ kg of saturated liquid, each having different specific properties denoted by suffixes g and f respectively but the same pressure and temperature. These specific properties can be read from the saturation table when the pressure or temperature of the wet vapour is known. Any specific property of the wet vapour can then be found by adding the corresponding extensive properties of the two constituents which together make up unit mass of wet vapour. Thus

$$u = (1 - x)u_f + xu_g$$
$$h = (1 - x)h_f + xh_g$$
$$s = (1 - x)s_f + xs_g$$

When the difference columns are given in the tables, i.e. properties denoted by the suffix fg, less arithmetic is required if the equations are expressed in the form

$$u = u_f + xu_{fg} \tag{8.6}$$
$$h = h_f + xh_{fg} \tag{8.7}$$
$$s = s_f + xs_{fg} \tag{8.8}$$

The physical interpretation of (8.7) say, is that the enthalpy of unit mass of wet vapour is equal to the enthalpy of unit mass of saturated liquid plus the increase of enthalpy entailed by the evaporation of a proportion x at constant pressure.

The specific volume of the wet vapour is similarly given by

$$v = (1 - x)v_f + xv_g \tag{8.9}$$

but this time the expression is modified in a different way. The term containing v_f can often be neglected, and in this case

$$v = xv_g \tag{8.10}$$

This simplification cannot be employed when the fraction x is small or the pressure high, but equation (8.10) will always be used unless otherwise stated.

The quantity of saturated vapour in unit mass of wet vapour, denoted by x, is referred to as the *dryness fraction*, or *quality*, of the vapour. The dryness fraction increases from 0 to 1 as the state of the substance changes from E to F in Fig. 8.5. It is a quantity which depends only on the state of the wet vapour, and it is therefore a property. In effect we have chosen to define the state of a wet vapour by (p, x).* If p and v are known, however, it is a simple matter to find x from equation (8.10) and then use (8.6), (8.7) and (8.8) to calculate the remaining properties.

---

*One practical reason for doing this is that in many circumstances the dryness fraction is more easily measured than the specific volume. A method of measuring the dryness fraction is described in section 10.3.

(ii) *Superheat table*

When the vapour is superheated, the two properties usually chosen as the independent variables for the table are pressure and temperature. At any given pressure a superheated vapour may have any temperature greater than the saturation value. Fig. 8.10 illustrates one method of tabulation which enables the values of v, u, h and s to be found at any state defined by a pair of values (p, t). The properties of superheated steam have been obtained for pressures up to 1000 bars and temperatures up to 800 °C. It will be noticed that the numbers in any vertical column cease when the saturation

p (t_s)		t	200	250	300	350	400	450 $\rightarrow$
↑	v_g	v	0·1317	0·1324	0·1520	0·1697	0·1865	0·2029	0·2191
15	u_g	u	2595	2597	2697	2784	2868	2952	3035
(198·3)	h_g	h	2792	2796	2925	3039	3148	3256	3364
	s_g	s	6·445	6·452	6·711	6·919	7·102	7·268	7·423
	v_g	v	0·0996		0·1115	0·1255	0·1386	0·1511	0·1634
20	u_g	u	2600		2681	2774	2861	2946	3030
(212·4)	h_g	h	2799		2904	3025	3138	3248	3357
	s_g	s	6·340		6·547	6·768	6·957	7·126	7·283
	v_g	v	0·0666		0·0706	0·0812	0·0905	0·0993	0·1078
30	u_g	u	2603		2646	2751	2845	2933	3020
(233·8)	h_g	h	2803		2858	2995	3117	3231	3343
	s_g	s	6·186		6·289	6·541	6·744	6·921	7·082
	v_g	v	0·0498			0·0588	0·0664	0·0733	0·0800
40	u_g	u	2602			2728	2828	2921	3010
(250·3)	h_g	h	2801			2963	3094	3214	3330
↓	s_g	s	6·070			6·364	6·584	6·769	6·935

Fig. 8.10   Table for superheated steam
(Units as in Fig. 8.9)

temperature corresponding to the pressure becomes greater than the value of the temperature at the head of the column. In abridged tables, where space saving is important, it is sometimes more convenient to head these columns with the number of degrees of superheat rather than the actual temperature of the vapour. There will then be no vacant spaces in the table.

Owing to the need for two independent variables, the superheat table is inevitably much more lengthy than the saturation table. For this reason the internal energy is often not included and must be found from $(h - pv)$.

(iii) *Compressed liquid table*

When the pressure of a liquid is higher than the saturation pressure corresponding to the temperature of the liquid, the state lies in the region to the left of MNc in Fig. 8.5. The pressure and temperature can be varied independently, and can therefore be considered to be the independent variables as in the case of a superheated vapour. Consequently the properties may be tabulated in the same way, i.e. as in Fig. 8.10. In fact it is possible to combine the compressed liquid and superheat tables by entering the compressed liquid data in the vacant spaces of Fig. 8.10. It is found, however, that the properties of a liquid change very little with pressure and it is often more convenient to list the compressed liquid values as deviations from the corresponding saturation values taken at the same temperature. Such a table is shown in Fig. 8.11. At ordinary pressures, say below 50 bars, deviations are very small and the properties v, u, h and s can be assumed to equal the saturation values v_f, u_f, h_f and s_f at the

	t	0·01	100	200	300	350	374·15
	p_s	0·006112	1·01325	15·55	85·92	165·4	221·2
p	$v_f \times 10^2$	0·1000	0·1044	0·1157	0·1404	0·1741	0·317
(t_s)	h_f	0	419	852	1345	1671	2084
	s_f	0	1·307	2·331	3·255	3·779	4·430
100	$(v-v_f) \times 10^2$	−0·0005	−0·0006	−0·0009	−0·0007		
(311·0)	$(h-h_f)$	+10	+7	+4	−2		
	$(s-s_f)$	0·000	−0·008	−0·013	−0·007		
221·2	$(v-v_f) \times 10^2$	−0·0011	−0·0012	−0·0020	−0·0051	−0·0107	0
(374·15)	$(h-h_f)$	+22	+17	+9	−12	−34	0
	$(s-s_f)$	+0·001	−0·017	−0·031	−0·053	−0·071	0
	$(v-v_f) \times 10^2$	−0·0023	−0·0024	−0·0042	−0·0117	−0·0298	−0·161
500	$(h-h_f)$	+49	+38	+23	−21	−94	−369
	$(s-s_f)$	0·000	−0·037	−0·068	−0·134	−0·235	−0·670
	$(v-v_f) \times 10^2$	−0·0044	−0·0044	−0·0075	−0·0191	−0·0427	−0·180
1000	$(h-h_f)$	+96	+76	+51	−17	−119	−415
	$(s-s_f)$	−0·007	−0·070	−0·124	−0·235	−0·385	−0·853

Fig. 8.11   Table for compressed water

(Units as in Fig. 8.9)

temperature of the compressed liquid. This is equivalent to assuming that curves such as 1–2 in Fig. 8.5 are vertical straight lines (i.e. that the liquid is incompressible), and that the internal energy of a compressed liquid is a function only of temperature. We shall usually make this assumption in subsequent chapters.

The properties of water in its liquid and vapour phases have been investigated more thoroughly than those of any other substance. They are of particular interest to engineers owing to the extensive use of steam as a working fluid in heat engines. Several international conferences have been held to correlate the results obtained in various countries and by various experimental techniques.* At the Sixth International Conference on the Properties of Steam (1963), agreement was reached on a set of values

*An interesting account of this work may be found in Ref. (21).

to each of which was attached a tolerance indicating the agreed margin of experimental error. These values comprise The International Skeleton Tables 1963. The tolerances are well within the degree of accuracy required for any engineering calculations. The National Engineering Laboratory, Scotland, has used these values in computing the full set of *N.E.L. Steam Tables* 1964, Ref. (1), which employ sufficiently small intervals to make accurate interpolation a simple matter. For student use, an abridged set of steam tables is included in *Thermodynamic and Transport Properties of Fluids*, Ref. (17). The numbers appearing in subsequent examples are taken from these abridged tables and, where interpolation is necessary, linear interpolation is always used.

8.4 Diagrams of properties

The characteristics of a liquid and vapour have already been described by referring to diagrams using p and v, or T and v, as coordinates. Diagrams with other pairs of properties as coordinates will be shown to be useful for various purposes in later chapters, and three such diagrams are described in this section, namely the T-s, h-s and p-h diagrams. Property charts do not entirely take the place of tables because values of the properties cannot be read from them with as great an accuracy as they are given in tables.

Fig. 8.12

(i) *Temperature-entropy diagram*

When analysing the characteristics of heat engine cycles it is often helpful to plot the cycle on a diagram having temperature and entropy as the coordinates. Fig. 8.12a illustrates this. If the processes in the cycle are reversible, the heat supplied during the cycle is given by

$$Q_{12} = \int_1^2 T \, ds$$

and the heat rejected by

$$Q_{21} = \int_2^1 T \, ds$$

144

These quantities are represented by the areas under the curves 1–2 and 2–1. It follows that the area enclosed by a reversible cycle represents the net heat supplied and hence, from the First Law, the net work done.

In its simplest form the *T-s* diagram consists of a series of constant pressure lines and the saturation curve. The general shape of the curves can be deduced by supposing that unit mass of liquid is heated reversibly at a constant pressure p. The heat added is equal to the increase of enthalpy (from the First Law), and therefore the amount required to raise the temperature of the liquid dT is $c_p \, dT$ [from equation (8.4)]. Consequently the increase of entropy is

$$ds = c_p \frac{dT}{T}$$

For the purpose of finding the approximate shape of the curve we may assume that c_p is constant, so that the increase of entropy between states A and B is

$$s_B - s_A = c_p \ln \frac{T_B}{T_A}$$

Here c_p denotes the mean specific heat of the liquid over the range of temperature T_A to T_B at the particular pressure p. If T_B is the saturation temperature corresponding to the pressure p, then AB in Fig. 8.12b is the constant pressure line up to the point of saturation. As the heat addition continues, the temperature remains constant at T_B until the saturated vapour state C is reached. The latent heat is equal to h_{fg} and the change of entropy s_{fg} is equal to h_{fg}/T_B. Thereafter the entropy increases in accordance with the equation

$$ds = c_p \frac{dT}{T}$$

where c_p is the specific heat of the superheated vapour. The constant pressure line therefore continues from C to D with much the same shape as in the liquid region.

Since h_{fg} falls with increase of pressure, the length BC decreases, and the family of constant pressure lines appears as in Fig. 8.12b. With water, the line representing a constant pressure of 0·006 112 bar cuts the temperature axis at 273·16 K, because this is the reference state at which the entropy is put equal to zero. Fig. 8.12b is out of proportion, because at any given temperature the change of entropy with pressure in the liquid region is in fact negligible. On any reasonable scale the constant pressure lines are so close together in the liquid region as to be indistinguishable, and for practical purposes they may be assumed to coincide with the saturated liquid line. Fig. 8.13 shows the *T-s* diagram for steam drawn to scale.

In view of the fact that the dryness fraction is used as one of the parameters defining the state of a wet vapour, it is useful to have lines of constant dryness fraction superimposed on the diagram. From equation (8.8) we have $x = (s - s_f)/s_{fg}$. Therefore points on the constant dryness fraction line of value x can be obtained by marking off, on each horizontal constant pressure line such as BC in Fig. 8.12b, a length equal to x(BC) from the saturated liquid end. The line of constant quality x is then found by joining up the points so obtained, as in Fig. 8.13.

It is also useful to have a general idea of the shape of constant volume and constant enthalpy lines in the wet and superheat regions. In the superheat region a constant

Fig. 8.13 Temperature-entropy diagram for steam

volume line will be of similar shape to a constant pressure line, since the change of entropy in a constant volume process is given by

$$ds = \left(\frac{dQ}{T}\right)_{rev} = c_v \frac{dT}{T}$$

The slope of a curve on the *T-s* diagram is dT/ds, which for a constant pressure line equals T/c_p and for a constant volume line equals T/c_v. In fact c_v is always less than c_p (see section 7.4 (iii) for a formal demonstration) so that at any given state the constant volume line is always steeper than the constant pressure line. In the wet region a constant volume line, of value v, say, can be plotted in the following way. At any given temperature or pressure, v_f and v_g can be found from the saturation table. Using the chosen value v, the corresponding value of the dryness fraction x can be found from equation (8.9). With x and either p or t known, it is possible to locate the point on the

146

diagram. If this is repeated for other temperatures, the series of points representing states having the volume v can be joined up to form a constant volume line. Such lines are found to be concave downwards in the wet region.

Finally, the general shape of a constant enthalpy line has been indicated in Fig. 8.13. In the wet region it may be located by the method already described for a constant-volume line, except that equation (8.7) is used instead of (8.9). In the superheat region, points of constant enthalpy can be located by finding the pressure, at various temperatures, which corresponds to the chosen value of h. This may be done by inter-polation from the superheat table. The fact that the resulting curve tends towards the horizontal as the amount of superheat increases can be deduced from a knowledge of the behaviour of a perfect gas. The more a vapour is superheated the more closely does it behave like a perfect gas, and we shall show in section 8.5 that the enthalpy of a perfect gas is a function only of temperature. Thus if the enthalpy remains con-stant so must the temperature, and hence the tendency of the curve to flatten out at high degrees of superheat. It is worth noting the fact that as the pressure falls at con-stant enthalpy, a wet vapour eventually becomes drier and if it is not too wet it is superheated; section 10.3 describes how this fact may be used in the measurement of the dryness fraction of a wet vapour.

(ii) *Enthalpy-entropy diagram*

The diagram in which the enthalpy and entropy are used as coordinates is known as the *Mollier diagram* after its proposer. It is very useful when analysing the performance of adiabatic steady-flow processes, such as occur in nozzles, diffusers, turbines and compressors, and a glance at equations (4.5) and (4.6) will show why. For example, the work done by a fluid flowing through a turbine is equal to the drop in enthalpy between inlet and outlet (assuming no change in kinetic energy). Consequently, if the thermodynamic state of the fluid at inlet and outlet can be located as points on the h-s diagram, the vertical distance between the two points is proportional to the work done. Adiabatic processes which are also reversible, i.e. isentropic processes, are represented by vertical lines on the diagram.

The Mollier diagram is shown in Fig. 8.14. It may be seen to be a distorted form of the T-s diagram—distorted in such a way that the constant enthalpy lines become straight and horizontal. The constant pressure lines in the liquid region merge into the saturated liquid line as before, and are straight in the wet region for the following reason. The increase of enthalpy from a saturated liquid state to any state of quality x at the same pressure is xh_{fg}. Since this is equal to the heat added in a reversible con-stant pressure process, the increase of entropy is xh_{fg}/T, where T is the saturation temperature corresponding to the given pressure. Thus the increase of entropy is directly proportional to the increase of enthalpy in the wet region, the slope of the constant pressure line $(dh/ds)_p$ being equal to T. The slope of successive constant pressure lines increases, because the saturation temperature increases with pressure. In the superheat region, the slope at any state is also equal to the temperature T at that state, because the increase of entropy is given by $ds = dh/T$. There is therefore

147

no discontinuity at the saturation vapour line: the constant pressure lines start with a slope determined by the saturation temperature and curve upwards as the temperature increases.

Constant dryness fraction lines can be added by dividing each straight line in the wet region in the ratio x to $(1 - x)$ as before, and they are shown by the dotted curves

Fig. 8.14 Enthalpy-entropy diagram for steam

in Fig. 8.14. Constant temperature lines are usually added in the superheat region; in the wet region they coincide with the constant pressure lines. As already explained in connection with the *T-s* diagram, the enthalpy becomes more nearly dependent on temperature alone as the amount of superheat increases. It follows, therefore, that on the *h-s* diagram constant temperature lines must tend towards the horizontal as shown in the figure. It should be noted that lines of constant degrees of superheat are sometimes used instead of lines of constant temperature, and that large-scale charts for practical use only show the most useful part of the diagram, enclosed by the dotted rectangle.

The fact already mentioned, that a reduction of pressure at constant enthalpy eventually results in the drying and superheating of a wet vapour, is clearly seen on the *h-s* diagram because the process is represented by a horizontal line.

(iii) *Pressure-enthalpy diagram*

A diagram showing the properties of a fluid on a pressure-enthalpy basis is of somewhat limited use, but as we shall refer to it later when considering refrigeration problems it deserves a brief mention here. Since pressure is one of the coordinates of this diagram, the main family of curves comprises those of constant temperature. In the liquid region they are substantially vertical (see Fig. 8.15) because the effect of pressure

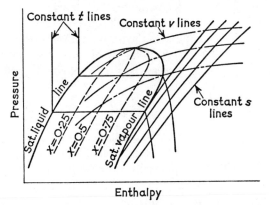

Fig. 8.15   Pressure-enthalpy diagram

on the enthalpy is negligible. They then run horizontally across to the saturated vapour line since they are also lines of constant pressure in the wet region. In the superheat region they fall steeply and approach the vertical for the same reason that they approach the horizontal on the enthalpy-entropy diagram. It is usual to add lines of constant volume in the wet and superheat regions, lines of constant dryness fraction in the wet region, and lines of constant entropy in the superheat region: these have the shape shown in Fig. 8.15.

8.5 Properties of a perfect gas

At pressures and temperatures usually obtaining in engineering applications many substances are to be found in a state lying to the right of curve MNf in Fig. 8.5. In this section we shall restrict our attention to two regions in this field: (a) the region of high temperature, where the temperature is in excess of twice the absolute critical temperature, and (b) the region of low pressure, where the pressure is one atmosphere or less. The so-called 'permanent' gases, oxygen, hydrogen, nitrogen, etc., all lie in the first region at ordinary working temperatures (the critical temperatures of the specified gases being 154·4 K, 33·3 K and 126·0 K respectively). The water vapour in the atmosphere lies in the second region, and is in fact superheated steam at very low pressure. The behaviour of substances in these two regions of states approximates very closely to that of a hypothetical substance known as a perfect gas. The perfect gas is a particularly useful concept because the properties are related by very simple equations. As before, we shall start by considering the relation between the primary properties p, v and T.

(i) *Equation of state of a perfect gas*

Let us imagine that the pressure, specific volume and temperature of a gas are measured for a large number of equilibrium states covering a wide range of these variables.

Fig. 8.16 *p-v-T data for gases*

When the quantity pv/T is calculated for each state, and plotted against pressure for various values of temperature, the curves of Fig. 8.16 are obtained. The curves all converge to a single point on the pv/T axis as the pressure is reduced to zero, and the same sort of family of curves is obtained for all gases. For any gas it is possible to write

$$\lim_{p \to 0} \left(\frac{pv}{T}\right) = A$$

but the constant A is different for each gas.

It is clear from Fig. 8.16 that any equation relating p, v and T which will express these results accurately will be complicated. However, for practical purposes, over the normal working range of pressure and temperature the equation of state can be expressed sufficiently accurately by the equation

$$pv = AT$$

If this equation were valid for all pressures and temperatures, the results in Fig. 8.16 would all fall on the horizontal dotted line. The hypothetical substance which has this characteristic equation is called a *perfect gas* or *ideal gas*. The lower the pressure and the higher the temperature, the more nearly does a real gas approximate to a perfect gas. This can be explained on the basis of a theory as to the microscopic structure of a gas (see section 8.6). The constant A is called the *specific gas constant*, or simply the *gas constant*, and henceforth it will be given its usual symbol R. The characteristic equation of a perfect gas therefore becomes

$$pv = RT \tag{8.11}$$

or if a mass m of gas is considered,

$$pV = mRT*$$

*This approach of showing that, as $p \to 0$, real gases tend towards perfect gas behaviour $pv = RT$, presupposes that experiments can be carried out with gases in which their absolute thermodynamic temperature can be measured in some independent way. In practice the gas thermometer is the only practical way of measuring thermodynamic temperature except at very low and high temperatures. To avoid circularity of argument, it is necessary to prove formally that a perfect gas temperature θ_g defined by $pv = R\theta_g$ is identical with the absolute thermodynamic temperature T, and this is done in section 7.6(i).

The product pV has the dimensions of energy, and in our system of units should be expressed in kJ. It is shown in equation (8.5) that when p is measured in bars (1 bar $= 10^5$ N/m^2) and V in m^3, pV must be multiplied by $10^5/10^3 = 100$ to obtain a result in kJ. The units of R which we shall employ are kJ/kg K.

Although the gas constant differs for different gases, the equation of state can be generalised to include all perfect gases by making use of an important law of chemistry. Before doing so, however, we must introduce another measure of a quantity of matter: it is the *amount of substance*, symbolised by n, and its unit is the *mole*. This was defined in 1972 as one of the base units of the SI by the following statement:

The mole is the amount of substance of a system which contains as many elementary entities as there are atoms in 0·012 kilogram of carbon 12. (When the mole is used, the elementary entities must be specified and may be atoms, molecules, ions, electrons, other particles, or specified groups of such particles.)

In this book the entities will always be groups of identical atoms or molecules, or groups of specified mixtures of atoms and/or molecules (as with 'air'), as will be clear from the context. Furthermore, it will always be convenient to work in *kilomoles*, the unit symbol being *kmol*.

The number of entities in 1 kmol of carbon 12, each of mass $m(^{12}\text{C})$, or by definition in 1 kmol of any other substance X having entities of mass $m(\text{X})$, is

$$N_A = \frac{12 \text{ kg}}{m(^{12}\text{C})} \text{ kmol}^{-1}$$

The mass per kmol of any substance, called the *molar mass M*, is therefore given by

$$M = N_A m(\text{X}) = \left[\frac{12 \text{ kg}}{m(^{12}\text{C})} \text{ kmol}^{-1}\right] \times m(\text{X}) = \left[\frac{m(\text{X})}{\frac{1}{12}m(^{12}\text{C})}\right] \frac{\text{kg}}{\text{kmol}}$$

Several important points can be made as follows.

(a) N_A is the *Avogadro constant* which, according to the latest measurements, is equal to $6\cdot022\,045 \times 10^{26}$ kmol^{-1}.

(b) The ratio $m(\text{X})/\frac{1}{12}m(^{12}\text{C})$ will be recognised as the definition of *relative atomic mass* A_r, or *relative molecular mass* M_r, depending upon whether X is an atomic or molecular species.

(c) The denominator $\frac{1}{12}m(^{12}\text{C})$ is the unit of measurement on the 'atomic weight' scale: it is called the *unified atomic mass unit* and is symbolised by u.

(d) u is numerically equal to $1/N_A$, and hence $1 \text{ u} = 1\cdot660\,565\,5 \times 10^{-27}$ kg.

(e) M is numerically equal to A_r or M_r but is not dimensionless: here it will be measured in kg/kmol.

(f) It follows from (e) that the mass m of an amount of substance n becomes

$$m = nM \tag{8.12}$$

Equation (8.11) can now be applied to a molar mass M, the volume of which is called the *molar volume* v_0, to yield

$$pv_0 = MRT \qquad (8.13a)$$

The fundamental law of chemistry required to generalise this equation of state is known as Avogadro's law. To explain the empirical fact that gases combine in simple integral ratios of volumes, Avogadro (1811) suggested that equal volumes of all gases contain the same number of molecules, when the volumes are measured under the same conditions of temperature and pressure. His argument can be found in any elementary textbook on chemistry. It is not exact in practice but, as with the equation of state, the accuracy of the law improves as the pressure of the gas is reduced. Now since a molar mass M contains the same number of molecules whatever the gas (i.e. the Avogadro constant), v_0 must be the same for all gases at the same p and T. It follows from equation (8.13a) that the product MR must be the same for all gases. This product is known as the *universal gas constant* and it will be symbolised by R_0. Equation (8.11) can therefore be written in a form which applies to all perfect gases, and approximately to all real gases, as

$$pv_0 = R_0 T \qquad (8.13b)$$

Or, if V is the volume of an amount of substance n,

$$pV = nR_0 T$$

Experiment shows that an average value for the volume of 1 kmol of the 'permanent' gases is $22 \cdot 4138 \, m^3$ at standard atmospheric pressure (1 atm $= 1 \cdot 013\ 25$ bar) and $0\ °C$. The value of R_0 is thus found to be

$$R_0 = \frac{pv_0}{T} = \frac{100 \times 1 \cdot 013\ 25 \times 22 \cdot 4138}{273 \cdot 15} = 8 \cdot 3144 \frac{kJ}{kmol\ K}$$

The gas constant for any particular gas can be obtained from

$$R = \frac{R_0}{M}$$

e.g. for oxygen

$$R = \frac{8 \cdot 3144}{32} = 0 \cdot 260 \, kJ/kg\ K$$

(ii) *Other property relations for a perfect gas*

We have already noted in section 8.2 that a knowledge of the characteristic equation is not, by itself, sufficient to enable the other thermodynamic properties to be determined. Some additional experimental data are necessary before u, h or s can be expressed in terms of two of the measurable properties p, v and T. It is possible to show hat this is so for the perfect gas without difficulty. We have seen that the combination of the First and Second Laws gives rise to equation (8.2), i.e.

$$ds = \frac{1}{T} du + \frac{p}{T} dv$$

This is a general relation between thermodynamic properties which must be satisfied by any substance. If we now apply this general equation to a perfect gas, the characteristic equation enables p to be eliminated to give

$$ds = \frac{1}{T} du + \frac{R}{v} dv \tag{8.14}$$

The expression for enthalpy is obtained more simply. From the definition of enthalpy we have

$$h = u + pv$$

which for a perfect gas becomes

$$h = u + RT \tag{8.15}$$

It is evident from (8.14) and (8.15) that we require an expression for u in terms of v and T; when this is known, u can be eliminated from each of these equations to give expressions for s and h in terms of T and v.

It so happens that there is a very simple expression for the internal energy of a perfect gas. Firstly, we may note that the internal energy is a function only of temperature, i.e.

$$u = \phi(T) \tag{8.16}$$

This can be shown to be a direct consequence of the characteristic equation of a perfect gas, but the proof is somewhat difficult and it has been dealt with in Chapter 7, section 7.6 (i). Although attempts have been made to see how far this result is true for a real gas, no accurate experimental technique has so far been suggested. Nevertheless an idealised experiment can be conceived as follows.

The apparatus consists of two vessels which may be put into communication with one another by opening a valve (Fig. 8.17). One vessel contains the gas under investigation and the other is evacuated. Let us imagine that we can prevent any heat from entering or leaving the system, i.e. the gas. The procedure is to measure the temperature of the gas, and open the valve so that the gas expands to fill the evacuated vessel. When equilibrium has been re-established, the temperature of the gas is again measured. Since the expansion is free, no work is done by the system or its surroundings. Moreover, we have assumed that the process is arranged to be adiabatic. It follows from the First Law that there can be no change of internal energy between the initial and final states. Both the pressure and specific volume have changed, so that if it is found that no change of temperature has occurred, we may deduce that the internal energy is a function only of temperature. That is, no matter what changes of pressure and volume occur, the internal energy remains the same if the temperature does not change.

Fig. 8.17 The Joule experiment

The experiment is impractical because it is not possible to ensure that the process is adiabatic. It is of course possible to reduce the heat exchange with the surrounding atmosphere to negligible proportions by lagging the apparatus, but the vessels themselves then constitute the effective surroundings of the gas which is the system under consideration. The first real attempt to perform the experiment was made by Joule in the 1840s, following up some experiments by Gay-Lussac. Joule submerged his vessels in a water bath, and measured the temperature of the water before and after the expansion. On observing no change, he inferred that the temperature of the gas had undergone no change either, and consequently was able to deduce the fact that $u = \phi(T)$. This relation is usually spoken of as Joule's Law. However, even if Joule had been able to use modern accurate thermometry, the heat capacity of the water bath is so very much larger than the heat capacity of the gas that a small change in gas temperature would not have been measurable.* The experiment cannot therefore be said to establish the law. In fact we now know that a real gas should exhibit a small drop in temperature after a free expansion, for reasons which will be discussed in the next section.

Accepting Joule's law as being true for a perfect gas (and as we have said, this can be proved), we can proceed to deduce the nature of the function in (8.16). From the definition of the specific heat at constant volume we have

$$\mathrm{d}u = (c_v \, \mathrm{d}T)_v$$

If u is a function only of temperature two consequences follow: (1) c_v must be either constant or a function only of temperature, and (2) the relation must apply to any infinitesimal process—not only to one at constant volume. The suffix v can therefore be dropped from the equation and we can write

$$\mathrm{d}u = c_v \, \mathrm{d}T \tag{8.17}$$

With real gases, c_v is found to vary mainly with temperature, and this variation is small enough for c_v to be treated as a constant if the range of temperature covered by the process is not too wide. It is customary, therefore, to define a perfect gas as one

*The Joule-Thomson experiment described in section 7.4(iv), which employs a steady-flow process, is not subject to the same criticism and is a more sensitive experiment. It shows that the enthalpy of a gas is a function only of temperature as the pressure under which the experiment is conducted is reduced, i.e. under conditions for which real gas behaviour approaches that of the perfect gas. It will be clear from the following discussion that $h = f(T)$ implies $u = \phi(T)$ and vice versa.

for which

(a) $pv = RT$

and

(b) $c_v = $ constant

The second part of the definition is not contained in the first; what is implied by (a) is that c_v is a function only of temperature. Any 'permanent' gas can be treated as a perfect gas defined by (a) and (b) with sufficient accuracy for many purposes. With c_v constant then, the change in internal energy between any two states at T_1 and T_2 is given by

$$u_2 - u_1 = c_v(T_2 - T_1) \qquad (8.18)$$

This equation applies to *any* process—whether reversible or irreversible—because the change in a property depends only on the end states, and in this case only on the initial and final temperatures.

It is evident that the additional experimental information, required to enable us to express u, h and s as functions of the primary properties, is the value of the specific heat at constant volume. If this is assumed to be known, suitable expressions for the enthalpy and entropy of a perfect gas can be found as follows. The differential form of (8.15) now becomes

$$dh = c_v \, dT + R \, dT$$
$$= (c_v + R) \, dT \qquad (8.19)$$

The enthalpy is therefore also a function of temperature alone. This expression can be further simplified, because from the definition of the specific heat at constant pressure we have

$$dh = (c_p \, dT)_p$$

Since h depends on temperature only, this must apply to *any* infinitesimal process undergone by a perfect gas, and the suffix p can be dropped. Thus

$$dh = c_p \, dT \qquad (8.20)$$

A relation between c_p, c_v and R immediately follows from (8.19) and (8.20). Thus

$$dh = (c_v + R) \, dT = c_p \, dT$$

and hence

$$c_p - c_v = R \qquad (8.21)$$

Since R is a positive quantity, c_p must always be greater than c_v. Equation (8.21) implies that *even when c_p and c_v vary with temperature, their difference must always be* *constant*. If a perfect gas is defined as one for which c_v is constant, then c_p must be constant also and the integration of (8.20) is simply

$$h_2 - h_1 = c_p(T_2 - T_1) \tag{8.22}$$

Finally, from (8.14) and (8.17) the entropy of a perfect gas is given by

$$ds = \frac{c_v}{T} dT + \frac{R}{v} dv$$

and hence

$$s_2 - s_1 = c_v \ln \frac{T_2}{T_1} + R \ln \frac{v_2}{v_1} \tag{8.23}$$

Unlike the internal energy and enthalpy, the entropy is a function of both temperature and volume. The relations between the entropy and the other two combinations of primary properties, p and T or p and v, can easily be obtained as follows. From the equation of state

$$\frac{p_1 v_1}{T_1} = \frac{p_2 v_2}{T_2}$$

Substituting for v_2/v_1 in (8.23) we have

$$s_2 - s_1 = c_v \ln \frac{T_2}{T_1} + R \ln \left(\frac{p_1}{p_2} \frac{T_2}{T_1} \right)$$

$$= (c_v + R) \ln \frac{T_2}{T_1} - R \ln \frac{p_2}{p_1}$$

and using (8.21) this reduces to

$$s_2 - s_1 = c_p \ln \frac{T_2}{T_1} - R \ln \frac{p_2}{p_1} \tag{8.24}$$

Similarly, by substituting for T_2/T_1 in (8.23) we get finally

$$s_2 - s_1 = c_v \ln \frac{p_2}{p_1} + c_p \ln \frac{v_2}{v_1} \tag{8.25}$$

It may be helpful if the characteristics of a perfect gas are summarised as follows:

(1) The equation of state is $pv = RT$.
(2) A consequence of (1) is that $u = \phi(T)$.
(3) From (2) and the definition of c_v it follows that $c_v = \phi(T)$ or is constant, and that $du = c_v \, dT$ for any infinitesimal process.
(4) From (3) and the definition of enthalpy it follows that $h = \phi(T)$.
(5) From (4) and the definition of c_p it follows that $dh = c_p \, dT$ for any infinitesimal process, and that $c_p - c_v = R$.

(6) If a perfect gas is defined as also having constant specific heats, the following relations are valid irrespective of the process joining the end states 1 and 2:

$$u_2 - u_1 = c_v(T_2 - T_1) \tag{8.18}$$

$$h_2 - h_1 = c_p(T_2 - T_1) \tag{8.22}$$

$$s_2 - s_1 = c_v \ln \frac{T_2}{T_1} + R \ln \frac{v_2}{v_1} \tag{8.23}$$

$$= c_p \ln \frac{T_2}{T_1} - R \ln \frac{p_2}{p_1} \tag{8.24}$$

$$= c_v \ln \frac{p_2}{p_1} + c_p \ln \frac{v_2}{v_1} \tag{8.25}$$

(iii) *Temperature-entropy diagram for a perfect gas*

A *T-s* diagram showing the properties of a perfect gas usually consists simply of a family of constant pressure lines, Fig. 8.18. It may be constructed in the following manner. First a reference state is chosen for which the entropy is considered to be zero, e.g. a state defined by (p_0, T_0). The increase of entropy from this state to any other state at p_0 but at some other temperature T, can be found from equation (8.24). Thus in general

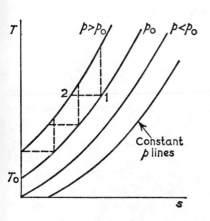

$$s_2 - s_1 = c_p \ln \frac{T_2}{T_1} - R \ln \frac{p_2}{p_1}$$

which in our case reduces to

$$s = c_p \ln \frac{T}{T_0} \tag{8.26}$$

The value of s can be found for several values of T, and the points so obtained can be joined up to form the constant pressure line p_0 on Fig. 8.18.

Fig. 8.18   Temperature-entropy diagram for a perfect gas

Now consider the isothermal compression process denoted by 1–2 on Fig. 8.18. The pressure increases from p_0 to p, and since T is constant equation (8.24) reduces to

$$s_2 - s_1 = -R \ln \frac{p}{p_0}$$

It is evident that $(s_2 - s_1)$ is the same at any temperature. This suggests that any constant pressure line p can be found by stepping off horizontal distances from the line p_0, each equal to $R \ln (p/p_0)$. In this way a series of constant pressure lines can

157

be rapidly constructed. It should be noticed that although the horizontal distance between any two constant pressure lines is the same at any temperature, the vertical distance increases with increase of temperature. This is due to the curvature of the constant pressure lines, and is clearly indicated by the dotted lines in the figure.

Constant volume lines could be added to the diagram, using equation (8.23). The general shape of the curves is the same except that they have a slightly steeper slope than the constant pressure lines at any given point. The slope of a constant pressure line can be found from the differential form of (8.26), i.e.

$$ds = \frac{c_p}{T} dT$$

and hence

$$\frac{dT}{ds} = \frac{T}{c_p}$$

Similarly the slope of a constant volume line is

$$\frac{dT}{ds} = \frac{T}{c_v}$$

Since $c_v < c_p$, the constant volume line must have the steeper slope.

8.6 Kinetic theory and the properties of real gases

Although thermodynamics is concerned solely with the behaviour of macroscopic systems, it is sometimes helpful to have in mind a mechanical model which exhibits the microscopic structure of the system. Such a model is particularly helpful when discussing the way in which real gases differ from a perfect gas.

(i) *Kinetic theory of gases*

On the kinetic theory of gases, a gas is conceived as a large number of minute hard particles, i.e. molecules, moving about at random with very high velocities. The pressure and temperature, which are terms used to describe the observed macroscopic phenomena, can then be explained in terms of the microscopic behaviour of the molecules. Thus, pressure is regarded as the average force per unit area on the walls of the container due to bombardment by the molecules, and the absolute temperature is regarded as being proportional to the mean of the squares of the velocities of the molecules, i.e. to their average kinetic energy of translation. Any problem concerning the behaviour of gases is thereby reduced to one of mechanics. The kinetic theory has been developed to the point where most of the observed phenomena can be explained and predicted with the use of classical mechanics.* At points where the theory apparently breaks down, the difficulties can be resolved by the use of quantum mechanics.

*For a clear account of the theory, see Ref. (14).

The concept of a perfect gas can be shown to be a consequence of making the implest assumptions about the molecules, and this part of the kinetic theory can be ound in any elementary textbook on physics. The main assumptions are as follows:

a) The molecules are perfectly elastic and perfectly rigid, implying that no momentum or time is lost during collisions with the wall of the containing vessel.

b) The volume occupied by the molecules is negligible compared with the total volume.

c) The attractive forces between adjacent molecules are negligible.

The first assumption seems reasonable because no loss of pressure is observed when a gas is kept in a closed vessel over a long period of time, and one can only conclude that no loss of momentum occurs as the result of a vast number of collisions with the wall. The second assumption seems reasonable because a gas can be so easily compressed, and the last because a gas readily expands to fill any space into which it is introduced. With these assumptions and the ordinary laws of mechanics, the characteristics of a perfect gas can be predicted quite simply, e.g. we can obtain the form of the characteristic equation, Joule's law and Avogadro's law.

If assumptions (b) and (c) are not strictly true, it is easy to see why real gases behave more like a perfect gas as the pressure is reduced and the temperature raised. At the lower pressures the molecules are on the average further apart, so that the volume occupied by the molecules is a smaller proportion of the total volume and the attractive forces are less. At the higher temperatures the molecules are moving past one another with higher velocities; they are near one another for a shorter period of time, and consequently the attractive forces have less effect.

By making more realistic assumptions, the behaviour of real gases can also be predicted. For example, more accurate forms of the characteristic equation can be deduced by allowing for the volume occupied by the molecules and the attractive forces between them. Van der Waals' equation provides a simple example, i.e.

$$\left(p + \frac{a}{v_2}\right)(v - b) = RT$$

R, a and b are constants which differ for different gases. The a/v^2 term takes account of the attractive forces, while b takes care of the volume of the molecules.

The departure from Joule's law is also easily explained by modifying assumptions (b) and (c). We know that the internal energy of a real gas depends slightly on volume as well as on temperature, and that a slight drop of temperature must accompany a free expansion under adiabatic conditions. Let us see if this is to be expected of a van der Waals gas. An expression for the internal energy of a gas having van der Waals' characteristic equation is derived in Chapter 7 (equation 7.46); in differential form it is

$$du = c_v \, dT + \frac{a}{v^2} \, dv$$

For an adiabatic free expansion the change in internal energy is zero, so that for this

159

process we have

$$c_v \, \mathrm{d}T = -\frac{a}{v^2} \, \mathrm{d}v$$

The constant a, indicating the forces of attraction between the molecules, is alway positive for gases. Therefore, since $\mathrm{d}v$ is positive in an expansion, $\mathrm{d}T$ must be negative The physical explanation becomes clear when it is remembered that the internal energy includes both the kinetic energy of the molecules and the potential energy due to the attractive forces (section 2.4). If the analogy of a stretched spring is borne in mind, it i evident that the potential energy is increased by the expansion because the molecule are further apart. Since the internal energy is unchanged, the kinetic energy of the molecules must have decreased, and this is reflected in a drop of temperature.

Before leaving the kinetic theory, the behaviour which it predicts for the specific heats of perfect gases is worth noting. According to this theory, the specific heat c should be constant and equal to the increase in the total kinetic energy of the molecule per degree rise of temperature. The value predicted for c_v varies with the complexit of the molecules; for example, c_v is $\frac{3}{2}R$ for a monatomic gas and $\frac{5}{2}R$ for a diatomic gas This difference is explained by the fact that monatomic gases possess translationa kinetic energy only, whereas more complex molecules possess energy of rotation also and all forms of energy are increased as the temperature is raised. For any given type of molecule, however, there is no reason why the increase in total energy per degree rise should be different at different temperatures.

For helium, of molecular weight 4, c_v becomes $(3/2)(8\cdot3144/4) = 3\cdot118$ kJ/kg K and this prediction is in excellent agreement with observation at all temperatures For nitrogen, a diatomic gas of molecular weight 28, c_v should be $(5/2)(8\cdot3144/28$ $= 0\cdot742$ kJ/kg K, which agrees with observation over a wide range of temperature up to about 400 K (Fig. 8.19). Above this temperature, however, c_v gradually increases owing to the fact that the molecules acquire vibrational energy as well a translational and rotational kinetic energy. This gradual rise cannot be predicted by the theory based on classical mechanics, although quantum theory does provide a satisfactory explanation. Quantum theory also suggests that at very low temperature c_v for polyatomic gases should gradually drop to the value $\frac{3}{2}R$ valid for monatomic gases, implying thereby that rotational motion of the molecules ceases at such temperatures. This has in fact been confirmed for hydrogen at 40 K, but for heavier molecules this is expected to happen much closer to absolute zero and experimental evidence is lacking.

Finally, it is worth noting the physical meaning which the concept of entropy acquires when a model based on the microscopic structure of the system is considered. We have seen that any irreversibility results in an increase in the entropy of an adiabatic system—this is the important 'principle of increasing entropy'. On the microscopic view, an irreversibility is always associated with an increase in the disorderly motion of the molecules. For example, during an adiabatic expansion of a gas, internal viscous friction will result in a reduction in work transfer because the

.ternal energy, i.e. random molecular energy, is higher in the final state than it would
.therwise have been. Similarly, in an adiabatic expansion through a nozzle viscous
ffects convert directional kinetic energy of a stream, which is fully available for doing
.ork on say a turbine blade, into such random energy. Or again, two gases when
.ixed represent a higher degree of disorder than when they are separated, and we
.all see in section 14.3 that there is an increase in entropy involved in an irreversible
.ixing process. In all natural processes irreversibilities are present, and there is a
.ndency to proceed towards a state of greater disorder. It is possible to establish a
.uantitative measure of disorder—in terms of a particular definition of the concept of

Fig. 8.19   Variation of c_v with temperature
for nitrogen at ordinary pressures

Fig. 8.20   Compressibility factor for
nitrogen

robability—and then to derive a relation between entropy and this concept of
robability We are then led to a statistical interpretation of the Second Law and to
 branch of the subject known as *statistical thermodynamics*.

.i) *Properties of real gases*

.ccurate equations of state are inconvenient for everyday calculations because they
.ad to unwieldy expressions. They are mostly used for the preparation of tables and
.iagrams of properties. When tables are not available, and for conditions where the
.erfect gas laws are too inaccurate, correction factors may be applied to the perfect
.as relations. Thus the equation of state, for example, can be written as

$$pv = ZRT \tag{8.27}$$

.here Z is the *compressibility factor*. The value of this factor depends upon the gas
.nd it is also a function of the pressure and temperature. The combination of Z
.= $\phi(p, T)$ and equation (8.27) is the complete characteristic equation. Charts have
.een prepared for the more common gases, showing Z plotted against pressure for
.arious temperatures. Fig. 8.20 shows the compressibility chart for nitrogen. The
.dvantages of plotting $Z = \phi(p, T)$, rather than $v = \phi(p, T)$ directly, are that (a) a
.uch smaller range of values is required and (b) it shows at once how far the gas
.eviates from a perfect gas. The disadvantage is that when v is one of the known

161

parameters of state a method of successive approximation is required to obtain the unknown property. For example, when v and T are given, p must be found from the perfect gas equation and a first approximation to Z obtained from the chart. Using this value of Z, a new value of p can be calculated from (8.27), a better approximation to Z obtained from the chart, and so on.

It so happens that the shape of the curves in the region to the right of the critical isothermal T_o in Fig. 8.5 is much the same for all gases although the scales are different. This suggests that the scales might be brought into line if the properties were quoted as fractions of the critical values, i.e. as

$$p_R = \frac{p}{p_c}, \quad v_R = \frac{v}{v_c} \quad \text{and} \quad T_R = \frac{T}{T_c}$$

p_R, v_R and T_R are referred to as the *reduced properties*. If the experimental p_R-v_R-T data for all gases were found to lie on the same set of curves, we could conclude that *any two gases at the same p_R and T_R have the same value of v_R*. This statement is known as the *law of corresponding states*. It can be expressed by saying that there is a functional relation

$$v_R = \phi(p_R, T_R)$$

which holds for all gases. It can be shown as follows that, if the law of corresponding states were true, it would be possible to compile a single set of curves expressing the variation of Z/Z_o with p_R and T_R which would hold for all gases. Thus since

$$Z = \frac{pv}{RT} = \frac{p_c v_c}{RT_o} \cdot \frac{p_R v_R}{T_R} = Z_c \left(\frac{p_R v_R}{T_R} \right)$$

then if $v_R = \phi(p_R, T_R)$ with ϕ the same for all gases, Z/Z_c must be a unique function of p_R and T_R. Unfortunately such a set of curves could not yield the necessary result that $Z \to 1$ as $p_R \to 0$ for all gases, because it is an experimental fact that Z_c differ for different gases. This is simply a way of saying that the law of corresponding states is not true.

Although too inaccurate to be used in this way, a modified form of the law is sometimes useful. The modified form asserts that $Z = f(p_R, T_R)$ with the function f the same for all gases. The *generalised compressibility chart* of Fig. 8.21 expresses the function f. Note that the chart implies that $Z_c = 0.22$ for all gases which is not the case, but at least it expresses the fact that $Z \to 1$ when $p_R \to 0$. Furthermore its use only involves a knowledge of p_c and T_c, and not of v_c which in general is not known accurately. [Critical data can be found from any tables of physical constants—see Ref. (33) for example.] Values of Z taken from the chart may deviate by as much as 10 per cent from the experimental value for a particular gas, but the chart is useful when very little experimental data are available for a substance, and also for showing when the assumption of perfect behaviour ($Z = 1$) is no longer sufficiently accurate for a given purpose. Its use suffers from the disadvantage previously mentioned in relation to Fig. 8.20, that when v is one of the known parameters of state, successive approximation is necessary to determine the unknown property.

162

We have briefly indicated the method of dealing with the properties of real gases by referring to the *p-v-T* relation as an example. We cannot here go into the equivalent method for determining values of the other thermodynamic properties. This whole topic is of particular concern to chemical engineers, and a much fuller treatment can be found in textbooks on chemical engineering thermodynamics.*

At the beginning of section 8.5 we suggested that a gas can be treated as perfect when the temperature is more than twice the critical temperature, or the pressure is one atmosphere or less. Fig. 8.21 shows the latter criterion to be valid at all temperatures, since for most gases a pressure of one atmosphere corresponds to a reduced pressure of under 0·1. The first of these rough guides ($T_R > 2$) is apparently not valid

Fig. 8.21   Generalised compressibility chart

for all pressures, and ceases to be true for a reduced pressure greater than about 5. For most gases, however, this corresponds to a much higher pressure than is generally encountered in engineering applications, e.g. the corresponding pressure for nitrogen is 170 bar. For our purposes we shall always be in a position to assume that a gas has the characteristic equation $pv = RT$.

The second part of the definition of a perfect gas—that the specific heats are constant—is not nearly such a good approximation. The slight variation of the specific heats with pressure can usually be neglected; indeed, in view of the decision to use the perfect gas equation of state, the specific heats must be treated as pure functions of temperature if we wish to be consistent [see section 8.5 (ii)]. Fig. 8.19 shows the variation of c_v with temperature for nitrogen. The corresponding curve for c_p can easily be obtained by using the relation (8.21) between the specific heats and the gas constant. If the approximate procedure of using a mean specific heat is not sufficiently accurate, some form of equation can be used such as

$$c_p = a + bT + cT^2$$

*See Ref. (4), Chapter 4, and also Ref. (15).

where a, b and c are constants. The change of enthalpy, for example, can then be obtained analytically as

$$h_2 - h_1 = \int_{T_1}^{T_2} (a + bT + cT^2)\, dT$$

Although air is strictly speaking a mixture of gases, its composition is sufficiently invariable for it to be treated as a single gas. Tables of properties of air which take account of the variation of c_p are available, Refs. (12) and (9). The tables of Keenan and Kaye cover the range for which the p-v-T data can be accurately represented by the perfect gas equation of state, so that the tables of internal energy and enthalpy require only one independent parameter—namely temperature. The internal energy and enthalpy are arbitrarily put equal to zero for all states at absolute zero of temperature.

The entropy of a perfect gas, however, has been shown to be a function of two independent properties. To avoid complicating the table by introducing a second independent parameter, the entropy is not given directly for all states, but a new function ϕ is tabulated in its place. The differential form of equation (8.24) is

$$ds = \frac{c_p}{T}\, dT - \frac{R}{p}\, dp$$

When the reference state is chosen at absolute zero and some pressure p_0, the entropy at any state defined by (p_1, T_1) is given by

$$s_1 = \int_0^{T_1} \frac{c_p}{T}\, dT - R \int_{p_0}^{p_1} \frac{dp}{p}$$
$$= \int_0^{T_1} \frac{c_p}{T}\, dT - R \ln \frac{p_1}{p_0}$$

The change of entropy between any two states 1 and 2 is therefore given by

$$s_2 - s_1 = \int_0^{T_2} \frac{c_p}{T}\, dT - \int_0^{T_1} \frac{c_p}{T}\, dT - R \ln \frac{p_2}{p_1}$$

Finally, if we write

$$\phi = \int_0^{T} \frac{c_p}{T}\, dT$$

the change of entropy becomes

$$s_2 - s_1 = \phi_2 - \phi_1 - R \ln \frac{p_2}{p_1} \qquad (8.28)$$

The quantity ϕ is a function of temperature only and may therefore be tabulated in the same way as the internal energy and enthalpy. Equation (8.28) is used when the change of entropy between two states is required. We shall not make use of such tables in subsequent chapters; for our purpose it will be sufficiently accurate to use a mean specific heat.

164

9 Non-flow Processes

Non-flow processes have been classified and analysed in general terms in Chapter 3. The processes were characterised by (1) the volume, pressure or temperature remaining constant, (2) the heat transfer being zero, (3) the pressure and volume varying in such a way that pv^n is constant. Most non-flow processes of practical interest approximate to one, or a succession, of these processes.

In this chapter we shall apply the analysis to closed systems consisting of particular fluids. Each process will be dealt with assuming the fluid to be (a) a vapour and (b) a perfect gas. Tables of properties must be used in the first case, but in the second the analysis can be carried out using algebraic relations. Steam is used as an example of a vapour, and we shall make use of the abridged tables of Ref. (17). Air is considered as an example of a perfect gas, and the following values will be used for the specific heat at constant pressure and the gas constant:

$$c_p = 1 \cdot 005 \text{ kJ/kg K}$$

$$R = 0 \cdot 287 \text{ kJ/kg K}$$

It follows from equation (8.21) that

$$c_v = 0 \cdot 718 \text{ kJ/kg K}$$

For a summary of the relations between the properties of a perfect gas, which are valid for any process, whether reversible or irreversible, the reader is referred to the end of section 8.5 (ii).

9.1 Constant volume process

If there is no 'paddle work', the energy equation for any fluid undergoing a constant volume process is

$$Q = u_2 - u_1$$

When the fluid is a perfect gas this becomes

$$Q = c_v(T_2 - T_1)$$

For a perfect gas we also have the following relations between properties which are valid for any two states of the same volume. From the equation of state,

$$\frac{p_1}{T_1} = \frac{p_2}{T_2}$$

and from the definition of entropy,

$$s_2 - s_1 = \int_1^2 \left(\frac{dQ}{T}\right)_{\text{rev}} = \int_1^2 \frac{c_v}{T}\, dT = c_v \ln \frac{T_2}{T_1}$$

The latter equation can also be obtained by putting $v_2 = v_1$ in the general expression for the entropy change, equation (8.23).

Fig. 9.1a

Example 9.1 A fluid in a closed vessel of fixed volume, 0·14 m³, exerts a pressure of 10 bar at 250 °C. If the vessel is cooled so that the pressure falls to 3·5 bar, determine the final temperature, heat transferred, and change of entropy.

(a) *Steam* (Fig. 9.1a)
The first step is to find the mass of fluid in the system. Since t_1 is greater than the saturation temperature corresponding to p_1, the vapour is initially superheated. Using the superheat table we find $v_1 = 0\cdot2328$ m³/kg.
Since $V = 0\cdot14$ m³,

$$m = \frac{V}{v_1} = \frac{0\cdot14}{0\cdot2328} = 0\cdot6014 \text{ kg}$$

The vapour must be wet in its final state because mv_g at p_2 is $0\cdot6014 \times 0\cdot5241 = 0\cdot3152$ m³ which is greater than 0·14 m³. The temperature must therefore be the saturation temperature corresponding to p_2, i.e. $t_2 = 138\cdot9$ °C.
The dryness fraction can be found from equation (8.10), i.e.

$$x_2 = \frac{V}{mv_{g2}} = \frac{0\cdot14}{0\cdot3152} = 0\cdot444$$

[N.B. If the accurate equation (8.9) is used x_2 becomes 0·443.]

166

From the superheat table,

$u_1 = 2711 \text{ kJ/kg}$

From the saturation table,

$u_2 = u_{f2} + x_2 u_{fg2} = 584 + 0.444(2549 - 584) = 1457 \text{ kJ/kg}$

Heat transferred,

$Q = m(u_2 - u_1) = 0.6014(1457 - 2711) = -754 \text{ kJ}$

From the superheat table, $s_1 = 6.926 \text{ kJ/kg K}$, and from the saturation table,

$s_2 = s_{f2} + x_2 s_{fg2} = 1.727 + 0.444 \times 5.214 = 4.042 \text{ kJ/kg K}$

Change of entropy,

$S_2 - S_1 = 0.6014(4.042 - 6.926) = -1.734 \text{ kJ/K}$

b) *Air* (Fig. 9.1b)

Fig. 9.1b

Mass of gas,

$$m = \frac{p_1 V}{R T_1} = \frac{100 \times 10 \times 0.14}{0.287 \times 523} = 0.933 \text{ kg}$$

Final temperature,

$$T_2 = \frac{p_2}{p_1} T_1 = \frac{3.5}{10} 523 = 183 \text{ K}$$

Heat transferred,

$Q = m c_v (T_2 - T_1)$

$= 0.933 \times 0.718(183 - 523) = -228 \text{ kJ}$

Change of entropy,

$$S_2 - S_1 = m c_v \ln \frac{T_2}{T_1}$$

$$= 0.933 \times 0.718 \ln \frac{183}{523} = -0.703 \text{ kJ/K}$$

9.2 Constant pressure process

For a reversible constant pressure process undergone by any fluid

$$W = p \int_1^2 dv = p(v_2 - v_1)$$

$$Q = (u_2 - u_1) + p(v_2 - v_1) = h_2 - h_1$$

When the fluid is a perfect gas

$$W = R(T_2 - T_1)$$

$$Q = c_p(T_2 - T_1)$$

For a perfect gas we also have the following relations between properties which are valid for any two states of the same pressure:

$$\frac{v_1}{T_1} = \frac{v_2}{T_2}$$

$$s_2 - s_1 = \int_1^2 \left(\frac{dQ}{T}\right)_{rev} = \int_1^2 \frac{c_p}{T} dT = c_p \ln \frac{T_2}{T_1}$$

The latter equation can also be obtained by putting $p_2 = p_1$ in the general expression for entropy change, equation (8.24).

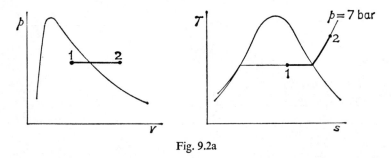

Fig. 9.2a

Example 9.2 0·2 kg of fluid, initially at a temperature of 165 °C, expands reversibly at a constant pressure of 7 bar until the volume is doubled. Find the final temperature, work done, and heat transferred in two cases: (a) when the fluid is steam with an initial dryness fraction of 0·7, (b) when the fluid is air.

(a) *Steam* (Fig. 9.2a)

Initial specific volume,

$$v_1 = x_1 v_{g1} = 0·7 \times 0·2728 = 0·191 \text{ m}^3/\text{kg}$$

Final specific volume,

$$v_2 = 2v_1 = 0·382 \text{ m}^3/\text{kg}$$

Since $v_2 > v_{g2}$ the vapour must be superheated in its final state. From the superheat table by interpolation, the final temperature is

$$t_2 = 300 + \frac{0\cdot382 - 0\cdot3714}{0\cdot4058 - 0\cdot3714}\, 50 = 315\,^{\circ}\text{C}$$

Work done,

$$W = mp(v_2 - v_1)$$
$$= 0\cdot2 \times 100 \times 7(0\cdot382 - 0\cdot191) = 26\cdot7 \text{ kJ}$$

From saturation table,

$$h_1 = h_{f1} + x_1 h_{fg1}$$
$$= 697 + 0\cdot7 \times 2067 = 2144 \text{ kJ/kg}$$

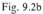

Fig. 9.2b

From superheat table, by interpolation

$$h_2 = 3060 + \frac{15}{50}\,(3164 - 3060) = 3091 \text{ kJ/kg}$$

Heat transferred,

$$Q = m(h_2 - h_1) = 0\cdot2(3091 - 2144) = 189 \text{ kJ}$$

b) *Air* (Fig. 9.2b)

Initial volume,

$$V_1 = \frac{mRT_1}{p_1} = \frac{0\cdot2 \times 0\cdot287 \times 438}{100 \times 7} = 0\cdot03592 \text{ m}^3$$

Final volume,

$$V_2 = 2V_1 = 0\cdot07184 \text{ m}^2$$

Final temperature,

$$T_2 = \frac{V_2}{V_1}\,T_1 = 876 \text{ K}$$

Work done,

$$W = p(V_2 - V_1) = 100 \times 7 \times 0\cdot03592 = 25\cdot1 \text{ kJ}$$

Heat transferred,

$$Q = mc_p(T_2 - T_1) = 0\cdot2 \times 1\cdot005 \times 438 = 88\cdot0 \text{ kJ}$$

169

9.3 Polytropic process

When a polytropic process is reversible, the work done is given by equation (3.12), i.e

$$W = \frac{(p_2 v_2 - p_1 v_1)}{1 - n} \tag{9.1}$$

Therefore, from the energy equation,

$$Q = (u_2 - u_1) + \frac{(p_2 v_2 - p_1 v_1)}{1 - n}$$

When the fluid is a perfect gas, these expressions can be written as

$$W = \frac{R}{1 - n}(T_2 - T_1)$$

$$Q = c_v(T_2 - T_1) + \frac{R}{1 - n}(T_2 - T_1)$$

$$= \left(c_v + \frac{R}{1 - n}\right)(T_2 - T_1)$$

Since $p_1 v_1{}^n = p_2 v_2{}^n$ and $pv = RT$, we also have the following relations between pro
perties which are valid whether the process is reversible or irreversible:

$$\frac{T_2}{T_1} = \left(\frac{p_2}{p_1}\right)^{(n-1)/n} \tag{9.2}$$

$$\frac{T_2}{T_1} = \left(\frac{v_2}{v_1}\right)^{1-n} \tag{9.3}$$

The change of entropy is best found from first principles as in the following example

Example 9.3 0·9 kg of fluid, initially at a pressure of 15 bar and a temperature of 250 °C
expand reversibly and polytropically to 1·5 bar. Find the final temperature, work done, hea
transferred, and change of entropy, if the index of expansion is 1·25.

(a) *Steam* (Fig. 9.3a)

From the superheat table,

$$v_1 = 0·152 \text{ m}^3/\text{kg}$$

Final specific volume,

$$v_2 = v_1 \left(\frac{p_1}{p_2}\right)^{1/n} = 0·1520 \times 10^{1/1·25} = 0·959 \text{ m}^3/\text{kg}$$

v_{g2} is 1·159 m³/kg so that the vapour is wet in the final state, and the temperature is th
saturation value corresponding to 1·5 bar, i.e. $t_2 = 111·4$ °C.

Dryness fraction,

$$x_2 = \frac{0 \cdot 959}{1 \cdot 159} = 0 \cdot 827$$

Work done,

$$W = \frac{m}{1 - n}(p_2 v_2 - p_1 v_1)$$

$$= \frac{100 \times 0 \cdot 9}{-0 \cdot 25}\{(1 \cdot 5 \times 0 \cdot 959) - (15 \times 0 \cdot 152)\} = 303 \text{ kJ}$$

From the superheat table,

$$u_1 = 2697 \text{ kJ/kg}$$

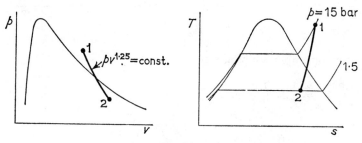

Fig. 9.3a

From the saturation table,

$$u_2 = u_{f2} + x_2 u_{fg2} = 467 + 0 \cdot 827(2519 - 467) = 2164 \text{ kJ/kg}$$

Heat transferred,

$$Q = m(u_2 - u_1) + W$$
$$= 0 \cdot 9(2164 - 2697) + 303 = -177 \text{ kJ} \quad \cdot$$
$$s_1 = 6 \cdot 711 \text{ kJ/kg K}$$
$$s_2 = s_{f2} + x_2 s_{fg2} = 1 \cdot 434 + 0 \cdot 827 \times 5 \cdot 789 = 6 \cdot 222 \text{ kJ/kg K}$$

Change of entropy,

$$S_2 - S_1 = 0 \cdot 9(6 \cdot 222 - 6 \cdot 711) = -0 \cdot 440 \text{ kJ/K}$$

(b) *Air* (Fig. 9.3b)

Final temperature,

$$T_2 = T_1 \left(\frac{p_2}{p_1}\right)^{(n-1)/n} = 523 \left(\frac{1}{10}\right)^{0 \cdot 25/1 \cdot 25} = 330 \text{ K}$$

Work done,

$$W = \frac{mR}{1 - n}(T_2 - T_1) = \frac{0 \cdot 9 \times 0 \cdot 287}{-0 \cdot 25}(330 - 523) = 199 \text{ kJ}$$

Heat transferred,

$$Q = mc_v(T_2 - T_1) + W$$

$$= 0.9 \times 0.718(330 - 523) + 199 = 74.3 \text{ kJ}$$

Fig. 9.3b

Change of entropy, from first principles,

$$ds = \left(\frac{dQ}{T}\right)_{\text{rev}} = \frac{1}{T}du + \frac{p}{T}dv$$

$$= \frac{c_v}{T}dT + \frac{R}{v}dv \text{ (for perfect gas)}$$

$$s_2 - s_1 = c_v \ln\frac{T_2}{T_1} + R \ln\frac{v_2}{v_1}$$

$$= c_v \ln\frac{T_2}{T_1} + \frac{R}{1-n}\ln\frac{T_2}{T_1} \text{ [using (9.3)]}$$

$$= \left(c_v + \frac{R}{1-n}\right)\ln\frac{T_2}{T_1}$$

$$S_2 - S_1 = 0.9\left(0.718 + \frac{0.287}{-0.25}\right)\ln\frac{330}{523} = 0.178 \text{ kJ/K}$$

9.4 Adiabatic process

For a fluid undergoing an adiabatic process the energy equation reduces to

$$W = -(u_2 - u_1)$$

When the fluid is a perfect gas

$$W = -c_v(T_2 - T_1)$$

For a perfect gas undergoing a reversible adiabatic (i.e. isentropic) process, we can deduce the following relations between properties:

$$0 = c_v \, dT + p \, dv \quad \text{(First Law)}$$

$$R \, dT = p \, dv + v \, dp \quad \text{(equation of state)}$$

Eliminating dT from these two equations, we obtain

$$0 = \left(1 + \frac{c_v}{R}\right) p \, dv + \frac{c_v}{R} v \, dp$$

$$= c_p p \, dv + c_v v \, dp \quad \text{(since } R = c_p - c_v\text{)}$$

Writing $c_p/c_v = \gamma$, this reduces to

$$\gamma \frac{dv}{v} + \frac{dp}{p} = 0$$

and on integration,

$$\gamma \ln v + \ln p = \text{constant}$$

$$pv^\gamma = \text{constant} \tag{9.4}$$

Since we also have $pv/T = \text{constant}$ we can eliminate v and p in turn to give

$$Tp^{(1-\gamma)/\gamma} = \text{constant} \quad \text{or} \quad \frac{T_2}{T_1} = \left(\frac{p_2}{p_1}\right)^{(\gamma-1)/\gamma} \tag{9.5}$$

$$Tv^{\gamma-1} = \text{constant} \quad \text{or} \quad \frac{T_2}{T_1} = \left(\frac{v_2}{v_1}\right)^{1-\gamma} \tag{9.6}$$

It is evident from (9.4) that, for a perfect gas, the reversible adiabatic process is a special case of the reversible polytropic process, and (9.5) and (9.6) can be obtained by putting $n = \gamma$ in (9.2) and (9.3). γ is called the *index of isentropic expansion* or *compression*. As in the case of the specific heats, for real gases γ varies with temperature and pressure (mainly with the former), but a mean value can be used for most purposes. If we put $n = \gamma$ in the expression for the work done in a polytropic process we have

$$W = \frac{R}{1 - \gamma} (T_2 - T_1)$$

which can easily be shown to be identical with the expression already derived—thus

$$\frac{R}{1 - \gamma} (T_2 - T_1) = \frac{c_p - c_v}{1 - c_p/c_v} (T_2 - T_1) = -c_v(T_2 - T_1)$$

In a reversible adiabatic process $s_2 = s_1$, and equation (9.4) could have been obtained by putting this result in the general expression (8.25). If this is done we have

$$c_p \ln \frac{v_2}{v_1} + c_v \ln \frac{p_2}{p_1} = 0$$

and hence

$$\frac{p_2}{p_1} = \left(\frac{v_1}{v_2}\right)^\gamma$$

Since the process is reversible, the foregoing relation applies to any pair of states between the end states, and consequently throughout the process

$$pv^\gamma = \text{constant}$$

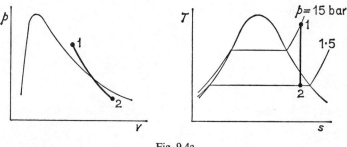

Fig. 9.4a

Example 9.4 Assuming the process in Example 9.3 is isentropic (instead of polytropic with index 1·25) find the final temperature and the work done.

(a) *Steam* (Fig. 9.4a)

To find the final state we make use of the fact that $s_2 = s_1$, and from the superheat table, $s_1 = 6\cdot711$ kJ/kg K. Since $s_1 = s_2 < s_{g2}$, the vapour must be wet in the final state, and therefore $t_2 = 111\cdot4\,°C$.

From

$$s_1 = s_2 = s_{f2} + x_2 s_{fg2}$$

we have

$$x_2 = \frac{6\cdot711 - 1\cdot434}{5\cdot789} = 0\cdot912$$

From the superheat table

$$u_1 = 2697 \text{ kJ/kg}$$

and from the saturation table

$$u_2 = u_{f2} + x_2 u_{fg2}$$
$$= 467 + 0\cdot912 \times 2052 = 2338 \text{ kJ/kg}$$

Work done,

$$W = -m(u_2 - u_1) = -0\cdot9(2338 - 2697) = 323 \text{ kJ}$$

(b) *Air* (Fig. 9.4b)

For air,

$$\gamma = \frac{1 \cdot 005}{0 \cdot 718} = 1 \cdot 40$$

The final temperature,

$$T_2 = T_1 \left(\frac{p_2}{p_1}\right)^{(\gamma-1)/\gamma} = 523 \left(\frac{1}{10}\right)^{1/3 \cdot 5} = 271 \text{ K}$$

Work done,

$$W = -mc_v(T_2 - T_1) = -0 \cdot 9 \times 0 \cdot 718(271 - 523) = 163 \text{ kJ}$$

Fig. 9.4b

For a vapour, there is no simple and unique relation between p and v which applies to every isentropic expansion and compression. Nevertheless it is sometimes convenient to assume an approximate relation of the polytropic form $pv^n = $ constant; n then becomes the index of isentropic expansion or compression although it is not necessarily equal to the ratio of the specific heats. Indeed, if part of the process occurs in the wet region, where $c_p = \infty$, this ratio has no meaning. The value of n can be found, as in the following example, by substituting the values of p and v at the end states in the expression

$$p_1v_1^n = p_2v_2^n$$

There is no unique value for n; it varies with the end states. As a rough approximation, however, the following average values of n may be used for steam:

$n = 1 \cdot 035 + 0 \cdot 1x$, when $0 \cdot 7 < x < 1 \cdot 0$ where x is the initial dryness fraction.*

$n = 1 \cdot 135$, when most of the process occurs in the wet region with the steam initially saturated or slightly superheated.

$n = 1 \cdot 30$, when the process occurs mostly in the superheat region.

*This is known as Zeuner's equation.

Example 9.5 Steam, initially dry saturated, expands isentropically from a pressure of 15 bar to 0·15 bar. Find the index of isentropic expansion.

$$s_2 = s_1 = s_{g1} = 6\text{·}445 \text{ kJ/kg K}$$

$$x_2 = \frac{s_2 - s_{f2}}{s_{fg2}} = \frac{6\text{·}445 - 0\text{·}755}{7\text{·}254} = 0\text{·}784$$

$$v_1 = v_{g1} = 0\text{·}1317 \text{ m}^3/\text{kg}$$

$$v_2 = x_2 v_{g2} = 0\text{·}784 \times 10\text{·}06 = 7\text{·}887 \text{ m}^3/\text{kg}$$

If we can write $p_1 v_1{}^n = p_2 v_2{}^n$, then

$$15 \times 0\text{·}1317^n = 0\text{·}15 \times 7\text{·}887^n$$

from which

$$n = 1\text{·}125$$

Fig. 9.5

This approximate method of treating the isentropic expansion or compression of a vapour as polytropic will be found useful in Chapter 18. Although it must be remembered that the isentropic process is only strictly polytropic when the fluid is a perfect gas with constant specific heats, when the vapour is *steam* and *superheated throughout the process*, the approximation is very close to actual behaviour as is indicated by the following. It has been found that over a wide field in the superheat region the *u-p-v* data are related by the simple equation

$$u - B = \frac{1}{n-1} p(v - C) \tag{9.7}$$

where the empirical constants n, B and C are

$$n = 1\text{·}3, \quad B = 1943 \text{ kJ/kg}, \quad C = -0\text{·}000\ 175 \text{ m}^3/\text{kg}$$

This equation was originally proposed by H. L. Callendar. By inspection of the superheat table it can be seen that, at pressures below 100 bar, C is always less than 1 per cent of v. For many engineering purposes C can be ignored and thus

$$u - B = \frac{1}{n-1} pv \tag{9.8}$$

From equation (9.8) it is possible to deduce rigorously* that the following set of relations must hold for an *isentropic* process:

$$pv^n = \text{constant}, \quad p/T^{n/(n-1)} = \text{constant},$$

$$Tv^{(n-1)} = \text{constant}, \quad pv/T = \text{constant} \tag{9.9}$$

n is the constant having the value 1·3, and in this context is the index of isentropic expansion or compression. The last equation in (9.9) does not imply that the perfect

*This is a useful exercise for students who have studied Chapter 7—the main equations required are (7.13), (7.14), (7.23) and (7.24).

gas equation is applicable, because it only applies to an isentropic process and the constant pv/T is different for different isentropic processes (i.e. it depends on the value of s).

Finally we may note that since $h = u + pv$, equation (9.8) can be written

$$h - B = \frac{n}{n-1} pv \tag{9.10}$$

It follows that for superheated steam any change in enthalpy is given very closely by

$$h_2 - h_1 = \frac{n}{n-1}(p_2 v_2 - p_1 v_1)$$

And for an *isentropic* process, for which we have seen that $p_2 v_2{}^n = p_1 v_1{}^n$, this becomes

$$h_2 - h_1 = \frac{n}{n-1} p_1 v_1 \left[\left(\frac{p_2}{p_1} \right)^{(n-1)/n} - 1 \right]$$

$$= (h_1 - B) \left[\left(\frac{p_2}{p_1} \right)^{(n-1)/n} - 1 \right] \tag{9.11}$$

[Relations (9.9) and (9.11) are particularly useful when dealing with supersaturated states, discussed in section 18.2, which are not covered by steam tables.]

9.5 Isothermal process

For any fluid undergoing a reversible isothermal process

$$Q = T(s_2 - s_1)$$

and hence from the energy equation,

$$W = T(s_2 - s_1) - (u_2 - u_1)$$

When the fluid is a perfect gas (for which u is a function only of temperature), $u_2 = u_1$ and it follows from the energy equation that

$$Q = W$$

Making use of the equation of state, we can find W from $\int p \, dv$ if the process is reversible. Thus for a reversible isothermal process undergone by a perfect gas,

$$Q = W = \int_1^2 p \, dv = RT \int_1^2 \frac{dv}{v} = RT \ln \frac{v_2}{v_1} \tag{9.12}$$

For a perfect gas we also have the following relations between properties which are valid for any two states of the same temperature:

$$p_1 v_1 = p_2 v_2$$

$$s_2 - s_1 = \frac{1}{T} \int_1^2 (dQ)_{\text{rev}} = R \ln \frac{v_2}{v_1} \quad \text{[from equation (9.12)]}$$

The latter equation can also be obtained by putting $T_2 = T_1$ in the general expression for the entropy change, equation (8.23). Since $p_1 v_1 = p_2 v_2$, equation (9.12) can be written alternatively as

$$Q = W = -RT \ln \frac{p_2}{p_1}$$

and hence also

$$s_2 - s_1 = -R \ln \frac{p_2}{p_1}$$

It is evident that for a perfect gas the reversible isothermal process is another special case of a reversible polytropic process, i.e. one for which $n = 1$. Note that we cannot put $n = 1$ in equation (9.1) to obtain the work done—the reason being that the integral of dv/v^n is $\ln v$ when $n = 1$.

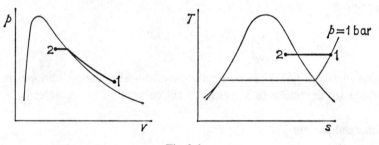

Fig. 9.6a

Example 9.6 A fluid, initially at 155·5 °C and 1 bar, is compressed reversibly and isothermally to a state where the specific volume is 0·28 m³/kg. Find the change of internal energy, change of entropy, heat transferred and work done, per kg of fluid.

(a) *Steam* (Fig. 9.6a)

From the superheat table, by interpolation,

$$u_1 = 2583 + \frac{5·5}{50} \, 76 = 2591 \text{ kJ/kg}$$

Since 0·28 m³/kg is less than the specific volume of saturated vapour at 155·5 °C, the vapour must be wet in its final state. Hence,

$$x_2 = \frac{0·28}{v_{g2}} = \frac{0·28}{0·3427} = 0·817$$

$$u_2 = u_{f2} + x_2 u_{fg2} = 655 + 0·817 \times 1910 = 2215 \text{ kJ/kg}$$

Change of internal energy,

$$u_2 - u_1 = 2215 - 2591 = -376 \text{ kJ/kg}$$

From the superheat table, by interpolation,

$$s_1 = 7\cdot614 + \frac{5\cdot5}{50}\,0\cdot220 = 7\cdot638 \text{ kJ/kg K}$$

Also,

$$s_2 = s_{f2} + x_2 s_{fg2} = 1\cdot897 + 0\cdot817 \times 4\cdot893 = 5\cdot895 \text{ kJ/kg K}$$

Change of entropy,

$$s_2 - s_1 = 5\cdot895 - 7\cdot638 = -1\cdot743 \text{ kJ/kg K}$$

Heat transferred,

$$Q = T(s_2 - s_1) = 428\cdot5(-1\cdot743) = -747 \text{ kJ/kg}$$

Work done,

$$W = Q - (u_2 - u_1) = -747 + 376 = -371 \text{ kJ/kg}$$

Fig. 9.6b

(b) *Air* (Fig. 9.6b)

Change of internal energy is zero.
 Initial specific volume,

$$v_1 = \frac{RT}{p_1} = \frac{0\cdot287 \times 428\cdot5}{100 \times 1\cdot0} = 1\cdot230 \text{ m}^3/\text{kg}$$

Change of entropy,

$$s_2 - s_1 = R \ln \frac{v_2}{v_1} = 0\cdot287 \ln \frac{0\cdot28}{1\cdot23} = -0\cdot425 \text{ kJ/kg K}$$

Heat transferred,

$$Q = T(s_2 - s_1) = 428\cdot5(-0\cdot425) = -182 \text{ kJ/kg}$$

Work done,

$$W = Q = -182 \text{ kJ/kg}$$

9.6 Irreversibility and the free expansion

We note in section 5.4 (i) that most practical non-flow processes can be assumed reversible. Nevertheless it is worth reviewing our position when the processes we have considered are irreversible.

(a) Constant volume, constant pressure, polytropic, and isothermal processes: when these are irreversible we must know either W or Q in addition to the end states before the process is completely determined, i.e. before only one unknown is left in the energy equation.

(b) Adiabatic process: when this is irreversible the work done can be predicted if the end states are known, but when it is reversible only *one* property in the final state need be known. The adiabatic process differs from the others in that only one property in the final state can be varied independently; the final state then depends upon the amount of irreversibility present.

To illustrate the principle of increasing entropy, we shall consider as a final example the Joule free expansion for a perfect gas [see section 8.5 (ii)]. In this case both Q and W are zero and the entropy of the gas must increase because the process is irreversible and adiabatic.

Fig. 9.7

Example 9.7 Air, initially at a pressure and temperature of 3·5 bar and 15 °C, is enclosed in a vessel 0·06 m³ in volume. The air is allowed to expand into an evacuated vessel also 0·06 m³ in volume. If the process is adiabatic, find the change in entropy of the air.

The mass of air in the system is given by

$$m = \frac{p_1 V_1}{R T_1} = \frac{100 \times 3\cdot5 \times 0\cdot06}{0\cdot287 \times 288} = 0\cdot254 \text{ kg}$$

From the energy equation, since $W = Q = 0$, $u_2 - u_1 = 0$. If air can be treated as a perfect gas, $T_2 - T_1 = 0$ also. To find the change of entropy, we can either use the general expression (8.23) or, if we wish to work directly from the equation defining entropy, $ds = (dQ/T)_{rev}$, we can adopt the following procedure. The irreversible process can be 'replaced' by any suitable reversible process which joins the two given end states defined by ($V_1 = 0.06$, $T_1 = 288$) and ($V_2 = 0.12$, $T_2 = 288$) and for which Q can be easily calculated. The obvious choice is a reversible isothermal process: then using equation (9.12) we have

$$s_2 - s_1 = \frac{1}{T} \int_1^2 (dQ)_{\text{rev}} = R \ln \frac{v_2}{v_1}$$

Since $\dfrac{v_2}{v_1} = \dfrac{V_2}{V_1} = 2$, we get finally

$$S_2 - S_1 = 0.254 \times 0.287 \ln 2 = 0.0505 \text{ kJ/K}$$

10   Flow Processes

In Chapter 4 it was shown how the heat and work crossing the boundary during a steady-flow process can be related to the properties of the fluid at the inlet and outlet of an open system. This relation, the steady-flow energy equation (4.2), was applied to some important open systems but without reference to any particular fluid. As in Chapter 9, we shall again consider the fluid to be either steam or air and illustrate the applications with numerical examples. As pointed out in section 5.4 (i), the irreversibility due to viscous friction cannot be neglected unless the velocity of flow is very small. In many cases it is necessary to calculate the unknown quantity assuming the process to be reversible, and multiply the result by a *process efficiency*. Various process efficiencies are introduced in the following examples. A second artifice for dealing with irreversibility in adiabatic steady-flow processes is also mentioned—this involves treating the process as a polytropic process.

Finally, this chapter ends with the consideration of multi-stream processes and nonsteady-flow processes.

10.1 Boiler and condenser

Since the work done in a boiler or condenser is zero, the steady-flow energy equation reduces to

$$Q = (h_2 - h_1) + \tfrac{1}{2}(C_2^2 - C_1^2)$$

The velocity of flow in the inlet and outlet pipes is small, and, as the following example will show, the change in kinetic energy can be neglected. The velocity in the boiler or condenser itself is even smaller, so that the effect of friction can be neglected and the process can be regarded as internally reversible. This implies that there is no pressure drop due to friction, and the pressure can be assumed constant throughout the system.

In the following example of a steam boiler plant, it is assumed that before the steam leaves the system it is passed through a separate bank of tubes, the *superheater*, situated in the stream of furnace gases. The superheater is not shown in Fig. 4.4.

Example 10.1   1500 kg of steam are to be produced per hour at a pressure of 30 bar with 100 K of superheat. The feed water is supplied to the boiler at a temperature of 40 °C. Find the rate at which heat must be supplied, assuming typical values for the velocity at inlet and outlet: 2 m/s in the feed pipe and 45 m/s in the steam main.

182

In section 8.3 (iii) we agreed to assume that the enthalpy of a compressed liquid is equal to the enthalpy of saturated liquid at the same temperature. Therefore, from the saturation table, $h_1 = 167\cdot5$ kJ/kg. The saturation temperature at 30 bar is 233·8 °C, and the outlet temperature is therefore 333·8 °C. From the superheat table, by interpolation,

$$h_2 = 2995 + \frac{33\cdot8}{50}\,122 = 3077 \text{ kJ/kg}$$

The heat required per kg of steam is

$$Q = (h_2 - h_1) + \tfrac{1}{2}(C_2{}^2 - C_1{}^2)$$

$$= (3077 - 167\cdot5)\left[\frac{\text{kJ}}{\text{kg}}\right] + \frac{1}{2}(45^2 - 2^2)\left[\frac{1}{\text{kg}} \times \frac{\text{kg m}^2}{s^2} \text{ or } \frac{\text{N m}}{\text{kg}}\right] \times \frac{1}{10^3}\left[\frac{\text{kJ}}{\text{N m}}\right]$$

$$= (2909 + 1\cdot01) \text{ kJ/kg} = 2910 \text{ kJ/kg}$$

Rate of heat transfer required

$$mQ = \frac{1500}{3600}\,2910 = 1213 \text{ kJ/s}$$

It is easy to see that the change of kinetic energy (1·01 kJ/kg) can be neglected.

10.2 Adiabatic steady-flow processes

For the reasons given in sections 4.2 (ii) and (iii), the processes in a nozzle, diffuser, turbine and rotary compressor can be assumed adiabatic. The energy equation for the nozzle or diffuser, where $W = 0$, becomes

$$\tfrac{1}{2}(C_2{}^2 - C_1{}^2) = (h_1 - h_2)$$

And for the turbine or rotary compressor, when the velocity at inlet is approximately the same as that at outlet, the energy equation reduces to

$$W = (h_1 - h_2)$$

The state of the fluid is usually known at the inlet of these systems but only one property, e.g. the pressure, can be arbitrarily fixed at the outlet. To fix the final thermodynamic state of the fluid, and hence determine h_2, it is necessary to assume the process to be reversible; the flow is then isentropic and we can use the fact that $s_2 = s_1$. This has been explained in section 6.6, and the subsequent examples should make the point clear.

When the fluid is a perfect gas, the foregoing special forms of the energy equation become:

nozzle and diffuser, $\quad \tfrac{1}{2}(C_2{}^2 - C_1{}^2) = c_p(T_1 - T_2)$

turbine and rotary compressor, $\quad W = c_p(T_1 - T_2)$

Also, from equation (8.24) we have, for a perfect gas,

$$s_2 - s_1 = c_p \ln \frac{T_2}{T_1} - R \ln \frac{p_2}{p_1}$$

This equation relates the difference in entropy between states 1 and 2 to the pressure and temperature of the gas at these states. The fact that the gas is flowing during any process has no effect on relations between its *thermodynamic* properties, and the equation can therefore be applied to the thermodynamic states of a perfect gas at the inlet and outlet of an open system. For any isentropic steady-flow process $(s_2 - s_1)$ is zero, and we shall arrive at the same relation between p and T that we deduced for the isentropic non-flow process, i.e. equation (9.5),

$$\frac{T_2}{T_1} = \left(\frac{p_2}{p_1}\right)^{(\gamma-1)/\gamma}$$

This, together with $p_1 v_1/T_1 = p_2 v_2/T_2$, yields

$$p_1 v_1^{\gamma} = p_2 v_2^{\gamma} \quad \text{and} \quad \frac{T_2}{T_1} = \left(\frac{v_2}{v_1}\right)^{1-\gamma}$$

Thus the relations between the thermodynamic properties of a perfect gas in its initial and final states are the same for both isentropic steady-flow and non-flow processes.

Since the velocity of flow is very high in a nozzle, diffuser, turbine or rotary compressor, the effect of friction cannot be neglected. Nevertheless the laws of thermodynamics do not enable quantitative predictions to be made for irreversible processes. One method of accounting for the effect of viscous friction is to calculate the unknown quantity assuming the process to be reversible and then to multiply the result by a *process efficiency* to obtain a more realistic estimate. The value of the process efficiency must be determined from tests carried out on a similar system. For nozzles it is usual to define the efficiency in terms of the outlet kinetic energy as

$$\text{nozzle efficiency,} \quad \eta_N = \frac{C_2^2}{(C_2')^2} \tag{10.1}$$

C_2 is the actual outlet velocity, while C_2' is the outlet velocity which would have been achieved had the final pressure been reached isentropically. A prime will always be used to denote the result obtained by assuming the process to be isentropic.

Diffusers are designed to obtain the maximum possible pressure rise at the expense of a given reduction in velocity, and the process efficiency is often defined by

$$\text{diffuser efficiency,} \quad \eta_D = \frac{p_2 - p_1}{p_2' - p_1} \tag{10.2}$$

Here p_2 is the actual outlet pressure, and p_2' is the pressure which would result from an isentropic process leading to the same outlet velocity.

Suitable process efficiencies for turbines and rotary compressors are respectively

$$\eta_T = \frac{W}{W'} \quad \text{and} \quad \eta_C = \frac{W'}{W} \tag{10.3}$$

When the fluid can be treated as a perfect gas, and the change in kinetic energy between inlet and outlet can be ignored, these become

$$\eta_T = \frac{T_1 - T_2}{T_1 - T_2'} \quad \text{and} \quad \eta_C = \frac{T_2' - T_1}{T_2 - T_1}$$

The way in which these efficiencies are used will be illustrated in the following examples. Apart from the fact that process efficiencies are always defined in such a way that they are numerically less than unity, their definitions are quite arbitrary—they should be regarded merely as empirical correction factors. Since all the foregoing process efficiencies are defined by comparing the actual process with an isentropic process, they may be called *isentropic efficiencies*. (Note, however, that the term 'adiabatic efficiency' is sometimes used.)

Fig. 10.1a

Example 10.2 A fluid expands from 3 bar to 1 bar in a nozzle. The initial velocity is 90 m/s, the initial temperature 150 °C, and from experiments on similar nozzles it has been found that the isentropic efficiency is likely to be 0·95. Estimate the final velocity.

(a) *Steam* (Fig. 10.1a)

From the superheat table,

$$h_1 = 2762 \text{ kJ/kg}, \quad s_2' = s_1 = 7 \cdot 078 \text{ kJ/kg K}$$

Since $s_2' < s_{g2}$, the vapour must be wet on leaving the nozzle after a reversible expansion. Hence

$$x_2' = \frac{s_2' - s_{f2}}{s_{fg2}} = \frac{7 \cdot 078 - 1 \cdot 303}{6 \cdot 056} = 0 \cdot 954$$

$$h_2' = h_{f2} + x_2' h_{fg2} = 417 + 0 \cdot 954 \times 2258 = 2580 \text{ kJ/kg}$$

185

From the energy equation

$$\frac{(C_2')^2 - 90^2}{2 \times 10^3} = (2762 - 2580)$$

$$(C_2')^2 = 0 \cdot 372 \times 10^6 \ \text{m}^2/\text{s}^2$$

$$C_2^2 = \eta_N (C_2')^2 = 0 \cdot 95 \times 0 \cdot 372 \times 10^6 \ \text{m}^2/\text{s}^2$$

$$C_2 = 595 \ \text{m/s}$$

(b) *Air* (Fig. 10.1b)

After a reversible expansion the final temperature would be

$$T_2' = T_1 \left(\frac{p_2}{p_1}\right)^{(\gamma-1)/\gamma}$$

Assuming $\gamma = 1 \cdot 40$ we have

$$T_2' = 423 \left(\frac{1}{3}\right)^{0 \cdot 4/1 \cdot 4} = 309 \ \text{K}$$

From the energy equation,

$$\frac{(C_2')^2 - 90^2}{2 \times 10^3} = 1 \cdot 005(423 - 309)$$

$$(C_2')^2 = 0 \cdot 229 \times 10^6 \ \text{m}^2/\text{s}^2$$

Fig. 10.1b

$$C_2^2 = 0 \cdot 95 \times 0 \cdot 229 \times 10^6 \ \text{m}^2/\text{s}^2$$

$$C_2 = 466 \ \text{m/s}$$

Example 10.3 A fluid enters a turbine at the rate of 14 kg/s with an initial pressure and temperature of 3 bar and 150 °C. If the final pressure is 1 bar and the isentropic efficiency of the turbine is 0·85, find the power developed and the change of entropy between inlet and outlet.

p_1, T_1 and p_2 are the same as in the previous example, and Figs. 10.1a and b, which show changes in the *thermodynamic* properties only, also apply to this case.

(a) *Steam*

Using the results of Example 10.2, we have

$$h_1 = 2762 \ \text{kJ/kg}, \quad x_2' = 0 \cdot 954, \quad h_2' = 2580 \ \text{kJ/kg}$$

From the energy equation,

$$W' = (h_1 - h_2') = 182 \ \text{kJ/kg}$$

$$W = \eta_T m W' = 0 \cdot 85 \times 14 \times 182 = 2170 \ \text{kJ/s or kW}$$

To find the change of entropy we must first determine the actual final state, i.e. state 2.

$$h_1 - h_2 = \eta_T(h_1 - h_2') = 0{\cdot}85 \times 182 = 154{\cdot}7 \text{ kJ/kg}$$
$$h_2 = 2762 - 154{\cdot}7 = 2607 \text{ kJ/kg}$$
$$x_2 = \frac{2607 - 417}{2258} = 0{\cdot}970$$
$$s_2 = 1{\cdot}303 + 0{\cdot}970 \times 6{\cdot}056 = 7{\cdot}177 \text{ kJ/kg K}$$

Since $s_1 = 7{\cdot}078$ kJ/kg K, the change of entropy is

$$s_2 - s_1 = 0{\cdot}099 \text{ kJ/kg K}$$
$$S_2 - S_1 = 14 \times 0{\cdot}099 = 1{\cdot}39 \text{ kJ/s K}$$

(b) *Air*

From Example 10.2 (b), $T_2' = 309$ K,
  Actual work done,

$$W = \eta_T m c_p (T_1 - T_2')$$
$$= 0{\cdot}85 \times 14 \times 1{\cdot}005(423 - 309) = 1363 \text{ kJ/s or kW}$$

The actual final temperature can be found from

$$T_1 - T_2 = \eta_T(T_1 - T_2')$$
$$T_2 = 423 - 0{\cdot}85 \times 114 = 326 \text{ K}$$

The values of p_1, T_1, p_2 and T_2 can be substituted in the appropriate general expression (8.24) for the change of entropy. Alternatively, we can replace the irreversible path 1–2, Fig. 10.1b, by a reversible path consisting of the isentropic process 1–2′ and the reversible constant pressure process 2′–2. We then have

$$s_2 - s_1 = s_2 - s_2' = \int_{2'}^{2} \left(\frac{dQ}{T}\right)_{\text{rev}} = c_p \ln \frac{T_2}{T_2'}$$
$$= 1{\cdot}005 \ln \frac{326}{309} = 0{\cdot}0538 \text{ kJ/kg K}$$
$$S_2 - S_1 = 14 \times 0{\cdot}0538 = 0{\cdot}753 \text{ kJ/s K}$$

It is worth mentioning one other method of accounting for irreversibility in an adiabatic flow process. The actual thermodynamic path of all these adiabatic processes may be regarded as being approximately polytropic in form, i.e. we may assume that

$$pv^n = \text{constant}$$

The polytropic expression can only have a meaning if p and v are regarded as average values at any point in the process because, although we normally assume that every elemental mass of fluid has the same properties at inlet and outlet of an open system, not all elements necessarily pass through exactly the same series of states between inlet and outlet.

The value of the index of an irreversible adiabatic expansion or compression is not equal to γ even when the fluid is a perfect gas; n is an empirical constant depending on the amount of friction. If n is known from previous experiments on a similar system, an unknown final state can then be predicted by making use of the equation

$$p_1 v_1^n = p_2 v_2^n$$

Or, when the fluid is a gas, we can also use

$$\frac{T_2}{T_1} = \left(\frac{p_2}{p_1}\right)^{(n-1)/n}$$

Once the final state has been determined, the energy equation, which applies both to reversible and irreversible processes, can then be used in the usual way.

To conclude this section, one further topic must be mentioned. In Chapter 11 we shall have occasion to find the work transferred when a *liquid* undergoes an isentropic steady-flow process. We have seen that when the change in kinetic energy can be neglected, the energy equation for this process reduces to

$$W = -(h_2 - h_1)$$

The enthalpy change can be found with sufficient accuracy, without the aid of a compressed liquid table, by making use of the relation between thermodynamic properties, equation (6.13), viz.

$$T \, ds = dh - v \, dp$$

In an isentropic process this reduces to

$$dh = v \, dp \tag{10.4}$$

When applied to liquids, which can be assumed incompressible at ordinary pressures, equation (10.4) can be integrated directly to give

$$h_2 - h_1 \simeq v(p_2 - p_1)$$

where v is the average specific volume of the liquid over the range of pressure p_1 to p_2. It follows that for an isentropic steady-flow expansion or compression of a liquid

$$W \simeq -v(p_2 - p_1) \tag{10.5}$$

The use of equation (10.5) will be illustrated in Chapter 11.

10.3 Throttling process

The throttling process is an adiabatic steady-flow process, but it was not included in the previous section because this was concerned with adiabatic processes which ideally should be isentropic. The throttling process is *essentially* irreversible—it is a means of reducing the pressure of a fluid by intentionally introducing friction into the flow.

In section 4.2 (iv) we noted that the throttling process was such that

$$h_1 = h_2$$

When the fluid is a perfect gas it follows that

$$T_1 = T_2$$

(The fact that a slight change of temperature is observed when the fluid is a real gas has been noted in section 7.4 (iv), and it indicates that the enthalpy of a real gas is not a function only of temperature.)

When a gas or vapour is throttled, the specific volume downstream of the restriction is greater than it is upsteam, i.e. $v_2 > v_1$. If the restriction is in a pipe of uniform diameter, the continuity equation (4.3) implies that $C_2 > C_1$. Nevertheless, the following example shows that for ordinary flow velocities the kinetic energy terms in the energy equation are negligible and for practical purposes $h_2 = h_1$.

Example 10.4   Air flows at the rate of 2·3 kg/s in a 15 cm diameter pipe. It has a pressure of 7 bar and a temperature of 95 °C before it is throttled by a valve to 3·5 bar. Find the velocity of the air downstream of the restriction, and show that the enthalpy is essentially the same before and after the throttling process. Also find the change of entropy.

Fig. 10.2 illustrates the process. The temperature actually falls as the fluid velocity increases through the restriction, and then rises because most of this increase in kinetic energy is dissipated by friction. The initial specific volume is

$$v_1 = \frac{RT_1}{p_1} = \frac{0\cdot287 \times 368}{100 \times 7} = 0\cdot151 \text{ m}^3/\text{kg}$$

and the initial velocity is

$$C_1 = \frac{mv_1}{A} = \frac{2\cdot3 \times 0\cdot151 \times 4}{\pi(0\cdot15)^2} = 19\cdot65 \text{ m/s}$$

Similarly, the final velocity is

$$C_2 = \frac{m}{A}\frac{RT_2}{p_2} = \frac{2\cdot3 \times 4}{\pi(0\cdot15)^2}\frac{0\cdot287 T_2}{100 \times 3\cdot5} = 0\cdot1067 T_2 \text{ m/s}$$

Fig. 10.2

Since Q and W are zero, the energy equation reduces to

$$h_1 + \tfrac{1}{2}C_1^2 = h_2 + \tfrac{1}{2}C_2^2$$

For a gas this becomes

$$c_p T_2 + \frac{(0\cdot1067 T_2)^2}{2 \times 10^3} = c_p 368 + \frac{19\cdot65^2}{2 \times 10^3}$$

$$1\cdot005 T_2 + 5\cdot69 \times 10^{-6} T_2^2 = 1\cdot005 \times 368 + 0\cdot193$$

Without further calculation it is evident that the kinetic energy terms can be ignored and T_2 is effectively equal to T_1. Hence the final velocity is

$$C_2 = 0.1067 \times 368 = 39.3 \text{ m/s}$$

Since the process is adiabatic and irreversible, the fluid must undergo an increase of entropy as it flows through the system. The initial and final thermodynamic states could be joined by a reversible, non-flow, isothermal process, and hence,

$$s_2 - s_1 = \int_1^2 \left(\frac{dQ}{T}\right)_{\text{rev}} = -R \ln \frac{p_2}{p_1}$$

$$= -0.287 \ln \frac{3.5}{7} = 0.199 \text{ kJ/kg K}$$

$$S_2 - S_1 = 2.3 \times 0.199 = 0.458 \text{ kJ/s K}$$

Before leaving the throttling process, we may note how use is made of it in the *throttling calorimeter* for the determination of the dryness fraction of a wet vapour. Normally the state of a fluid is most conveniently determined by measuring the intensive properties p and T, but for a wet vapour these are not independent properties. The throttling calorimeter, illustrated in Fig. 10.3, enables a wet state 1 to be determined by measuring p and T alone. A steady stream of wet vapour is sampled from the main flow and passed through a small orifice in a well lagged chamber. The vapour pressure, initially p_1, falls to some value p_2 as a result of the throttling process. Since $h_2 = h_1$, it is evident from the h-s diagram (Fig. 8.14) that if the pressure drop is sufficient and the vapour not too wet, the sampled vapour will be superheated after the expansion. The thermodynamic state can therefore be determined by measuring the independent properties p_2 and T_2. The enthalpy h_2 can then be found from the superheat table. If the initial pressure p_1 is also measured, and the corresponding values of h_f and h_{fg} are taken from the saturation table, the dryness fraction is found from

Fig. 10.3   Throttling calorimeter

$$x_1 = \frac{h_1 - h_{f1}}{h_{fg1}} = \frac{h_2 - h_{f1}}{h_{fg1}}$$

The process must of course be continued long enough for the calorimeter to reach a steady temperature before readings are taken. The chief difficulty is in ensuring that the sample is representative of the main flow—unfortunately the liquid accumulates on the wall of the pipe and the sample tends to be drier than the main stream as a whole.

0.4 Isothermal steady-flow process

t was pointed out in section 4.2 that a reciprocating compressor (or expander) can
»e regarded as an open system undergoing a steady-flow process, provided that a
·eceiver is included between the cylinder and measuring section to damp out the
»ulsations. The fluid flows through a reciprocating machine at a comparatively slow
·ate and there is time for an appreciable amount of heat to be exchanged with the
·urroundings. It is possible to imagine a limiting case in which the temperature of the
luid remains constant throughout its passage through the system. This is the case
onsidered here.

Since we have postulated low velocities, the energy equation reduces to

$$Q - W = (h_2 - h_1)$$

When the fluid is a perfect gas $(h_2 - h_1)$ is zero, because the enthalpy is a function
»nly of temperature. For the isothermal expansion or compression of a perfect gas
·e therefore have

$$Q = W$$

f the process is assumed reversible, these equations can be evaluated to find the work
lone when the states of the fluid are known at inlet and outlet. Thus, when the fluid
·s a vapour, $(h_2 - h_1)$ and $(s_2 - s_1)$ can be found from tables of properties and, since
he temperature T is constant, Q can be calculated from $T(s_2 - s_1)$. Hence W can be
ound from the energy equation. When the fluid is a perfect gas a simple expression
·or W can be deduced in terms of the temperature T and the inlet and outlet pressures
·_1 and p_2. Thus putting $T = $ constant in (8.24),

$$s_2 - s_1 = -R \ln \frac{p_2}{p_1}$$

·nd hence

$$W = Q = -RT \ln \frac{p_2}{p_1} \tag{10.6}$$

Equation (10.6) is in fact the same expression that was obtained for the work done
·when a perfect gas undergoes a reversible *non-flow* isothermal process (section 9.5).
The same equality of work does not hold when the fluid is a vapour because in a non-
low process

$$Q - W = (u_2 - u_1)$$

Only when the fluid is a perfect gas are the flow and non-flow isothermal works the
·ame, because only then does $(u_2 - u_1) = (h_2 - h_1)$, both being zero.

Owing to the comparatively low fluid velocities, the effect of fluid friction in a
·eciprocating machine is small. Hence the question of introducing a process efficiency
·o take account of irreversibility in this type of process has not arisen in practice.

There remains the question of how to predict the work done when the heat exchange is not sufficient to maintain the temperature constant, but consideration of this wi be left to Chapter 16.

Example 10.5 Recalculate the work and heat transfers for Example 9.6 assuming that tl isothermal compression occurs in a steady-flow process.

The thermodynamic states before, during, and after compression are identical with thos of the non-flow process of Example 9.6. The processes for steam and air are therefore de picted respectively by Figs. 9.6a and b, and some of the required property values can be take from that example.

(a) *Steam*

The heat transferred is the same as for the non-flow process, i.e. $Q = -747$ kJ/kg. From th energy equation the work done is

$$W = Q - (h_2 - h_1)$$

From the superheat table, by interpolation

$$h_1 = 2777 + \frac{5 \cdot 5}{50} \, 99 = 2788 \text{ kJ/kg}$$

x_2 has been found to be 0·817, and thus

$$h_2 = h_f + x_2 h_{fg}$$
$$= 417 + 0 \cdot 817 \times 2258 = 2262 \text{ kJ/kg}$$

Hence

$$W = -747 - (2262 - 2783) = -226 \text{ kJ/kg}$$

(b) *Air*

The heat transferred is the same as for the non-flow process, i.e. $Q = -182$ kJ/kg. Also since $(h_2 - h_1) = (u_2 - u_1)$, both being zero, the work done equals the heat transferred as fo the non-flow process.

10.5 Multi-stream steady-flow processes

Reference to section 4.1 will show that, provided the streams do not react chemically with one another, there is nothing inherent in the derivation of the steady-flow energy equation (4.2) to prevent its application to an open system in which more than one stream of fluid enters and leaves the system. The enthalpy, kinetic energy and potential energy terms associated with each stream can be summed at the inlet and outlet respectively, to give the following equation:

$$Q - W = \sum_{\text{out}} m(h + \tfrac{1}{2}C^2 + gz) - \sum_{\text{in}} m(h + \tfrac{1}{2}C^2 + gz)$$

Also, for conservation of mass we have the equation

$$\sum_{\text{out}} m = \sum_{\text{in}} m$$

One simple example of a multi-stream system is given here, but other more important cases can be found in Chapter 14.

Example 10.6 Steam is to be condensed by direct injection of cold water. The steam enters the condenser at a rate of 450 kg/h with a dryness fraction 0·9 and a pressure 1 atm. The estimated heat loss from the condenser to the surroundings is 8500 kJ/h. If the cold water enters with a temperature of 15 °C, and the mixture of condensate and cooling water is to leave at 95 °C, determine the rate of flow of cooling water required.

Fig. 10.4 illustrates the system. The energy equation, neglecting the kinetic and potential energy terms, becomes

$$Q = m_3 h_3 - (m_1 h_1 + m_2 h_2)$$

and for conservation of mass,

$$m_3 = m_1 + m_2$$

From the saturation table,

$$h_1 = h_{f1} + x_1 h_{fg1} = 419{\cdot}1 + 0{\cdot}9 \times 2256{\cdot}7 = 2450 \text{ kJ/kg}$$
$$h_2 = 62{\cdot}9 \text{ kJ/kg}, \quad h_3 = 398{\cdot}0 \text{ kJ/kg}$$

Therefore

$$-8500 = (450 + m_2)\,398 - (450 \times 2450 + m_2\,62{\cdot}9)$$
$$m_2 = 2730 \text{ kg/h}$$

Fig. 10.4

10.6 Nonsteady-flow processes

The only nonsteady-flow processes we shall consider are those undergone by the fluid in an open system as it changes from one equilibrium state to another. The appropriate energy equation has been derived in section 4.3, viz. equation (4.9). If we neglect the potential energy terms, as we have done with the steady-flow processes, the equation becomes

$$Q - W = (m''u'' - m'u') + \Sigma \delta m_2(h_2 + \tfrac{1}{2}C_2{}^2) - \Sigma \delta m_1(h_1 + \tfrac{1}{2}C_1{}^2)$$

Q and W refer to the quantities of heat and work crossing the boundary of the open system during the time it takes for the system to change from one equilibrium state to another. The term $(m''u'' - m'u')$ refers to the change of internal energy of the system between these two equilibrium states, and the remaining two terms refer to the summation of the quantities of enthalpy and kinetic energy leaving and entering the system with each element of fluid. We shall now apply this equation to two particular cases.

(i) *Filling a reservoir from an infinite source of fluid*

Fig. 10.5 illustrates the system. An air bottle of volume V contains air at pressure p' and temperature T'. It is to be filled from a compressed air line maintained at a constant pressure p_1 and temperature T_1. We require to find the mass and temperature of the air in the bottle after the valve has been opened to allow the pressure in the bottle to rise to the supply pressure p_1.

We may assume that when the valve is opened the air enters the bottle so rapidly that the process is adiabatic. If we draw the boundary in such a way that the entrance to the open system is in the main supply line, the thermodynamic properties of the fluid at station 1 are constant throughout the process, and C_1 is negligible. The velocity will be high in the locality of the valve—but local transformations of energy within the open system do not concern us. Also, no fluid leaves the system, and W is zero. The energy equation therefore reduces to

$$(m''u'' - m'u') = h_1 \Sigma \delta m_1$$

Since mass is conserved

$$(m'' - m') = \Sigma \delta m_1$$

and hence

$$(m''u'' - m'u') = h_1(m'' - m') \qquad (10.7)$$

Fig. 10.5

In our example the fluid is air, which can be treated as a perfect gas. Consequently (10.7) can be written as

$$c_v(m''T'' - m'T') = c_p T_1(m'' - m')$$

Moreover, we also have

$$\frac{m''}{m'} = \frac{p''V/RT''}{p'V/RT'} = \frac{p''/T''}{p'/T'} \quad \text{and} \quad p'' = p_1$$

Substituting these in the energy equation, we get

$$c_v \left(\frac{p_1/T''}{p'/T'} T'' - T' \right) = c_p T_1 \left(\frac{p_1/T''}{p'/T'} - 1 \right)$$

Since p', T', p_1 and T_1 are given, this equation can be used to find T''. Having found T'' we can find m'' from

$$m'' = \frac{p_1 V}{RT''}$$

Note that in the foregoing treatment we have assumed the internal energy to be zero at absolute zero. The actual reference state, however, is unimportant. Equation (10.7) can be written as

$$m''(u''_r + u_0) - m'(u'_r + u_0) = (u_{r1} + u_0 + p_1v_1)(m'' - m')$$

and hence

$$(m''u''_r - m'u'_r) + u_0(m'' - m') = h_{r1}(m'' - m') + u_0(m'' - m')$$

Here the suffix r is used to denote specific internal energy relative to the specific internal energy u_0 at some reference state 0. Since the terms containing u_0 vanish, the reference state is obviously arbitrary.

The following numerical example illustrates the treatment of the problem when the fluid is not a perfect gas.

Example 10.7   A vessel of volume 0·085 m³ contains steam at 1 bar and 0·8 dryness fraction. The vessel is momentarily connected to a steam main in which the steam is at 20 bar and 260 °C. If the pressure in the vessel is allowed to rise to the main pressure, estimate the final state and mass of the steam in the vessel.

Making the appropriate assumptions, the energy equation (10.7) can be used, i.e.

$$(m''u'' - m'u') = h_1(m'' - m')$$

From the superheat table, by interpolation,

$$h_1 = 2904 + \left(\frac{10}{50} \times 121\right) = 2928 \text{ kJ/kg}$$

From the saturation table,

$$u' = 417 + 0·8 \times 2089 = 2088 \text{ kJ/kg}$$

$$m' = \frac{V}{x'v'_g} = \frac{0·085}{0·8 \times 1·694} = 0·0627 \text{ kg}$$

$$m'' = \frac{V}{v''} = \frac{0·085[m^3]}{v''}$$

Substituting these values in the energy equation we get

$$\left(\frac{0·085[m^3]}{v''}u'' - 0·0627[kg] \times 2088\left[\frac{kJ}{kg}\right]\right) = 2928\left[\frac{kJ}{kg}\right]\left(\frac{0·085[m^3]}{v''} - 0·0627[kg]\right)$$

or

$$\frac{u''/v''}{[kJ/m^3]} = \frac{2928}{v''/[m^3/kg]} - 620$$

195

The final pressure p'' is 20 bar, but no exact algebraic relation between u'', v'' and p'' is known. A graphical method can be used to solve the equation as follows:

20 bar and °C	300	350	400
$v''/[\mathrm{m^3/kg}]$	0·1255	0·1386	0·1511
$u''/[\mathrm{kJ/kg}]$	2 744	2 861	2 946
$(u''/v'')/[\mathrm{kJ/m^3}]$	21 900	20 600	19 500
$\{2928[\mathrm{m^3/kg}]/v''-620\}$	22 700	20 500	18 800

By plotting the last two lines of figures against temperature as in Fig. 10.6, it is found that the steam has a temperature of 343 °C in the final state. That the final temperature in the vessel is higher than the temperature in the steam main is due to the fact that the steam is recompressed adiabatically in the vessel after being throttled.

The final mass of steam in the vessel can now be found. At 20 bar and 343 °C, by interpolation from the superheat table,

$$v'' = 0.1255 + \frac{43}{50}\,0.0131 = 0.1368 \text{ m}^3/\text{kg}$$

Hence

$$m'' = \frac{V}{v''} = \frac{0.085}{0.1368} = 0.621 \text{ kg}$$

Fig. 10.6

[N.B. If the approximate relation between p, v and u given by equation (9.7) is used, $v'' = 0.135 \text{ m}^3/\text{kg}$ and $m'' = 0.63 \text{ kg}$.]

(ii) *Maximum work attainable by allowing a reservoir of fluid to discharge via a turbine*

This example has already been introduced in general terms in section 4.3, and Fig 4.10 illustrates the system. By making suitable assumptions, the energy equation was shown to reduce to

$$W = (m'u' - m''u'') - \Sigma\delta m_2 h_2$$

The maximum work will be obtained when the expansion through the turbine is reversible. Since we are assuming the process to be adiabatic, the additional assumption of reversibility implies that the process is isentropic. In the following numerical example we shall consider the fluid to be air.

196

Example 10.8 A tank of volume 0.3 m^3 is filled with air at a pressure and temperature of 35 bar and 40 °C. The air is allowed to discharge through a turbine into the atmosphere, until the pressure in the tank has fallen to the atmospheric level of 1 bar. Find the maximum amount of work that could be delivered by the turbine.

The air in the tank progressively decreases in pressure and temperature from p' and T' to p'' and T''. But since the pressure p_2 at the outlet is constant, i.e. atmospheric, and the expansion is isentropic, the temperature T_2 at the outlet must also remain constant during the process. This can be made clear by the following argument. Consider the mass of air which remains in the tank at the end of the expansion and treat it as a closed system. This mass, originally occupying only a fraction of the volume of the tank, expands isentropically during the process. If at any instant the pressure and temperature is p and T, then we have

$$\frac{T}{T'} = \left(\frac{p}{p'}\right)^{(\gamma-1)/\gamma}$$

But p and T must also represent the pressure and temperature of the air as it enters the turbine at this particular instant, and since the expansion through the turbine is also isentropic,

$$\frac{T_2}{T} = \left(\frac{p_2}{p}\right)^{(\gamma-1)/\gamma}$$

Multiplying this pair of equations we get

$$\frac{T_2}{T'} = \left(\frac{p_2}{p'}\right)^{(\gamma-1)/\gamma}$$

T' and p' are the given initial conditions in the tank, and p_2 is constant, i.e. atmospheric pressure. Therefore T_2 must be constant also. Moreover, since $p'' = p_2$, the final temperature of the air in the tank must be equal to T_2.

From the conservation of mass, and the fact that T_2 is constant, the energy equation becomes

$$W = c_v(m'T' - m''T'') - c_p T_2(m' - m'')$$

where

$$T'' = T_2 = T'\left(\frac{p_2}{p'}\right)^{(\gamma-1)/\gamma} = 313\left(\frac{1}{35}\right)^{0.4/1.4} = 113 \text{ K}$$

$$m' = \frac{p'V}{RT'} = \frac{100 \times 35 \times 0.3}{0.287 \times 313} = 11.68 \text{ kg}$$

$$m'' = \frac{p''V}{RT''} = \frac{100 \times 1 \times 0.3}{0.287 \times 113} = 0.925 \text{ kg}$$

Hence

$$W = 0.718(11.68 \times 313 - 0.925 \times 113) - 1.005 \times 113(11.68 - 0.925)$$

$$= 1330 \text{ kJ}$$

11   Vapour Power Cycles

One common method of producing mechanical power employs the transfer of heat from a reservoir to a working fluid which is taken through a thermodynamic cycle The cycles considered in this chapter have two characteristics in common: (a) the working fluid is a condensable vapour which is in the liquid phase during part of the cycle, and (b) the cycle consists of a succession of steady-flow processes, with each process carried out in a separate component specially designed for the purpose. Each component constitutes an open system, and all the components are connected in series so that as the fluid circulates through the power plant each fluid element passes through a cycle of mechanical and thermodynamic states (see section 4.4).

To simplify the analysis, we shall assume that the change in kinetic and potential energy of the fluid between entry and exit of each component is negligible compared with the change in enthalpy. This implies that for the present purpose the energy equation can be written as

$$Q - W = h_2 - h_1$$

We shall also assume, in the main, that the processes are reversible. Cycles composed of reversible processes are called *ideal cycles*.

After discussing suitable criteria for comparing their effectiveness, various ideal cycles are analysed in turn. The Carnot cycle is considered first, and the Rankine cycle is then presented as a logical modification of this to meet certain practical considerations. Finally, other modifications used in modern power plant are described.

Although the working fluid usually employed is steam, because it is cheap and chemically stable, the principles presented here apply equally to any condensable vapour. The chapter ends with a discussion of the characteristics required of an ideal working fluid.

11.1 Criteria for the comparison of cycles

The choice of power plant for a given purpose is determined largely by considerations of operating cost and capital cost. The former is primarily a function of the overall efficiency of the plant, while the latter depends mainly on its size and complexity. In general the efficiency can always be improved by adding to the complexity of the plant, so that a suitable compromise between low operating and capital costs must be reached.

Consider the efficiency of the plant first. Fig. 11.1 illustrates the essential features of the type of power plant considered here. Until the 1950s the high temperature reservoir had always been a stream of hot gases produced by the continuous combustion of a hydocarbon fuel in air. Now an increasing number of nuclear reactors are being brought into operation, and also solar heat sources are being used in suitable locations [see Ref. (28)]. These new sources of energy do not affect the main discussion here, which is concerned with factors which govern the performance of the vapour cycles themselves. Whatever the source of energy, the *overall thermal efficiency* of a vapour cycle plant is suitably measured by the proportion of latent energy in the fuel which is converted into useful mechanical work. The overall efficiency can be expressed as the product of two efficiencies: (a) the combustion (or nuclear reactor) efficiency which expresses the proportion of latent energy in the fuel transferred as heat to the working fluid, and (b) the cycle efficiency which expresses the proportion of this heat which is converted into mechanical work. In this chapter we are concerned with the comparison of various cycles and, except for considerations raised in sections 11.6 and 11.9, the 'source' efficiency may be assumed to be independent of the type of cycle used. Combustion efficiency itself is discussed in detail in section 15.6.

Fig. 11.1   The essential elements of a vapour power plant

Let us now recapitulate briefly the reasons why the cycle efficiency is not unity.

(a) *Thermodynamic*: The Second Law expresses the fact that even in the best power cycle some heat must be rejected. This best form of cycle was shown in section 6.1 to be one in which: (1) all the heat supplied is transferred while the working fluid is at a constant upper temperature T_a, and all the heat rejected leaves while the working fluid is at a constant lower temperature T_b, and (2) all the processes are reversible. The efficiency of such a cycle is $(T_a - T_b)/T_a$, irrespective of the working fluid. If the cycle, even though reversible, is such that the temperature of the working fluid changes during a heat transfer process, the cycle efficiency will be less than $(T_a - T_b)/T_a$, where T_a and T_b now refer to the maximum and minimum temperatures reached by the working fluid (section 6.3).

(b) *Practical*: All real processes are irreversible, and irreversibilities reduce the cycle efficiency (section 5.5).

It is a simple matter to calculate the efficiency of a cycle when all the processes are assumed to be reversible. The result is known as the *ideal cycle efficiency*. By introducing process efficiencies as defined in Chapter 10 it is also possible to estimate the actual cycle efficiency. The ratio of the actual cycle efficiency to the ideal cycle efficiency is called the *efficiency ratio*.

Some cycles are more sensitive to irreversibilities than others. That is, two cycles may have the same ideal cycle efficiencies, but after allowing for the same process efficiencies we find that their actual cycle efficiencies are markedly different. A high ideal cycle efficiency is not, therefore, by itself, a good indication of whether or not the cycle will provide a power plant of high overall efficiency. An additional criterion which indicates the cycle's sensitivity to irreversibilities is required. Such a criterion is furnished by the *work ratio* r_w. Any power cycle consists of processes of both positive and negative work transfer, and r_w is defined as the ratio of the net work to the positive work done in the cycle. Irreversibilities decrease the positive work and increase the negative work. If the ideal negative work is only slightly less than the ideal positive work, i.e. if r_w is only slightly greater than zero, quite a small amount of component inefficiency is sufficient to reduce the net work output to zero and so reduce the actual

	Ideal		With $\eta_T = \eta_C = 0.9$	
	Cycle 1	Cycle 2	Cycle 1	Cycle 2
+ve Q	120	120	120	120
+ve W	100	40	90	36
−ve W	61	1	67·8	1·1
net W	39	39	22·2	34·9
r_w	0·390	0·975	0·247	0·969
η	0·325	0·325	0·185	0·291

Fig. 11.2 The effect of work ratio

cycle efficiency to zero. On the other hand, as the work ratio approaches unity, the same amount of component inefficiency will have a much smaller effect on the net work output and hence on the actual cycle efficiency. This point is illustrated in Fig 11.2, by taking two ideal cycles involving positive work from a turbine and negative work from a compressor; the cycles have the same heat input and efficiency, thus producing the same net work, but with radically different work ratios. It is seen that when process efficiencies for the turbines and compressors are taken into account [see equation (10.3)], the efficiency of the cycle having a low work ratio drops much more than that of the cycle having a high work ratio.

Summarising, we may say that *a high ideal cycle efficiency together with a high work ratio provides a reliable indication that the real power plant will have a good overall efficiency.*

The next consideration is some criterion which will indicate the relative size of plant for a given power output; the complexity of the plant will be obvious from inspection of the ideal cycle. Here again the work ratio sometimes serves as a useful criterion. The component producing the positive work has to provide the power for the component requiring the negative work, and therefore a work ratio approaching unity indicates that these components will be of least possible size for a given net power output. Most practical vapour cycles, however, have work ratios very near unity: the work ratio is then not very informative and a more direct criterion of size is necessary.

In general, the size of the components will depend on the amount of working fluid which has to be passed through them. A more direct indication of relative sizes of steam plant is therefore provided by the *specific steam consumption* (s.s.c.), i.e. the mass flow of steam required per unit power output.[*] It is usually expressed in kg/kW h: if the numerical value of the net work output per unit mass flow is $W/[kJ/kg] = W/[kW\,s/kg]$ the s.s.c. can be found from

$$\text{s.s.c.} = \frac{1}{W} = \left(\frac{1}{W/[kJ/kg]}\right)\left[\frac{kg}{kW\,s}\right] \times 3600\left[\frac{s}{h}\right] = \left(\frac{3600}{W/[kJ/kg]}\right)\frac{kg}{kW\,h}$$

11.2 Carnot cycle

One ideal cycle in which the heat is taken in at a constant upper temperature (T_a) and rejected at a constant lower temperature (T_b) is that suggested by Sadi Carnot, Ref. (2). It consists of two reversible isothermal processes at T_a and T_b respectively, connected by two reversible adiabatic (isentropic) processes. When the working fluid is a condensable vapour, the two isothermal processes are easily obtained by heating and

Fig. 11.3   The Carnot cycle

cooling at constant pressure while the fluid is a wet vapour. The cycle is represented on the *T-s* diagram for steam in Fig. 11.3, and a diagrammatic sketch of the plant accompanies it.

Saturated water in state 1 is evaporated in a boiler at constant pressure to form saturated steam in state 2, the heat added being

$$Q_{12} = h_2 - h_1$$

[*]If the effect of different working fluids is being investigated, account must be taken of their different densities because it is strictly the volume flow which is relevant.

The steam is expanded isentropically to state 3 while doing work in a turbine or reciprocating engine.* The work done is

$$W_{23} = -(h_3 - h_2)$$

After expansion the steam is then partially condensed at constant pressure while heat

$$Q_{34} = h_4 - h_3$$

is rejected. Condensation is stopped at state 4 where $s_4 = s_1$. Finally the steam is compressed isentropically in a rotary or reciprocating compressor to state 1, the work required being

$$W_{41} = -(h_1 - h_4)$$

The calculation of the work and heat transfers in the separate processes can be carried out by the methods given in Chapter 10.

Example 11.1 Calculate the heat and work transfers, cycle efficiency, work ratio and steam consumption of a Carnot cycle (Fig. 11.3), using steam between pressures of 30 and 0·04 bar.

From tables we find that at 30 bar $T_1 = T_2 = 507·0$ K and

$$h_1 = h_f = 1008 \text{ kJ/kg}, \quad h_2 = h_g = 2803 \text{ kJ/kg}$$

Putting $s_4 = s_1$ and $s_3 = s_2$ (see section 10.2), we find that at the condenser pressure of 0·04 bar, at which $T_3 = T_4 = 302·2$ K,

$$x_3 = 0·716, \quad \text{and} \quad x_4 = 0·276$$

Hence from $h = h_f + x h_{fg}$ we have

$$h_3 = 121 + x_3 2433 = 1863 \text{ kJ/kg}$$
$$h_4 = 121 + x_4 2433 = 793 \text{ kJ/kg}$$

The turbine work is

$$W_{23} = -(h_3 - h_2) = 940 \text{ kJ/kg}$$

and the compressor work is

$$W_{41} = -(h_1 - h_4) = -215 \text{ kJ/kg}$$

The heat transfers in the boiler and condenser respectively are

$$Q_{12} = (h_2 - h_1) = 1795 \text{ kJ/kg}$$

and

$$Q_{34} = (h_4 - h_3) = -1070 \text{ kJ/kg}$$

The net work from the cycle is therefore

$$W = W_{23} + W_{41} = 725 \text{ kJ/kg}$$

and this can be seen to equal $Q_{12} + Q_{34}$.

---

*For the remainder of this chapter we shall refer to this component as a turbine. The operation of turbines and reciprocating engines is described in detail in Chapters 19 and 16, and the circumstances in which one or other is preferable are mentioned in section 16.4.

The cycle efficiency is

$$\eta = \frac{W}{Q_{12}} = \frac{725}{1795} = 0{\cdot}404$$

This must, of course, also be given by

$$\eta = \frac{T_1 - T_3}{T_1} = \frac{507{\cdot}0 - 302{\cdot}2}{507{\cdot}0} = 0{\cdot}404$$

The work ratio is

$$r_w = \frac{W}{W_{23}} = \frac{725}{940} = 0{\cdot}771$$

and the specific steam consumption is

$$\text{s.s.c.} = \frac{3600}{725} = 4{\cdot}97 \text{ kg/kWh}$$

The wider the range of temperature, the more efficient becomes the cycle. The lowest possible temperature of the condensing steam is governed by two factors. Firstly, by the temperature of the sink of heat—atmosphere, ocean or river—for which an average figure of 15 °C may be assumed.* Secondly, by the temperature difference required for the heat transfer process. The rate of heat transfer is proportional to the surface area across which the heat flows, and to the temperature difference between the cooled and cooling fluid (see Part IV). In practice the required rate of heat transfer can only be obtained with a reasonable size of condenser if a temperature difference of 10 to 15 K is provided. The lowest practical condensing temperature in the cycle is therefore 25–30 °C, the corresponding pressure being 0·032–0·042 bar. The condenser pressure of 0.04 bar used in Example 11.1, for which the saturation temperature is 29·0 °C, will be used in all the other examples in this chapter. To show how important it is to keep the condenser pressure as low as possible, the variation of cycle efficiency with condenser pressure has been calculated for a constant boiler pressure of 30 bar and the results are presented in Fig. 11.4.

The maximum possible temperature of the working fluid, in power plants of the kind considered in this chapter, is governed by the strength of the materials available for the highly stressed parts of the plant, e.g. boiler tubes or turbine blades. This *metallurgical limit* may be assumed to be 600–650 °C for steam plant at the present time, the exact figure depending upon the life required of the plant. For a Carnot cycle operating with steam in the wet region, the highest possible temperature is that corresponding to the critical state: $t_c = 374{\cdot}15$ °C and $p_c = 221{\cdot}2$ bar. Modern materials cannot, therefore, be used to their best advantage with this cycle when steam

*Only natural sinks of heat need be considered—the work required to produce an artificial sink by refrigeration will, ideally, be equal to the additional work obtained from the power cycle. With real processes the refrigeration work would always be greater.

is the working fluid. Fig. 11.5 shows the variation of cycle efficiency and steam consumption with boiler pressure, for a constant condenser pressure of 0·04 bar. It is evident that in the higher pressure range there is only a small gain in cycle efficiency for increased boiler pressure. The cost of the boiler, pipe lines and turbine casing

Fig. 11.4   Variation of Carnot η with condenser pressure

increases with boiler pressure. Moreover the specific consumption, and hence the size of plant, increases after a certain pressure is reached. Evidently the most economical pressure must be determined by reaching a compromise between low operating cost and low capital cost.

Fig. 11.5   Variation of Carnot η and s.s.c. with boiler pressure

   In the previous section we pointed out that the ideal cycle efficiency is not by itself a good guide to the probable overall efficiency of the actual plant. The work ratio for the cycle in Example 11.1 is, by comparison with other types of vapour cycle, quite low (0·771), and we must expect irreversibilities to have an appreciable effect on the cycle efficiency and steam consumption. The following example shows the effect of assuming values less than unity for the process efficiencies of the turbine and compressor when the boiler and condenser pressures are 30 and 0·04 bar as before.

Example 11.2   Recalculate Example 11.1 with isentropic efficiencies of 0·80 for the compression and expansion processes (Fig. 11.6), to estimate the actual cycle efficiency and steam consumption.

The actual turbine work is

$$W_{23} = -(h_3 - h_2) = -0 \cdot 80(h'_3 - h_2) = 752 \text{ kJ/kg}$$

and the actual compressor work is

$$W_{41} = -(h_1 - h_4) = -\frac{(h'_1 - h_4)}{0 \cdot 80} = -269 \text{ kJ/kg}$$

The net work is therefore

$$W = W_{23} + W_{41} = 483 \text{ kJ/kg}$$

The enthalpy in state 1 is

$$h_1 = h_4 - W_{41} = 1062 \text{ kJ/kg}$$

and hence the heat transfer to the boiler is

$$Q_{12} = (h_2 - h_1) = 1741 \text{ kJ/kg}$$

The thermal efficiency is

$$\eta = \frac{W}{Q_{12}} = 0 \cdot 277$$

Fig. 11.6

and the specific steam consumption is

$$\text{s.s.c.} = \frac{3600}{483} = 7 \cdot 45 \text{ kg/kW h}$$

The previous examples show that the cycle efficiency is reduced by irreversibilities from 0·404 to 0·277, and that the steam consumption is increased from 4·97 to 7·45 kg/kW h. We shall show that other forms of cycle having work ratios nearer unity are much less affected by irreversibilities.

11.3 Rankine cycle

There are two major reasons why the Carnot cycle is not used in practice. Firstly, because it has a low work ratio. Secondly, because of practical difficulties associated with the compression. It would be difficult to control the condensation process so that it stopped at state 4, and then carry out the compression of a very wet vapour efficiently. The liquid tends to separate out from the vapour and the compressor would have to deal with a non-homogeneous mixture. Moreover the volume of the fluid is high and the compressor would be comparable in size and cost with the turbine. It is comparatively easy, on the other hand, to condense the vapour completely, and compress the liquid to boiler pressure in a small feed pump. The resulting cycle, shown in Fig. 11.7, is known as the *Rankine cycle*.

205

It is evident without calculation that the efficiency of this cycle will be less than that of the Carnot cycle operating between the same temperatures, because all the heat supplied is not transferred at the upper temperature. Some heat is added while the temperature of the liquid varies from T_5 to T_1. It is also evident, from the comparative areas of the two cycles, that the net work output per unit mass of steam is greater in the Rankine cycle [in section 8.4 (i) it was shown that the area $= \oint(T ds)_{rev} = \oint dQ = W$]. It follows that the steam consumption is less and the work ratio is greater. Thus, in spite of a lower ideal cycle efficiency, the actual cycle efficiencies may not be so different, and the size of the Rankine plant will certainly be much smaller. The following example verifies these deductions.

Fig. 11.7   The Rankine cycle

Example 11.3   Calculate the cycle efficiency, work ratio, and steam consumption of the Rankine cycle in Fig. 11.7, working between the same pressures as the Carnot cycle in Example 11.1.

Estimate the actual cycle efficiency and steam consumption when the isentropic efficiencies of the expansion and compression processes are each 0·80.

(a) *Ideal cycle*

As in Example 11.1,

$$h_2 = 2803 \text{ kJ/kg} \quad \text{and} \quad h_3 = 1863 \text{ kJ/kg}$$

And the turbine work is as before

$$W_{23} = -(h_3 - h_2) = 940 \text{ kJ/kg}$$

Using equation (10.5), the compression work W_{45} is found as follows.

$$W_{45} = -(h_5 - h_4) \approx -v_f(p_5 - p_4)$$
$$= -0·001(30 - 0·04) \times 100 = -3 \text{ kJ/kg}$$

Since $h_4 = h_f = 121 \text{ kJ/kg}$, $h_5 = 124 \text{ kJ/kg}$. The heat supplied is

$$Q_{52} = (h_2 - h_5) = 2803 - 124 = 2679 \text{ kJ/kg}$$

Therefore

$$\eta = \frac{W}{Q_{52}} = \frac{940 - 3}{2679} = 0.350$$

$$r_w = \frac{W}{W_{23}} = \frac{940 - 3}{940} = 0.997$$

$$\text{s.s.c.} = \frac{3600}{940 - 3} = 3.84 \text{ kg/kW h}$$

Fig. 11.8

(b) *Actual cycle* (Fig. 11.8)

Actual expansion work is

$$W_{23} = 0.80 \times 940 = 752 \text{ kJ/kg}$$

and the actual compression work is

$$W_{45} = -\frac{3}{0.80} = -4 \text{ kJ/kg}$$

The enthalpy at state 5 now becomes

$$h_5 = h_4 - W_{45} = 125 \text{ kJ/kg}$$

and the heat supplied is

$$Q_{52} = (h_2 - h_5) = 2803 - 125 = 2678 \text{ kJ/kg}$$

Therefore

$$\eta = \frac{752 - 4}{2678} = 0.279$$

$$\text{s.s.c.} = \frac{3600}{752 - 4} = 4.81 \text{ kg/kW h}$$

The first point to note from the previous example is that the compressor work, or *feed pump term,* is very small compared with the other energy transfers. When the boiler pressure is comparatively low, a good approximation to the ideal cycle efficiency will be obtained if it is neglected. The approximate calculation yields the following results for the ideal cycle of the previous example:

$$W \approx W_{23} = 940 \text{ kJ/kg}$$
$$Q_{52} \approx (h_2 - h_4) = 2682 \text{ kJ/kg}$$
$$\eta \approx 0.350 \quad \text{and} \quad \text{s.s.c.} \approx 3.83 \text{ kg/kW h}$$

This confirms the fact that the points representing states 4 and 5 almost coincide on the *T-s* diagram if it is drawn to any reasonable scale [see section 8.4 (i)].

The second point worth noting is that the difference between the performance of the ideal and actual cycles is much smaller than in the case of the Carnot cycle working between the same pressures and with the same component efficiencies. This is emphasised by the following table, in which the actual cycle performances are compared with the ideal values. For the assumed operating conditions the actual cycle efficiency of the Rankine plant (0·279) is slightly greater than that of the Carnot plant (0·277).

Cycle	r_w	Efficiency ratio	$\dfrac{\text{Actual s.s.c.}}{\text{Ideal s.s.c.}}$
Carnot	0·771	0·686	1·50
Rankine	0·997	0·797	1·25

Fig. 11.9 shows the results of a series of ideal cycle calculations with various boiler pressures and a constant condenser pressure of 0·04 bar. The cycle efficiency does not increase continuously with boiler pressure up to the critical pressure, but reaches a maximum when the pressure is about 160 bar. The reason is that the latent heat, and therefore the proportion of the heat supplied which is transferred at the upper temperature, becomes smaller as the pressure increases. Sooner or later the effective average temperature at which the heat is added starts to decrease and the efficiency falls.

Fig. 11.9 Variation of Rankine η and s.s.c. with boiler pressure

For a high efficiency it is essential to keep the condenser pressure as low as possible. However, where space is restricted as in a steam locomotive, or where no cooling water is available, the condenser is omitted and an open cycle is used (see section 4.4). When the boiler feed water temperature is 100 °C, the ideal cycle efficiency is then identical with that of a closed cycle plant condensing at 1 atm. The condition of the fluid is the same at entry to the boiler in the two cases, and the fact that the fluid entering the boiler is not the same as that leaving the turbine makes no difference. If the boiler feed water temperature is less than 100 °C the efficiency will be reduced because more 'sensible heat' must be added in the boiler.

208

It has already been remarked that the metallurgical limit cannot be approached when the steam leaves the boiler in a saturated condition. But by placing in the combustion gases a separate bank of tubes (the superheater) leading saturated steam away from the boiler, it is possible to raise the steam temperature without at the same

Fig. 11.10   The Rankine cycle with superheat

time raising the boiler pressure. The Rankine cycle then appears as in Fig. 11.10, from which it is evident that the average temperature at which heat is supplied is increased by superheating and hence the ideal cycle efficiency is increased. There is very little change in susceptibility to irreversibilities, the work ratio being so very near unity in the unsuperheated Rankine cycle that the slight increase obtained by superheating

Fig. 11.11   Effect of superheating on the Rankine cycle

makes very little difference. Nevertheless, the specific steam consumption is markedly reduced, the net work per unit mass of steam being much greater, so that the added complexity of a superheater is compensated by a reduction in the size of other components. Fig. 11.11a shows the variation of ideal cycle efficiency and steam consumption with superheat temperature up to 800 °C. There is evidently still much to be gained by improving high-temperature materials. Unlike the efficiency of the unsuperheated cycle, the efficiency of the superheated cycle increases continuously with pressure. Fig. 11.11b illustrates this for a cycle with 0·04 bar condenser pressure and superheating to 600 °C—the metallurgical limit for long-life plant.

There is an additional reason why superheating is always used in practice. In the unsuperheated cycle of Example 11.3 the dryness fraction at the turbine exhaust is 0·716. Inspection of the *T-s* diagram shows that with higher boiler pressures the turbine exhaust would become even wetter. This is a very undesirable feature, because droplets in the steam erode the blading and reduce the turbine isentropic efficiency—see section 19.2 (iv). In practice the dryness fraction at the turbine exhaust is not allowed to fall below about 0·88. Superheating increases the dryness fraction, although with the present metallurgical limit in the range of 600–650 °C it is not always by itself sufficient to maintain the dryness fraction above 0·88. For example, when the steam enters the turbine at 160 bar and 600 °C, the turbine exhaust dryness fraction after isentropic expansion to 0·04 bar is 0·772; irreversibilities in the turbine tend to increase the dryness fraction but they would not bring the value up to 0·88. With supercritical cycles (i.e. when 2345 in Fig. 11.10 is a supercritical isobar) this problem is even more severe; the remedy lies in the use of reheat cycles discussed in the next section.

Fig. 11.11b shows the η and s.s.c. curves extended beyond the critical pressure of 221·2 bar. It would appear that the use of pressures beyond 160 bar, let alone supercritical pressures, is hardly justified. However, when reheating and regenerative feed heating, described in the next two sections, are applied to such cycles, it becomes worth using such extreme pressures.

11.4 Reheat cycle

With the reheat cycle the expansion takes place in two turbines. The steam expands in the high-pressure turbine to some intermediate pressure, and is then passed back to yet another bank of tubes in the boiler where it is reheated at constant pressure, usually to the original superheat temperature.* It then expands in the low-pressure turbine to the condenser pressure. The cycle appears as in Fig. 11.12, and the following example shows the effect of reheating on the ideal cycle efficiency and specific consumption. A boiler pressure of 30 bar has been assumed for the sake of comparison with the other examples in this chapter, although this is well below the most economical pressure for which to design a reheat cycle plant.

Example 11.4 Find the ideal cycle efficiency and steam consumption of a reheat cycle (Fig. 11.12) operating between pressures of 30 and 0·04 bar, with a superheat temperature of 450 °C. Assume that the first expansion is carried out to the point where the steam is dry saturated and that the steam is reheated to the original superheat temperature. The feed pump term may be neglected.

---

*It has been suggested that as an alternative to the conventional reheat scheme, it is better to reheat the steam near the turbine via a liquid metal circulating between the boiler and reheater. The appreciable pressure drop in the lengthy, conventional, reheat circuit is avoided, but this gain is partially offset by the extra temperature drop required for the liquid metal circuit.

From tables

$$h_2 \approx h_1 = 121 \text{ kJ/kg}, \quad h_5 = 3343 \text{ kJ/kg}, \quad s_5 = 7\cdot082 \text{ kJ/kg K}$$

To find the intermediate reheat pressure p_6, we find from the saturation table the pressure at which $s_g = s_5$. This gives

$$p_6 = p_7 = 2\cdot3 \text{ bar, and hence } h_6 = 2713 \text{ kJ/kg}$$

From the superheat table

$$h_7 = 3381 \text{ kJ/kg} \quad \text{and} \quad s_7 = 8\cdot310 \text{ kJ/kg K}$$

At the turbine outlet

$$x_8 = 0\cdot980 \quad \text{and} \quad h_8 = 2505 \text{ kJ/kg}$$

Fig. 11.12   The reheat cycle

The total quantity of heat transferred to the steam in the boiler is

$$Q_{25} + Q_{67} = (h_5 - h_2) + (h_7 - h_6)$$
$$= 3222 + 668 = 3890 \text{ kJ/kg}$$

and the total turbine work is

$$W_{56} + W_{78} = -(h_6 - h_5) - (h_8 - h_7)$$
$$= 630 + 876 = 1506 \text{ kJ/kg}$$

The cycle efficiency is therefore

$$\eta = \frac{W_{56} + W_{78}}{Q_{25} + Q_{67}} = \frac{1506}{3890} = 0\cdot387$$

and the specific steam consumption is

$$\text{s.s.c.} = \frac{3600}{1506} = 2\cdot39 \text{ kg/kW h}$$

This example, as well as the other examples in this chapter, can be solved more quickly by means of an *h-s* chart. This is illustrated in Fig. 11.13. The chart usually covers only the superheat region and the two-phase region near the saturated vapour line, and it cannot therefore be used for finding h_1, h_2 and h_3. It enables one, however, to find h_5, h_6, h_7 and h_8 more quickly than by means of tables.

Comparing the foregoing results with those of Examples 11.1 and 11.3, and the appropriate values from Fig. 11.11a (all cycles with a boiler pressure of 30 bar and a condenser pressure of 0·04 bar), we arrive at the following table:

Cycle (ideal)	Carnot	Rankine (no superheat)	Rankine (superheat to 450 °C)	Reheat (to 450 °C)
η	0·404	0·350	0·375	0·387
s.s.c.	4·97	3·84	2·98	2·39

It can be seen that reheating reduces the steam consumption appreciably, as might be expected from consideration of the area of the cycle on the *T-s* diagram which equals the net work done per unit mass of steam. This reduction is a particularly important consideration in higher pressure cycles because it implies a smaller boiler—an expen-

Fig. 11.13

Fig. 11.14   The effect of reheat pressure on the cycle efficiency

sive item in high-pressure plant. However, this decrease in size cannot completely offset the disadvantage of the added complexity, and it must be stressed that the main reason for reheating is to avoid too wet a condition in the turbine.

It is clear that reheating makes only a little difference to the ideal cycle efficiency. Whether it increases or decreases the efficiency, in comparison with the Rankine cycle having the same conditions in the boiler and condenser, depends upon the point in the expansion at which reheating is carried out. The reheat cycle can be regarded as a combination of the Rankine cycle and the cycle *abcd* (or *a′b′c′d*) in Fig. 11.14. If the latter has a lower efficiency than the Rankine cycle, the efficiency of the combination will be lower also. The effective average temperature at which the heat is added in the cycle *abcd* may well be lower than that in the Rankine cycle if the intermediate pressure is too low.

Cycles using supercritical pressures have been coming into operation in recent years. With such cycles the problem of moisture in the turbine is even more severe, and the use of reheating is essential. So far, it has been possible to avoid the additional complication of reheating at more than one point in the expansion. Typical modern supercritical and subcritical ideal cycles are compared in the table below, where

	Superheat to 600 °C		Superheat and reheat to 600 °C	
	160 bar	350 bar	160 bar	350 bar
η	0·453	0·468	0·468	0·485
s.s.c./[kg/kW h]	2·31	2·37	1·76	1·80
x_{exit}	0·772	0·708	0·954	0·871

reheating is assumed to occur when the steam is just saturated ($x_a = 1$ in Fig. 11.14). Although the increase in efficiency resulting from the use of supercritical pressures seems small, it should not be discounted. A one per cent increase in efficiency for a cycle which has an actual efficiency of, say, 40 per cent is equivalent to $1/0\cdot4 = 2\cdot5$ per cent saving in fuel: a saving well worth having. Nevertheless it is true to say that a marked improvement in efficiency cannot be obtained by using higher pressures: it can only be achieved by developing new high-temperature alloys or other materials for turbine blading and superheater tubing and thereby raising the metallurgical limit.

11.5 Regenerative cycle

The efficiency of the unsuperheated cycle (Fig. 11.7) is less than the Carnot efficiency $(T_a - T_b)/T_a$, because some of the heat supplied is transferred while the temperature of the working fluid varies from t_5 to t_1. If some means could be found to transfer this heat *reversibly* from the working fluid in another part of the cycle, then all the heat supplied from an *external* source would be transferred at the upper temperature and the Carnot cycle efficiency would therefore be achieved. A cycle in which this artifice is used to raise the efficiency is called a *regenerative cycle*.

One method which is theoretically possible is indicated in Fig. 11.15. The feed water, at t_6, is passed into passages in the turbine casing at the point where the expanding steam has a temperature only infinitesimally greater than t_6. The water flows in a direction opposite to that of the expanding steam, and is thereby heated to a temperature only infinitesimally smaller than t_2 before it passes into the boiler. At all points the heat transfer is effected by an infinitesimal temperature difference and the process is therefore reversible. Most of the expansion in the turbine is no longer an adiabatic process, and it follows the path 2–3–4. Since the heat rejected in process 2–3 equals that supplied in process 6–1, the shaded areas on the T-s diagram of Fig. 11.15 must be equal. The heat supplied from the external source is equal to the area under line 1–2, and the heat rejected to the external sink equals the area under line 4–5. Inspection

Fig. 11.15   A regenerative cycle

of the areas involved shows that not only are the efficiencies of the regenerative cycle and the Carnot cycle 1278 equal, but the steam consumptions also. On the other hand the work ratio of the regenerative cycle, requiring the negligible pump work $-(h_6 - h_5)$, is very much higher than that of the Carnot cycle and is comparable to that of the Rankine cycle.

Such a regenerative cycle is impractical for two reasons. Firstly, it would be impossible to design a turbine which would operate efficiently as both turbine and heat exchanger. Secondly, the steam expanding through the turbine would have an impracticably low dryness fraction. For these reasons the basic idea of regeneration is applied in a somewhat different way—instead of taking the feed water to the turbine, some of the turbine steam is taken to the feed water.

The simplest approximation to a practical regenerative cycle is shown in Fig. 11.16. Consider 1 kg of steam in state 2 passing from the boiler into the turbine. This steam is expanded to an intermediate state 3, at which point a quantity y kg is bled off and taken to a *feed water heater*. The remaining $(1 - y)$ kg is expanded to condenser

214

pressure and leaves the turbine in state 4. After condensation to state 5, the $(1 - y)$ kg of water is compressed in the first feed pump to the bleeding pressure $(p_6 = p_3)$. It is then mixed in the feed water heater with y kg of bled steam in state 3, and the 1 kg of mixture leaves the heater in state 7. A second feed pump compresses the water to boiler pressure, state 1. In this cycle the average temperature at which the heat is added from an *external source* is higher than in the Rankine cycle, and therefore the efficiency is increased. The cycle is not a true ideal cycle, because the mixing process between steam at state 3 and water at state 6 is irreversible. In the limit, as the number of bleedings and feed water heaters is increased to infinity, the mixing processes become reversible, all the heat from the external source is supplied at the upper temperature, and the cycle will therefore have the Carnot efficiency.

Fig. 11.16   A simple regenerative cycle using one feed water heater

Since a regenerative cycle of this kind is really a combination of cycles undergone by different quantities of fluid, there is some difficulty in portraying it on a *T-s* diagram. Fig. 11.16 must be interpreted as implying that $(1 - y)$ kg is carried through cycle 56712345, while y kg follows the path 71237.

Example 11.5   Find the cycle efficiency and steam consumption of a regenerative cycle with one feed heater, if the steam leaves the boiler dry saturated at 30 bar and is condensed at 0·04 bar (Fig. 11.16).

It is first necessary to select the pressure p_3 at which the steam is to be bled off. It is shown later that the cycle efficiency is a maximum when the temperature t_3 is approximately halfway between t_2 and t_4, i.e.

$$t_3 = \frac{t_2 + t_4}{2} = \frac{233 \cdot 8 + 29 \cdot 0}{2} = 131 \cdot 4 \,°\text{C}$$

and hence

$$p_3 = 2 \cdot 8 \text{ bar}$$

We can now calculate the correct mass y to be bled off at p_7. Consider the heater as an adiabatic open system (Fig. 11.17). No work is done and the energy equation reduces to

$$0 = 1h_7 - yh_3 - (1 - y)h_6$$

Putting $h_6 = h_5$, we find that

$$y = \frac{h_7 - h_5}{h_3 - h_5}$$

From tables

$$h_5 = 121 \text{ kJ/kg} \quad \text{and} \quad h_7 = 551 \text{ kJ/kg}$$

Also

$$h_2 = 2803 \text{ kJ/kg} \quad \text{and} \quad s_2 = 6{\cdot}186 \text{ kJ/kg K}$$

Putting $s_2 = s_3 = s_4$, we find

$$x_3 = 0{\cdot}846, \quad h_3 = 2388 \text{ kJ/kg}$$

$$x_4 = 0{\cdot}716, \quad h_4 = 1863 \text{ kJ/kg}$$

Fig. 11.17 The feed water heater as an open system

Hence

$$y = \frac{551 - 121}{2388 - 121} = 0{\cdot}1897$$

The heat transferred in the boiler is

$$Q_{12} = (h_2 - h_1) = 2803 - 551 = 2252 \text{ kJ/kg}$$

The turbine work is

$$W_{23} + W_{34} = -1(h_3 - h_2) - (1 - y)(h_4 - h_3)$$
$$= 415 + 425 = 840 \text{ kJ/kg}$$

Therefore

$$\eta = \frac{W_{23} + W_{34}}{Q_{12}} = \frac{840}{2252} = 0{\cdot}373$$

and the specific steam consumption is

$$\text{s.s.c.} = \frac{3600}{840} = 4{\cdot}29 \text{ kg/kW h}$$

In Fig. 11.18 the cycle efficiency is plotted against bleed temperature, all the other conditions remaining constant. It can be seen that the cycle efficiency is a maximum when the bleed temperature is approximately equal to the mean of the boiler and condenser temperatures, and that a slight deviation from this value does not affect the efficiency significantly.

Comparing the results of Example 11.5 with those of the ideal unsuperheated Rankine cycle in Example 11.3, the efficiency has been increased from 0·350 to 0·373 with the use of only one feed heater. On the other hand the specific consumption has been increased from 3·84 to 4·29 kg/kW h. In practice the best return for the additional

apital cost is obtained with several feed heaters, the exact number depending on the
team conditions. The proper choice of the bleed pressures is a matter of lengthy
ptimisation calculations which cannot be entered into here. The *open heater*, in which
he bled steam is freely mixed with the feed water, is not often used because of the large
umber of feed pumps required (i.e. number of heaters + 1). Fig. 11.19 illustrates a
lant using three heaters of the *closed* type, in which the heat is transferred through
anks of tubes. For example, considering heater 1, the bled steam in state 5 drains
rom the feed heater in state 14 as a separate stream; this is different from the open
eater arrangement of Fig. 11.16. The drain in state 14 can be pumped back into the
eed line at point 1 by means of a small drain or drip pump, but a more common
rrangement is to cascade it back through throttle valves, via the lower pressure feed
eaters, to the condenser. The advantage of doing this is that only one feed pump is

Fig. 11.18   Effect of bleed temperature on the cycle efficiency

equired; a similar train of open heaters would require four feed pumps in the feed
vater line. The throttling of the drains is accompanied in practice by the formation of
ome steam, and this can lead to rapid erosion in the cascade line. To avoid this ero-
ion, the drains are throttled into large flash chambers (not shown) which reduce the
elocity and therefore the erosion.

Considering the *T-s* diagram in Fig. 11.19, the feed water leaving the feed pump in
tate 10 is successively heated to temperatures t_{11}, t_{12} and t_1 which, ideally, are equal to
he saturation temperatures t_{17}, t_{15} and t_{13}* respectively. In practice $t_{11} < t_{17}$, $t_{12} < t_{15}$,
 $< t_{13}$, the differences usually being referred to as the *terminal temperature differences*

---

*It might appear feasible, where the bled steam is initially superheated and condensed in a counter-flow
eater, for the exit temperature of the feed water to attain the superheat temperature, e.g. $t_1 = t_5$ in heater
. This ideal cannot be achieved, however, because the difference between the sensible enthalpy capacities of
he bled superheated steam and the feed water would necessitate heat transfer from colder steam to hotter feed
ater over the temperature range $t_{13} \rightarrow t_5$. Although the condition $t_1 = t_5$ is impossible, the heater could
e designed so that the ideal feed temperature t_1 is slightly above t_{13}, but this possibility will be neglected
 what follows.

of the heaters. Similarly the condensed bled steam should drain at temperature $t_{14} = t_{12}$, $t_{16} = t_{11}$ and $t_{18} = t_{10}$. Such ideal conditions can only be achieved with closed heaters having infinitely large heat exchange areas. The following example which is concerned with the ideal cycle, assumes such limiting conditions to exist.

Fig. 11.19   A regenerative cycle with three feed heaters

Example 11.6 Calculate the ideal cycle efficiency of a regenerative cycle using three closed heaters as shown in Fig. 11.19. The steam leaves the boiler at 30 bar superheated to 450 °C, and the condenser pressure is 0·04 bar. Choose the bleed pressures so that the temperature difference $(t_2 - t_9)$ is divided into approximately equal steps. (Such a choice of bleed pressures makes the efficiency of the ideal cycle approximately a maximum.)

Making the temperature differences $(t_2 - t_{13})$, $(t_{13} - t_{15})$, $(t_{15} - t_{17})$, $(t_{17} - t_9)$, approximately equal, the bleed pressures become 11, 2·8 and 0·48 bar. All the relevant enthalpies in the plant can now be found. Before, during and after expansion in the turbine these are

$$h_4 = 3343 \text{ kJ/kg}, \quad h_5 = 3049 \text{ kJ/kg}$$
$$h_6 = 2749 \text{ kJ/kg}, \quad h_7 = 2458 \text{ kJ/kg}$$
$$h_8 = 2133 \text{ kJ/kg}$$

In finding the enthalpies in the feed line, the following assumptions will be made. (a) The feed pump term is negligible, i.e. $h_9 \approx h_{10}$. (b) In throttling the condensed bled steam, which is a process of equal initial and final enthalpy, the state after throttling lies approximately on the saturation line; e.g. in throttling from 14 to 15, $h_{14} = h_{15}$ will be identical with h_f corresponding to pressure p_8. (c) The enthalpy of the compressed liquid in the feed line is approximately equal to that of saturated liquid at the same temperature, e.g. $h_{12} \approx h_{15}$. With these assumptions we can write

$$h_{10} = h_{18} = h_{19} = h_9 = 121 \text{ kJ/kg}$$
$$h_{11} = h_{16} = h_{17} = 336 \text{ kJ/kg}$$
$$h_{12} = h_{14} = h_{15} = 551 \text{ kJ/kg}$$
$$h_1 = h_{13} = 781 \text{ kJ/kg}$$

To determine the correct amounts of steam to be bled for each heater per kg of steam leaving the boiler, an energy equation can be written down for each heater.

1st heater: $1h_1 + y_a h_{14} - y_a h_5 - 1h_{12} = 0$

$$y_a = \frac{h_{13} - h_{15}}{h_5 - h_{15}} = 0.0921$$

2nd heater: $1h_{12} + (y_a + y_b)h_{16} - y_b h_6 - 1h_{11} - y_a h_{15} = 0$

$$y_b = (1 - y_a)\frac{h_{15} - h_{17}}{h_6 - h_{17}} = 0.0809$$

3rd heater: $1h_{11} + (y_a + y_b + y_c)h_{18} - y_c h_7 - (y_a + y_b)h_{17} - 1h_{10} = 0$

$$y_c = (1 - y_a - y_b)\frac{h_{17} - h_9}{h_7 - h_9} = 0.0761$$

(Always start with the highest pressure heater so as to have only one unknown at each stage of calculation.)

We can now calculate the heat and work transfers for the cycle. The heat added in the boiler is

$$Q_{1,4} = (h_4 - h_1) = 2562 \text{ kJ/kg}$$

and the heat rejected in the condenser is

$$Q_{8,9} = 1h_9 - (1 - y_a - y_b - y_c)h_8 - (y_a + y_b + y_c)h_{19} = -1511 \text{ kJ/kg}$$

The work done in the turbine is

$$W_{4,8} = -1(h_5 - h_4) - (1 - y_a)(h_6 - h_5) - (1 - y_a - y_b)(h_7 - h_6) -$$
$$(1 - y_a - y_b - y_c)(h_8 - h_7) = 1051 \text{ kJ/kg}$$

[As a check, $W_{4,8}$ must also be given by

$$W_{4,8} = Q_{1,4} + Q_{8,9} = 1051 \text{ kJ/kg]}$$

The cycle efficiency is

$$\eta = \frac{W_{4,8}}{Q_{1,4}} = 0.410$$

and the specific consumption is

$$\text{s.s.c.} = \frac{3600}{1051} = 3.43 \text{ kg/kWh}$$

The ideal efficiency and steam consumption for the Rankine cycle corresponding to the cycle 92489 in Fig. 11.19 can be found from Fig. 11.11a to be 0·375 and 2·98 kg/kW h. For the regenerative cycle these become 0·410 and 3·43 kg/kW h. Apart from the marked improvement in cycle efficiency, regeneration has the advantage of easing the difficulties in the design of the blading at the low pressure end of the turbine. With the Rankine cycle this blading has to pass a very large volume flow and the blades must be undesirably long. By bleeding, the volume flow at the low pressure end is considerably reduced; this is indicated by the stepped outline of the turbine in Fig. 11.19. The size of the condenser is also reduced, which partially compensates for the additional feed heating components.

It is necessary to point out that if the isentropic efficiency of the turbine is not unity, points 5, 6, 7, 8 on the *T-s* diagram will not be vertically below point 4. To estimate the actual cycle efficiency with an irreversible expansion it is necessary to know the path of the expansion (i.e. 'line of condition'), and this is considered in section 19.2 (v).

With modern steam conditions the best compromise between additional capital cost and improved efficiency is obtained when a train of about eight feed heaters is used. It is common to include one open heater at a moderate pressure which incorporates a de-aerator, because dissolved air leads to corrosion in the boiler, and is detrimental to heat transfer and the maintenance of a high vacuum in the condenser. There will normally be two closed heaters at pressures below the open heater, with the condensed bleed cascading back into the condenser. At pressures above the open heater there will be about three closed heaters whose condensate cascades back into the open heater. A description of modern plant and a more detailed thermodynamic analysis of regenerative cycles can be found in Refs. (18) and (34) respectively.

11.6 The economiser and air-preheater

So far we have been comparing the performance of the various cycles on the basis of cycle efficiency, and have assumed in effect that the combustion process can be carried out with equal efficiency in all plants. This is not, in fact, the case. Without discussing combustion efficiency in detail, it can be appreciated that if the combustion gases leave the boiler plant at a temperature higher than atmospheric, some of the

Fig. 11.20   Diagrammatic sketch of a boiler plant with (a) an economiser, (b) an air preheater

latent energy in the fuel is being wasted. Useful work could still be obtained by operating a cycle between the exhaust gas temperature and atmospheric temperature. With the Carnot cycle, the lowest temperature at which the exhaust gases can leave the plant is the upper temperature of the cycle. With the Rankine cycle, on the other hand, the exhaust gases could be reduced, ideally, to the temperature of the water leaving the feed pump. A reduction in exhaust gas temperature is accomplished in practice by passing the feed water through a bank of tubes, the *economiser*, situated in the chimney.* The flow is in the direction indicated in Fig. 11.20a, so that the hottest gases are transferring heat to the hottest water, and the temperature difference between the fluids is kept to a minimum throughout the process.

It would appear that the advantage of higher ideal efficiency possessed by the Carnot cycle is at least partly diminished by its inability to abstract all the energy available

*One limitation is that the products of combustion must not be cooled below their dew point (see section 14.5), otherwise condensation containing corrosive acids may cause serious damage to the economiser tubes: in practice 150 °C is considered to be the safe lower limit.

in the products of combustion—a further argument in favour of the practical superiority of the Rankine cycle. The regenerative cycle suffers from the same disadvantage as the Carnot cycle in this respect. If an economiser is added to the cycle shown in Fig. 11.19, the minimum temperature to which the combustion gases can be cooled is t_1. When a large number of feed heaters is used, the feed water entering the boiler is nearly at boiling point and an economiser becomes superfluous.

An alternative method of reducing the temperature of the combustion gases, which can be used with all the cycles, is known as *air-preheating*. The air-preheater is a bank of tubes in the chimney through which the combustion air passes before reaching the grate, Fig. 11.20b. The combustion gases are cooled while the temperature of the air is increased prior to combustion. Since the reactants start with a higher temperature, less fuel need be burnt to arrive at the same products temperature. The net result is an increase in combustion efficiency comparable to that obtained with an economiser. Air-preheating is often used in conjunction with an economiser in Rankine and reheat plants, but the degree of preheating cannot be carried as far as in the regenerative plant where no economiser is fitted.

One practical consequence of cooling the combustion gases is that only a small amount of natural draught can be obtained in the chimney. The heat-exchanging banks of tubes, grate and chimney require a fair pressure gradient to maintain the necessary air flow for combustion. For this reason a forced draught, produced by a fan, is always used in modern boiler plant. The power expended on driving the fan is small compared with the advantage gained by improving the combustion efficiency.

Enough has been said to indicate that a complete evaluation of an external-combustion power cycle must take account of the interdependence between the combustion and cycle efficiencies. The adoption of means to improve one efficiency may easily lead to the reduction of the other. A formal analysis of the plant as a whole is possible using the concept of availability introduced in section 7.9, and Refs. (12) and (13) of Part I provide a good introduction to this approach.

11.7 Binary vapour cycle

The greatest proportion of the heat supplied to the Rankine cycle is added during evaporation of the liquid, whereas the amount added to the feed water, and during superheating, is relatively small. Thus a major improvement in the cycle efficiency can only be achieved if the evaporation itself can be carried out at a temperature corresponding to the metallurgical limit of the turbine. The limitation imposed on the efficiency of a steam turbine plant is therefore not imposed so much by the type of cycle employed as by the properties of the fluid used. Steam has a relatively low critical temperature (374·15 °C) and the critical pressure is very high (221·2 bar). It is desirable to find a fluid whose critical temperature lies well above the metallurgical limit of about 600 °C, and whose vapour pressure is moderate at this temperature. Mercury satisfies these requirements fully. At a vapour temperature of 600 °C the saturation pressure is merely 23 bar.

The use of mercury by itself has the disadvantage that at the usual temperatures of heat rejection (25–30 °C) its vapour pressure is extremely low and its specific volume is enormous. This means that an impracticably high vacuum would have to be maintained in a condenser of impracticable size. A further disadvantage of mercury is its low latent heat (about one-eighth the latent heat of steam), implying a high specific mercury consumption.

Fig. 11.21   The binary vapour cycle

In order to use the whole temperature range, from the metallurgical limit down to atmospheric, mercury is always used in conjunction with steam in a two-stage or *binary cycle*. This is illustrated in Fig. 11.21. Mercury is evaporated in the mercury boiler and is expanded to a pressure of about 0·1 bar, the mercury condenser performing the function of a steam boiler. Preheating of the feed water, and the superheating of the steam after evaporation, is done by the combustion gases of the main mercury boiler. The method of calculating the cycle efficiency is illustrated in the following example.

Example 11.7 Calculate the ideal efficiency of a binary vapour cycle. The steam cycle operates between pressures of 30 and 0·04 bar, and uses a superheat temperature of 450 °C. The mercury cycle works between pressures of 14 and 0·1 bar, the mercury entering the turbine in a dry saturated condition (Fig. 11.21).

It is first necessary to calculate the heat and work transfers for each cycle per kg of respective fluid circulated. The enthalpies at the relevant points in the cycle are found in the usual way from steam and mercury tables.

H_2O *Cycle*: Heat added to the steam cycle in the mercury condenser is

$$Q_{3,4} = h_4 - h_3 = 1795 \text{ kJ/kg}$$

Heat added from the external source is

$$Q_{2,3} + Q_{4,5} = (h_3 - h_2) + (h_5 - h_4)$$
$$= (1008 - 121) + (3343 - 2803)$$
$$= 1427 \text{ kJ/kg}$$

The heat rejected in the condenser is

$$Q_{6,1} = (h_1 - h_6) = (121 - 2133) = -2012 \text{ kJ/kg}$$

The turbine work is

$$W_{5,6} = -(h_6 - h_5) = -(2133 - 3343) = 1210 \text{ kJ/kg}$$

Hg *Cycle*: The heat added to the cycle is

$$Q_{8,10} = (h_{10} - h_8) = (362 \cdot 55 - 34 \cdot 33) = 328 \cdot 22 \text{ kJ/kg}$$

After isentropic expansion in the mercury turbine we find

$$x_{11} = 0 \cdot 7286 \quad \text{and} \quad h_{11} = 248 \cdot 93 \text{ kJ/kg}$$

The turbine work is

$$W_{10,11} = -(h_{11} - h_{10}) = 113 \cdot 62 \text{ kJ/kg}$$

The heat rejected in the condenser is

$$Q_{11,7} = (h_7 - h_{11}) = -214 \cdot 60 \text{ kJ/kg}$$

It is now possible to determine the correct flow of z kg of mercury per kg of steam. Writing down the energy equation for the 'mercury condenser/steam boiler', we have

$$z(h_7 - h_{11}) + 1(h_4 - h_3) = 0$$

Hence

$$z = \frac{1795}{214 \cdot 60} = 8 \cdot 364 \text{ kg Hg/kg } H_2O$$

Calculating the total energy transfers on the basis of 1 kg steam circulating, we get for the heat added from the external source

$$Q_A = zQ_{8,10} + 1(Q_{2,3} + Q_{4,5})$$
$$= 2745 + 1427 = 4172 \text{ kJ/kg } H_2O$$

nd the net work done, neglecting the feed pump terms, is

$$W = zW_{10,11} + 1W_{5,6}$$
$$= 950 + 1210 = 2160 \text{ kJ/kg H}_2\text{O}$$

As a check, we can find the total heat rejected from

$$W - Q_A = 2160 - 4172 = -2012 \text{ kJ/kg H}_2\text{O}$$

his must be equal to the heat rejected in the steam condenser $Q_{6,1}$.)
The ideal cycle efficiency is

$$\eta = \frac{W}{Q_A} = \frac{2160}{4172} = 0.518$$

Although the efficiency in the foregoing example represents a marked improvement ·n that of any other ideal cycle we have considered, only few binary vapour plants ave been built. Mercury is highly poisonous and attacks many metals in common se. Moreover, its inability to 'wet' boiler and condenser surfaces leads to difficulties 1 the design of these components. With the use of remote controls, welded construc- ion, special alloys, and 'wetting agents', these difficulties have been largely solved. here remains, however, the almost prohibitive cost of the necessary quantity of 1ercury. So far no satisfactory substitute for mercury as a 'topper' has been ound, and the principle of the binary cycle is much more likely to be used in the orm of a combined gas and steam cycle. This is described in section 12.5: here nergy in the exhaust gases from a gas turbine plant is used to produce steam for Rankine cycle plant.

Fig. 11.22   A simple heating plant

1.8 Back-pressure and extraction turbines

n many industrial processes, as well as in central heating, heat has to be supplied at a noderately low temperature. A common method of supplying such heat is to evaporate vater at the appropriate temperature and pressure, transfer the heat to the process luid by condensing the steam in a heat exchanger, and then return the condensate o the boiler. This is illustrated in Fig. 11.22.

In all such plants, coal or oil is used as the source of energy, and the energy is made vailable at a temperature far in excess of that of the process steam. To adopt such a scheme is very wasteful from the thermodynamic point of view, because the large temperature difference between the products of combustion and the process steam is not used to produce work. In most industrial plants requiring process heat, electrical or

Fig. 11.23 A back-pressure turbine plant

mechanical power is required at the same time. It is more economical, therefore, to utilise the temperature difference between the metallurgical limit and the process steam temperature for the production of work, and obtain the process heat from the heat rejected in the steam condenser. Such a plant, called a *total-energy plant*, is illustrated in Fig. 11.23. A turbine exhausting into a condenser at a relatively high pressure, in which the rejected heat is employed usefully, is called a *back-pressure turbine*.

Fig. 11.24 Comparison of a steam and mercury back-pressure cycle for a high process temperature

It is clear that, apart from the thermodynamic economy achieved in combining the power and heating plant, the capital cost is also reduced. A single boiler serves two purposes, and the usual condenser of the power plant becomes superfluous as it is replaced by the process heat exchanger.

If the process temperature is fairly high, it may not be worthwhile to use a back-pressure turbine, because the power produced is then very small [Fig. 11.24a]. In

such a case it may still be worthwhile to use a mercury back-pressure turbine, the
advantage of the latter becoming obvious when Figs. 11.24a and b are compared.

The use of a back-pressure turbine is practical if the power and heat requirements
are fairly steady and well matched. If the power requirements are larger than those
that can be obtained from a simple back-pressure turbine, an *extraction*, or *pass-out*,
turbine can be used. In this alternative scheme some of the steam is expanded to the
lowest possible temperature of about 29 °C, the remainder being extracted from the
turbine at the correct intermediate temperature (Fig. 11.25). The extracted steam is
condensed in the process heat exchanger and finally returned to the boiler. This
scheme enables variations in power and heat requirements to be easily accommodated
by controlling the amount of extracted steam.

Fig. 11.25   A pass-out or extraction turbine plant

With the new emphasis on energy conservation, serious consideration is now
being given to the use of back-pressure or pass-out turbine plant for power
stations near cities. The rejected heat would be conveyed to buildings by hot
water mains. Up to now, when capital and maintenance costs are included, the
economic advantage over individual heating appliances has been marginally
against such schemes in the U.K. The saving in fuel consumption, however,
would be very considerable: see Ref. (38) for example.

11.9 Steam cycles for nuclear power plant

Nuclear power plant has been developed mainly on what can be called conventional
lines, by using the nuclear reactor as a source of heat for a steam cycle. Some effort
is being directed towards the direct conversion of nuclear energy into electrical
energy, e.g. via the magnetohydrodynamic generator briefly described in Chapter 20,
but these efforts still have to bear fruit. Here we shall discuss briefly some of the consequences arising from the application of nuclear energy sources to steam cycles.
Although the principles discussed in the preceding sections apply also to nuclear
plant, there exist a number of difficulties peculiar to such plant. They arise largely
from present restrictions on the maximum permissible temperature which 'fuel'
elements in a reactor are allowed to reach, and from the use of a secondary fluid to

Fig. 11.26 Simple nuclear power plant

transfer energy from them to the steam. There is every prospect that further development of reactor materials will ease these restrictions and therefore remove at least some of these peculiar difficulties.

Fig. 11.26a shows diagrammatically a simple scheme employing a secondary fluid which circulates through the reactor and transfers heat to the steam boiler. In early reactors the permissible maximum temperature T_{max} of the secondary fluid leaving the reactor was below the critical temperature of steam, and it might appear that the cycle shown in Fig. 11.26b would have been the obvious choice. However, quite apart from the difficulty of excessive moisture in the turbine which such a choice would entail, it should be remembered that the secondary fluid falls in temperature as it flows through the heat exchanger, and therefore the saturation temperature of the steam must be well below T_{max}. This is most easily appreciated on the $T–H$ diagram of Fig. 11.27a, where H is the product of mass flow and specific enthalpy of the respective fluids. ΔH must be the same for the two curves, it being equal to the total quantity of heat transferred. The saturation temperature is chosen so that at

Fig. 11.27

the so-called *pinch point* the secondary fluid temperature is marginally higher than the saturation temperature, and it can be seen that under these conditions a certain amount of superheating is possible.

The secondary fluid temperature drop, for a specified heat transfer, can be varied by varying the mass flow as indicated in Fig. 11.27b. A higher mass flow enables a higher saturation temperature to be used for the power cycle, and thus a higher steam-cycle efficiency to be obtained, but it also entails a higher circulating pump power and it is the net (turbine − pump) work which is important. It is obvious that the steam conditions cannot be optimised on the basis of steam-cycle efficiency alone, but must be considered in conjunction with the pumping power of the secondary fluid circuit. An analogous interaction of cycle and combustion efficiency was discussed in section 11.6, but such interactions are not nearly as critical as those in the nuclear reactor. In all British plant of commercial size the secondary fluid chosen has been a

Fig. 11.28 Dual-pressure cycle

gas, namely compressed CO_2, and for a gas the pumping power is large and increases rapidly with rate of mass flow. A liquid metal would require much less pumping power and have very good heat transfer properties, but would raise numerous difficulites with regard to operational safety and capital cost. It should be appreciated that the number of possible secondary fluids is limited because, as with any material in the reactor core, the nuclear characteristics play an overiding role.

A secondary fluid which changes phase (e.g. boiling water) undergoes no temperature drop as it condenses in the heat exchanger and would enable the highest possible saturation temperature to be used for the steam cycle. But it carries the risk of 'burn-out' of the reactor fuel elements [see section 22.8 (ii)] and it requires from the designer a precise knowledge of boiling heat transfer which until recently has been lacking. A further difficulty which arises with a boiling fluid in a reactor derives from the fact that the nuclear reactions in the fuel elements are appreciably affected by the ratio of vapour to liquid surrounding the elements, and very close control of the boiling process is essential. One attraction of using boiling heat transfer to steam is that the secondary fluid circuit can be dispensed with by passing the steam directly into the turbine, although care has to be taken to avoid steam leakage from any part of the circuit because it may be contaminated with radioactivity.

Nothing very final can be said about the best choice of cycle, because reactor materials and metallurgical limits are still changing rapidly. At present, British practice is to use a *dual-pressure* cycle, a simplified version of which is shown in Fig. 11.28. It is seen to incorporate regenerative feed heating to improve cycle efficiency, and other refinements (not shown) are mainly concerned with the reduction of moisture by

Fig. 11.29   Ideal fluids for a vapour power cycle

reheating, moisture extraction or some method of steam drying. Comparing Figs. 11.27a and 11.28b, it can be seen that the dual-pressure cycle improves the cycle efficiency by reducing the average temperature difference between the secondary fluid and steam, or in other words by increasing the average temperature at which heat is supplied to the working fluid in the cycle. Further information on steam cycles for nuclear power plant can be found in Ref. (35).

11.10 The ideal working fluid

Apart from the Carnot cycle, which was shown to be impractical, all the cycles discussed in this chapter have efficiencies and specific consumptions which are dependent upon the properties of the working fluid. Although for reasons of cost and chemical stability steam is almost always used as the working fluid in vapour cycles, its behaviour is far from ideal. It is worth considering briefly the characteristics that would be desirable in an ideal working fluid. A T-s diagram for one such fluid is sketched in Fig. 11.29a, and its main features are as follows.

(a) The critical temperature is well above the metallurgical limit. Superheating is therefore superfluous and most of the heat can be added at the upper temperature. The saturation pressure at the metallurgical limit is moderate to reduce the capital and maintenance costs of the plant.

(b) The specific heat of the liquid is small, i.e. the saturation line is steep. The heat required to ʹring the liquid to boiling point is then also small, again increasing the proportion added at the upper temperature.

(c) The latent heat is high, implying a low specific consumption and therefore smaller planᵗ for a given power output. The size of plant is also reduced if the fluid has a high density.

(d) The saturated vapour line is very steep, so that the dryness fraction after the expansion can be maintained above 0·9 without recourse to superheating.

(e) The saturation pressure at the lower temperature of about 29 °C is slightly higher than atmospheric, so that no vacuum is needed in the condenser. This avoids leakage of air into the system, the undesirability of which is discussed in sections 14.4 and 22.8 (i).

No single fluid has been found which possesses all these desirable thermodynamic characteristics, and yet at the same time is cheap, chemically stable, non-toxic, and non-corrosive.

It will be apparent from what was said about cycles for nuclear power plant in section 11.9, that the desirable characteristics of the working fluid will depend to some extent on the nature of the source of heat. The foregoing ideal working fluid is desirable only if the source is of essentially constant temperature. When the source is a hot gas undergoing a marked drop in temperature as it gives up its energy, the desirable *T-s* diagram of the working fluid is as shown in Fig. 11.29b. A cycle using a supercritical pressure, shown by the chain dotted lines, then enables the external irreversibility due to the temperature difference between source and working fluid to be reduced to a minimum. In this case it would be desirable for the critical temperature of the working fluid to be *below* the metallurgical limit. A more comprehensive discussion of the desirable characteristics of working fluids can be found in Ref. (39). (A quantitative assessment of the effect of external irreversibilities can best be made using the concept of availability discussed in section 7.9.)

12 Gas Power Cycles

This chapter is concerned with cycles which form the basis of power plant in which the working fluid is a gas. The gas is normally air, or the products of combustion of fuel and air. Such power plant may be classified into two main groups—gas turbine engines and reciprocating engines. The former are analogous to steam plant in that the individual processes are steady-flow processes carried out in separate components. The latter involve non-flow processes accomplished in a cylinder fitted with a reciprocating piston.

With few exceptions, only the ideal cycles which form the basis of these engines will be discussed in this chapter. Gas turbine cycles follow on naturally from the previous discussion of vapour power cycles; they are therefore considered first, although in a historical survey the order of presentation would be reversed.

12.1 Internal-combustion engines and air-standard cycles

In the previous chapter it was seen that a practical vapour power cycle can be devised consisting of two isentropic processes and two constant pressure processes, i.e. the Rankine cycle. This cycle has a good ideal efficiency, since most of the heat supplied is transferred at the upper temperature and the whole of the heat rejected is transferred at the lower temperature. Furthermore, the work ratio is very near unity, and therefore the efficiency is not greatly affected by irreversibilities. Let us now consider the probable effect on the performance of the cycle of using a gas as the working fluid; the cycle is then known as the *Joule* (or *Brayton*) *cycle*. The p-v and T-s diagrams for the cycle appear as in Fig. 12.1. A diagrammatic sketch of the plant—known as a closed cycle gas turbine—is also shown. The steady-flow constant pressure processes during which heat is transferred are no longer constant temperature processes and the ideal efficiency must therefore be appreciably less than the Carnot efficiency based upon the maximum and minimum temperatures of the cycle, i.e. $\eta < (T_3 - T_1)/T_3$. Also, the negative compressor work, $c_p(T_1 - T_2)$, is an appreciable proportion of the positive expansion work, $c_p(T_3 - T_4)$, so that the work ratio is considerably less than unity. The cycle has therefore a poorer efficiency than the Rankine cycle and it is much more susceptible to irreversibilities. Clearly, if such a cycle is to find favour, it must do so on other grounds.

Bearing in mind the fact that the source of energy is normally provided by the combustion of fuel in air, it can be seen that there is an advantage to be gained by using a

Fig. 12.1   The Joule cycle applied to the closed cycle gas turbine plant

gas cycle if the gas is air, since the fuel can be burnt directly in the working fluid. This idea of *internal combustion* leads to the open-cycle gas turbine plant of Fig. 12.2, in which the heater of the closed cycle is replaced by an internal-combustion chamber, and the cooler becomes superfluous because the turbine exhaust gases are rejected

Fig. 12.2   The internal-combustion gas turbine

to the atmosphere. The plant is thus much less bulky and expensive than an equivalent vapour plant with its large boiler and condenser. These are very important advantages in the case of power plant for transport, where small size and low weight are important requirements, or for peak-load generating sets where the capital cost is more important than the running cost.

The internal-combustion turbine is particularly suitable for aircraft propulsion

233

when used in conjunction with either an air-screw or propelling nozzle [see section 18.4 (i)]. When a propeller-turbine [Fig. 18.15b] is used, the net shaft work (W_3 + W_{12}) in Fig. 12.2 is simply supplied to the air-screw. If propulsion is by jet, the turbine is required to supply merely the compressor work and it uses only part of the expansion to atmospheric pressure, from 3 to 5 say. The remaining expansion, from 5 to 4, occurs in the propulsion nozzle, and the exit velocity can be calculated by the method of Example 10.2 (b).

It should be appreciated that an open cycle plant with internal combustion depart considerably from the conventional thermodynamic idea of a heat engine. In the thermodynamic sense no heat is involved at all, and the cycle is replaced by a continuous, adiabatic, steady-flow process during which the chemical energy in the fuel is converted into work in a more direct way Power plant of this kind are called *internal combustion engines*, and Fig. 12.3 illustrates the essential features common to all such plant (cf. Fig. 11.1). The combustion process is now an integral part of the whole, and it is no longer possible to treat the overall efficiency of the plant as the product of a cycle efficiency and a combustion efficiency as was done in the previous chapter. A full analysis of the performance of internal-combustion engines is therefore complex, and it involves considerations of the combustion process itself (see Chapters 15 and 17). Fortunately a great deal of useful information can be obtained from a simplified treatment.

Fig. 12.3

Let us compare the closed and open gas turbine plants of Fig. 12.1 and Fig. 12.2 a little more closely. Firstly, the net work outputs differ mainly because the specific heats and mass flows of the gases passing through the turbines are not the same—in the closed plant the fluid is air and in the open plant the fluid consists of products of combustion. Secondly, in Fig. 12.1 the fuel is consumed externally while heat is transferred to raise the temperature of the air from T_2 to T_3, whereas in Fig. 12.2 the fuel is consumed internally while air at T_2 changes to combustion products at T_3. Since the major proportion of the working fluid is nitrogen in both types of plant, it is not surprising that the quantities of work and of fuel consumed are not widely different. As a first approximation, useful deductions about the performance of internal-combustion engines can be made from equivalent gas cycles using air as the working fluid. These equivalent cycles are called *air-standard cycles*, and their efficiencies are called *air-standard efficiencies*.

The following sections will only deal with the performance of air-standard cycles It will always be assumed that the air has constant specific heats, and that the kinetic energy of the fluid is the same at inlet and outlet of each component. The discussion will not, however, be limited to cycles which are *ideal* in the sense of being reversible. but will also include the effect of irreversibilities, i.e. friction, in the compressor and the turbine.

12.2 Simple gas turbine cycle

Expressions for the efficiency and work ratio of the Joule cycle shown in Fig. 12.1 can be found in the following way. The compressor and turbine work transfers are respectively:

$$W_{12} = -(h_2 - h_1) = -c_p(T_2 - T_1)$$

and

$$W_{34} = -(h_4 - h_3) = -c_p(T_4 - T_3)$$

The heat supplied during the cycle is

$$Q_{23} = (h_3 - h_2) = c_p(T_3 - T_2)$$

and the cycle efficiency is therefore

$$\eta = \frac{W_{12} + W_{34}}{Q_{23}} = \frac{-(T_2 - T_1) - (T_4 - T_3)}{(T_3 - T_2)} \tag{12.1}$$

The cycle temperatures can be related to the *pressure ratio* $r_p = p_2/p_1 = p_3/p_4$. For isentropic compression and expansion

$$T_2 = T_1 r_p^{(\gamma-1)/\gamma} \quad \text{and} \quad T_3 = T_4 r_p^{(\gamma-1)/\gamma}$$

Inserting these values in (12.1), the ideal air-standard efficiency of the Joule cycle is found to be

$$\eta = 1 - \left(\frac{1}{r_p}\right)^{(\gamma-1/\gamma)} \tag{12.2}$$

The work ratio is

$$r_w = \frac{W_{12} + W_{34}}{W_{34}} = \frac{-(T_2 - T_1) - (T_4 - T_3)}{-(T_4 - T_3)}$$

On introducing the pressure ratio r_p, this expression can be reduced to

$$r_w = 1 - \frac{T_1}{T_3} r_p^{(\gamma-1)/\gamma} \tag{12.3}$$

The efficiency of the ideal cycle is a function of the pressure ratio only, while the work ratio, and therefore the cycle's susceptibility to irreversibilities, depend also upon T_1 and T_3. The efficiency of the real plant will therefore depend on these temperatures as well as the pressure ratio. Evidently T_1 should be as low as possible and T_3 as high as possible. In practice T_1 is limited to atmospheric temperature (say 288 K at sea level). T_3 is fixed by the metallurgical limit, i.e. the maximum temperature which the highly stressed parts of the turbine can withstand. The value depends upon the life required of the plant and the heat resistant alloy available, but for the purpose of the present discussion we may assume a limit of 1000 K. This is rather higher than the conservative figure chosen for steam plant in the previous chapter, but for most applications of gas turbines a somewhat shorter working life is acceptable and it is possible to employ alloys which are too expensive to use in a large boiler.

Having fixed T_1 and T_3, the maximum pressure ratio theoretically possible is given by

$$(r_p)_{max} = \left(\frac{T_3}{T_1}\right)^{\gamma/(\gamma-1)} \tag{12.4}$$

A glance at Fig. 12.4 shows that this is not a practical value, because at this pressure ratio the compressor and turbine works are equal and there is no net work output.

Fig. 12.4   Fig. 12.5

The net output is also zero at a pressure ratio of unity, so that there will be some intermediate pressure ratio at which the work output is a maximum. The net output per unit mass flow, i.e. the *specific work output*,* is given by

$$W = -c_p(T_2 - T_1) - c_p(T_4 - T_3)$$

$$= -c_p T_1 \left\{ r_p^{(\gamma-1)/\gamma} - 1 \right\} - c_p T_3 \left\{ \frac{1}{r_p^{(\gamma-1)/\gamma}} - 1 \right\} \tag{12.5}$$

Treating T_1 and T_3 as constant, differentiating with respect to r_p, and equating dW/dr_p to zero, it can be shown that for maximum work output

$$r_p = \left(\frac{T_3}{T_1}\right)^{\gamma/2(\gamma-1)}, \quad \text{or} \quad r_p = \sqrt{(r_p)_{max}} \tag{12.6}$$

With T_1 at 288 K and T_3 at 1000 K, the optimum pressure ratio is 8·8, and with this value the mass flow of air for a given work output, and hence the size of plant, becomes a minimum. Fig. 12.5 shows the variation of ideal efficiency with pressure ratio

---

*In gas turbine practice it is usual to use the specific work output as a criterion of plant size, rather than its reciprocal which would be the equivalent of the specific steam consumption used in the previous chapter.

obtained from equation (12.2). Evidently the gain in efficiency obtained by using a pressure ratio greater than 8·8 is not very large, and would probably not be worth the accompanying increase in size of plant.

As soon as compressor and turbine inefficiencies are introduced, as indicated in Fig. 12.6, the importance of a high work ratio becomes obvious. With a low value of T_3, the difference between the turbine work and compressor work becomes very small, and a slight decrease in the former and increase in the latter is sufficient to reduce the work output, and cycle efficiency, to zero. General expressions for cycle efficiency and specific work output become rather unwieldy when isentropic component efficiencies are taken into account, and to show the effect of variation of r_p and T_3 on the performance it is easier to make a series of calculations of the type shown in the following example.

Example 12.1 Calculate the efficiency and specific work output of a simple gas turbine plant operating on the Joule cycle. The maximum and minimum temperatures of the cycle are 1000 K and 288 K respectively, the pressure ratio is 6, and the isentropic efficiencies of the compressor and turbine are 85 and 90 per cent respectively.

The isentropic temperature changes in the compressor and turbine are found from

$$T_2' = 288 \times 6^{0.4/1.4} = 288 \times 1.6685 = 481 \text{ K}$$

and

$$T_4' = \frac{1000}{6^{0.4/1.4}} = \frac{1000}{1.6685} = 599 \text{ K}$$

Fig. 12.6

Hence the actual temperature after compression is given by

$$T_2 - T_1 = \frac{T_2' - T_1}{\eta_c} = \frac{481 - 288}{0.85} = 227 \text{ K}$$

or

$$T_2 = 515 \text{ K}$$

the actual temperature after expansion is given by

$$T_3 - T_4 = \eta_t (T_3 - T_4') = 0.90(1000 - 599)$$
$$= 361 \text{ K}$$

or

$$T_4 = 639 \text{ K}$$

The compressor and turbine works are

$$W_{12} = -c_p(T_2 - T_1) = -1.005(515 - 288) = -228 \text{ kJ/kg}$$
$$W_{34} = -c_p(T_4 - T_3) = -1.005(639 - 1000) = 363 \text{ kJ/kg}$$

and the heat added is

$$Q_{23} = c_p(T_3 - T_2) = 1.005(1000 - 515) = 487 \text{ kJ/kg}$$

237

Therefore the cycle efficiency is

$$\eta = \frac{W_{12} + W_{34}}{Q_{23}} = \frac{135}{487} = 0.277$$

and the specific work output is

$$W = W_{12} + W_{34} = 135 \text{ kJ/kg} \quad \text{or} \quad \text{kW per kg/s of air}$$

The results of calculations similar to the foregoing, but for different values of r_p and T_3, are given in Fig. 12.7. The efficiency of the irreversible cycle is not only a function of r_p, but also of T_3. There is an optimum pressure ratio for maximum cycle efficiency as well as an optimum pressure ratio for maximum specific work output

Fig. 12.7   The simple cycle with $T_1 = 288$ K, $\eta_c = 0.85$, $\eta_t = 0.90$

(although these optimum pressure ratios are not the same). The latter pressure ratio would probably be adopted in an actual plant, because it results in the smallest plant and the efficiency curve is fairly flat in this region. The ideal efficiency curve is also shown in Fig. 12.7, to emphasise the marked reduction in efficiency caused by irreversibilities in the compressor and turbine. The following section deals with the more important modifications to the simple cycle which may be adopted to improve both the ideal efficiency and work ratio.

12.3 Gas turbine cycles with heat exchange, intercooling and reheating

(i) *Simple cycle with heat exchange*

With normal values of the pressure ratio and turbine inlet temperature, the turbine outlet temperature is always above the compressor outlet temperature. An improvement in performance can therefore be effected by the addition of a *heat exchanger* which transfers heat from the gases leaving the turbine to the air before it enters the combustion chamber. The equivalent air standard cycle then appears as in Fig. 12.8. By arranging for the heat exchange to occur in a counter-flow heat exchanger (see section 24.1), it is theoretically possible for the temperature of the compressed air to be raised from T_2 to $T_x = T_4$, while the gas leaving the turbine is cooled from T_4 to $T_y = T_2$. At each point in the heat exchange process the heat is being transferred over an infinitesimal temperature difference, so that the process can be regarded as reversible and the cycle as ideal.

The effect of the heat exchanger is to reduce the amount of heat required from an external source to

$$Q_{x3} = c_p(T_3 - T_4) \tag{12.7}$$

The net work output and work ratio are unchanged, but it can easily be shown that the ideal air-standard efficiency now becomes

$$\eta = 1 - \frac{T_1}{T_3} r_p^{(\gamma-1)/\gamma} \tag{12.8}$$

Fig. 12.8

In this case, the lower the pressure ratio the higher the efficiency, the maximum value being $(T_3 - T_1)/T_3$ when $r_p = 1$. This is the Carnot efficiency based upon the maximum and minimum temperatures of the cycle. This result is not surprising because, as r_p is reduced to unity, the external heat transfers tend to take place more nearly at the upper and lower temperatures (see Fig. 12.4). The efficiency curves, shown in Fig. 12.9a, meet the curve for the simple cycle where the pressure ratio is such

Fig. 12.9   Effect of heat exchange

239

that $T_2 = T_4$, because at this point the heat exchanger becomes superfluous. Equating equations (12.8) and (12.2), it is apparent that this occurs when

$$r_p = \left(\frac{T_3}{T_1}\right)^{\gamma/2(\gamma-1)}$$

i.e. at the optimum pressure ratio for maximum work output, see equation (12.6). Clearly, if a heat exchanger is to be used, a pressure ratio somewhat less than the optimum must be adopted. Fig. 12.9a emphasises a fact which is obvious from the T-s diagram, i.e. that the higher the value of T_3 the greater is the advantage to be obtained from the addition of a heat exchanger.

In practice the heat exchanger is never 'perfect', and the actual temperature T_x reached by the compressed air is always less than T_4. This results in a major change in the shape of the efficiency curves because, as r_p tends to unity and the net work output tends to zero, the external heat supplied no longer tends to zero. The result is that at $r_p = 1$ the efficiency has a value of zero, instead of the Carnot efficiency as with the ideal heat exchanger. An actual efficiency curve therefore appears as shown in Fig. 12.9b.

(ii) *Intercooling and reheating*

The addition of a heat exchanger improves the ideal efficiency but does not improve the work ratio. The latter may be increased by either reducing the compressor work or increasing the turbine work.

Consider the compression work first. The curvature of the constant pressure lines on the T-s diagram is such that the vertical distance between them decreases in

Fig. 12.10

the direction of the arrow in Fig. 12.10. Therefore the further to the left the compression 1–2 takes place, the smaller is the work required to drive the compressor. State 1 is determined by the atmospheric pressure and temperature, but if the compression is carried out in two stages, 1–3 and 4–5, with the air cooled at constant pressure p_i between the stages, some reduction in compression work can be obtained. The sum of the temperature rises $(T_3 - T_1)$ and $(T_5 - T_4)$ is clearly less than the temperature rise $(T_2 - T_1)$.

Ideally, it is possible to cool the air to atmospheric temperature, i.e. $T_4 = T_1$, and in this case the intercooling is said to be *complete*. With isentropic compression and complete intercooling the compression work is

$$W = -c_p(T_3 - T_1) - c_p(T_5 - T_4)$$

$$= -c_p T_1 \left\{ \left(\frac{p_i}{p_1}\right)^{(\gamma-1)/\gamma} - 1 \right\} - c_p T_1 \left\{ \left(\frac{p_2}{p_i}\right)^{(\gamma-1)/\gamma} - 1 \right\}$$

The saving in work will depend on the choice of the intercooling pressure p_i. By equating dW/dp_i to zero, the condition for minimum work is found to be

$$p_i = \sqrt{(p_1 p_2)} \qquad\qquad (12.9)$$

Hence

$$\frac{p_i}{p_1} = \frac{p_2}{p_i} = r_{pi}$$

and

$$r_{pi} = \sqrt{\left(\frac{p_2}{p_1}\right)} = \sqrt{r_p} \qquad (12.10)$$

Thus for minimum compressor work, the compression ratios and work inputs for the two stages are equal.

The compression work can be reduced further by increasing the number of stages and intercoolers, but the additional complexity and cost make more than two or three stages uneconomic. It is possible to generalise the expression for the minimum compression work to cover n stages and show that the pressure ratios in all stages must be equal.

Fig. 12.11

From here it is only a small step to the idea of *reheating* during the expansion. The principle of reheating has been mentioned previously in section 11.4, but there the main object was to avoid excessive moisture in the steam turbine, and the work ratio, already near unity in the Rankine cycle, is practically unaffected. Reheating is employed in gas turbine plant principally to increase the work ratio, and so increase the specific work output and decrease the effect of component losses. Fig. 12.11 illustrates the relevant part of the cycle showing an expansion in two stages with reheating to the metallurgical limit, i.e. $T_9 = T_6$. The turbine work is increased from W_{67} to

$$W_{6.8} + W_{9.10} = -c_,(T_8 - T_6) - c_p(T_{10} - T_9)$$

It is possible to show that with isentropic expansion the optimum intermediate pressure, this time for *maximum* work, is given by

$$p_i = \sqrt{(p_6 p_7)} \quad \text{or} \quad r_{pi} = \sqrt{r_p} \qquad\qquad (12.11)$$

Reheating can also be extended to more than two stages, although this is seldom done in practice and with open-cycle plant a limit is set by the oxygen available for combustion.

Although intercoolers and reheaters improve the work ratio, these devices by themselves can lead to a decrease of ideal cycle efficiency. This is because the heat supplied is increased as well as the net work output. The full advantage is only reaped if a heat exchanger is also included in the plant, as shown in Fig. 12.12. The additional heat required for the colder air leaving the compressor can then be obtained from the hotter exhaust gases, and there is a gain in ideal cycle efficiency as well as work ratio.

It is worth considering briefly what happens if the idea of multi-stage compression and expansion, with intercooling, reheating and heat exchange, is carried to its logical conclusion. Fig. 12.13 shows a cycle with a large number of such stages. It is evident that with an infinite number of stages this cycle would have all its heat addition at the

Fig. 12.12   The gas turbine plant with intercooling, heat exchange and reheating

upper temperature T_3, and all the heat rejection at its lower temperature T_1. The compression and expansion processes become isothermal, as shown by the dotted lines, and the efficiency of the cycle must be equal to the Carnot efficiency, i.e. $(T_3 - T_1)/T_3$. This cycle is called the *Ericsson cycle*.

Fig. 12.13   The Ericsson cycle

12.4   Closed cycle gas turbine

In section 12.1 it was implied that a gas turbine working on the closed cycle (Fig. 12.1) had little to recommend it in comparison with the equivalent steam plant. Nevertheless one or two closed cycle plants have been constructed and this type deserves a brief mention here. It appears that it will not be long before nuclear reactors (section 11.9) are capable of operating at the high temperatures required for gas turbine cycles. When considering the use of a nuclear reactor in conjunction with a gas turbine, it would seem desirable to pass the gas from the compressor directly through the reactor and thus avoid the complication of a secondary fluid circuit and additional heat exchangers. If so, however, it is essential to think in terms of a closed cycle, because it would be dangerous to reject continuously into the atmosphere gas which may have been contaminated by radioactivity. There are several attractive possibilities peculiar to the closed cycle gas turbine which, if realised in practice, may make it a serious competitor of steam plant for nuclear power stations. These possibilities may be summarised as follows.

1. The system may be pressurised, i.e. the general pressure level of the cycle may be raised, so that the size of all the components can be reduced for the same mass flow.
2. It is possible to use a gas having more favourable properties than air. A simple calculation with the aid of equation (12.2) shows that the cycle efficiency is markedly improved if the value of γ is increased from 1·4 to 1·67, which is the value for all monatomic gases such as helium and argon. Helium has the further attraction of a much higher thermal conductivity than air, resulting in the use of smaller heat exchangers.
3. With a pressurised system it is possible to accommodate changes in load by varying the mass flow of fluid in the circuit instead of by reducing the turbine inlet temperature as in the open cycle plant. The efficiency at part load will then be closer to the value obtained when running at the design condition.

Before leaving the subject of gas turbines it is worth emphasising that the processes are carried out in separate components which can be multiplied and linked together in a very wide variety of ways to suit different applications. No attempt has been made here to indicate the actual diversity which is possible. For this kind of information, and for methods of estimating the performance of actual gas turbine plant as opposed to air-standard cycles, the reader is referred to more specialised books, e.g. Refs. (3) and (10). The remaining sections in this chapter will be devoted to a survey of the air-standard cycles which form the basis of reciprocating internal-combustion engines.

12.5 Combined gas and steam cycles

The advantage of a binary cycle in making the best use of the available temperature difference between source and sink was pointed out in section 11.7. Because the exhaust gases leave a gas turbine at a moderately high temperature (about 600 K) it is possible to use the energy therein to form steam for a Rankine cycle plant. Such combined power plants are often referred to as 'Cogas' or 'Stag' plants, and an increasing number are being built for base load electricity generation. A dual-pressure steam cycle might be used to make the best use of the low-grade heat, as explained in section 11.9 on cycles for nuclear power plant and depicted in Fig. 11.28. The only change is that the turbine exhaust gases replace the reactor coolant in the figure.

Because of the restriction on the maximum temperature which the turbine blades can withstand, gas turbines operate with a very high air/fuel ratio. One consequence is that additional fuel can be burnt in the ample oxygen remaining in the exhaust gases. An alternative to using the complex dual-pressure steam cycle, therefore, is to burn fuel in the steam boiler and use a higher pressure, simple steam cycle. Low-grade fuel can be used for this part of the cycle. Which solution is preferred in practice depends upon the effect of the relative capital cost of the different types of boiler upon the cost of the electricity produced. Combined cycle plant efficiencies of about 45 per cent are possible.

It should be obvious from the foregoing that gas turbines can also be used in conjunction with waste-heat boilers to produce process steam or hot water for use in factories: when used in this way the term 'total-energy' plant is used.

12.6 Reciprocating engine cycles

So far in the discussion of vapour and gas power cycles we have thought of the cycle as a series of steady-flow processes, each being carried out in a separate component. The turbine has therefore always been regarded as the natural power producing component. As will be seen in Chapter 19, the turbine becomes very small for low power outputs, i.e. for small mass flows of working fluid, and the effect of viscous friction becomes relatively great. Consequently, for a small plant the process efficiency of the turbine is very low. It is usually only practicable to think in terms of a turbine plant if the power required is at least several hundred kilowatts.

The combination of steady-flow components into a power plant is not the only method of obtaining mechanical power from the combustion of fuel. For small powers it is usually preferable to employ a cycle consisting of a succession of non-flow processes. A given mass of working fluid can be taken through a series of processes in a cylinder fitted with a reciprocating piston. Such non-flow processes are much more nearly reversible than the rapid flow processes occurring in a turbine or rotary compressor [see section 5.4 (i)].

A second advantage of the reciprocating engine is that the maximum permissible temperature of the working fluid is much higher than it is in a turbine plant (≈ 2800 K compared with ≈ 1000 K in a gas turbine). This is because the cylinder of the intermittently working reciprocating engine is only exposed to the peak temperature of the fluid for a small fraction of the duration of a complete cycle. The cylinder temperature therefore never even approaches this peak value. It follows that metallurgical considerations are much less stringent in a reciprocating engine. The chief disadvantage of a reciprocating engine is the comparatively low rate at which it can handle the working fluid. This feature makes it excessively bulky and heavy when large power outputs are required. There is of course a range of power output where reciprocating and turbine plant are competitive, and their relative merits will be more fully appreciated after a reading of Chapters 17 and 19.

Like the normal type of gas turbine plant, the practical reciprocating engine does not use an external heat source or employ a true thermodynamic cycle, but derives its energy from the direct combustion of fuel in the working fluid. The engine takes in air and fuel and rejects products of combustion as depicted in Fig. 12.3. The actual processes occurring in reciprocating engines will be described in Chapter 17, and here we shall only consider the equivalent air-standard cycles with external heat addition and rejection taking the place of internal combustion and rejection of hot exhaust gases. There are three important air-standard cycles which form the basis of all practical reciprocating engines; they are the *Otto*, *Diesel*, and *dual* or *mixed* cycles. A few other cycles have been used and will be mentioned briefly, but so far they have not attained any importance.

12.7 The Otto, Diesel and mixed cycles

The *Otto cycle* forms the basis of spark-ignition and high speed compression-ignition engines. The four non-flow processes comprising a complete cycle are shown in Fig. 12.14. They may be imagined to occur in a cylinder fitted with a reciprocating piston having a swept volume equal to $m(v_1 - v_2)$, where m is the mass of fluid in the cylinder. The processes are as follows.

1-2   Air is compressed isentropically through a volume ratio v_1/v_2, known as the *compression ratio* r_v.
2-3   A quantity of heat Q_{23} is added at constant volume until the air is in state 3.
3-4   The air is expanded isentropically to the original volume.
4-1   Heat Q_{41} is rejected at constant volume until the cycle is completed.

The efficiency of the cycle is

$$\eta = \frac{W}{Q_{23}} = \frac{Q_{23} + Q_{41}}{Q_{23}}$$

Assuming constant specific heats for the air, and considering unit mass of fluid, the heat transfers are

$$Q_{23} = c_v(T_3 - T_2)$$
$$Q_{41} = c_v(T_1 - T_4)$$

Consequently

$$\eta = 1 - \frac{T_4 - T_1}{T_3 - T_2}$$

For the two isentropic processes,

$$\frac{T_2}{T_1} = \frac{T_3}{T_4} = r_v{}^{\gamma-1}$$

By algebraic manipulation we then have

$$1 - \frac{T_2}{T_3} = 1 - \frac{T_1}{T_4}, \quad \text{or} \quad \frac{T_4 - T_1}{T_3 - T_2} = \frac{T_4}{T_3}$$

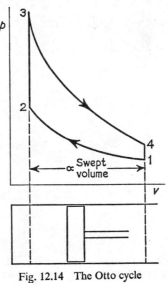

Fig. 12.14   The Otto cycle

Hence the ideal cycle efficiency becomes

$$\eta = \frac{T_3 - T_4}{T_3} \tag{12.12}$$

or in terms of the compression ratio,

$$\eta = 1 - \frac{1}{r_v^{\gamma-1}} \tag{12.13}$$

The maximum possible efficiency based upon the maximum and minimum temperatures of the cycle, i.e. the Carnot efficiency, is $(T_3 - T_1)/T_3$. Equation (12.12) indicates that the cycle efficiency is less than this, since $T_4 > T_1$. This result is to be expected since the heat addition and rejection do not take place at the upper and lower temperatures. Equation (12.13) emphasises the important fact that the efficiency depends only on the compression ratio and not upon the peak temperature T_3. The specific work output, on the other hand, does increase with increase of T_3.

In the *Diesel cycle** the heat addition occurs at constant pressure instead of constant volume. The cycle is shown in Fig. 12.15 and consists of the following processes.

1-2   Air is compressed isentropically through the compression ratio $r_v = v_1/v_2$.
2-3   Heat Q_{23} is added while the air expands at constant pressure to volume v_3. At state 3 the heat supply is cut off and the volume ratio v_3/v_2 may conveniently be called the *cut-off ratio* r_c.
3-4   The air is expanded isentropically to the original volume.
4-1   Heat Q_{41} is rejected at constant volume until the cycle is completed.

The heat transfers in this cycle are

$$Q_{23} = c_p(T_3 - T_2)$$

$$Q_{41} = c_v(T_1 - T_4)$$

Hence the efficiency is

$$\eta = \frac{Q_{23} + Q_{41}}{Q_{23}} = 1 - \frac{T_4 - T_1}{\gamma(T_3 - T_2)}$$

The efficiency can be expressed in terms of r_v and r_c in the following way.

$$\frac{T_2}{T_1} = r_v^{\gamma-1}, \quad \text{and hence} \quad T_1 = T_2 \left(\frac{1}{r_v}\right)^{\gamma-1}$$

Fig. 12.15   The Diesel cycle

---

*The name is strictly speaking a misnomer because Diesel's efforts were originally directed towards building an engine which would have the Carnot cycle as its equivalent air-standard cycle. Ackroyd Stuart was the first to build a successful engine working on the so-called Diesel cycle.

246

For the constant pressure process

$$\frac{T_3}{T_2} = r_c, \quad \text{and hence} \quad T_3 = T_2 r_c$$

Also

$$\frac{T_4}{T_3} = \left(\frac{v_3}{v_4}\right)^{\gamma-1} = \left(\frac{v_3}{v_2}\frac{v_2}{v_4}\right)^{\gamma-1} = \left(\frac{r_c}{r_v}\right)^{\gamma-1}$$

and hence

$$T_4 = T_2 r_c \left(\frac{r_c}{r_v}\right)^{\gamma-1}$$

Substituting for T_1, T_3 and T_4 in the expression for the efficiency, we have

$$\eta = 1 - \frac{1}{r_v^{\gamma-1}}\left\{\frac{r_c^{\gamma} - 1}{\gamma(r_c - 1)}\right\} \tag{12.14}$$

Evidently the efficiency of the Diesel cycle depends upon r_c, and hence upon the quantity of heat added, as well as on the compression ratio r_v. Since the term in brackets is always greater than unity (except for the trivial case where $r_c = 1$ and there is no heat addition), the Diesel cycle always has a lower efficiency than the Otto cycle of the same compression ratio (Fig. 12.16). This is not a very significant result because, as will be explained in Chapter 17, practical engines based upon the Diesel cycle can employ higher compression ratios than those based on the Otto cycle.

The behaviour of many reciprocating engines is more adequately represented by the

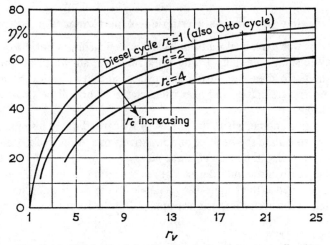

Fig. 12.16   Efficiency of the Diesel cycle for various cut-off ratios

dual or *mixed cycle* shown in Fig. 12.17. In this cycle part of the heat addition occurs during a constant volume process and the remainder during a constant pressure process. Writing $r_v = v_1/v_2$, $r_c = v_4/v_3$ and $r_p = p_3/p_2$, it is possible to show that the efficiency is given by

$$\eta = 1 - \frac{1}{r_v^{\gamma-1}} \frac{r_p r_c^\gamma - 1}{(r_p - 1) + \gamma r_p(r_c - 1)} \tag{12.15}$$

Fig. 12.18 shows the three cycles on the *T-s* diagram. They have been drawn for the case where both the compression ratios and heat inputs are the same for each. The quantity of heat rejected can be seen from the areas to be least in the Otto cycle and

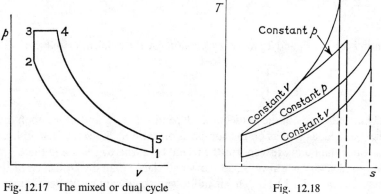

Fig. 12.17   The mixed or dual cycle   Fig. 12.18

greatest in the Diesel cycle. Without recourse to equations (12.13), (12.14) and (12.15) therefore, it can be seen that the cycle efficiencies decrease in the order—Otto, mixed, Diesel.

12.8 Mean effective pressure as a criterion of performance

In section 11.1 it was explained that the ideal cycle efficiency is not the only criterion by which a cycle should be judged. The work ratio was shown to be another useful criterion of the performance of steady-flow cycles, since it indicates the susceptibility of the cycle to irreversibilities and also whether the plant will be small or large per unit power output.

In a reciprocating engine, where the processes are carried out in a single component, it is not so easy to isolate the positive and negative work in the cycle. For this reason another criterion, called the *mean effective pressure*, is usually preferred to the work ratio when comparing air-standard cycles of reciprocating engines. The mean effective pressure p_m, is defined as the height of a rectangle on the *p-v* diagram having the same length and area as the cycle. This is illustrated in Fig. 12.19, from which it is apparent that

$$p_m(v_1 - v_2) = \oint p \, dv = W \tag{12.16}$$

where W is the net work output per unit mass of fluid. p_m can be regarded as that con-stant pressure which, by acting on the piston over one stroke, can produce the net work of the cycle. The mean effective pressure, unlike the work ratio, is not dimen-

Fig. 12.19

sionless, and it will here be expressed in bars. A method of evaluating the mean effective pressures of actual engines is described in section 16.2.

From equation (12.16) it is evident that a cycle with a large mean effective pressure will produce a large work output per unit swept volume, and hence an engine based on this cycle will be small for a given work output. Irreversibilities due to viscous friction are usually small in non-flow processes, unless a considerable amount of turbulence is artificially introduced as in certain types of compression-ignition engine. What is of considerable significance in reciprocating engines is *mechanical* friction, which results in the useful shaft power being appreciably less than the work done by the fluid on the piston. Since mechanical friction decreases with decrease in engine size, a large mean effective pressure implies that a smaller fraction of the net work of the cycle will be dissipated in this manner.*

It follows that the mean effective pressure is a useful criterion for the comparison of reciprocating engine cycles in that it indicates relative engine size, and also how far the actual engine efficiency will depart from the cycle efficiency. As an illustration of this consider the modification to the Diesel cycle shown in Fig. 12.20. In this case

Fig. 12.20

*Mechanical friction also depends to some extent on the peak pressure and the average absolute pressure level during the cycle—higher pressures leading to greater bearing losses, particularly if lubrication is imperfect. It has been suggested that a better criterion for susceptibility to mechanical friction loss is provided by the ratio of mean effective pressure to peak pressure. This criterion, which is dimensionless, is truly analogous to the work ratio.

the rejection of heat occurs during a constant pressure process instead of during a constant volume process. The cycle efficiency is improved, because the heat rejection occurs at a lower average temperature (T_5 being lower than T_4). The additional work resulting from this modification, however, is obtained at the expense of a disproportionate increase in piston stroke, i.e. the mean effective pressure is reduced. The actual engine based upon this cycle would therefore not necessarily have a higher efficiency than one based upon the Diesel cycle, and it would certainly be a larger engine. The modified cycle is in fact the Joule cycle, described previously in section 12.1, which has found acceptance only as a basis for gas turbine plant. The following example shows the relative magnitudes of the mean effective pressure of the Diesel and Joule cycles.

Example 12.2 1 kg of air is taken through (a) a Diesel cycle (Fig. 12.15) and (b) a Joule cycle (Fig. 12.1). Initially the air is at 288 K and 1 atm. The compression ratio v_1/v_2 for both cycles is 15, and the heat added is 1850 kJ and in each case. Calculate the ideal cycle efficiency and mean effective pressure for each cycle.

For both cycles:

The initial volume of the air is

$$v_1 = \frac{RT_1}{p_1} = \frac{0 \cdot 287 \times 288}{1 \cdot 01325 \times 100} = 0 \cdot 816 \text{ m}^3$$

After compression

$$T_2 = T_1 \left(\frac{v_1}{v_2}\right)^{\gamma-1} = 288 \times 15^{0 \cdot 4} = 288 \times 2 \cdot 954 = 851 \text{ K}$$

and

$$v_2 = \frac{0 \cdot 816}{15} = 0 \cdot 054 \text{ m}^3$$

The temperature rise during heat addition is given by

$$Q_{23} = c_p(T_3 - T_2)$$
$$= 1 \cdot 005(T_3 - 851) = 1850 \text{ kJ/kg}$$

and hence

$$T_3 = 2692 \text{ K}$$

For the Diesel cycle:

$$\frac{v_4}{v_3} = \frac{v_4}{v_2} \frac{v_2}{v_3} = 15 \times \frac{851}{2692} = 4 \cdot 742$$

and hence

$$T_4 = T_3 \left(\frac{v_3}{v_4}\right)^{\gamma-1} = 2692 \left(\frac{1}{4 \cdot 742}\right)^{0 \cdot 4} = 1444 \text{ K}$$

The heat rejected is

$$Q_{41} = c_v(T_1 - T_4) = 0 \cdot 718(288 - 1444) = -830 \text{ kJ/kg}$$

and the net work is

$$W = Q_{23} + Q_{41} = 1020 \text{ kJ/kg}$$

Therefore the efficiency is

$$\eta = \frac{W}{Q_{23}} = \frac{1020}{1850} = 0.551$$

The mean effective pressure is

$$p_m = \frac{W}{v_1 - v_2} = \frac{1020}{(0.816 - 0.054)100} = 13.4 \text{ bar}$$

For the Joule cycle:

$$T_4 = T_3 \left(\frac{v_3}{v_4}\right)^{\gamma-1} = 2692 \left(\frac{1}{15}\right)^{0.4} = 911 \text{ K}$$

and

$$v_4 = \frac{RT_4}{p_4} = \frac{0.287 \times 911}{1.01325 \times 100} = 2.580 \text{ m}^3$$

The heat rejected is

$$Q_{41} = c_p(T_1 - T_4) = 1.005(288 - 911) = -626 \text{ kJ/kg}$$

and the net work is

$$W = Q_{23} + Q_{41} = 1224 \text{ kJ/kg}$$

The efficiency is therefore

$$\eta = \frac{W}{Q_{23}} = \frac{1224}{1850} = 0.662$$

The mean effective pressure is

$$p_m = \frac{W}{v_4 - v_2} = \frac{1224}{(2.580 - 0.054)100} = 4.85 \text{ bar}$$

Note that the net work could have been calculated from $\oint p \, dv$, and the efficiency from equations (12.2) and (12.14) respectively.

12.9 Cycles having the Carnot efficiency

The Otto, Diesel and mixed cycles all have efficiencies less than the Carnot efficiency based upon the maximum and minimum temperatures in the cycle. The Carnot cycle itself is quite unsuitable as a basis for a reciprocating engine using a 'perfect gas' as a working fluid because the mean effective pressure of the cycle is very small. This should be apparent from the thin appearance of the cycle on the *p-v* diagram in Fig. 12.21. There are other cycles having the Carnot efficiency which do not suffer from this defect, and although they are of little practical significance at the present time, they are worthy of a brief mention here.

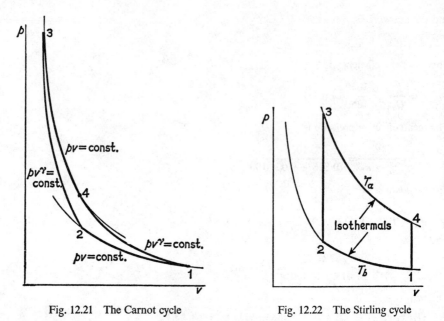

Fig. 12.21   The Carnot cycle   Fig. 12.22   The Stirling cycle

Consider the cycle shown in Fig. 12.22, consisting of two constant volume processes and two isothermals—it is called the *Stirling cycle*. The heat supplied during process 2–3 is equal in quantity to the heat rejected during process 4–1. Furthermore the temperature of the fluid varies between the same limits during these two processes. It is therefore theoretically possible for the heat rejected, Q_{41}, to be returned to the working fluid as Q_{23}. Ideally, this heat transfer can be accomplished reversibly in a *regenerator*, which consists essentially of a matrix of wire gauze or small tubes. Fig. 12.23 illustrates the principle of operation. A temperature gradient from T_a to T_b is maintained along the matrix. The working fluid enters the matrix in state 4, transfers heat to the matrix, and leaves in state 1; each element of the matrix is raised in temperature by an infinitesimal amount. Then working fluid in state 2 passes in the reverse direction, cooling each element of the matrix by an infinitesimal amount, and leaving in state 3. At no time need the working fluid and matrix differ more than infinitesimally in temperature, so that the entire heat transfer process is carried out reversibly.

With a perfect regenerator, the only heat added from an external source during the cycle is that which is transferred during the isothermal process 3–4 (i.e. at T_a), and the only heat rejected to an external sink is the quantity transferred during the isothermal process 1–2 (i.e. at T_b). It follows that the efficiency of the cycle is $(T_a - T_b)/T_a$. The mean effective pressure of this cycle is much greater than that of the Carnot cycle, and is in fact comparable to that of the Otto cycle. Engines which have been built to operate on the Stirling cycle are in fact heat engines in the sense of Fig. 12.1, with external combustion. The main reason why such engines have had little success is the

Fig. 12.23   The regenerator   Fig. 12.24   The Ericsson cycle

difficulty of designing an efficient regenerator of reasonable size (air being a poor conductor of heat) which can operate at a sufficiently high temperature. One end of the regenerator matrix must be continuously maintained at the upper temperature, and the maximum permissible temperature is therefore subject to metallurgical limitations: it cannot be as high as in reciprocating internal-combustion engines. Developments in high-temperature alloys, and improved knowledge of heat transfer processes, have led to new attempts to construct a 'Stirling engine'. These are described in Ref. (24), and efficiencies as high as those achieved in compression-ignition engines have been claimed for experimental engines.* Since there are no restrictions on the type of fuel that can be used, such engines may be useful for military applications and for small power plant in developing countries. Furthermore, external combustion systems can be designed to give exhausts with low noxious emissions, and increasing concern with environmental pollution may lead to the development of Stirling engines for road vehicles. They would then have to compete with attempts to build Rankine cycle plant for the same purpose.

A similar cycle—the *Ericsson cycle*—consists of two constant pressure processes and two isothermals (Fig. 12.24). In this case the heat rejected during one constant pressure process is returned via a regenerator to the working fluid during the other constant pressure process, and again the cycle has the Carnot efficiency. This is in fact the same cycle as that depicted in Fig. 12.13—and it was originally proposed as a basis for a reciprocating engine. The Ericsson cycle suffers from the same drawbacks as the Stirling cycle.

*A very successful air liquefaction plant working on the reversed Stirling cycle has also been developed and is now widely used.

13　Heat Pump and Refrigeration Cycles

In a power cycle, heat is received by the working fluid at a high temperature and rejected at a low temperature, while a net amount of work is done *by* the fluid. Cycles can be conceived in which the reverse happens, heat being received at a low temperature and rejected at a high temperature, while a net amount of work is done *on* the fluid. The latter are called *heat pump* or *refrigeration* cycles. The term 'heat pump' is usually applied to a machine whose principal purpose is to supply heat at an elevated temperature, and the term 'refrigerator' to one whose purpose is the extraction of heat from a cold space. This distinction in terminology is arbitrary because a heat pump and refrigerator are identical in principle and it is possible to use one machine to fulfil the function of a heat pump and refrigerator simultaneously. For example, one machine can provide a cold space for food storage and also supply heat to a domestic hot water tank. When no distinction is necessary we shall refer to the cycle as a refrigeration cycle.

Practical refrigeration cycles are composed of flow processes, each process being carried out in a separate component. As in Chapter 11, we shall again assume that changes in potential and kinetic energy between the inlet and outlet of each component are negligible. The working fluids are almost always in the liquid or vapour phase. Air is occasionally used, but only in air-conditioning plant where the air fulfils the dual function of the refrigerating fluid and air-conditioning medium.

13.1 Reversed Carnot cycle and performance criteria

If the series of processes which make up a reversible power cycle are plotted on a *p-v* or *T-s* diagram, the enclosed area is traced out in a clockwise sense, indicating that the net work done is positive. The negative net work of a reversible refrigeration cycle is proportional to an area traced out by processes in an anticlockwise sense. A reversed Carnot cycle, using a wet vapour as a working fluid, is shown in Fig. 13.1. Vapour is compressed isentropically from a low pressure and temperature (state 1) to a higher pressure and temperature (state 2), and is passed through a condenser in which it is condensed at constant pressure to state 3. The fluid is then expanded isentropically to its original pressure (state 4), and is finally evaporated at constant pressure to state 1.

The criterion of performance of the cycle, expressed as the ratio output/input, depends upon what is regarded as the output. In a refrigerator we endeavour to extract the maximum amount of heat Q_{41} in the evaporator for a net expenditure of work W. Therefore the *coefficient of performance* of a refrigerator is defined as

$$c.p._{ref} = -\frac{Q_{41}}{W} \tag{13.1}$$

Since W is negative and Q_{41} positive with our sign convention, the negative sign is introduced to make c.p._{ref} a positive number. In a heat pump we are concerned to

Fig. 13.1   The refrigeration or heat pump cycle

obtain the maximum amount of heat Q_{23} from the condenser for the net expenditure of work W, and therefore the coefficient of performance of a heat pump is defined by

$$c.p._{hp} = \frac{Q_{23}}{W} \tag{13.2}$$

The relation between these two coefficients of performance can be established by applying the First Law. Thus

$$(Q_{41} + Q_{23}) - W = 0$$

and hence

$$c.p._{hp} = c.p._{ref} + 1 \tag{13.3}$$

Inspection of the areas representing Q_{41}, Q_{23} and W on the T-s diagram (Fig. 13.1) shows that c.p._{ref} may be greater than unity, and that c.p._{hp} must always be greater than unity—the reason why the term 'efficiency' is not employed.

From equation (13.2) it is evident that c.p._{hp} is the reciprocal of the efficiency of a power cycle. The c.p._{hp} of a reversed Carnot cycle, operating between an upper temperature T_a and lower temperature T_b, must therefore be equal to $T_a/(T_a - T_b)$. Equation (13.3) then shows that the c.p.$_{ref}$ of a reversed Carnot cycle is $T_b/(T_a - T_b)$

Both coefficients become smaller as the temperature range of the cycle is increased. The upper cycle temperature must be 10 K or so above the temperature of the atmosphere (refrigerator) or hot water tank (heat pump), so that the required rate of heat transfer can be accomplished in a condenser of economic size. Similarly the lower cycle temperature must be below the temperature of the cold space (refrigerator) or surroundings (heat pump) to enable the evaporator to be of economic size. Evidently these necessary temperature differences (i.e. external irreversibilities) place a limit on the coefficient of performance attainable.

A typical temperature range over which the working fluid of a heat pump might operate is from 5 °C to 60 °C. The ideal coefficient of performance of a heat pump operating on a reversed Carnot cycle under these conditions is therefore $333/(60 - 5)$ = 6·05. That is, 6·05 times the quantity of work supplied is delivered as heat at the required temperature. If the heat pump is driven by a power plant operating with an actual overall efficiency of about 0·30, the heat delivered at the required temperature is twice the heat supplied from the high-temperature source of the power cycle. This ratio might be termed the 'thermal advantage' of the combined plant. It would appear that there is much to be gained by burning fuel in a heat engine coupled to a heat pump, instead of using the combustion as a direct source of heat for warming a building. This is only to be expected: the latter procedure involves a considerable degradation of energy, because the quantity of heat is allowed to flow through a large temperature range without producing work. To arrive at the advantage of about 2, however, we have assumed the use of a perfect heat pump. In practice the ideal cycle used has a lower coefficient of performance than the reversed Carnot cycle, and irreversibilities reduce the coefficient of performance still further. The real advantage of such a plant would therefore be smaller than 2. Whether in any proposed installation the advantage will be worth the outlay in plant must be determined by careful economic analysis of fuel, capital and maintenance costs.

13.2 Practical refrigeration cycles

Using arguments similar to those in Chapter 5, it is possible to show that all reversible cycles operating between the same two reservoirs have the same coefficient of performance, i.e. that of the reversed Carnot cycle. One has only to assume the opposite to be true and show that this contradicts the Second Law. Practical refrigerators, however, use a different form of cycle which, although having a lower ideal c.p., has other features to commend it.

A major simplification in the plant is achieved by dispensing with an expansion machine and using a simple throttle valve to obtain the reduction in pressure (Fig. 13.2). Thus the end states 3 and 4 of the expansion process lie on a constant enthalpy line instead of an isentropic. The effect of this modification can be seen by comparing the *T*-*s* diagrams of Figs. 13.1 and 13.2. The net work required is now equal to the

ompression work, because no work is obtained from the expansion process. The heat
ransfer in the condenser is unaffected, but the heat extracted in the evaporator is
diminished. Both c.p.$_{ref}$ and c.p.$_{hp}$ are therefore reduced by the modification. Also,
he rate of flow of working fluid is increased *when the cycle is used for refrigeration*,
because the heat extracted per unit mass of refrigerant (Q_{41}) is reduced. Q_{41}, known
as the *refrigeration effect*, is a useful criterion for comparing cycles on the basis of
ize of plant required for a given *duty* (heat extracted per unit time).

Although the heat transfers still take place at the upper and lower temperatures of
he cycle, the throttling process introduces an irreversibility which makes the cycle
irreversible as a whole. The coefficient of performance is then no longer independent

Fig. 13.2   A refrigeration cycle incorporating a throttle valve

of the working fluid as it is for the reversed Carnot cycle. By and large, the properties
required of an ideal working fluid for heat pumps and refrigerators are similar to
those required for an engine fluid and listed in section 11.10. The principal modifica-
ions are that: (a) the critical temperature need only be above atmospheric temperature
or hot water temperature) and not above the metallurgical limit, and (b) the satura-
on pressure should be above atmospheric pressure at the lowest evaporator tempera-
ure likely to be required. (b) is desirable so that any leak in the circuit will lead to an
fflux of refrigerant (easily detectable), and not an influx of atmospheric air. The
atter contains moisture which would ice up the throttle valve. Fluids in common use
re ammonia (NH_3), methyl chloride (CH_3Cl), sulphur dioxide (SO_2), and fluorine
ompounds such as Freon-12 (CF_2Cl_2). Carbon dioxide (CO_2), which has been widely
sed in the past, has high vapour pressures at normal temperatures (requiring ex-
ensive pipe work) and a relatively low critical temperature, and it has mostly been
isplaced by the other fluids. Abridged tables of properties of ammonia and Freon-12
re given in Ref. (17), and comprehensive tables for all commonly used refrigerants
an be found in Ref. (27).

Example 13.1 Calculate the refrigeration effect and coefficient of performance for the refrigeration cycle shown in Fig. 13.2, when the fluid is ammonia and the upper and lower temperatures are 30 °C and −15 °C respectively. Find also the corresponding values for a reversed Carnot cycle operating between the same temperatures.

From the ammonia tables we have

$$h_4 = h_3 = h_f = 323 \cdot 1 \text{ kJ/kg}$$
$$h_2 = h_g = 1468 \cdot 9 \text{ kJ/kg}, \quad s_2 = 4 \cdot 984 \text{ kJ/kg K}$$

From $s_1 = s_2$, we have $x_1 = 0 \cdot 889$, and hence

$$h_1 = h_f + x_1 h_{fg} = 1280 \cdot 7 \text{ kJ/kg}$$

The refrigeration effect is

$$Q_{41} = (h_1 - h_4) = 957 \cdot 6 \text{ kJ/kg}$$

and the net work expended is

$$W = W_{12} = -(h_2 - h_1) = -188 \cdot 2 \text{ kJ/kg}$$

As a refrigerator the coefficient of performance is

$$\text{c.p.}_{\text{ref}} = -\frac{957 \cdot 6}{-188 \cdot 2} = 5 \cdot 09$$

For the reversed Carnot cycle,

$$\text{c.p.}_{\text{ref}} = \frac{273 - 15}{30 - (-15)} = 5 \cdot 73$$

$s_3 = 1 \cdot 204 \text{ kJ/kg K}$, and from $s_4 = s_3$ we have $x_4 = 0 \cdot 147$, and hence $h_4 = 305 \cdot 6 \text{ kJ/kg}$. The refrigeration effect is therefore

$$Q_{41} = (h_1 - h_4) = 975 \cdot 1 \text{ kJ/kg}$$

The foregoing example shows that the refrigeration effect is not significantly altered by replacing the expansion machine with a throttle valve, but that the coefficient of performance is noticeably reduced—from 5·73 to 5·09. The work recovered in the expander of the reversed Carnot plant is, however, quite small compared with the compression work, e.g. 17·5 versus −188·2 kJ/kg. In an actual plant using this ideal cycle much of the expansion work would be dissipated by mechanical friction. The substitution of a throttling expansion does not, therefore, result in a significant change in the performance of the real cycle, and it constitutes a worthwhile mechanical simplification of the plant.

A practical ideal cycle differs in two other respects from the reversed Carnot cycle. Firstly, the compression is usually carried out in the superheat region. Not only would it be difficult to arrange for the evaporation to cease at state 1 (Fig. 13.2), but any liquid refrigerant passing into the compressor would tend to wash away the lubricating oil. This is obviously undesirable if, as is usual, the compressor is a reciprocating or positive-displacement rotary type with rubbing surfaces. Moreover the oil would be carried to the evaporator where it might form a film on the tube surfaces and impair the heat transfer process. It is usual, therefore, to transfer a saturated or slightly super-heated vapour to the inlet of the compressor. The second modification is that the condensed liquid is often subcooled before entering the throttle valve. As already observed, most of the heat transferred in the condenser must be rejected by virtue of an appreciable temperature difference, to keep the size of condenser within reasonable limits. It is therefore possible to subcool the liquid after condensation to within a few degrees of the surrounding temperature.

Both these modifications are illustrated in Fig. 13.3a, and it is clear that they both tend to increase the refrigeration effect. The former also increases the compression work, $-(h_2 - h_1)$, and the net effect is a slight decrease in the coefficient of perform-ance as illustrated by the following example. It was suggested in section 8.4 (iii) that

Fig. 13.3   A practical simple refrigeration cycle

a *p-h* chart is a useful aid when considering refrigeration cycles. Fig. 13.3b shows the cycle on such a diagram, from which it appears that the refrigeration effect, heat rejected, and work required can be read directly from the *h*-axis. This method is used in the example.

Example 13.2   The cycle in Example 13.1 is modified so that the refrigerant enters the com-pressor as a saturated vapour, and is subcooled to 18 °C before entering the throttle valve. Find the refrigeration effect and c.p.$_\text{ref}$. What would the result be if Freon-12 were the refrig-erant instead of ammonia?

Fig. 13.4a shows the cycle located on the *p-h* chart for ammonia, from which the following values are obtained

$h_1 = 1427 \text{ kJ/kg}, \quad h_2 = 1660 \text{ kJ/kg}$
$h_3 = h_4 = 266 \text{ kJ/kg}$

The refrigeration effect is

$Q_{41} = (h_1 - h_4) = 1161 \text{ kJ/kg}$

The net work required is

$$W = -(h_2 - h_1) = -233 \text{ kJ/kg}$$

and the coefficient of performance is

$$\text{c.p.}_\text{ref} = -\frac{Q_{41}}{W} = -\frac{1161}{-233} = 4.98$$

Similarly Fig. 13.4b shows the cycle on a p-h chart for Freon-12, from which we obtain:
The refrigeration effect

$$Q_{41} = 181.0 - 53.0 = 128.0 \text{ kJ/kg}$$

and the coefficient of performance

$$\text{c.p.}_\text{ref} = -\frac{128.0}{-24.8} = 5.16$$

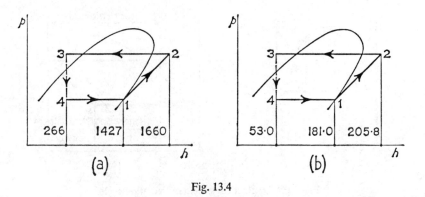

Fig. 13.4

From the previous example we see that Freon-12 gives a slightly higher c.p. than ammonia. On the other hand the refrigeration effect *per unit mass* of ammonia is about 9 times larger than for Freon-12. A comparison of refrigeration effects per unit *volume* of refrigerant is more useful, because the displacement of the compressor and the size of the plant in general depends mainly on this value. Freon is about 5 times denser than ammonia, so that the advantage of ammonia in this respect is less than 2. Other factors also govern the choice of refrigerant. Ammonia, for example, attacks many common metals and is very hygroscopic—two disadvantages from which Freon-12 does not suffer.

Comparing the results of Examples 13.1 and 13.2, it can be seen that the dry compression and subcooling increase the refrigeration effect of ammonia from 958 to 1161 kJ/kg, while only reducing the coefficient of performance from 5.09 to 4.98.

All the values in the examples have been calculated on the assumption that the cycles are composed of ideal processes. The adjective 'ideal' here does not imply that a process is reversible—the throttling process is essentially irreversible—it simply means that there are no unintended effects. In practice the compression is neither adiabatic nor reversible, the throttling process is not perfectly adiabatic, and the processes of condensation and evaporation are accompanied by small pressure drops due to friction in the pipes. Such deviations from the ideal lead to a reduction in the coefficient of performance, and the values obtained here must be regarded only as an indication of the best that can be achieved with the given cycle conditions and given refrigerant.

As with power cycles, a better performance can always be obtained at the expense of additional complexity. Only one example of a more complex cycle will be mentioned here—the multi-compression system with flash chamber, shown in Fig. 13.5. The object of this layout is to reduce the disadvantageous effect of the useless vaporisation which occurs in the throttling expansion. In the simple plant, this flash vapour passes through the evaporator without extracting any heat from the cold source, and work must be supplied to compress this vapour, together with the 'useful' vapour, through

Fig. 13.5   A refrigerator with two-stage compression and a flash chamber

the full range of pressure. In the two-stage plant, the liquid is expanded to an intermediate pressure in the flash chamber, and the useless flash vapour is passed to the inlet of the second-stage compressor. The liquid in the flash chamber is expanded in a second throttle valve to the final pressure and passes to the evaporator in the usual way. Less work is required, because the flash vapour removed at the intermediate pressure need only be compressed over a part of the pressure range of the cycle. For a given amount of heat rejection, a reduction of work implies an increase in heat extraction (First Law), and therefore the coefficient of performance is improved.

13.3 Water refrigerator

So far no mention has been made of water as a refrigerant. It is quite unsuitable for use in the type of plant we have been discussing because of its very low vapour pressure and high specific volume at low temperatures. At temperatures encountered in air-conditioning plant, however, it is possible to maintain the degree of vacuum required. Even so, at 7 °C the vapour pressure is about 0·01 bar, and the corresponding specific volume is about 129 m³/kg. The large volume flow implied by this has led to the use of an ejector pump in place of the usual compressor for this type of plant. The ejector pump is illustrated in Fig. 13.6a. Steam is expanded in a nozzle to form a high-speed jet at low pressure, which entrains the vapour to be extracted from the vacuum chamber. The combined stream from the mixing chamber is then diffused in the divergent part of the venturi until the required exhaust pressure is reached. Such a pump has no moving parts, but it has a very low efficiency.

Fig. 13.6b shows how the pump is used in the refrigeration plant. Water in the evaporator at low pressure vaporises and is pumped, together with the motive steam, into the condenser. A portion of the condensate is throttled into the evaporator, and

Fig. 13.6 Application of an ejector pump to a steam refrigeration cycle

the remainder is pumped back to the steam boiler. If waste steam is available from an existing boiler plant, the poor efficiency of the ejector pump is of no consequence, and the refrigeration plant itself is cheap to construct and operate.

13.4 Absorption refrigerators

The type of plant we have considered in previous sections is known as *vapour-compression plant*, and in such plant the work of compression is comparable in magnitude with the other energy transfers in the cycle. The compression work can be reduced considerably, however, if the vapour is dissolved in a suitable liquid before compression. After compression the vapour can be drawn off, and then throttled and evaporated in the usual way. This principle is employed in *absorption refrigerators*.

An ammonia absorption refrigerator is illustrated in Fig. 13.7. The processes of condensation (2–3), throttling (3–4), and evaporation (4–1), occur in conventional components. After evaporation to state 1, the vapour is first passed into an absorber

262

and dissolved in water. The water temperature is only a little above atmospheric, and at such a temperature the solubility of ammonia vapour is very high. The process of solution is accompanied by a rejection of heat Q_s to the surroundings (just as the process of driving off the vapour from the solution requires heat absorption). The liquid solution is then compressed to the required pressure and heated, first in a heat exchanger, and then in the generator which is maintained at a temperature T_g. At the higher temperature the solubility of ammonia in water is appreciably smaller, and the heat received Q_g results in the removal of vapour from solution. The generator may be heated by steam coils, gas, or electricity. The ammonia vapour is passed to the condenser to complete the main cycle, while the remaining weak solution is returned to the absorber via a throttle valve. The heat exchanger is merely used to reduce the quantities of heat Q_g and Q_s.

Fig. 13.7   The ammonia absorption refrigerator

The work required to compress the liquid solution is very much smaller than that required for compressing a corresponding amount of ammonia vapour, but on the other hand heat has to be supplied to the generator from an external source. The analysis of the compression process requires a knowledge of the thermodynamic properties of ammonia-in-water solutions and cannot be given here.* Nevertheless, by concentrating on the essential nature of the cycle, simple thermodynamic principles can be applied to show that there is a limit to the coefficient of performance which can be achieved with any given set of temperature levels.

Energy is transferred in the form of heat at three temperature levels: (1) atmospheric temperature T_o, at which heat is rejected in the condenser and absorber, (2) the temperature T_c at which heat is taken from the cold chamber, and (3) the temperature T_g, at

*An example of such an analysis can be found in Ref. (11).

which heat is received in the generator. A small quantity of energy is also transferred as work in the pump, but this is negligible compared with the other energy transfers. It is possible to imagine an arrangement of reversible machines (Fig. 13.8) performing a function equivalent to that of the absorption plant. Firstly, a reversible heat engine receives a quantity of heat Q_g at T_g and rejects heat at T_a while producing a quantity of work W_{ga} with an efficiency

$$\frac{W_{ga}}{Q_g} = \frac{T_g - T_a}{T_g}$$

Secondly, a reversible refrigerator receives a quantity of heat Q_c at T_c and rejects heat at T_a while absorbing a quantity of work W_{ca}. The coefficient of performance of the refrigerator is

$$-\frac{Q_c}{W_{ca}} = \frac{T_c}{T_a - T_c}$$

Fig. 13.8 A 'reversible equivalent' of an absorption refrigerator

If W_{ga} is made equal to $-W_{ca}$, this plant will be equivalent to an absorption refrigerator for which the pump work is zero.

The coefficient of performance of the combined plant acting as a refrigerator can be defined as Q_c/Q_g, which on combining the previous two equations becomes

$$\text{c.p.}_{\text{ref}} = \frac{Q_c}{Q_g} = \frac{T_c(T_g - T_a)}{T_g(T_a - T_c)} \tag{13.4}$$

Although this limiting coefficient is greater than unity with normal operating temperatures, the properties of ammonia-in-water solutions are such than an ammonia absorption refrigerator has a coefficient of performance less than unity. These refrigerators are rarely used unless waste heat is available from some existing source.

A special form of ammonia absorption refrigerator, in which the pump is dispensed with entirely, is widely used for domestic purposes where the power consumption (electricity or gas) is so small that efficiency is not a prime consideration.* Having no moving parts at all, the refrigerator requires little maintenance. Its operation depends on the principle that if a liquid is exposed to an inert atmosphere, which is not saturated with the vapour of the liquid, some of the liquid will evaporate and so produce a cooling effect. The total pressure is effectively uniform throughout the plant, but the partial pressure of the ammonia vapour varies, being high in the condenser and low in the evaporator. This variation is accomplished by concentrating a gas (hydrogen) in the parts of the circuit where the ammonia vapour pressure is required to be low. There being no material change in total pressure during the cycle, circulation of the fluid can be maintained by convection currents set up by density gradients.

*The refrigerator described here was invented by B. von Platen and C. G. Munters when studying at the Royal Institute of Technology in Stockholm. It is often called the *Electrolux* refrigerator after the firm who first developed it commercially.

A diagrammatic sketch of the more essential components of this type of absorption refrigerator is shown in Fig. 13.9. The ammonia-in-water solution is boiled in the generator and rises to the separator. From here the water, while cooling, returns to the absorber through a liquid trap, and the ammonia vapour rises further to the condenser. There it loses heat to the surroundings, and the ammonia condensate flows through a liquid trap into the evaporator coils. In the evaporator there is some hydrogen

Fig. 13.9   Absorption refrigerator with no moving parts

present, and the ammonia evaporates at a low temperature corresponding to its low partial pressure. The mixture of hydrogen and ammonia flows down through a heat exchanger to the absorber, and the ammonia vapour returns into the cold water solution. From the absorber the hydrogen rises back to the evaporator, while the ammonia-in-water solution flows into the generator to complete the cycle. Thus there are three essential circuits to this plant: (a) the main ammonia circuit, (b) the subsidiary water circuit, and (c) the subsidiary hydrogen circuit. The actual plant is in fact rather more complicated than Fig. 13.9 suggests, and the thermodynamic analysis of the cycle is difficult.

13.5 Gas cycles

When a gas is used as a refrigerating fluid it is essential to employ an expander because the temperature remains substantially unchanged by throttling. The processes in the cooler and heater, which replace the condenser and evaporator of a vapour compression machine, are usually constant pressure, but not constant temperature, processes.

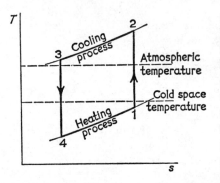

Fig. 13.10 A simple refrigeration cycle using air

The ideal cycle is in fact a reversed Joule cycle as shown in Fig. 13.10. The temperature at state 3 must be somewhat above atmospheric for a refrigerator, and at state 1 somewhat below the cold space temperature. Consequently a refrigerator using a gas as a working fluid is less efficient than one using a vapour, because its operating temperature range is very much wider, i.e. from t_4 to t_2 instead of from t_1 to t_3. It is also much bulkier than a vapour plant of the same duty, because a gas requires relatively larger surface areas for a given heat transfer. For these reasons gases are never used nowadays as working fluids for refrigerators, when the sole object is refrigeration.

When cool air is required for air-conditioning, however, it is sometimes more convenient to cool the air directly than by the means indicated in section 13.3. The air then combines the function of refrigerating fluid and air-conditioning medium. A typical modern application is illustrated by Fig. 13.11 and the following example.

Fig. 13.11 An aircraft air-conditioning unit

Example 13.3 An air conditioning unit of a pressurised aircraft receives its air from the jet engine compressor at a pressure of 1·22 bar. The atmospheric pressure and temperature are 0·227 bar and 217 K respectively.

266

The air conditioning unit consists of a freewheeling compressor and turbine mounted on one shaft (Fig. 13.11). The air coming into the unit is compressed to a pressure p, cooled at constant pressure, and expanded in the turbine to a pressure of 1 bar and temperature of 280 K. Assuming that all processes are reversible, calculate the pressure p and the temperature of the air at exit from the cooler.

The engine compressor intake state 1 is

$$p_1 = 0.227 \text{ bar}, \quad T_1 = 217 \text{ K}$$

and the air bleed state 2 is

$$p_2 = 1.22 \text{ bar}$$

$$T_2 = T_1 \left(\frac{p_2}{p_1}\right)^{(\gamma-1)/\gamma} = 217 \left(\frac{1.22}{0.227}\right)^{0.4/1.4} = 217 \times 1.6168 = 351 \text{ K}$$

The freewheeling compressor outlet temperature is

$$T_3 = T_2 \left(\frac{p}{p_2}\right)^{0.4/1.4}$$

and hence this compressor work is

$$W_{23} = -(h_3 - h_2) = -c_p(T_3 - T_2) = -c_p T_2 \left\{ \left(\frac{p}{p_2}\right)^{0.4/1.4} - 1 \right\}$$

The required turbine outlet state 5 is

$$p_5 = 1 \text{ bar}, \quad T_5 = 280 \text{ K}$$

The cooler outlet temperature is therefore

$$T_4 = T_5 \left(\frac{p}{p_5}\right)^{0.4/1.4}$$

and the turbine work is

$$W_{45} = -(h_5 - h_4) = -c_p(T_5 - T_4) = -c_p T_5 \left\{ 1 - \left(\frac{p}{p_5}\right)^{0.4/1.4} \right\}$$

Since the turbine is only to produce sufficient work to drive the compressor,

$$W_{23} = -W_{45}$$

and hence the equation from which the unknown pressure p can be found is

$$-c_p T_2 \left\{ \left(\frac{p}{p_2}\right)^{0.4/1.4} - 1 \right\} = c_p T_5 \left\{ 1 - \left(\frac{p}{p_5}\right)^{0.4/1.4} \right\}$$

By trial and error the cooler pressure is

$$p = 3.06 \text{ bar}$$

The required cooler exit temperature is

$$T_4 = T_5 \left(\frac{3.06}{1}\right)^{0.4/1.4} = 385 \text{ K}$$

Another circumstance in which a gas cycle is used is when the object is to liquefy the gas. The simplest liquefaction process was first developed by Linde for liquefying air, and it is illustrated in Fig. 13.12. Air at atmospheric pressure, state 1, is compressed to state 2, and is then cooled at constant pressure back to atmospheric temperature, state 3, in a water cooler. The air is cooled further to state 4 in a heat exchanger before being throttled to state 5. In this state the liquid portion 5_f is drained off and the vapour portion 5_g is taken to the heat exchanger and back to the compressor. To keep the mass of air in the system constant, make-up air, equal in mass to the liquid air drained off, is introduced into the system in state 1.

It is evident that when starting such a plant there is no cold gas in the system, and the throttling process will follow path 3–6. But provided that the gas is below its inversion temperature [see section 7.4 (iv)], a little cooling will occur. This cooling

Fig. 13.12   A simple gas liquefaction plant

effect is cumulative, as it is utilised in the heat exchanger, and the throttling curve 3–6 will shift towards path 4–5. The ultimate steady position of state 4 depends on the rate at which liquid is drained off, and on the rate at which heat leaks into the system.

When the gas to be liquefied has an inversion temperature below atmospheric, it is necessary to precool the gas below its inversion temperature with a conventional refrigerator, or to replace the throttle valve with an expansion machine. The reason why a throttle valve has been preferred to an expansion machine is that it is difficult to construct a machine which will work satisfactorily at very low temperatures. Using new materials, suitable expanders have been built in recent years, and much better plant efficiencies can now be obtained. The subject of low-temperature engineering—called *cryogenics*—is of increasing importance but it is outside the scope of this book. A survey of recent work in this field can be found in Ref. (13).

14   Properties of Mixtures

Any homogeneous mixture of gases can be regarded as a single substance if the constituents do not react chemically with one another and are in a fixed proportion by weight. The properties of such a mixture can be determined experimentally just as for a single substance, and they can be tabulated or related algebraically in the same way. The composition of air can be assumed invariable for most purposes and air is usually treated as a single substance. As air is such a common fluid, its properties have been determined by direct measurement. For less common mixtures, however, it is necessary to have some method of deducing the properties of the mixture from those of its constituents. This chapter describes how this can be done in the case of mixtures of perfect-gases, and mixtures of gases and vapours at low pressure. More refined methods, for use when the gases or vapours cannot be assumed to be perfect, are given in textbooks on chemical thermodynamics [e.g. Ref. (4)].

Steam condensers, air-conditioning plant, and cooling towers, provide suitable examples of systems involving mixtures of gases and vapours at low pressure, and the thermodynamic analysis of the processes which occur in these systems is given here. The special nomenclature associated with the study of moist atmospheric air (hygrometry) is introduced, and the chapter ends with a brief description of the hygrometric chart.

14.1 An empirical law for mixtures of gases

Any mixture of gases can be imagined to be separated into its constituents in such a way that each occupies a volume equal to that of the mixture and each is at the same temperature as the mixture. An important empirical law can then be stated as follows:

The pressure and internal energy of a mixture of gases are respectively equal to the sums of the pressures and internal energies of the individual constituents when each occupies a volume equal to that of the mixture at the temperature of the mixture.

When stated with reference to pressure alone, this is the proposition known as *Dalton's law*. The extended law has been referred to as the *Gibbs-Dalton law* and we shall adopt this usage.

As an immediate deduction from the Gibbs-Dalton law it may be shown that an equivalent statement can be made about the enthalpy of a mixture. Thus, the Gibbs-Dalton law can be expressed symbolically by

$$p = \sum p_i]_{V.T} \quad \text{and} \quad U = \sum U_i]_{V.T} \tag{14.1}$$

269

Suffix i refers to a typical constituent i, and the suffixes V and T outside the square bracket indicate that the property must be evaluated when the constituent alone occupies the volume V of the mixture at the temperature T of the mixture. By definition the enthalpy H is $(U + pV)$ and therefore, using (14.1), the enthalpy of a mixture becomes

$$H = \sum U_i]_{V,T} + V \sum p_i]_{V,T}$$

But

$$H_i]_{V,T} = (U_i + p_i V)]_{V,T} = U_i]_{V,T} + V p_i]_{V,T}$$

and therefore

$$H = \sum H_i]_{V,T} \tag{14.2}$$

The Gibbs-Dalton law has been determined from experiments on mixtures of gases. It has been found that the accuracy of the law improves as the pressure of the mixture is decreased, and consequently it may be inferred that it is strictly true for mixtures of perfect gases. This inference is confirmed by the fact that the law can be deduced from the postulates of the simple kinetic theory of gases. Briefly, if the size of the molecules is negligible compared with their distance apart and there are no attractive forces between them, each constituent will contribute its share of pressure and internal energy as if it were the only gas in the vessel; and this in fact is what the law implies.

In the following section we shall deduce the consequences of the Gibbs-Dalton law for a mixture of perfect gases, and show how the properties of such a mixture can be determined from the properties of the constituents. Arguments based on the kinetic theory can tell us what to expect. According to this theory the perfect gas relations [equations (8.18) to (8.25)] are theoretical deductions based only on the laws of mechanics and thermodynamics. Although the values of the constants R, c_v and c_p, depend upon the particular gas, i.e. on the mass and structure of the molecules, the *form* of the relations is the same for all perfect gases. For any mixture of perfect gases, therefore, we may also expect to be able to write:

$$pv = RT \tag{14.3}$$

$$u - u_0 = c_v(T - T_0) \tag{14.4}$$

$$h - h_0 = c_p(T - T_0) \tag{14.5}$$

$$s - s_0 = c_p \ln \frac{T}{T_0} - R \ln \frac{p}{p_0} \tag{14.6}$$

In this case R, c_v and c_p are specific constants for the particular mixture. The problem of determining the behaviour of a mixture of perfect gases then reduces to one of obtaining expressions for R, c_v and c_p in terms of the gas constants and specific heats of the constituents. Once the mixture constants have been determined, the analysis of both non-flow and flow processes can be carried out just as though the fluid were a single gas.

14.2 Mixtures of perfect gases

First we shall show formally that equations (14.3) to (14.6) follow from the application of the Gibbs-Dalton law to mixtures of perfect gases. If a mixture occupies a volume V at a temperature T, the pressure p_i of any constituent when it alone occupies the volume V at temperature T is given by

$$p_i V = n_i R_0 T$$

p_i is known as the *partial pressure* of the constituent i, and n_i is the *amount of substance* of the constituent in the volume V (here measured in kmol). By writing an equation of the above form for each constituent and summing the equations we have

$$V \Sigma p_i = R_0 T \Sigma n_i$$

Remembering that the number of kilomoles is proportional to the number of molecules, and that the number of molecules remains unchanged during any process which does not involve a chemical reaction, the number of kilomoles of mixture n is given by

$$n = \Sigma n_i$$

Also, from Dalton's law we have

$$p = \Sigma p_i]_{V,T}$$

Consequently for the mixture we can write

$$pV = n R_0 T$$

The molar mass M of the mixture, from equation (8.12), is

$$M = \frac{m}{n} \tag{14.7}$$

where m is the mass of the mixture. We therefore have

$$pV = \frac{m}{M} R_0 T = mRT$$

where $R = R_0/M$ is the specific gas constant for the mixture. Per unit mass of mixture this becomes equation (14.3), viz.

$$pv = RT$$

Thus the characteristic equation of the mixture has the same form as that for any individual gas.

An expression for the specific gas constant of a mixture, in terms of the gas constants of the constituents, can now be found.

$$R = \frac{R_0}{M} = \frac{n}{m} R_0$$

Substituting $n = \Sigma n_i = \Sigma(m_i/M_i)$ and $M_i = (R_0/R_i)$, this becomes

$$R = \frac{R_0}{m} \Sigma m_i \frac{R_i}{R_0} = \Sigma \frac{m_i}{m} R_i \qquad (14.8)$$

m_i/m is the *mass fraction* of the constituent i having the gas constant R_i.

It has been stated in section 8.5 (ii), and proved in section 7.6 (i), that the internal energy of any substance having the characteristic equation of a perfect gas must be a function only of temperature. Having shown that (14.3) holds for a mixture, it follows that the internal energy of the mixture must be a function only of the temperature of the mixture. We can therefore apply the equation

$$dU = mc_v\, dT$$

to the mixture as well as to each constituent. But from the Gibbs-Dalton law, equation (14.1), we have

$$U = \Sigma U_i]_{V.T}$$

Since the internal energy depends only on temperature, the suffix V can be omitted when applying the law to perfect gases. In differential form the equation therefore becomes

$$mc_v\, dT = \Sigma m_i c_{vi}\, dT$$

and hence the specific heat at constant volume for the mixture is given by

$$c_v = \Sigma \frac{m_i}{m} c_{vi} \qquad (14.9)$$

If the specific heats of the constituents can be treated as constants, then c_v for the mixture is constant also and we arrive at (14.4),

$$u - u_0 = c_v(T - T_0)$$

Another consequence of the characteristic equation of a perfect gas is that

$$c_p - c_v = R$$

and this relation must also hold for the mixture. The specific heat at constant pressure of a mixture of perfect gases is therefore given by

$$c_p = \Sigma \frac{m_i}{m} R_i + \Sigma \frac{m_i}{m} c_{vi} = \Sigma \frac{m_i}{m} c_{pi} \qquad (14.10)$$

The expected equation (14.5) then follows from the definition of enthalpy, and (14.6) from the laws of thermodynamics, just as in the case of a single gas, and there is no need to repeat the argument here.

From the foregoing paragraphs it is evident that for any process from some reference state 0 to some other state, the changes in u, h and s can be found by calculating R,

c_v and c_p for the mixture [using equations (14.8), (14.9) and (14.10)], and then substituting these values in equations (14.4), (14.5) and (14.6). Another approach, which is particularly useful when dealing with mixtures of gases and vapours, follows more immediately from the Gibbs-Dalton law. Thus from equation (14.1) we can write,

$$U - U_0 = \sum U_i]_T - \sum U_i]_{T_0}$$

or

$$m(u - u_0) = \sum m_i u_i]_T - \sum m_i u_i]_{T_0} = \sum m_i c_{vi}(T - T_0) \tag{14.11}$$

Similarly, from equation (14.2) we get

$$m(h - h_0) = \sum m_i h_i]_T - \sum m_i h_i]_{T_0} = \sum m_i c_{pi}(T - T_0) \tag{14.12}$$

These two equations express the fact that the changes in internal energy and enthalpy of a mixture can be calculated from the sum of the changes in internal energy and enthalpy of the constituents.

The equivalent equation for the change in entropy cannot be deduced in such a direct manner because there is no explicit reference to entropy in our statement of the Gibbs-Dalton law. We may proceed by substituting (14.8) and (14.10) in (14.6) to give

$$m(s - s_0) = \sum m_i c_{pi} \ln \frac{T}{T_0} - \sum m_i R_i \ln \frac{p}{p_0} \tag{14.13}$$

Now for each constituent

$$p_i V = m_i R_i T$$

and for the mixture

$$pV = mRT$$

By division it is evident that p_i/p is constant during any process for which the composition (i.e. each mass fraction) remains constant. Hence

$$\frac{p_i}{p} = \frac{p_{i0}}{p_0} \quad \text{or} \quad \frac{p}{p_0} = \frac{p_i}{p_{i0}}$$

Substituting this result in (14.13) we get

$$m(s - s_0) = \sum m_i \left(c_{pi} \ln \frac{T}{T_0} - R_i \ln \frac{p_i}{p_{i0}} \right) \tag{14.14}$$

Equation (14.14) is akin to (14.11) and (14.12), merely differing in form because, unlike u or h, s is a function of both temperature *and* pressure.

Remembering that p_i is the pressure of a constituent at the volume and temperature of the mixture, (14.14) can be written alternatively as

$$m(s - s_0) = \sum m_i s_i]_{V,T} - \sum m_i s_i]_{V_0,T_0}$$

273

If the entropy of each constituent is put equal to zero at the state (V_0, T_0), then we can say that

$$ms = \sum m_i s_i]_{V,T} \quad \text{or} \quad S = \sum S_i]_{V,T} \tag{14.15}$$

Evidently the Gibbs-Dalton law implies that *the entropy of a mixture of perfect gases is the sum of the entropies of the constituents evaluated when each occupies alone the volume of the mixture at the temperature of the mixture.*

Example 14.1 A mixture of gases has the following composition by mass:

$$N_2 \ 60\%; \ O_2 \ 30\%; \ CO_2 \ 10\%$$

0·9 kg of mixture, initially at 1 bar and 27 °C, are compressed to 4 bar. The compression can be assumed to be a reversible non-flow polytropic process with a polytropic index of 1·2. Find the work done, heat transferred and change of entropy. The values of c_p for the constituents are

$$N_2 \ 1·040; \ O_2 \ 0·918; \ CO_2 \ 0·846 \text{ kJ/kg K}$$

The constants for the mixture are:

$$R = \sum \frac{m_i}{m} R_i = 0·60 \, \frac{8·314}{28} + 0·30 \, \frac{8·314}{32} + 0·10 \, \frac{8·314}{44} = 0·275 \text{ kJ/kg K}$$

$$c_p = \sum \frac{m_i}{m} c_{pi} = 0·60 \times 1·04 + 0·30 \times 0·918 + 0·10 \times 0·846 = 0·984 \text{ kJ/kg K}$$

$$c_v = c_p - R = 0·984 - 0·275 = 0·709 \text{ kJ/kg K}$$

The final temperature is given by

$$T_2 = T_1 \left(\frac{p_2}{p_1}\right)^{(n-1)/n} = 300 \left(\frac{4}{1}\right)^{0·2/1·2} = 378 \text{ K}$$

The work done is

$$W = \frac{mR(T_2 - T_1)}{1 - n} = \frac{0·9 \times 0·275(378 - 300)}{-0·2} = -96·5 \text{ kJ}$$

The heat transferred is

$$Q = mc_v(T_2 - T_1) + W = 0·9 \times 0·709(378 - 300) - 96·5 = -46·7 \text{ kJ}$$

The change of entropy is

$$S_2 - S_1 = m \left(c_p \ln \frac{T_2}{T_1} - R \ln \frac{p_2}{p_1}\right)$$

$$= 0·9 \left(0·984 \ln \frac{378}{300} - 0·275 \ln \frac{4}{1}\right) = -0·138 \text{ kJ/K}$$

It is worth noting another empirical law, variously known as the *law of partial volumes, Amagat's law* or *Leduc's law*, which states that:

The volume of a mixture of gases is equal to the sum of the volumes of the individual constituents when each exists alone at the pressure and temperature of the mixture.

For perfect gases this law will now be shown to be a consequence of Dalton's law: for real gases this law and Dalton's law are only approximately true and one cannot be deduced from the other. If a constituent i is placed in an empty vessel having the volume of the mixture V, at the mixture temperature T, it will exert the partial pressure p_i. Keeping the temperature constant and increasing the pressure to the pressure of the mixture p, the volume of this constituent reduces to

$$V_i = \frac{p_i}{p} V$$

V_i is the *partial volume* of the constituent by definition. If all such equations, for each constituent, are added, we have

$$\Sigma V_i = \frac{V}{p} \Sigma p_i$$

But $\Sigma p_i = p$, from Dalton's law, and hence

$$V = \Sigma V_i]_{p,T} \tag{14.16}$$

Sometimes problems concerning mixtures are simplified if the analysis is carried through in terms of *mole fractions* rather than mass fractions. The following relations are then useful. For each constituent we can write

$$p_i V = n_i R_0 T$$

and for the mixture

$$pV = n R_0 T$$

By division, and writing $n_i/n = x_i$ for the mole fraction of the constituent i, we have

$$p_i = x_i p \tag{14.17}$$

We also have

$$pV_i = n_i R_0 T$$

and by a similar argument

$$V_i = x_i V \tag{14.18}$$

The molar mass of the mixture is given by $M = m/n$, and hence

$$Mn = \Sigma m_i = \Sigma M_i n_i$$

275

or

$$M = \Sigma x_i M_i \tag{14.19}$$

If required, the gas constant for the mixture can then be found from $R = R_0/M$.

The concept of a *molar heat*, i.e. specific heat per unit amount of substance, is also useful and widely used. A molar heat will be denoted by C to distinguish it from the specific heat c, and the units will be kJ/kmol K. Since by definition $C = Mc$, an expression for the molar heat at constant volume equivalent to (14.9) can be found by combining this equation with (14.7). Thus

$$C_v = M\Sigma \frac{m_i}{m} c_{vi} = \Sigma M \frac{m_i}{m} \frac{C_{vi}}{M_i}$$

$$= \Sigma \frac{n_i}{n} C_{vi} \quad \text{or} \quad \Sigma x_i C_{vi} \tag{14.20}$$

Similarly it can be shown that

$$C_p = \Sigma x_i C_{pi} \tag{14.21}$$

Further, from the definition of C,

$$C_p - C_v = M(c_p - c_v) = MR = R_0 \tag{14.22}$$

When an analysis of a gas mixture is carried out experimentally, e.g. by using an Orsat apparatus (section 15.3), the analysis by volume is usually obtained. The volume V of a sample of gas is first measured at some fixed pressure and temperature, usually atmospheric. Then each constituent is absorbed in turn by a suitable chemical reagent and the reduction in volume is noted. The volumes are always measured at the same pressure and temperature, so that any reduction in volume will be equal to the partial volume V_i of the constituent in the original sample. The result is expressed as a volumetric analysis, i.e. a set of values of V_i/V, and it follows from equation (14.18) that this is also an analysis by mole fractions. The mass fractions, and hence the gravimetric analysis, can be deduced from the volumetric analysis if the molar masses of the constituents are known. Thus

$$\frac{m_i}{m} = \frac{n_i M_i}{\Sigma n_i M_i} = \frac{n_i M_i/n}{\Sigma(n_i M_i/n)} = \frac{x_i M_i}{\Sigma x_i M_i}$$

And finally, substituting V_i/V for x_i from equation (14.18) we have

$$\frac{m_i}{m} = \frac{(V_i/V)M_i}{\Sigma(V_i/V)M_i} \tag{14.23}$$

Example 14.2 A mixture of gases at a temperature of 150 °C has a pressure of 4 bar. A sample is analysed, and the volumetric analysis is found to be

$$CO_2 \ 14\%; \ O_2 \ 5\%; \ N_2 \ 81\%$$

Determine the gravimetric analysis and partial pressures of the gases in the mixture.

If 2·3 kg of mixture are cooled at constant pressure to 15 °C find the final volume.

Using equation (14.23), the procedure for finding the gravimetric analysis can be set out as follows:

i	V_i/V	M_i	$(V_i/V)M_i$	m_i/m
CO_2	0·14	44	6·16	0·202
O_2	0·05	32	1·60	0·053
N_2	0·81	28	22·68	0·745
	1·00		30·44	1·000

The gravimetric analysis appears in the last column. Note that according to (14.18) the second column also gives the mole fractions, and according to (14.19) the final result 30·44 in the fourth column is the molar mass of the mixture. From (14.17) the partial pressures are:

CO_2 $p_i = 0·14 \times 4 = 0·56$ bar
O_2 $p_i = 0·05 \times 4 = 0·20$ bar
N_2 $p_i = 0·81 \times 4 = 3·24$ bar

The amount of substance in 2·3 kg is

$$n = \frac{2·3}{30·44} = 0·0756 \text{ kmol}$$

and the final volume of the mixture is

$$V = \frac{nR_0T}{p} = \frac{0·0756 \times 8·314 \times 288}{100 \times 4} = 0·453 \text{ m}^3$$

Fig. 14.1

14.3 The mixing process

So far we have considered methods of dealing with gases which are already mixed; in this section we shall consider the mixing process itself. The simplest case is the adiabatic mixing of perfect gases which are initially at the same pressure and temperature.

Imagine an insulated vessel in which two gases are separated by a partition, Fig. 14.1. The pressure and temperature of each gas is p and T, and the masses are m_a and m_b. On removing the partition, the gases diffuse into one another and each gas expands to fill the total volume. No work is done, and we are considering an adiabatic mixing process, so that the internal energy must remain unchanged. It follows that the temperature of the mixture will be the same as the original temperature of the constituents. The volumes initially occupied by the gases are, by definition, the partial volumes associated with a mixture of them at pressure p and temperature T. Conversely, from the law of partial volumes and the fact that the temperature T is unchanged, the pressure of the mixture must be the same as the initial pressure of the constituents. Both the pressure and temperature of the mixture are therefore the same

277

as the original pressure and temperature of the constituents, although the pressure which each gas exerts is now reduced to the corresponding partial value p_a or p_b. Each gas behaves as though the other were not present, and can be considered to have undergone a free expansion. Evidently this diffusion process is irreversible, and it follows from the principle of increasing entropy that the entropy of this isolated system must have increased. The initial entropies of the separate gases, each reckoned from the same datum state (p_0, T_0), are given by equation (8.24) as

$$S_a = m_a c_{pa} \ln \frac{T}{T_0} - m_a R_a \ln \frac{p}{p_0}$$

$$S_b = m_b c_{pb} \ln \frac{T}{T_0} - m_b R_b \ln \frac{p}{p_0}$$

As expressed by equation (14.15), the entropy of the mixture is equal to the sum of the entropies of the constituents, evaluated at the volume and temperature of the mixture. These entropies, from equation (14.14), are

$$S_b' = m_a c_{pa} \ln \frac{T}{T_0} - m_a R_a \ln \frac{p_a}{p_0}$$

$$S_a' = m_b c_{pb} \ln \frac{T}{T_0} - m_b R_b \ln \frac{p_b}{p_0}$$

The total change of entropy for the whole system is therefore

$$(S_a' + S_b') - (S_a + S_b) = -m_a R_a \ln \frac{p_a}{p} - m_b R_b \ln \frac{p_b}{p}$$

Since p_a and p_b are each less than p, this is a positive quantity indicating an increase of entropy.*

If the temperatures of the gases are initially different, there will be an additional increase of entropy due to the irreversible redistribution of internal energy which occurs within the system. The following example illustrates this more general case, and also the use of molar heats.

Example 14.3 Two containers, each $1\cdot 4 \ m^3$ in volume, are isolated from one another by a valve. One contains oxygen at 7 bar and 150 °C, and the other carbon dioxide at 2 bar and 15 °C. The valve is opened and the gases are allowed to mix. Assuming the process is adiabatic, determine the equilibrium temperature and pressure of the mixture, and the change of

---

*It is interesting to note that if the two gases become more and more alike the change of entropy on mixing does not reduce to zero in the limiting case of two identical gases. Yet there is no change of entropy when a mass $(m_a + m_b)$ of gas exists in a closed vessel at fixed p and T although the molecules are continuously 'diffusing' through the volume. This contradiction is known as *Gibbs' paradox*. It has been resolved by Bridgman [Ref. (2), Part I] in terms of our inability even in principle to distinguish between the two parts of the gas when the molecules are identical: and if there is a discontinuity in the physical operations implicit in the description of the problem it is reasonable to expect a discontinuity in the result of the analysis.

entropy associated with the system as a whole. The values of the molar heats at constant volume C_v for the constituents are

O_2 21·0; CO_2 28·3 kJ/kmol K

For convenience let the oxygen be denoted by suffix a and the carbon dioxide by suffix b. Since $n = pV/R_0T$, the number of kilomoles of each gas is

$$n_a = \frac{100 \times 7 \times 1\cdot4}{8\cdot314 \times 423} = 0\cdot279 \text{ kmol}, \quad n_b = \frac{100 \times 2 \times 1\cdot4}{8\cdot314 \times 288} = 0\cdot117 \text{ kmol}$$

The total numbers of kilomoles in the mixture is

$$n = n_a + n_b = 0\cdot396 \text{ kmol}$$

Since W and Q are zero, the total internal energy of the system must remain constant. Initially the internal energy, reckoned from zero at T_0, is

$$n_a C_{va}(T_a - T_0) + n_b C_{vb}(T_b - T_0)$$

and finally it is $nC_v(T - T_0)$, where T is the equilibrium temperature and C_v is the molar heat of the mixture. But according to equation (14.20),

$$nC_v = n_a C_{va} + n_b C_{vb}$$

Hence, on equating the initial and final internal energy, we get

$$(n_a C_{va} + n_b C_{vb})(T - T_0) = n_a C_{va}(T_a - T_0) + n_b C_{vb}(T_b - T_0)$$

or

$$n_a C_{va}(T - T_a) + n_b C_{vb}(T - T_b) = 0$$
$$0\cdot279 \times 21\cdot0(T - 423) + 0\cdot117 \times 28\cdot3(T - 288) = 0$$
$$T = 374 \text{ K}$$

The equilibrium pressure of the mixture is

$$p = \frac{nR_0T}{V} = \frac{0\cdot396 \times 8\cdot314 \times 374}{100 \times 2\cdot8} = 4\cdot40 \text{ bar}$$

The change in entropy can be found by summing the changes associated with each constituent. Thus the change in entropy of oxygen, in terms of T and V, is

$$\Delta S_a = n_a \left(C_{va} \ln \frac{T}{T_a} + R_0 \ln \frac{V}{V_a} \right)$$

$$= 0\cdot279 \left(21\cdot0 \ln \frac{374}{423} + 8\cdot314 \ln \frac{2\cdot8}{1\cdot4} \right) = 0\cdot886 \text{ kJ/K}$$

And similarly for carbon dioxide,

$$\Delta S_b = 0\cdot117 \left(28\cdot3 \ln \frac{374}{288} + 8\cdot314 \ln \frac{2\cdot8}{1\cdot4} \right) = 1\cdot540 \text{ kJ/K}$$

The total change in entropy is an increase of

$$\Delta S = \Delta S_a + \Delta S_b = 2\cdot426 \text{ kJ/K}$$

It should not be supposed from the foregoing that it is impossible *in principle* for gases to be mixed in a reversible manner. To see how this can be done it is necessary first to introduce the concept of a *semipermeable membrane*, i.e. a membrane which is permeable to one gas only and impermeable to the other gases in the mixture. This is an idealised piece of apparatus, like the frictionless pulley or the non-conducting wall. A membrane of palladium foil which, at red heat, is permeable to hydrogen but practically impermeable to all other common gases, is an example which closely approximates to the ideal. The gas to which the membrane is permeable will continue to pass through it until the partial pressure of that gas is the same on each side; *membrane equilibrium* is then said to exist.

Consider the imaginary device of Fig. 14.2, suggested by Planck. A cylinder is divided into two compartments of equal volume by a membrane permeable to gas *a*, but impermeable to a second gas *b*. Two pistons are coupled together, one of which is

Fig. 14.2   Reversible mixing process

impermeable to both gases and the other permeable only to gas *b*. The initial configuration is as shown in Fig. 14.2a, with the pistons to the right and the gases in separate compartments at the same temperature. As the pistons are moved to the left the gases diffuse through the semipermeable membranes until eventually the two gases are mixed in the left-hand compartment leaving a vacuum in the other. It remains to show that this mixing process is reversible.

We shall suppose (1) that the motion of the pistons is infinitely slow so that membrane equilibrium exists throughout the process, and (2) that the whole system is maintained at constant temperature. Under these conditions the volumes and temperatures of the separate gases, which are initially equal, are also the same as the volume and temperature of the final mixture. Therefore if p_a and p_b are the initial pressures of the gases, they will also be the partial pressures of the gases when mixed. At any intermediate point in the process, if the piston area is A, the net force acting on the left-hand piston is

$$A\{(p_a + p_b) - p_b\} = Ap_a$$

tending to move it to the left. And the net force on the other piston is

$$A(p_a - 0) = Ap_a$$

tending to move it to the right. It follows that the piston combination is always acted upon by balanced forces. Since only an infinitesimal force is required to set the pistons in motion in either direction, the process is reversible. No work is done, there is no change in temperature and therefore of internal energy, and so according to the First Law no heat crosses the boundary. Thus the surroundings are not affected by the process.

We shall have occasion to refer again to the concept of the semipermeable membrane and the reversible mixing process in the next chapter.

14.4 Gas and saturated vapour mixtures

When a liquid is introduced into an evacuated vessel of greater volume than the liquid, some of it will evaporate. The more rapidly moving molecules of liquid escape from the surface to the space above, and at first few of these return to the liquid. As more and more molecules fill the vacant space and the vapour pressure increases, so a greater proportion find their way back into the liquid. If the vessel is maintained at a constant temperature, eventually equilibrium will be established between the rates of evaporation and condensation, and the pressure will remain constant. The space above the liquid then contains the maximum possible amount of vapour at the given temperature. The vessel is filled with saturated liquid and saturated vapour, and the pressure is the saturation vapour pressure corresponding to the given temperature. At any instant, only the behaviour of the molecules in the immediate vicinity of the surface governs the rates of evaporation and condensation. Therefore the saturation pressure is independent of the relative masses of liquid and vapour present and depends solely on the temperature. An increase of temperature permits more molecules to overcome the attractive forces of the liquid molecules, and a new equilibrium is established at a higher pressure. If the temperature is increased sufficiently, all the liquid evaporates and the vapour becomes superheated.

We have already seen that, provided the pressure is not too high, each gas in a gaseous mixture can be considered to behave as though the others were not present. This is also approximately true if one of the constituents is a vapour. Thus in the case we have been considering, if the vessel is filled with a gas prior to the introduction of the liquid, the gas will have a negligible effect on the final state of equilibrium between the liquid and its vapour. The presence of the gas molecules will retard the establishment of equilibrium, but eventually the vapour will exert the saturation vapour pressure corresponding to the temperature of the mixture. This saturation pressure is the partial pressure of the vapour in the mixture, and the total pressure in the vessel is the sum of the partial pressures of the vapour and the gas. Under these conditions the gas (or mixture of gases) is said to be 'saturated with vapour', although strictly speaking it is the vapour which is saturated. Note that the liquid is subjected to the vapour *and* gas pressure and so is strictly speaking a compressed liquid. However, as pointed out in section 8.3 (iii), the properties of a compressed liquid are nearly the same as those of saturated liquid at the same temperature. For practical purposes, therefore, both the liquid and vapour may be regarded as saturated.

Example 14.4   A closed vessel contains 0·1 kg of wet steam having a dryness fraction of 0·1, and 0·15 kg of nitrogen. The temperature of the mixture is 95 °C. Find the volume of the vessel and the pressure of the mixture. Also find the temperature to which the mixture must be raised for all the water to be evaporated, and the corresponding pressure of the mixture.

There are 0·01 kg of saturated steam and 0·09 kg of saturated water in the vessel. It is immaterial whether the liquid is a single mass or is distributed through the vapour as droplets. Hitherto we have treated such a mixture as wet steam, but it is more convenient here to consider the two phases separately. When considering mixtures in general, we use suffixes to denote the substances involved. In this chapter we shall use s for the vapour phase in the mixture (saturated in this example because liquid is present) and w for the liquid phase (saturated because vapour is present). From the saturation table at 95 °C, $p_s = 0·8453$ bar and $v_s = 1·982$ m³/kg, and p_s is the partial pressure of the steam in the mixture. The volume of the vessel must be equal to the volume of the water plus the volume of the mixture of steam and nitrogen. But the volume of the mixture of steam and nitrogen can be found from the volume of the steam alone at the partial pressure of the steam and the temperature of the mixture. Therefore the volume of the vessel is

$$V = m_w v_w + m_s v_s = (0·09 \times 0·001) + (0·01 \times 1·982)$$
$$= 0·00009 + 0·01982 = 0·0199 \text{ m}^3$$

The volume of the steam itself, i.e. 0·01982 m³, is equal to the volume of the nitrogen at the partial pressure of the nitrogen p_N and the temperature of the mixture. Therefore

$$p_N = \frac{m_N R_N T}{V_s} = \frac{0·15 \times 8·314 \times 368}{100 \times 28 \times 0·01982} = 8·28 \text{ bar}$$

The pressure of the mixture is

$$p = p_s + p_N = 9·125 \text{ bar}$$

When the water is completely evaporated,

$$v_s = \frac{V}{m_s} = \frac{0·0199}{0·1} = 0·199 \text{ m}^3/\text{kg}$$

From the saturation table, the corresponding vapour pressure is, by interpolation,

$$p_s = 10 - \frac{0·199 - 0·1944}{0·2149 - 0·1944} 1 = 9·78 \text{ bar}$$

and similarly the temperature required is found to be 179 °C or 452 K. The volume occupied by the nitrogen is now 0·0199 m³ and therefore the corresponding partial pressure is

$$p_N = \frac{m_N R_N T}{V} = \frac{0·15 \times 8·314 \times 452}{100 \times 28 \times 0·0199} = 10·11 \text{ bar}$$

The final pressure of the mixture is

$$p = p_s + p_N = 9·78 + 10·11 = 19·9 \text{ bar}$$

We have seen that for perfect gases the internal energy, enthalpy and entropy of a mixture are respectively equal to the sums of the internal energies, enthalpies and entropies of the constituents, evaluated when each exists alone at the volume and temperature of the mixture. This result holds well for real gases, and with somewhat less accuracy for mixtures of gases and saturated vapours. When a liquid phase is present the internal energy, for example, of a mixture of liquid, vapour and gas can be found by considering the system in two parts: (a) the saturated vapour and gas mixture, and (b) the saturated liquid. The internal energy of the vapour-gas mixture can be determined by applying the Gibbs-Dalton law, and the internal energy of the liquid can be found from tables in the usual way. The total internal energy of the system is equal to the sum of the internal energies of the two parts. Such an approach can always be adopted when dealing with extensive properties.

Example 14.5 Find the heat that must be added during the process of evaporating the water in Example 14.4. Assume that for nitrogen $c_v = 0.747$ kJ/kg K

Applying the non-flow energy equation to this constant volume process, we have

$$Q = U_2 - U_1$$

The initial internal energy of the mixture is

$$U_1 = m_{w1}u_{w1} + m_{s1}u_{s1} + m_N u_{N1}$$

where u_{w1} and u_{s1} are found from the saturation tables of u_f and u_g at t_1. Similarly the final internal energy is

$$U_2 = m_{s2}u_{s2} + m_N u_{N2}$$

where $m_{s2} = m_{w1} + m_{s1}$ and u_{s2} is u_g at t_2. It follows that

$$Q = (m_{s2}u_{s2} - m_{w1}u_{w1} - m_{s1}u_{s1}) + m_N c_v N(t_2 - t_1)$$

Now the initial and final temperatures are $t_1 = 95\ ^\circ$C and $t_2 = 179\ ^\circ$C; and the masses are $m_{s1} = 0.01$ kg, $m_{w1} = 0.09$ kg, $m_N = 0.15$ kg. Using the saturation table, we find that the heat transferred is

$$Q = \{(0.1 \times 2583) - (0.09 \times 399) - (0.01 \times 2500)\} + \{0.15 \times 0.747(179 - 95)\}$$
$$= 207\ \text{kJ}$$

The higher the pressure the less accurate does the foregoing treatment become, but in a great many applications of practical interest to engineers the pressure is atmospheric or less and the results are sufficiently accurate. The steam condenser provides an important example of such a case and is worth considering in some detail. We shall consider the most common type—the shell and tube condenser—where steam flows over a bank of tubes through which cold water is pumped continuously. We have seen in section 11.2 that the temperature of the condensing steam may be as low as 25 °C. At this temperature the corresponding saturation pressure is 0·032 bar, i.e. a vacuum of 736 mm of mercury. With a pressure of this order in the low pressure side of the plant, it is impossible to prevent some air from leaking into the system through glands, or air dissolved in the boiler feed water from coming out of solution. The condenser is therefore always filled with a mixture of water, steam and air.

We may now consider the effect of the air on the design and performance of the condenser. Firstly, it might be thought that because the back pressure on the turbine is increased (it is equal to the saturation pressure of the steam plus the partial pressure of the air), the work output of the turbine will be reduced. As the subsequent numerical example will show, this effect is negligible for normal rates of air leakage. Secondly, for the reasons given at the end of section 22.8 (i), the air reduces the rate of heat transfer per unit area per unit temperature difference between steam and cooling water. Thus either the surface area of tubing or the temperature difference must be increased for a given condenser duty. The most obvious consequence of air leakage is the need for a pump of sufficient capacity to extract the air continuously and so maintain the vacuum in the condenser. Since the air cannot be separated from the steam with which it is mixed, this entails a loss of potential condensate. Furthermore, the presence of air results in undercooling, i.e. the condensate is cooled below the saturation temperature of the incoming steam. As the condensate is usually returned to the boiler, a reduction of condensate temperature necessitates an increased supply of heat in the boiler.

To see how undercooling can arise, suppose that steam enters the condenser in a dry saturated condition and that with every kilogramme of steam there are m_a kg of air. m_a is of course a small fraction. Consider the behaviour of a mass of steam-air mixture occupying a volume equal to the volume of 1 kg of steam. The pressure at entry is equal to the sum of the partial pressure of the air and the saturation pressure of the steam entering the condenser. Since the velocity of steam flow is small, the pressure may be assumed constant throughout the condenser. As the mass of mixture we are considering flows through the condenser, some of the steam condenses and the volume each kilogramme occupies is reduced. Since the volume of the remaining vapour is also the volume occupied by m_a kg of air, the partial pressure of the air increases. But the total pressure is constant, and an increase in the partial pressure of the air must be accompanied by a decrease in the saturation pressure, and hence temperature, of the steam. The temperature of the condensate therefore falls progressively below that of the incoming steam as condensation proceeds. The following example should help to make this clear.

Example 14.6 A condenser is to deal with steam of dryness fraction 0·95 and temperature 38 °C, at the rate of 5000 kg/h. If the estimated air leakage is 5 kg/h, determine the amount of vapour carried away with the air for condensate temperatures of 37, 35, 32 and 28 °C. Find also the capacity of the air extraction pump in each case.

The pressure in the condenser is first found from the inlet conditions. At 38 °C, from the saturation table,

$$p_s = 0·06624 \text{ bar} \quad \text{and} \quad v_s = 21·63 \text{ m}^3/\text{kg}$$

There are 5 kg of air to $0.95 \times 5000 = 4750$ kg of vapour, and therefore the mass of air, m_a, in a mass of mixture having a volume of 21.63 m^3 is $5/4750 = 1/950$ kg. Hence the partial air pressure

$$p_a = \frac{m_a R_a T}{v_s} = \frac{0.287 \times 311}{100 \times 950 \times 21.63} = 0.00004 \text{ bar}$$

The partial pressure of the air at entry is therefore negligible, and the total condenser pressure p is 0.0662 bar. At 37 °C, by interpolation from the saturation table,

$$p_s = 0.0628 \text{ bar} \quad \text{and} \quad v_s = 22.8 \text{ m}^3/\text{kg}$$

$$p_a = p - p_s = 0.0662 - 0.0628 = 0.0034 \text{ bar}$$

The mass of air per kg of vapour is

$$m_a = \frac{p_a v_s}{R_a T} = \frac{100 \times 0.0034 \times 22.8}{0.287 \times 310} = 0.08713$$

Consequently the mass of vapour per kg of air is $1/m_a = 11.48$ kg. Since 5 kg of air must be extracted per hour, this result implies that 57.4 kg of steam are extracted per hour with the air. Each kg of steam occupies 22.8 m^3, so that the capacity required of the air pump is

$$57.4 \times 22.8 = 1310 \text{ m}^3/\text{h}$$

Similar calculations made for other condensate temperatures yield the following results:

Condensate temperature/°C	37	35	32	28
m_s/m_a	11·5	3·53	1·59	0·828
Air pump capacity/[m^3/h]	1310	447	235	152

In the foregoing example, the water flow lost from the feed circuit varies from 57.4 to 4.14 kg/h as the condensate temperature is reduced from 37 °C to 28 °C. Neglecting the effect of the change in quantity of condensate, the additional heat which must be supplied in the boiler as a result of the reduced temperature of the feed water is $5000 \times 4.19(37 - 28) = 189\,000$ kJ/h (taking the specific heat of water as 4.19 kJ/kg K). Evidently a compromise must be reached between the conflicting requirements of low feed water loss and high condensate temperature. When plotted as in Fig. 14.3, the results show that in our example there is little to be gained in this respect by designing for a lower condensate temperature than about 32 °C. The size of air pump required is also relatively unaffected by a further reduction of temperature.

It is possible, by special design, to reduce the feed water loss and air pump capacity without excessive undercooling. For example, before leaving the condenser the air-steam mixture can be drawn over a section of tubes, the *air-cooler*, which is shielded from the main flow of steam and condensate. Most of the condensing process occurs with very little undercooling and the remainder is carried out in the air-cooler at a lower temperature. The effect of the small amount of low temperature feed water upon the temperature of the hot-well is negligible, little vapour is lost, and a smaller air pump can be used. Further improvement in performance is obtained with the 'regenerative' condenser. Fig. 14.4 illustrates a central-flow condenser of this type

Fig. 14.3

wherein some of the steam is allowed to pass unimpeded to the bottom, and in passing upwards meets the cooler condensate falling through the nest of tubes. The condensate is thereby reheated to a temperature approaching that of the steam at inlet. The air, saturated with steam, is withdrawn from the centre of one end of the condenser, after passing over the air-cooling section of tubes.

Fig. 14.4   Cross-section of central-flow condenser

Fig. 14.5

The shell and tube condenser is evidently an example of the multi-stream open system introduced in section 10.5. To determine the mass flow of cooling water required, the steady-flow energy equation may be applied as in the following example.

Example 14.7 A condensate temperature of 37 °C is to be adopted for the condenser considered in the previous example. Separate condensate and air pumps are to be employed and an air-cooling section of tubes is to reduce the temperature of the air pump suction to 32 °C. Calculate the mass flow of cooling water required if the temperature rise of the cooling water is to be 11 K.

Fig. 14.5 illustrates the system. From the previous example we have

$$m_{a1} = n_{a3} = 5 \text{ kg/h}, \quad m_{s1} = 4750 \text{ kg/h}, \quad m_{w1} = 250 \text{ kg/h}$$

The amount of steam extracted with the air is 1·59 kg per kg of air at 32 °C. Therefore $n_{s3} = 7·95$ kg/h. Also

$$m_{w2} = (m_{s1} + m_{w1}) - m_{s3} = 4992 \text{ kg/h}$$

The energy equation, neglecting kinetic energy terms, becomes

$$Q = (m_{a3}h_{a3} + m_{s3}h_{s3} + m_{w2}h_{w2}) - (m_{a1}h_{a1} + m_{s1}h_{s1} + m_{w1}h_{w1})$$
$$= m_a c_{pa}(t_3 - t_1) + (m_{s3}h_{s3} - m_{s1}h_{s1}) + (m_{w2}h_{w2} - m_{w1}h_{w1})$$

Using the saturation table, and taking $c_{pa} = 1·005$ kJ/kg K,

$$Q = 5 \times 1·005(32 - 38) + \{(7·95 \times 2559·3) - (4750 \times 2570·1)\} +$$
$$\{(4992 \times 154·9) - (250 \times 159·1)\} = -11·46 \times 10^6 \text{ kJ/h}$$

Therefore, for a temperature rise of 11 K, the cooling water flow is

$$\frac{11·46 \times 10^6}{11 \times 4·19} = 0·249 \times 10^6 \text{ kg/h}$$

What may be called the 'thermodynamic analysis' of a condenser has now been completed. Such an analysis cannot yield any information about the actual size, i.e. surface area of tubing, required for a given duty. Before this can be estimated, a study of the actual mechanism of heat transfer processes is necessary. Part IV is devoted to this topic.

14.5 Hygrometry (or psychrometry)

The previous section described a method of dealing with a mixture of saturated vapour and gas. If the vapour is superheated, however, the procedure can be simplified. When the pressure is sufficiently low for the Gibbs-Dalton law to be valid, a superheated vapour can be treated as a perfect gas. A mixture of superheated vapour and gas can then be regarded as a mixture of perfect gases and the method of section 14.2 is applicable. Moist atmospheric air is an example of a mixture which can be treated in this way.

Normally the atmosphere is not saturated with water vapour but is a mixture of air and superheated vapour. Many industrial processes, e.g. in the paper and textile industries, require close control of the vapour content of the atmosphere as well as its temperature. The air-conditioning of buildings for comfort, the design of cooling towers, and the science of meteorology, are other instances which involve a study of the properties of moist air. This study has been given the name of *hygrometry* (or *psychrometry*). Hygrometry makes use of certain technical terms which arise either from the need to specify the composition of the mixture or from the method of measuring the composition. These technical terms must now be introduced.

Specific humidity (or *moisture content*) ω is the ratio of the mass of water vapour m_s to the mass of dry air m_a in any given volume V of mixture. Thus

$$\omega = \frac{m_s}{m_a} = \frac{m_s/V}{m_a/V} = \frac{v_a}{v_s} \tag{14.24}$$

where v_a and v_s are the specific volumes of the dry air and vapour in the mixture. Treating the water vapour as a perfect gas, and writing p_s for the partial pressure of the vapour, we have

$$p_s V = m_s R_s T$$

Similarly for the dry air

$$p_a V = m_a R_a T$$

It follows that

$$\omega = \frac{R_a p_s}{R_s p_a}$$

Now $R_a = 0.2871 \text{ kJ/kg K}$ and $R_s = \dfrac{8\cdot3144}{18\cdot015} = 0.4615 \text{ kJ/kg K}$. Also, $p_a = p - p_s$, where p is the barometric pressure. Therefore,

$$\omega = 0.622 \frac{p_s}{p_a} = 0.622 \frac{p_s}{p - p_s} \tag{14.25}$$

Relative humidity ϕ is the ratio of the actual partial pressure p_s of the vapour, to the partial pressure p_g of the vapour when the air is saturated at the same temperature.*
Thus

$$\phi = \frac{p_s}{p_g} \tag{14.26}$$

and it is often expressed as a percentage. The ease with which the atmosphere takes up moisture from any surface depends on how far the air is short of being saturated rather than upon the absolute vapour content. The relative humidity is therefore of greater significance than the specific humidity when industrial drying processes or air-conditioning problems are being considered. Alternative expressions for ϕ can

*Since in this chapter suffix s is used to denote the constituent 'steam', we shall use suffix g to refer to the saturated state.

be found if the vapour is treated as a perfect gas. Thus, multiplying the numerator and denominator of equation (14.26) by $V/R_s T$ we have

$$\phi = \frac{p_s V/R_s T}{p_g V/R_s T} = \frac{m_s}{(m_s)_{\text{sat}}}$$

And, dividing m_s and $(m_s)_{\text{sat}}$ by V, we have

$$\phi = \frac{m_s/V}{(m_s)_{\text{sat}}/V} = \frac{v_g}{v_s} \tag{14.27}$$

where v_s is the actual specific volume of the vapour in the mixture, and v_g is the specific volume of the vapour when the air is saturated at the same temperature.

Dew-point is the temperature to which unsaturated air must be cooled at constant pressure for it to become saturated (or for condensation to begin). For the vapour in the mixture at temperature T we have

$$p_s = \frac{m_s R_s T}{V}$$

and for the mixture

$$p = \frac{mRT}{V}$$

Therefore,

$$\frac{p_s}{p} = \frac{m_s R_s}{mR}$$

Fig. 14.6

Since the ratio m_s/m remains unchanged as the unsaturated mixture is cooled at constant pressure, p_s must remain constant. The cooling process undergone by the vapour constituent is illustrated in Fig. 14.6. The partial pressure of the vapour at the dew-point is the saturation pressure p_{gd} corresponding to the dew-point temperature, and hence $p_s = p_{gd}$. Once the dew-point has been determined, for example by cooling a metal surface and observing its temperature when dew begins to form on it, the relative humidity can be found from equation (14.26), i.e.

$$\phi = \frac{p_{gd}}{p_g}$$

An alternative way of finding the relative humidity is by means of the *wet- and dry-bulb temperatures*. If a stream of unsaturated air flows past a thermometer having a wetted sleeve of cotton or linen around the bulb, the temperature recorded will be less than the actual temperature of the air. The temperature falls owing to evaporation from the wetted sleeve, and as a result there will be a transfer of heat from the air to the wetted sleeve. The thermometer reading reaches a steady value, called the wet-bulb temperature, when the rate of heat transfer balances the loss of energy due to vaporisation. The actual temperature of the air is sometimes called the dry-bulb temperature to emphasize the distinction. When the air is unsaturated the wet-bulb temperature lies between the dry-bulb and dew-point temperatures. The lower the relative humidity of the air, the more rapid is the evaporation from the wet-bulb and the larger becomes the *wet-bulb depression*, i.e. the difference between the dry- and wet-bulb temperatures. When the air is saturated, the wet-bulb, dry-bulb and dew point temperatures are identical.

Since the evaporation, and therefore the wet-bulb temperature, depends on the equilibrium established between the heat and mass transfer rates to and from the wetted sleeve, any slight draught increases the wet-bulb depression. It is found, however, that although the wet-bulb temperature falls as the air velocity is increased to about 2 m/s, it thereafter remains sensibly constant for velocities up to about 40 m/s. Provided that the air velocity is kept within this range, it is possible to relate the relative humidity to the wet- and dry-bulb temperatures alone. The equations are

Fig. 14.7   Wet- and dry-bulb hygrometer

complicated and the relation is best expressed in the form of tables as in Ref. (29). Fig. 14.7 illustrates a simple form of wet- and dry-bulb hygrometer, in which the air stream is produced by a small fan. The sling hygrometer is a commonly used portable type, and this consists of wet- and dry-bulb thermometers mounted on a frame which can be whirled through the air by hand. A brief description of these and other methods of measuring humidity can be found in Ref. (31).

We have seen that the relative humidity can be deduced from measurements of either the dew-point or wet- and dry-bulb temperatures. If required, the specific humidity can be determined from equations relating it to the relative humidity. Thus combining (14.24) and (14.27) we have

$$\phi = \omega \frac{v_g}{v_a} \tag{14.28}$$

or by combining (14.25) and (14.26),

$$\phi = \frac{\omega(p - p_s)}{0 \cdot 622 p_g} \tag{14.29}$$

Example 14.8 The pressure and temperature of the air in a room is 1 bar and 28 °C. If the relative humidity is found to be 30 per cent, find (a) the partial pressure of the water vapour and the dew-point, (b) the specific volume of each constituent, and (c) the moisture content (specific humidity).

From the saturation table, at 28 °C $p_g = 0.03778$ bar. Since $\phi = (p_s/p_g)$, $p_s = 0.3 \times 0.03778 = 0.01133$ bar. But $p_s = p_{gd}$ (Fig. 14.6), and the saturation temperature corresponding to 0.01133 bar is 8.8 °C: this temperature is the dew-point. [When dealing with saturated steam at very low pressures the temperature table of Ref. (17) will be used because the intervals are smaller than in the pressure table and linear interpolation is more accurate.] For the vapour

$$v_s = \frac{R_s T}{p_s} = \frac{0.462 \times 301}{100 \times 0.01133} = 122.7 \text{ m}^3/\text{kg}$$

Similarly for the dry air,

$$v_a = \frac{R_a T}{(p - p_s)} = \frac{0.287 \times 301}{100(1.0 - 0.0113)} = 0.874 \text{ m}^3/\text{kg}$$

The specific humidity is therefore, from equation (14.24),

$$\omega = \frac{v_a}{v_s} = \frac{0.874}{122.7} = 0.00712$$

The application of hygrometric principles to processes which occur in air-conditioning plant and cooling towers may now be considered.

(i) *Air-conditioning plant*

Air-conditioning* involves a regulated supply of air at some desired temperature and relative humidity. A typical plant is illustrated in Fig. 14.8. In winter, heating coils are in operation to warm the air. If this reduces the relative humidity to an uncomfortably low level, water is sprayed into the air at a controlled rate. In summer, when the air is cooled, the relative humidity increases unless some form of dehumidifying process is introduced. The combined cooling and dehumidifying is conveniently accomplished by the process illustrated in Fig. 14.9 which shows what happens to the vapour in the mixture. Cooling coils reduce the temperature of the incoming air to the dew-point and condensation commences. Sufficient cooling surface is provided for the temperature to fall to the dew-point corresponding to the required *exit* condition of the air. The air leaving the cooling coils is therefore saturated and at a lower temperature than is required finally. Simple heating at constant pressure to the final temperature will then reduce the relative humidity to the desired value. This is best explained by a numerical example.

*See Ref. (5) for a detailed treatment of this subject.

Fig. 14.8   Typical air-conditioning plant

Example 14.9   On a summer's day the air has a pressure of 1 bar, a temperature of 35 °C, and a relative humidity of 90 per cent. An air-conditioning plant is to deliver 30 m³/min at a temperature of 20 °C and with a relative humidity of 55 per cent. This is to be accomplished by cooling the air to the dew-point of the delivery air and reheating it. An axial-flow fan, situated before the cooler, absorbs 1·1 kW. Find the temperature to which the air must be cooled by the cooling coils, and the heat transfer rates required in the cooler and heater. Assume that the whole of the intake consists of atmospheric air, and that the pressure remains constant throughout the process at 1 bar.

Fig. 14.9   Combined cooling and dehumidifying

The temperature of the air leaving the cooler, at plane 2 in Fig. 14.10, can be found by considering the process which occurs in the heater. Here the composition of the mixture remains unchanged, so that the partial pressure of the vapour must be constant between planes 2 and 3. Hence

$$p_{s2} = p_{s3} = \phi_3 p_{g3} = 0.55 \times 0.02337 = 0.01285 \text{ bar}$$

At plane 2 the air has just left the cooler where condensation has occurred and the air is saturated with vapour. The temperature must therefore be the saturation temperature corresponding to p_{s2}, i.e. $t_2 = 10.7$ °C.

292

The energy equation may be applied between planes 2 and 3 to find the rate of heat transfer
in the heater. Neglecting changes in kinetic energy, we have

$$Q_{23} = (m_{a3}h_{a3} + m_{s3}h_{s3}) - (m_{a2}h_{a2} + m_{s2}h_{s2})$$

And for conservation of mass,

$$m_{a3} = m_{a2} \quad \text{and} \quad m_{s3} = m_{s2}$$

At the low partial pressures encountered in hygrometric work, a glance at the superheat table
will show that the enthalpy is not a function of pressure and the value of h_{s3} at 20 °C could

Fig. 14.10

be obtained by interpolating along the line of triple-point pressure values. It is more convenient, however, to treat the vapour above the saturation temperature t_2 as a perfect gas,
and write the equation as

$$Q_{23} = m_{a3}c_{pa}(t_3 - t_2) + m_{s3}c_{ps}(t_3 - t_2)$$

Over the range of temperature met with in air-conditioning plant, an average value for c_{ps}
is 1·86 kJ/kg K and for c_{pa} is 1·005 kJ/kg K [see Ref. (17)]. The mass of vapour in 30
m³/min of mixture is

$$m_{s3} = \frac{p_{s3}V}{R_s T} = \frac{100 \times 0·01285 \times 30}{0·462 \times 293} = 0·285 \text{ kg/min}$$

Similarly, the mass of air is

$$m_{a3} = \frac{(p - p_{s3})V}{R_a T} = \frac{100(1·0 - 0·01285)30}{0·287 \times 293} = 35·22 \text{ kg/min}$$

The heat transferred in the heater is therefore

$$Q_{23} = \{35·22 \times 1·005(20 - 10·7)\} + \{0·285 \times 1·86(20 - 10·7)\}$$
$$= 329·2 + 4·9 = 334·1 \text{ kJ/min}$$

To find the heat Q_{12} transferred in the cooler, the energy equation is applied to the open
system bounded by planes 1-2-4. Thus

$$Q_{12} = (m_{a2}h_{a2} + m_{s2}h_{s2} + m_{w4}h_{w4}) - (m_{a1}h_{a1} + m_{s1}h_{s1}) + W$$

This equation can be rearranged, remembering from conservation of mass that $m_{a3} =
m_{a2} = m_{a1} = m_a$ and $m_{s1} = m_{s2} + m_{w4}$, to give either

$$Q_{12} = m_a c_{pa}(t_2 - t_1) + \{(m_{s2}h_{s2} + m_{w4}h_{w4}) - m_{s1}h_{s1}\} + W \tag{a}$$

or

$$Q_{12} = m_a c_{pa}(t_2 - t_1) + m_{s2} c_{ps}(t_2 - t_1) + m_{w4}(h_{w4} - h_{s1}) + W \qquad \text{(b}$$

Now $m_a = 35 \cdot 22$ kg/min, and $m_{s2} = m_{s3} = 0 \cdot 285$ kg/min. m_{s1} can be found from the inle state. Thus

$$p_{s1} = \phi_1 p_{g1} = 0 \cdot 90 \times 0 \cdot 056\ 29 = 0 \cdot 050\ 66 \text{ bar}$$

The inlet specific humidity, from equation (14.25), is

$$\omega_1 = 0 \cdot 622 \frac{p_{s1}}{p_1 - p_{s1}} = 0 \cdot 622 \frac{0 \cdot 050\ 66}{1 \cdot 0 - 0 \cdot 050\ 66} = 0 \cdot 0332 \text{ kg steam/kg dry air}$$

and hence

$$m_{s1} = \omega_1 m_a = 0 \cdot 0332 \times 35 \cdot 22 = 1 \cdot 169 \text{ kg/min}$$

$$m_{w4} = m_{s1} - m_{s2} = 1 \cdot 169 - 0 \cdot 285 = 0 \cdot 884 \text{ kg/min}$$

The apparent advantage of using the second expression (b) for Q_{12} is that the enthalpy term involving m_{s2} can be easily calculated from the perfect gas approximation for steam (as was indeed done when calculating Q_{23}). But when some of the steam or water changes phase in the process, there is very little saving in labour and the use of expression (a reduces the chance of errors from algebraic manipulation. All enthalpies must then be calculated relative to the triple point datum of the steam tables. This approach will be adopted here.

The temperature of the condensed vapour will be equal to t_2, and hence $h_{w4} = 44 \cdot 9$ kJ/kg from the h_f column of the saturation table. The enthalpy $h_{s2} = h_g$ at $10 \cdot 7\,°C = 2519 \cdot 8$ kJ/kg. The enthalpy h_{s1} can be found from

$$h_{s1} = (h_g \text{ at } p_{s1}) + c_{ps}(t_1 - t_g \text{ at } p_{s1})$$
$$= 2561 \cdot 3 + 1 \cdot 86(35 - 33 \cdot 1) = 2565 \text{ kJ/kg}$$

Alternatively, h_{s1} can be found interpolating along any low-pressure line in the superhea table, because h is almost independent of p. Choosing the closest pressure to p_{s1}, viz $0 \cdot 05$ bar,

$$h_{s1} = 2561 + \frac{35 - 32 \cdot 9}{50 - 32 \cdot 9}(2594 - 2561) = 2565 \text{ kJ/kg}$$

The energy equation for the cooler therefore becomes

$$Q_{12} = 35 \cdot 22 \times 1 \cdot 005(10 \cdot 7 - 35) + \{(0 \cdot 285 \times 2519 \cdot 8 + 0 \cdot 884 \times 44 \cdot 9)$$
$$- 1 \cdot 169 \times 2565\} - 1 \cdot 1 \times 60 = -3166 \text{ kJ/min}$$

Here again, as we noted at the end of our analysis of condensers, the thermodynamic analysis only enables the required rates of heat transfer to be found. Further informa- tion is required before the necessary size of the cooling and heating coils can be determined. To reduce the quantities of heat involved, and therefore the size of the heat transfer surfaces, and yet maintain the freshness in the room, it is usual to draw about one-third of the total air supply from the atmosphere and to recirculate the remainder from the rooms (see Fig. 14.8).

(ii) *Cooling towers*

Steam power stations, refrigerating plant, and many other industrial processes, require large quantities of cooling water. If the plant is not situated near a natural source such as a river, it is necessary to recool the water after use and recirculate it. One of the most effective methods is to use the principle of evaporative cooling in a cooling tower. The warm water is sprayed into the tower near the top and allowed to fall through a packing of wooden slats which break up the stream and provide a large wetted surface to facilitate evaporation. A current of air passes up the tower, either naturally as in a chimney, or induced by a fan as shown in Fig. 14.11: the warm water is cooled, mainly by evaporation, while the air is raised in temperature and saturated, or nearly saturated, with water vapour. The water may be cooled to a temperature below that of the air entering the tower—ideally to the wet-bulb temperature of the air. The rate of cooling falls off as this equilibrium condition is approached and the increase in size of tower which would be required to obtain the last possible few degrees of cooling is uneconomic. Generally, the tower is designed for the outlet water temperature to be about 8 °C greater than the wet-bulb temperature of the atmosphere.

Although some of the cooling water is lost in the evaporative process, a cooling tower can be much smaller than a simple heat-exchanger wherein the heat is transferred from the water to the air through a dividing wall. Moreover, with a heat exchanger the water temperature can only be made to approach the dry-bulb temperature of the air, not its wet-bulb temperature.

For an introduction to the theory of design of cooling towers, and further references, the reader may turn to Ref. (16). The design is a problem in both heat and mass transfer from droplets and liquid films, and the detailed analysis to determine the volume of packing required for a given cooling duty is outside the scope of this book.

Example 14.10   Air enters the base of a natural-draught cooling tower at the rate of 1100 m³/min with a pressure of 1 bar, a temperature of 15 °C, and a relative humidity of 65 per cent. The water enters the tower at 38 °C and leaves at 17 °C. If the air leaves the tower at 32 °C in a saturated condition, find the mass flow of water entering the tower and the percentage loss of water by evaporation.

Fig. 14.12 illustrates the system. At the air inlet section 1 we have

$$p_{s1} = \phi_1 p_{g1} = 0.65 \times 0.01704 = 0.01108 \text{ bar}$$

$$m_{s1} = \frac{p_{s1}V}{R_s T} = \frac{100 \times 0.01108 \times 1100}{0.462 \times 288} = 9.16 \text{ kg/min}$$

$$m_{a1} = \frac{(p - p_{s1})V}{R_a T} = \frac{100(1.0 - 0.01108)1100}{0.287 \times 288} = 1316 \text{ kg/min}$$

Fig. 14.11   Induced-draught cooling tower

Fig. 14.12   Natural-draught cooling tower

At the air outlet section 2 we have

$$p_{s2} = p_{g2} = 0.04754 \text{ bar}$$

$$\omega_2 = 0.622 \frac{p_{s2}}{p - p_{s2}} = \frac{0.622 \times 0.04754}{(1.0 - 0.04754)} = 0.03105 \text{ kg}_s/\text{kg}_a$$

Since $m_{a2} = m_{a1}$

$$m_{s2} = \omega_2 m_{a1} = 0.03105 \times 1316 = 40.86 \text{ kg/min}$$

Water is lost by evaporation at the rate of

$$m_{s2} - m_{s1} = 40.86 - 9.16 = 31.7 \text{ kg/min}$$

Neglecting changes of kinetic energy and heat loss from the tower to atmosphere, the energy equation is

$$(m_{a2}h_{a2} + m_{s2}h_{s2} + m_{w4}h_{w4}) - (m_{a1}h_{a1} + m_{s1}h_{s1} + m_{w3}h_{w3}) = 0$$

Or in more convenient form,

$$m_a c_{pa}(t_2 - t_1) + (m_{s2}h_{s2} - m_{s1}h_{s1}) + (m_{w4}h_{w4} - m_{w3}h_{w3}) = 0$$

From conservation of mass,

$$m_{s2} + m_{w4} = m_{s1} + m_{w3}, \text{ and hence } m_{w4} = m_{w3} - 31.7 \text{ kg/min}$$

From saturation table, since t_{g1} at p_{s1} is 8.48 °C,

$$h_{s1} = h_{g1} + c_{ps}(t_1 - t_{g1})$$
$$= 2516 + 1.86(15 - 8.48) = 2528 \text{ kJ/kg}$$

and

$$h_{s2} = h_{g2} = 2559.3 \text{ kJ/kg}, \quad h_{w3} = h_{f3} = 159.1 \text{ kJ/kg}$$
$$h_{w4} = h_{f4} = 71.3 \text{ kJ/kg}$$

Finally, taking $c_{pa} = 1 \cdot 005$ kJ/kg K, we get

$$1316 \times 1 \cdot 005(32 - 15) + \{(40 \cdot 86 \times 2559) - (9 \cdot 16 \times 2528)\} +$$
$$\{(m_{w3} - 31 \cdot 7)71 \cdot 3 - m_{w3}159 \cdot 1\} = 0$$

The water flow at inlet, m_{w3}, is consequently 1160 kg/min. The percentage loss by evaporation is

$$\frac{m_{s2} - m_{s1}}{m_{w3}} \, 100 = 2 \cdot 7\%$$

14.6 Hygrometric chart

Calculations involved in the type of problem considered in section 14.5 can be considerably simplified if use is made of the hygrometric chart, Fig. 14.13. This chart enables the enthalpy of moist atmospheric air, of any relative humidity, to be read off directly. The enthalpy of the mixture is expressed as kJ per kg of *dry* air in the mixture. If this quantity is denoted by H, then

$$H = h_a + \omega h_s$$

Although h_a is fixed solely by the temperature of the air (dry-bulb temperature t), the value of ωh_s depends upon the specific humidity and both the pressure and temperature

Dry bulb temperature °C

Fig. 14.13   Hygrometric chart

of the air. Thus, equation (14.25) indicates that when ω and p are known the partial pressure of the vapour p_s is determined. If in addition the temperature t is known, then h_s is fixed. The chart is drawn for a fixed barometric pressure so that ω and t can be regarded as the only independent variables. These form the ordinate and abscissa scales respectively, and lines of constant enthalpy H are plotted on the diagram.

Equation (14.29) shows that, for a given barometric pressure, the relative humidity ϕ is a function of ω, p_s and p_g. p_g is the saturation pressure corresponding to the dry-bulb temperature t, and we have already seen that p_s is a function of ω for a fixed barometric pressure. It follows that ϕ is also a function of ω and t, and therefore lines of constant relative humidity can be added to the diagram. Since humidity measurements are so often made in terms of wet- and dry-bulb temperatures, lines of constant wet-bulb temperature are also included. Note that the wet-bulb temperature at any state along the saturation line ($\phi = 100\%$) is equal to the dry-bulb temperature, and also the dew-point, at that state. Lines of constant dew-point are horizontal because they are also lines of constant specific humidity. The constant wet-bulb temperature lines slope down towards the right of the diagram and follow very closely the lines of constant enthalpy.

20   35
Dry bulb temperature °C

Fig. 14.14

Finally, lines of constant specific volume of the dry air in the mixture, v_a, are usually included. v_a is a function of the dry-bulb temperature t and the partial pressure of the air $(p - p_s)$. But we have seen that p_s is a function of ω and p, and so, for a given barometric pressure p, v_a is a function of t and ω.

The way in which the hygrometric chart can be used to simplify the solution of a problem will be indicated briefly by referring to Example 14.9 on the air-conditioning plant of Fig. 14.10. The process undergone by the mixture can be represented as in Fig. 14.14, points 1 and 3 being located from the given values of t and ϕ at the inlet and outlet of the system. The temperature to which the air must be cooled, t_2, can be read directly from the chart. Before applying the energy equation, the next step must be to find the mass flow of dry air because all the 'specific' quantities on the chart refer to unit mass of dry air. Thus, v_{a3} is read from the chart, and hence m_a can be found from V_3/v_{a3} where V_3 is the volume flow of mixture delivered.

The energy equation for the heating process, i.e. between planes 2 and 3 in Fig. 14.10, can be written in the convenient form:

$$Q_{23} = m_a(H_3 - H_2)$$

since the values of H are in kJ/kg of dry air in the mixture, and m_a does not change between planes 2 and 3. Both H_3 and H_2 can be read from the chart, and so Q_{23} can be determined directly.

Similarly, the energy equation for the cooling and condensing process can be written as

$$Q_{12} = m_a(H_2 - H_1) + m_{w4}h_{w4} + W$$

H_2 and H_1 can be read from the chart, W is given, and h_{w4} can be found from the saturation table in the usual way; it remains to find m_{w4}. For conservation of mass,

$$m_{s1} = m_{s2} + m_{w4}$$

or,

$$m_{w4} = m_a(\omega_1 - \omega_2)$$

Both ω_1 and ω_2 can be read from the chart, and therefore m_{w4}, and thence Q_{12}, can be rapidly calculated.

Although a hygrometric chart is drawn for a particular barometric pressure, it can be used without much loss of accuracy for any reasonable variation in this pressure. For example, the chart of Ref. (6) is drawn for a pressure of 1 atm, but it can be used without sensible error between 0·9 and 1·1 bar. All the information conveyed by the chart can also be obtained in the form of tables such as those of Ref. (29).

15 Combustion Processes

Previous chapters have been wholly concerned with the behaviour of fluids which remain in the same chemical state throughout the processes considered. Here we shall deal with the thermodynamic aspects of a particular type of process involving chemical reactions, namely combustion. The term combustion refers to the fairly rapid reaction, usually accompanied by a flame, which occurs between a fuel and an oxygen carrier such as air. The molecules of fuel and air have a certain amount of energy stored in the bonds between their constituent atoms, a form of internal energy which we have not had to consider hitherto because it remains constant for all processes not involving chemical reactions. In the new molecules formed by a reaction this 'chemical' energy is at a lower level and the energy thus released can be transferred to the surroundings in the form of heat. Combustion is therefore said to be an *exothermic* reaction.

After discussing the relevant combustion equations, which relate the masses of reactants and products, methods of applying the energy equation to the combustion process are examined and the efficiency of such a process is defined. Finally, the idea of a reversible combustion process is introduced, and a method of estimating the effect of dissociation is described.

15.1 Fuels

Most common fuels consist mainly of hydrogen and carbon, whether the fuel be solid (e.g. coal), liquid (e.g. petroleum) or gaseous (e.g. coal gas). We shall assume that the chemical analysis of the fuel is known.* For solid and liquid fuels the analysis is usually quoted as a percentage by mass of each chemical *element* in the fuel—it is termed an *ultimate* analysis. The analysis of a gaseous fuel, on the other hand, is usually given in terms of the percentage by volume of each gas or type of hydrocarbon present.

In coal, the combustible elements are chiefly carbon (C) and hydrogen (H) with a minor quantity of sulphur (S); the incombustibles are nitrogen (N), moisture (H_2O) and small quantities of minerals which form the ash after combustion. Typical ultimate analyses of the three main types of coal—anthracite, bituminous coal, and lignite—are given in Fig. 15.1. Since there are numerous varieties of each type of coal, the figures can only serve as a rough guide. Moreover the analyses refer to dry coals:

---

*See Ref. (19) for complete data on typical fuels.

the moisture content is as low as 1 per cent in anthracite, varies from 2 to 10 per cent
in the bituminous coals, and is as high as 15 per cent in lignite.

Petroleum oils are complex mixtures of a large number of different hydrocarbons.
These hydrocarbons have been classified into groups, the most common being:
paraffins (having a molecular formula of the form C_nH_{2n+2}), olefins and naphthenes

Coal			C	H	O	N+S	ash
Anthracite	.	.	90·27	3·00	2·32	1·44	2·97
Bituminous	.	.	74·00	5·98	13·01	2·26	4·75
Lignite	.	.	56·52	5·72	31·89	1·62	4·25

Fig. 15.1   Ultimate analyses of coals

(C_nH_{2n}),* and aromatics (C_nH_{2n-6}). Although analysis in terms of actual compounds
is difficult and sometimes impossible, the ultimate analysis is a simple matter and
Fig. 15.2 gives typical examples. The analyses are very similar and the choice of fuel
for a particular application depends mainly on other characteristics such as viscosity,
ignition temperature, knocking tendency (in petrol engines), coking tendency (in
gas turbines), and so on.

	Motor petrol	Vaporising oil	Kerosene	Diesel oil (gas oil)	Light fuel oil	Heavy fuel oil
C	85·5	86·8	86·3	86·3	86·2	86·1
H	14·4	12·9	13·6	12·8	12·4	11·8
S	0·1	0·3	0·1	0·9	1·4	2·1

Fig. 15.2   Ultimate analyses of petroleum oils

Most hydrocarbons in petroleum deposits occur naturally as a liquid, but a few
exist in the gaseous phase at atmospheric temperature. Methane (a paraffin, CH_4) is
the most common natural gas—it is alternatively known as marsh gas and fire-damp.
The most widely used gaseous fuels, however, are formed by heating coal in various
ways: three typical compositions are given in Fig. 15.3. Hydrogen and carbon monoxide are easily isolated, but the various hydrocarbons are usually grouped into two
types representing respectively an average composition of the paraffins in the gas
(assumed to have the formula CH_4) and an average of the remainder (C_nH_m).

In the following sections concerned with typical combustion calculations, we shall
neglect the effect of minor elements, such as sulphur, in the fuel. Although for the
purpose of setting out mass and energy balances this is permissible, it should not be
thought that the presence of a small quantity of a substance is always unimportant.
From some points of view its presence (or absence) may well determine the choice of

*Although olefins and naphthenes have the same C/H ratio, they are distinguished by an important
difference in molecular structure.

fuel for a particular application, e.g. because of corrosive compounds formed by the combustion of an impurity.

	H_2	CO	'CH$_4$'	C_nH_m	O_2	N_2	CO_2
Coal gas	49·4	18·0	20·0	C_4H_8 2·0	0·4	6·2	4·0
Producer gas . . .	12·0	29·0	2·6	C_2H_4 0·4	—	52·0	4·0
Blast furnace . . .	2·0	27·0	—	—	—	60·0	11·0
North Sea gas	—	—	93·0	C_2H_6 4·8	—	2·0	0·2

Fig. 15.3 Volumetric analyses of gaseous fuels

15.2 Chemical equations—conservation of mass

The first step in the description of a combustion process is the formulation of the chemical equation which shows how the atoms of the reactants are rearranged to form the products. Before the chemical equation can be written it is necessary to know the number of atoms of each element in the molecules of the reactants and products. In any purely chemical, as opposed to nuclear reaction, the process is simply one of rearrangement of the atoms to form new molecules, and the total number of atoms of each element is unchanged. A chemical equation expresses the principle of the conservation of mass in terms of the conservation of atoms.

Consider the simple chemical equation expressing the complete combustion of carbon and oxygen to carbon dioxide:

$$C + O_2 \rightarrow CO_2 \tag{15.1}$$

This equation states that one atom of carbon will combine with one molecule of oxygen (2 atoms of O) to produce one molecule of carbon dioxide (1 atom of C + 2 atoms of O). The number of atoms of each element is the same on either side of the equation, and the arrow represents the direction of the reaction.

Equation (15.1) also represents the *amounts of substance* involved in the reaction and can be written as

$$1 \text{ kmol C} + 1 \text{ kmol O}_2 \rightarrow 1 \text{ kmol CO}_2 \tag{15.2}$$

Since the *molar masses* of the substances are numerically equal to the *relative atomic* or *molecular masses*,* equation (15.2) can be converted into a mass

*For a discussion of the relation between amount of substance, molar mass and relative atomic or molecular mass, see section 8.5(i).

Various 'atomic weight' scales have been used in the past. A chemical scale was fixed when the existence of three isotopes of oxygen of differing mass was not known: it was formed by allotting the number 16 to a 'mean' oxygen atom. This proved unsatisfactory because the proportions of oxygen isotopes vary in nature. When nuclear studies became important a physical scale was adopted in which the number 16 was attached to the most common isotope of oxygen (^{16}O). In 1961 a new scale was adopted by both chemists and physicists based on a particular isotope of carbon (^{12}C) which was assigned the value 12. Values on the old chemical scale and new ^{12}C scale differ only by about 0·004 per cent. For combustion calculations whole number values (viz. C-12, O-16, H-1, N-14) are sufficiently accurate, and thus these changes in the basis of the relative atomic mass scale are insignificant.

equation as follows:

$$12 \text{ kg C} + 32 \text{ kg O}_2 \rightarrow 44 \text{ kg CO}_2 \tag{15.3}$$

Now all gases occupy equal volumes *per kmol* when reduced to the same pressure and temperature. Although this is only strictly true for perfect gases, it is substantially true for all gases and dry vapours at moderate pressures. If any of the reactants or products are in the solid or liquid phase, the volume occupied by them can be neglected in comparison with the volume of the gases present. It follows that equation (15.2) is also equivalent to

$$0 \text{ vol. C} + 1 \text{ vol. O}_2 \rightarrow 1 \text{ vol. CO}_2 \tag{15.4}$$

Equation (15.4) implies that the volume of the products is equal to the volume of the reactants, when measured at the same pressure and temperature.

If insufficient oxygen is present for all the carbon to burn to carbon dioxide, some will burn to carbon monoxide. The reaction is then symbolised by

$$C + \tfrac{1}{2}O_2 \rightarrow CO \tag{15.5}$$

Carbon monoxide can be subsequently burnt to carbon dioxide, the appropriate equation being

$$CO + \tfrac{1}{2}O_2 \rightarrow CO_2 \tag{15.6}$$

and it is for this reason that the combustion of carbon to carbon monoxide is said to be 'incomplete', and to carbon dioxide 'complete'.

Equations similar to (15.1) to (15.4) can be written for the combustion of hydrogen, the other main constituent of a hydrocarbon fuel:

$$
\begin{aligned}
H_2 \quad &+ \quad \tfrac{1}{2}O_2 \quad \rightarrow \quad H_2O \\
1 \text{ kmol } H_2 &+ \tfrac{1}{2} \text{ kmol } O_2 \rightarrow \quad 1 \text{ kmol } H_2O \\
2 \text{ kg } H_2 \quad &+ \quad 16 \text{ kg } O_2 \rightarrow \quad 18 \text{ kg } H_2O \\
1 \text{ vol. } H_2 \quad &+ \quad \tfrac{1}{2} \text{ vol. } O_2 \rightarrow 1 \text{ vol. } H_2O \text{ (vap)}
\end{aligned} \tag{15.7}
$$

The abbreviation (vap) indicates that the water vapour has not been condensed.

In most engineering combustion systems the necessary oxygen is obtained by mixing the fuel with air.* For our purposes it will be sufficiently accurate to use the

---

*The chief exception is the rocket motor, described in section 18.4(ii). In this case the necessary oxygen is obtained from liquid oxygen, liquid ozone, or some oxygen carrier such as hydrogen peroxide or nitric acid.

following figures for the composition of air:

	O_2	N_2
Volumetric analysis %	21	79
Gravimetric analysis %	23·3	76·7

Here the symbol N is used to denote not only the nitrogen, but all the inert gases present in air which remain unchanged throughout the reaction. The molar mass of N_2 can be taken as 28 kg/kmol, and that of air as 29 kg/kmol.

The following two examples, concerning a liquid and gaseous fuel respectively, show how chemical equations may be used (a) to predict the analysis of the products of combustion when the composition of the fuel and the air/fuel ratio are known and (b) to determine the *stoichiometric* (or *correct*) air/fuel ratio. A stoichiometric mixture is one in which there is theoretically just sufficient oxygen to burn all the combustible elements in the fuel completely. In practical combustion systems the time available for the reaction is limited and excess air must be provided to ensure complete combustion. This is not only because of imperfect mixing of the fuel and air, but also because the inert gas molecules obstruct the reaction between the active molecules of fuel and oxygen.

The analysis of the products of combustion may be quoted on a volume or a mass basis. In either case, if the water vapour is assumed to be present as it is in the hot exhaust gases from an engine, the analysis is called a *wet* analysis. When the vapour is assumed to be condensed and removed, it is called a *dry* analysis.

Example 15.1 Determine the stoichiometric air/fuel ratio for a petrol approximating to hexane C_6H_{14}. Hence deduce the chemical equation if the petrol is burnt in 20 per cent excess air, and the wet volumetric analysis of the products (a) if all the water vapour is present, and (b) if the products are cooled to an atmospheric pressure and temperature of 1 bar and 15 °C. Determine also the dry volumetric analysis.

Finally estimate the chemical equation if only 80 per cent of the air required for stoichiometric combustion is provided.

Since the combustion is theoretically complete with a stoichiometric mixture, the products will consist only of CO_2, H_2O and the N_2 brought in with the air. The amount of oxygen required can be found by first balancing the carbon and hydrogen atoms on each side of the equation and then summing the number of atoms of oxygen in the products.

$$C_6H_{14} + 9\tfrac{1}{2}O_2 \rightarrow 6CO_2 + 7H_2O$$
$$86 \text{ kg } C_6H_{14} + 304 \text{ kg } O_2 \rightarrow 264 \text{ kg } CO_2 + 126 \text{ kg } H_2O$$

Since air contains 23·3 per cent by mass of oxygen, we have

$$\text{stoichiometric air/fuel ratio} = \frac{304 \times 100}{86 \times 23\cdot3} = 15\cdot17$$

If there is 20 per cent excess air, the equation, including the nitrogen, becomes

$$C_6H_{14} + 1\cdot2\left\{9\tfrac{1}{2}O_2 + \left(\frac{79}{21}\right)9\tfrac{1}{2}N_2\right\} \rightarrow 6CO_2 + 7H_2O + (0\cdot2)9\tfrac{1}{2}O_2 + 1\cdot2\left(\frac{79}{21}\right)9\tfrac{1}{2}N_2$$

The amount of substance of the products is

$$6(CO_2) + 7(H_2O) + 1 \cdot 9(O_2) + 42 \cdot 89(N_2) = 57 \cdot 79 \text{ kmol}$$

Now the mole fraction of a constituent equals the volume fraction [equation (14.18)]. The wet volumetric analysis if all the water vapour is present is therefore

(a) $CO_2 \dfrac{6 \times 100}{57 \cdot 79} = 10 \cdot 38\%;$ $H_2O \dfrac{7 \times 100}{57 \cdot 79} = 12 \cdot 11\%;$

   $O_2 \dfrac{1 \cdot 9 \times 100}{57 \cdot 79} = 3 \cdot 29\%;$ $N_2 \dfrac{42 \cdot 89 \times 100}{57 \cdot 79} = 74 \cdot 22\%$

The partial pressure of saturated water vapour at 15 °C is 0·01704 bar. The mole fraction of the water vapour in the mixture, from equation (14.17), is

$$x_s = \frac{0 \cdot 01704}{1 \cdot 0} = 0 \cdot 01704$$

If y is the number of kilomoles of water vapour remaining in the products after cooling, the total amount of substance becomes $\{(57 \cdot 79 - 7) + y\}$ kmol and therefore

$$\frac{y}{50 \cdot 79 + y} = 0 \cdot 01704, \quad \text{from which} \quad y = 0 \cdot 88$$

The total amount of substance of products under these conditions is 51·67 kmol in the vapour state, and the volumetric analysis becomes

(b) CO_2 11·61%; H_2O 1·70%; O_2 3·68%; N_2 83·01%

Merely cooling the products does not remove the water vapour entirely. If all the vapour is assumed to be removed, say by a drying agent, the total amount of substance is 50·79 kmol, and the dry analysis is

CO_2 11·81%; O_2 3·74%; N_2 84·45%

When insufficient air is supplied for complete combustion, some of the carbon will burn to carbon monoxide, and there may also be some free hydrogen. Insufficient information is given for the determination of the chemical equation. Hydrogen, however, has a greater affinity for oxygen than has carbon, and if the mixture is not too rich in fuel it is reasonable to assume that all the hydrogen will be burnt.* We can then write, for the rich mixture,

$$C_6H_{14} + 0 \cdot 8 \left\{ 9\tfrac{1}{2}O_2 + \left(\frac{79}{21}\right) 9\tfrac{1}{2}N_2 \right\} \rightarrow aCO_2 + bCO + 7H_2O + 0 \cdot 8 \left(\frac{79}{21}\right) 9\tfrac{1}{2}N_2$$

From the carbon balance,

$$6 = a + b$$

from the oxygen balance,

$$0 \cdot 8 \times 9\tfrac{1}{2} = a + b/2 + 7/2$$

*An accurate prediction of the products analysis for a mixture lacking sufficient oxygen for complete combustion can only be obtained by the method outlined in section 15.7. This makes use of additional information in the form of equilibrium constants for the various reactions.

Solving these simultaneous equations, we get

$$a = 2\cdot2, \quad \text{and} \quad b = 3\cdot8$$

The chemical equation is therefore

$$C_6H_{14} + 7\cdot6O_2 + 28\cdot6N_2 \rightarrow 2\cdot2CO_2 + 3\cdot8CO + 7H_2O + 28\cdot6N_2$$

Example 15.2 A gaseous fuel has the following percentage composition by volume:

$$CO \ 12\cdot6\%; \quad H_2 \ 41\cdot6\%; \quad CH_4 \ 26\cdot6\%; \quad O_2 \ 1\cdot9\%; \quad CO_2 \ 2\cdot6\%; \quad N_2 \ 14\cdot7\%$$

Find the wet volumetric and gravimetric analyses of the products of combustion when 10 per cent excess air is provided.

The solution can be set out in tabular form. The first table determines the oxygen required for stoichiometric combustion, and the amounts of CO_2 and H_2O produced thereby, per mol of fuel burnt. Note that the amount of oxygen in the fuel is taken into account when estimating the oxygen required from the air.

Reactants	kmol per kmol of fuel	Reaction	O_2 required	Products CO_2	Products H_2O
CO	0·126	$CO + \frac{1}{2}O_2 \rightarrow CO_2$	0·063	0·126	—
H_2	0·416	$H_2 + \frac{1}{2}O_2 \rightarrow H_2O$	0·208	—	0·416
CH_4	0·266	$CH_4 + 2O_2 \rightarrow CO_2 + 2H_2O$	0·532	0·266	0·532
O_2	0·019	—	−0·019	—	—
CO_2	0·026	—	—	0·026	—
N_2	0·147	—	—	—	—
	1·000		0·784	0·418	0·948

0·784 kmol of oxygen is required, per kmol of fuel burnt, for stoichiometric combustion. Since 10 per cent excess air is provided, 0·0784 kmol of oxygen is additionally present and remains unchanged throughout the process. The total amount of nitrogen in the products is

$$1\cdot1 \left(\frac{79}{21}\right) 0\cdot784 + 0\cdot147 = 3\cdot392 \ \text{kmol/kmol of fuel}$$

The analyses of the products by volume and by mass [see equation (14.23)] therefore become:

Products	kmol per kmol fuel (a)	% by volume	M [kg/kmol]	kg per kmol fuel $M \times (a)$	% by mass
CO_2	0·418	8·6	44	18·39	13·8
H_2O	0·948	19·6	18	17·06	12·8
O_2	0·078	1·6	32	2·50	1·9
N_2	3·392	70·2	28	94·98	71·5
	4·836	100·0		132·93	100·0

15.3 Experimental products analysis

When investigating the effectiveness of combustion plant it is often necessary to analyse the products experimentally. By taking samples at different stages of a reaction the combustion process can be studied, and the analyses can be used to discover lack of uniformity in fuel-air distribution. A products analysis also enables the air/fuel ratio to be computed in cases where direct measurement of the rate of air flow is not easily accomplished.

A brief description of various exhaust gas analysers can be found in Ref. (8). The most common method is by successive absorption of the CO_2, O_2 and CO as in the

Fig. 15.4   Orsat apparatus

Orsat apparatus, Fig. 15.4. The sample of exhaust gas is drawn into the measuring burette by lowering the levelling bottle containing water. This known volume of sample at atmospheric temperature and pressure is forced in succession into each of the reagent bottles which contain the absorbents. The volume is measured after each absorption process by returning the sample to the burette and bringing it to atmospheric pressure with the aid of the levelling bottle. The change in volume after a particular constituent has been absorbed is then the partial volume of that constituent in the original sample (see section 14.2). The gas which remains is assumed to be nitrogen.

The resulting analysis is by volume and is on a dry basis. Although most of the water vapour is condensed when the sample is first cooled to atmospheric temperature, the sample is nevertheless saturated with water vapour at this temperature throughout the analysis because the sample is contained in the measuring burette over water.

The reason why the Orsat apparatus yields a true dry analysis is therefore not obvious and it is left to the following example to show that this is so. A second example shows how the analysis can be used to compute the air/fuel ratio when the percentage by mass of carbon in the fuel is known.

Example 15.3 A sample of exhaust gases, when cooled to atmospheric temperature, has the following percentage composition by volume:

$$CO_2 \quad a; \qquad H_2O \quad b; \qquad O_2 \quad c; \qquad N_2 \quad d$$

Show that the Orsat analysis yields the true *dry* analysis which is evidently

$$CO_2 \quad \frac{100a}{100 - b}; \qquad O_2 \quad \frac{100c}{100 - b}; \qquad N_2 \quad \frac{100d}{100 - b}$$

Since the sample is in contact with water at a constant temperature throughout the analysis, the partial pressure of the water vapour, p_s, must remain constant at the saturation value. But from equation (14.17)

$$\frac{p_s}{p} = \frac{n_s}{n}$$

and p, the pressure of the mixture, is atmospheric. It follows that n_s/n is constant. Since n decreases as absorption of the constituents proceeds, n_s must decrease also, implying that some of the water vapour condenses.

a, b, c and d can be regarded as kmol of constituent per 100 kmol of mixture, so that initially

$$\frac{n_s}{n} = \frac{b}{100}$$

If n_{s1} represents the number of kmol of water vapour condensed during absorption of c kmol of CO_2, the mole fraction of vapour in the remaining mixture is

$$\frac{b - n_{s1}}{100 - (a + n_{s1})}$$

Hence

$$\frac{b}{100} = \frac{b - n_{s1}}{100 - (a + n_{s1})}, \quad \text{from which} \quad n_{s1} = \frac{ab}{100 - b}$$

The first percentage contraction of volume noted from the Orsat apparatus is therefore

$$a + n_{s1} = \frac{100a}{100 - b}$$

which is the percentage by volume of CO_2 in a *dry* sample.

Similarly, if n_{s2} is the number of kmol of water vapour condensed on absorption of the O_2 we have

$$\frac{b}{100} = \frac{b - (n_{s1} + n_{s2})}{100 - (a + n_{s1} + c + n_{s2})}$$

After substituting for n_{s1} this gives

$$n_{s2} = \frac{bc}{100 - b}$$

The second contraction noted from the Orsat apparatus is therefore

$$c + n_{s2} = \frac{100c}{100 - b}$$

which is the percentage by volume of O_2 in the dry sample.
   The remaining volume, consisting of N_2 and water vapour, is

$$100 - (a + n_{s1} + c + n_{s2})$$

After substituting for n_{s1} and n_{s2} this reduces to

$$\frac{100d}{100 - b}$$

which is the percentage by volume of N_2 in a dry sample.

Example 15.4   A petrol has the following ultimate analysis:

   C   83·7%;   H   16·3%

If the dry products analysis by volume is

   CO_2 11·8%; O_2 3·7%; N_2 84·5%

determine the air/fuel ratio used.

Method (i): *carbon balance*
Treating the volume fractions as mole fractions, 1 kmol of dry products contains
0·845 kmol of N_2. 0·845 kmol of N_2 are brought in with

$$0{\cdot}845 \times 28 \left(\frac{100}{76{\cdot}7}\right) = 30{\cdot}85 \text{ kg of air}$$

1 kmol of dry products contains 0·118 kmol of CO_2, which in turn contain $0{\cdot}118 \times 12$ kg
of carbon. This amount of carbon implies the combustion of

$$\frac{0{\cdot}118 \times 12}{0{\cdot}837} = 1{\cdot}692 \text{ kg of petrol}$$

Thus 1 kmol of dry products is produced when 30·85 kg of air are burnt with 1·692 kg of
petrol, i.e.

$$\text{air/fuel ratio} = \frac{30{\cdot}85}{1{\cdot}692} = 18{\cdot}2$$

   The foregoing method, using a carbon balance between reactants and products, is the most
satisfactory when carbon is the predominant element in the fuel. The result may be inaccurate,

however, if the exhaust gases are smoky, indicating unburnt carbon not included in the products analysis. An oxygen-hydrogen balance can be made as a check, although since the relevant quantities are relatively small the result is more open to error from inaccuracies of measurement.

Method (ii): *oxygen-hydrogen balance*

One kmol of dry products contains

$$32(0{\cdot}118 + 0{\cdot}037) = 4{\cdot}960 \text{ kg of } O_2$$

But the oxygen accompanying the nitrogen is

$$0{\cdot}845 \times 28 \left(\frac{23{\cdot}3}{76{\cdot}7}\right) = 7{\cdot}187 \text{ kg}$$

Hence $(7{\cdot}187 - 4{\cdot}960) = 2{\cdot}227$ kg of O_2 have combined with the hydrogen in the fuel to form water vapour. It follows from equation (15.7) that $2{\cdot}227/8$ kg of H_2 are involved, which corresponds to

$$\frac{2{\cdot}227}{8 \times 0{\cdot}163} = 1{\cdot}708 \text{ kg of petrol}$$

Therefore 30·85 kg of air are associated with 1·708 kg of fuel, i.e.

$$\text{air/fuel ratio} = \frac{30{\cdot}85}{1{\cdot}708} = 18{\cdot}1$$

With coal-burning appliances there is an additional complication because some of the combustion products will be in solid form. The analysis of the refuse is easily accomplished by heating a sample above its ignition temperature in a stream of oxygen. The loss of weight is measured and assumed to be carbon while the remainder is the ash. The following example shows how a complete mass balance can be compiled for a combustion process if the composition of the fuel, gaseous products and refuse are measured. In general the air will have a certain moisture content, which can be found by wet- and dry-bulb temperature measurements (sections 14.5 and 14.6), and this is also taken into account in the example.

Example 15.5 Determine the complete mass balance for a combustion process, on the basis of 1 kg of coal burnt, given the following data:

Fuel composition (fractions by mass)

C 0·69;   H 0·03;   O 0·04;   N 0·01; H_2O 0·10;   ash 0·13

Dry exhaust gas analysis (fractions by volume)

CO_2 0·146;   CO 0·020; O_2 0·055; N_2 0·779

Refuse analysis (fractions by mass)

C 0·40;   ash 0·60

Air: dry-bulb temperature 22·8 °C, wet-bulb temperature 15·5 °C.

a) The mass of refuse, m_R, per kg of fuel burnt can be found from an ash balance, i.e.

$0.6\, m_R = 0.13\,\text{kg}$

$m_R = 0.217\,\text{kg}$

b) The mass of dry gaseous products can be found from a carbon balance.

Basis: 1 kg of fuel.

Carbon in fuel	= 0.69	kg	or	0.0575 kmol
Carbon in refuse = 0.4 × 0.217	= 0.087	kg	or	0.0072 kmol
Carbon in gaseous products	= 0.603	kg	or	0.0503 kmol

Basis: 1 kmol of dry gaseous products.

Carbon in CO_2	= 0.146 kmol
Carbon in CO	= 0.020 kmol
Carbon in gaseous products	= 0.166 kmol

Ratio of dry gaseous products to fuel

$= 0.0503/0.166 = 0.303\ \text{kmol/kg}$

The dry gaseous products/kg of fuel are therefore:

$CO_2 = 0.146 \times 0.303 = 0.0442\ \text{kmol or} \times 44 = 1.95\ \text{kg}$
$CO\ \ = 0.020 \times 0.303 = 0.0061\ \text{kmol or} \times 28 = 0.17\ \text{kg}$
$O_2\ \ = 0.055 \times 0.303 = 0.0167\ \text{kmol or} \times 32 = 0.53\ \text{kg}$
$N_2\ \ = 0.779 \times 0.303 = 0.236\ \ \text{kmol or} \times 28 = 6.61\ \text{kg}$

$$\text{Total} = 9.26\ \text{kg}$$

c) The mass of dry air can be found from a nitrogen balance.
Basis: 1 kg of fuel.

Nitrogen in gaseous products $= 0.236\ \text{kmol (from b)}$
Nitrogen in fuel $= 0.01/28\ \ \ = \text{negligible}$
Nitrogen required from air $\ \ \ = 0.236\ \text{kmol}$
$\qquad\qquad$ Dry air supplied $= 0.236/0.79$
$\qquad\qquad\qquad\qquad\ = 0.299\ \text{kmol or} \times 29 = 8.67\ \text{kg}$

(d) The mass of water vapour in the air, found from the hygrometric chart or tables, is

specific humidity $= 0.0081\ \text{kg/kg of dry air}$

Therefore the mass of water vapour in the air/kg of fuel $= 0.0081 \times 8.67 = 0.07\ \text{kg}$

(e) The water vapour in the gaseous products can be found from a hydrogen balance.
Basis: 1 kg of fuel.

H_2O entering with air	= 0.07 kg
H_2O in fuel	= 0.10 kg
H_2O from combustion of H_2 = 0.03 × 9	= 0.27 kg

$$\text{Total} = 0.44\ \text{kg}$$

MASS BALANCE SUMMARY

Reactants			Products		
Fuel .	.	. 1·00 kg	Dry gases .	.	9·26 kg
Dry air	.	. 8·67 kg	Moisture .	.	0·44 kg
Moisture	.	. 0·07 kg	Refuse	.	. 0·22 kg
		9·74 kg			9·92 kg

The degree of balance obtained is an indication of the accuracy of the analyses of fuel and products of combustion.

15.4 First Law applied to combustion processes

The First Law of Thermodynamics applies to any system, and the non-flow and steady flow energy equations deduced from this law must be applicable to systems under going combustion processes. Before these equations can be used, however, data must be available from which we can find the change of internal energy or enthalpy when the chemical composition of the system changes as well as the thermodynamic properties.

Since the internal energy is a function of state, a change in internal energy is independent of the path of the process. Consider a non-flow combustion process, starting with a mixture of fuel and air at state (V_1, T_1) and ending with products at state (V_2, T_2). It can conveniently be regarded as consisting of the three successive processes shown in Fig. 15.5. The reactants are brought to a reference state (V_0, T_0) by some

Fig. 15.5

process *a-b* not involving a chemical reaction; the reaction *b-c* is then brought about in such a way that the products are at (V_0, T_0); and finally the products are brought to the final state (V_2, T_2) by some process *c-d*. We may write

$$(U_{P2} - U_{R1}) = (U_{P2} - U_{P0}) + (U_{P0} - U_{R0}) + (U_{R0} - U_{R1}) \tag{15.8}$$

where suffixes R and P refer to reactants and products respectively, and suffixes 1 and 2 refer to the initial and final states. The first and third terms on the right-hand side represent changes of internal energy in processes not involving chemical reaction, which can be evaluated by the methods applying to mixtures presented in Chapter 14. The middle term has to be obtained from additional experimental information because it involves a chemical reaction. However, by choosing a path which always involves the same pair of reference states at (V_0, T_0), only one piece of information is required for any given chemical reaction.

Because the process b-c occurs at constant volume, no work is done and the non-flow energy equation becomes

$$Q_{R0,P0} = (U_{P0} - U_{R0})$$

$(U_{P0} - U_{R0})$ can therefore be readily obtained by measuring the heat transferred in a constant-volume calorimetric process. When the chemical reaction is a combustion process, a temperature rise will be observed, and to return the system to its original temperature T_0 some heat must be transferred to the surroundings. Q will be negative and it must be concluded that there has been a reduction in internal energy of the system, even though two independent properties, V and T, are the same for the reactants and products. The explanation is that each molecule possesses potential energy stored in the bonds between the constituent atoms, and this energy—which we may term chemical energy—is smaller for the molecules of products than it is for the molecules of reactants. In the constant-volume process the energy is released to the surroundings as heat.

It has been found by experiment that at normal pressures this change of internal energy depends upon the temperature T_0 but not upon the volume V_0. Consequently the term $(U_{P0} - U_{R0})$ is called the *internal energy of combustion at* T_0, and it is symbolised by ΔU_0. In practice the suffix 0 is replaced by the actual reference temperature employed. Because of the experimental method of determining it, ΔU_0 is sometimes referred to as the *constant-volume heat combustion at* T_0. This name is not to be encouraged, however, because it is undesirable to associate the term 'heat' with the value of a property. Equation (15.8) can now be written

$$(U_{P2} - U_{R1}) = (U_{P2} - U_{P0}) + \Delta U_0 + (U_{R0} - U_{R1}) \tag{15.9}$$

In combustion processes the mixtures of reactants and products normally consist of gases or vapours at low pressure which can be assumed to behave as perfect gases, and possibly some liquids and solid substances. Using known specific heat data, and assuming no changes of phase occur during the processes a-b and c-d, the first and third terms on the right-hand side can therefore be found from

$$U_{P2} - U_{P0} = \sum_P m_i c_{vi} (T_2 - T_0)$$

$$U_{R0} - U_{R1} = \sum_R m_i c_{vi} (T_0 - T_1)$$

m_i is the mass, and c_{vi} the mean specific heat at constant volume, of the constituent i. If any constituent undergoes a change of phase in the temperature range $T_1 \rightarrow T_0$ or $T_0 \rightarrow T_2$ the u_{fg} data must also be known and included in the summation.

Similar considerations lead to the establishment of an equivalent equation for the change of enthalpy between reactants at T_1 and products at T_2, i.e.

$$(H_{P2} - H_{R1}) = (H_{P2} - H_{P0}) + \Delta H_0 + (H_{R0} - H_{R1}) \qquad (15.10)$$

This time the first and last terms on the right-hand side can be found from

$$(H_{P2} - H_{P0}) = \sum_P m_i c_{pi} (T_2 - T_0)$$

$$(H_{R0} - H_{R1}) = \sum_R m_i c_{pi} (T_0 - T_1)$$

where c_{pi} is the mean specific heat at constant pressure of the constituent i. In this case, if any constituent undergoes a change of phase, an h_{fg} term must be included in the summation. The term ΔH_0 is called the *enthalpy of combustion at T_0*, or the *constant-pressure heat of combustion at T_0*. It is usually found by measuring the heat rejected during a steady-flow combustion process carried out in such a way that the reactants and products are both at T_0, that no work is done, and that the change in kinetic energy is negligible. Under these circumstances the steady-flow energy equation reduces to

$$Q_{R0,P0} = (H_{P0} - H_{R0}) = \Delta H_0$$

Values of ΔU_0 and ΔH_0 are usually quoted for a reference temperature T_0 of 25 °C, and for one kilogram or one kilomole of the fuel. When determining ΔU_0 or ΔH_0 experimentally, it does not matter how much non-reacting material is present, e.g. nitrogen or excess oxygen. The change in internal energy or enthalpy of the non-reacting substances will be zero, because the products are brought back to the initial temperature. This point is of considerable practical importance since it means that a plentiful supply of oxygen can be used to ensure that all the combustible material is completely burnt.

When quoting values of ΔU_0 or ΔH_0 it is necessary to specify the phase of any individual reactant or product which may be in doubt. ΔH_0 for a liquid hydrocarbon, for example, is less than ΔH_0 for the same hydrocarbon in gaseous form: it is less by an amount equal to the latent heat of vaporisation at the reference temperature T_0. Or to take another example, when the fuel contains hydrogen the resulting H_2O in the products may be in the vapour or liquid phase. This point will be considered again in the next section. The conventional way of conveying the data is to give the combustion equation to which the value of ΔU_0 or ΔH_0 is applicable, e.g. ΔH_0 for hydrogen burning to water vapour is

$$H_2 + \tfrac{1}{2}O_2 \rightarrow H_2O \text{ (vap)}; \quad \Delta H_{25} = -120\,900 \text{ kJ/kg of } H_2$$

$$(\text{or} -241\,800 \text{ kJ/kmol of } H_2)$$

It is usual to make direct experimental determinations of ΔU_0 for solid or liquid fuels, and of ΔH_0 for gaseous fuels. Once either ΔU_0 or ΔH_0 has been determined experimentally, the other quantity can be calculated by making use of the following relation.

$$\Delta H_0 = H_{P0} - H_{R0} = (U_{P0} + p_{P0}V_{P0}) - (U_{R0} + p_{R0}V_{R0})$$
$$= \Delta U_0 + (p_{P0}V_{P0} - p_{R0}V_{R0})$$

For gaseous reactants and products we can write $pV = nR_0T$, and for solid and liquid constituents the pV terms are negligible compared with the internal energy term. The relation between ΔH_0 and ΔU_0 can therefore be written as

$$\Delta H_0 = \Delta U_0 + R_0 T_0 (n_P - n_R) \qquad (15.11)$$

where n_R and n_P are the numbers of kilomoles of *gaseous* reactants and products respectively. It follows that when the numbers of kilomoles of gaseous reactants and products are the same ΔH_0 and ΔU_0 are numerically equal. For most fuels the difference is quite small.

It is evident from equations (15.9) and (15.10) that in general we require a knowledge of the specific heats of the various reactants and products, in addition to values of ΔU_0 or ΔH_0. Values of c_p, tabulated against temperature, for the most common gases and some representative hydrocarbon vapours, are given in the abridged tables of Ref. (17). These values have been used in the numerical examples which follow. The mean specific heat for a range of temperature T_1 to T_2 is assumed to be equal to the specific heat at the temperature $(T_1 + T_2)/2$. The result would only be equal to the true mean specific heat, defined by $(h_2 - h_1)/(T_2 - T_1)$, if the specific heat varied linearly with temperature. Since the temperature ranges met with in combustion processes are often large, this procedure is not always sufficiently accurate. In such circumstances internal energy and enthalpy tables can be used. Refs. (12) and (7) tabulate u and h (in Btu/lb-mol) against temperature for the common gases, so that, for example, $(H_{P2} - H_{P0})$ can be found from

$$(H_{P2} - H_{P0}) = \sum_P n_i (h_{i2} - h_{i0})$$

It is worth emphasising that in such tables the internal energy or enthalpy is arbitrarily put equal to zero at some reference temperature (usually 0 K or 298·15 K). The tables can therefore only be used to find changes in internal energy and enthalpy during processes which do *not* involve a chemical reaction, because no account is taken of the particular form of internal energy which changes when a reaction occurs.

Example 15.6 When all the products of the gaseous hydrocarbon ethane (C_2H_6) are in the gaseous phase, the enthalpy of combustion at 25 °C is −47 590 kJ/kg. Find (a) the corresponding internal energy of combustion, and (b) the enthalpy of combustion at 540 °C.

Calculate the heat transferred when 0·2 kg of ethane is burnt at constant pressure in a cylinder containing 4·0 kg of dry air, the temperature of the reactants and products being 40 °C and 540 °C respectively.

The relevant mean specific heats at constant pressure, for the range 25 °C to 540 °C, are found from Ref. (17) at the mean temperature 555 K to be

C_2H_6 2·800; O_2 0·989; CO_2 1·049; $H_2O(vap)$ 1·987;

N_2 1·066 kJ/kg K

And for the range 25 °C to 40 °C

C_2H_6 1·788; O_2 0·919; N_2 1·040 kJ/kg K

(a) The stoichiometric combustion equation is

$$C_2H_6 + 3\tfrac{1}{2}O_2 \rightarrow 2CO_2 + 3H_2O$$

There are $4\tfrac{1}{2}$ kmol of reactants and 5 kmol of products per kmol of ethane, so that in equation (15.11) we have

$$(n_P - n_R) = \tfrac{1}{2} \text{ kmol}$$

Since ΔH_{25} is given in kJ/kg, and 1 kmol of ethane is equivalent to a mass of 30 kg

$$\Delta U_{25} = -47\,590 - \frac{8\cdot314 \times 298}{2 \times 30} = -47\,630 \text{ kJ/kg}$$

(b) From equation (15.10) we have

$$\Delta H_{540} = (H_{P540} - H_{R540})$$
$$= (H_{P540} - H_{P25}) + \Delta H_{25} + (H_{R25} - H_{R540})$$

In terms of mass the combustion equation is

$$30 \text{ kg } C_2H_6 + 112 \text{ kg } O_2 \rightarrow 88 \text{ kg } CO_2 + 54 \text{ kg } H_2O$$

On the basis of 1 kg of C_2H_6 we have

$$(H_{P540} - H_{P25}) = \sum_P m_i c_{pi}(540 - 25)$$

$$= \frac{515}{30}\{(88 \times 1\cdot049) + (54 \times 1\cdot987)\} = 3427 \text{ kJ/kg}$$

Similarly,

$$(H_{R25} - H_{R540}) = \sum_R m_i c_{pi}(25 - 540)$$

$$= -\frac{515}{30}\{(30 \times 2\cdot80) + (112 \times 0\cdot989)\} = -3345 \text{ kJ/kg}$$

$$\Delta H_{540} = 3427 - 47\,590 - 3345 = -47\,510 \text{ kJ/kg}$$

From (a) and (b) it should be noted that the difference between ΔH_0 and ΔU_0 is small, and that the values are not greatly affected by a change in reference temperature.

In the constant pressure combustion, the products are not returned to the initial temperature and we must take account of the change in enthalpy of the non-reacting substances.

The heat transferred is equal to the change of enthalpy if the expansion work can be regarded as $\int p \, dv$. Therefore

$$Q = (H_{P540} - H_{R40})$$
$$= (H_{P540} - H_{P25}) + \Delta H_{25} + (H_{R25} - H_{R40})$$

The stoichiometric combustion equation for 0·2 kg of ethane is

$$0\text{·}2 \text{ kg } C_2H_6 + 0\text{·}747 \text{ kg } O_2 \to 0\text{·}587 \text{ kg } CO_2 + 0\text{·}360 \text{ kg } H_2O$$

But 4 kg of air are provided, consisting of

$$4 \times 0\text{·}233 \text{ kg } O_2 + 4 \times 0\text{·}767 \text{ kg } N_2 = 0\text{·}932 \text{ kg } O_2 + 3\text{·}068 \text{ kg } N_2$$

Evidently there is some excess air, and the combustion equation for the specified mixture is

$$0\text{·}2 \text{ kg } C_2H_6 + 0\text{·}932 \text{ kg } O_2 + 3\text{·}068 \text{ kg } N_2 \to$$
$$0\text{·}587 \text{ kg } CO_2 + 0\text{·}360 \text{ kg } H_2O + 0\text{·}185 \text{ kg } O_2 + 3\text{·}068 \text{ kg } N_2$$

On the basis of 0·2 kg of ethane:

$$(H_{P540} - H_{P25}) = 515\{(0\text{·}587 \times 1\text{·}049) + (0\text{·}360 \times 1\text{·}987) +$$
$$(0\text{·}185 \times 0\text{·}989) + (3\text{·}068 \times 1\text{·}066)\} = 2464 \text{ kJ}$$
$$(H_{R25} - H_{R40}) = -15\{(0\text{·}2 \times 1\text{·}788) + (0\text{·}932 \times 0\text{·}919) +$$
$$(3\text{·}068 \times 1\text{·}040)\} = -66 \text{ kJ}$$

Therefore,

$$Q = 2464 - (0\text{·}2 \times 47\,590) - 66 = -7120 \text{ kJ}$$

In order to depict any combustion process on a U-T diagram, it is necessary to plot U-T curves for the reactants and products on a single diagram in the correct relative positions. In Fig. 15.6a the products curve has been drawn in an arbitrary position by putting $U = 0$ at $T = 0$, and the reactants curve has then been located above it a distance ΔU_0 at T_0.

During any combustion process 1–2 in a closed system, both work and heat may be transferred across the boundary in accordance with the First Law

$$Q - W = U_2 - U_1$$

There are two combustion processes of special interest. One which we have already mentioned is the 'calorimeter' process 1–3, carried out at constant volume and with $T_3 = T_1$. In this process the work is zero and the heat transferred is equal to the internal energy of combustion at T_1. This process yields the maximum possible transfer of energy to surroundings at T_1. The second process of special interest is adiabatic combustion at constant volume. Since both the heat and work transfers are zero, the internal energy remains constant. There is merely a redistribution of internal energy resulting in a rise of temperature, i.e. we may say that chemical energy is transformed into *sensible* internal energy. This process, 1–4 in Fig. 15.6a, results in the maximum possible temperature rise for any given fuel-air mixture.

When considering steady-flow processes, it is more convenient to depict these on an *H-T* diagram, Fig. 15.6b. The steady-flow 'calorimeter' process, in which the heat transferred is equal to the enthalpy of combustion, is represented by 1–5. Steady-flow adiabatic combustion, with zero work and no change of kinetic energy, is represented by 1–6; this process yields the maximum possible temperature rise for steady-flow combustion.

Fig. 15.6

Example 15.7 Liquid heptane (C_7H_{16}) has an enthalpy of combustion of $-44\,560$ kJ/kg at 25 °C, when the water in the products is in the vapour phase. This fuel is to be burnt adiabatically in a steady stream of air in stoichiometric proportions. If the initial temperature of the fuel and air is 15 °C, estimate the temperature of the products. The specific heat of liquid heptane is 2·30 kJ/kg K.

The stoichiometric combustion equation, omitting the nitrogen, is

$$C_7H_{16} + 11O_2 \rightarrow 7CO_2 + 8H_2O$$
$$100 \text{ kg } C_7H_{16} + 352 \text{ kg } O_2 \rightarrow 308 \text{ kg } CO_2 + 144 \text{ kg } H_2O$$

There will be 352(76·7/23·3) kg of N_2 also, on each side of the equation, so that for 1 kg of heptane the complete equation becomes

$$1 \text{ kg } C_7H_{16} + 3\cdot52 \text{ kg } O_2 + 11\cdot59 \text{ kg } N_2 \rightarrow 3\cdot08 \text{ kg } CO_2 + 1\cdot44 \text{ kg } H_2O + 11\cdot59 \text{ kg } N_2$$

The steady-flow energy equation reduces to

$$H_{Pt} - H_{R15} = 0$$

where t refers to the products temperature. Expanding this equation we have

$$(H_{Pt} - H_{P25}) + \Delta H_{25} + (H_{R25} - H_{R15}) = 0$$

The evaluation of this equation involves a method of successive approximation because the appropriate mean specific heat data for the first term depend upon the unknown temperature

318

t. From Ref. (17) we find that the mean specific heats at constant pressure for the range 15 °C to 25 °C are as follows:

  O_2 0·917; N_2 1·040; CO_2 0·838; H_2O(vap) 1·862 kJ/kg K

Therefore

$$(H_{R25} - H_{R15}) = \sum_R m_i c_{pi}(25 - 15)$$

$$= 10\{(1 \times 2\cdot30) + (3\cdot52 \times 0\cdot917) + (11\cdot59 \times 1\cdot040)\}$$

$$= 175\cdot8 \text{ kJ}$$

Still using these specific heats as a first approximation, we have

$$(H_{Pt} - H_{P25}) = \sum_P m_i c_{pi}(t - 25)$$

$$= (t - 25)\}(3\cdot08 \times 0\cdot838) + (1\cdot44 \times 1\cdot862) + (11\cdot59 \times 1\cdot040)\}$$

$$= 17\cdot32(t - 25) \text{ kJ}$$

Hence,

$$17\cdot32(t - 25) - 44\,560 + 175\cdot8 = 0$$

$$t \approx 2590 \,°C$$

As a second approximation, the mean specific heats for the range 2590 °C to 25 °C can be used in evaluating the products term. The relevant specific heats at the mean temperature of 1580 K are

  CO_2 1·336; H_2O (vap) 2·651; N_2 1·252 kJ/kg K

With these new values

$$(H_{Pt} - H_{P77}) = (t - 25)\{(3\cdot08 \times 1\cdot336) + (1\cdot44 \times 2\cdot651) + (11\cdot59 \times 1\cdot252)\}$$

$$= 22\cdot44(t - 25) \text{ kJ}$$

Hence,

$$22\cdot44(t - 25) - 44560 + 175\cdot8 = 0$$

$$t \approx 2003 \,°C$$

The procedure can be repeated until the desired accuracy is obtained, the accurate value being 2065 °C. This example illustrates the large effect which the variation of specific heats has on the final temperature reached in adiabatic combustion.

The actual temperature of the products from the adiabatic combustion of a stoichiometric mixture would not be as high as the value calculated in the manner of the previous example. The combustion would in fact be incomplete, partly for the reasons already given in section 15.2, and partly because at high temperatures the products are unstable. Some of the molecules split into their constituents, usually with an accompanying absorption of energy. This phenomenon is known as *dissociation*, and the method of estimating the final temperature when dissociation occurs is discussed in section 15.7.

Most fuels produce adiabatic combustion temperatures between 2200 and 2700 K when burnt stoichiometrically in air initially at atmospheric temperature. Higher temperatures can always be achieved, either by heating the air before combustion, or by enriching it with oxygen and so effectively reducing the mass of nitrogen present. The temperature will be increased in the latter case because the mass of the products, and hence their heat capacity, is reduced. The rise in temperature for a given release of chemical energy is therefore greater. It should be appreciated, however, that although very high temperatures may be desirable from a thermodynamic point of view, they cannot always be employed in power plant. In a gas turbine, for example, where the highly stressed parts of the turbine are in a continuous stream of hot gaseous products, serious overheating would occur unless a large quantity of excess air were provided to reduce the temperature at exit from the combustion chamber.

15.5 Calorific value of fuels

When dealing with energy transfers in combustion processes, many engineers prefer to relate these transfers to a *calorific value* of the fuel rather than to the properties ΔU_0 and ΔH_0. Calorific values are defined in terms of the number of heat units liberated when unit mass of fuel is burnt completely in a calorimeter under specified conditions. Since they are defined in terms of a quantity of heat, a rigorous definition must specify not merely the reference end states but also the process connecting these end states.

For a long time calorific values were used which were not uniquely defined, and it is only recently that some attempt has been made to remedy this. British Standard 526:1961 defines at some length the four calorific values listed below: their relation to ΔU_0 or ΔH_0 given in parentheses will be apparent from what follows.

(a) Gross (or higher) calorific value at constant volume, $Q_{gr,v}$ ($\approx -\Delta U_{25}$ with H_2O in products as liquid).
(b) Net (or lower) calorific value at constant volume, $Q_{net,v}$ ($\approx -\Delta U_{25}$ with H_2O in products as vapour).
(c) Gross (or higher) calorific value at constant pressure, $Q_{gr,p}$ ($\approx -\Delta H_{25}$ with H_2O in products as liquid).
(d) Net (or lower) calorific value at constant pressure, $Q_{net,p}$ ($\approx -\Delta H_{25}$ with H_2O in products as vapour).

'Gross' refers to calorific values obtained when the water in the products is in the liquid phase, and 'net' to those obtained with the water in the vapour phase. It is conventional to write the calorific value as a positive number, even though the heat is leaving the system undergoing the reaction.

For solid and liquid fuels the calorific value which can be found accurately by experiment is $Q_{gr,v}$, and the other three calorific values should be found by calculation from this value and the known composition of the fuel (see example 15.8). $Q_{gr,v}$ is found by burning a measured quantity of fuel in a bomb calorimeter. This consists essentially of a gas-tight steel vessel, or 'bomb', which is immersed in a conventional water calorimeter. The water equivalent of the calorimeter and bomb is found by conducting a preliminary experiment using an electrical resistance method of heating (or benzoic acid whose calorific value is accurately known). The bomb is then charged with the fuel sample, and with oxygen at a high pressure to ensure complete combustion. After ignition, the final steady temperature of the water in the calorimeter is noted—it is arranged to be not more than $3 \cdot 0 \pm 0 \cdot 5$ K above an initial temperature of $23 \cdot 5 \pm 8 \cdot 5$ °C. Although the temperature of the products is not exactly equal to that of the reactants, the numerical result will be substantially the same as the internal energy of combustion, ΔU_{25}, when the products contain liquid water only. A small quantity of water is always placed in the bomb initially, to saturate the volume, and any water formed by the combustion will therefore condense.

The calorific value which can be found accurately for a gaseous fuel is $Q_{gr,p}$, and again the other three calorific values should be found by calculation. $Q_{gr,p}$ is most conveniently measured by burning a mixed stream of gas and air at atmospheric pressure and temperature, under steady flow conditions, in a gas calorimeter. The hot gases flow over cooling coils through which water flows steadily in the opposite direction. With this counterflow arrangement the products leave the calorimeter at approximately atmospheric temperature and the water vapour produced is condensed. The calorific value of the gas is calculated from the mass flow and temperature rise of the cooling water, and the value is substantially the same as $-\Delta H_{25}$ for a reaction yielding water in the liquid phase.

The reasons why the calorific values so obtained are not exactly equal to their $-\Delta U_{25}$ or $-\Delta H_{25}$ equivalents, arise from the conditions under which the experiments are conducted. For example, the calorific value includes small amounts of heat transfer resulting from some of the products dissolving in the water present in the reaction vessel. Or again, the reactants and products are not both exactly at the reference temperature of 25 °C. Example 15.6 illustrated the fact that quite a large change in reference temperature has only a small effect on ΔH_0. It will be appreciated, therefore, that the differences referred to here are insignificant, except for the most exacting scientific work when in any case it is recommended that ΔH_{25} or ΔU_{25} be used in preference to calorific values. We may also infer from example 15.6 that the difference between a calorific value at constant pressure and the corresponding value at constant volume will be small although, as will be shown in the following example 15.8, the difference between gross and net values cannot be ignored. If m_w is the mass of water produced per kilogram of fuel burnt, then

$$(Q_{gr,v} - Q_{net,v}) = m_w\, u_{fg}$$

and for the constant pressure values

$$(Q_{gr,p} - Q_{net,p}) = m_w\, h_{fg}$$

u_{fg} and h_{fg} should be taken from steam tables at the reference temperature of 25 °C.

Example 15.8 The calorific value $Q_{gr,v}$ of kerosene, whose ultimate analysis is 86 per cent carbon and 14 per cent hydrogen, was found by experiment to be 46 890 kJ/kg. Calculate the other three calorific values.

1 kg of fuel produces $0.86 \times 44/12 = 3.15$ kg of CO_2

and $0.14 \times 18/2 = 1.26$ kg of H_2O

At 25 °C, $u_{fg} = 2304$ kJ/kg and $h_{fg} = 2442$ kJ/kg

$$Q_{net,v} = Q_{gr,v} - 1.26 \times 2304 = 46\,890 - 2903 = 43\,987 \text{ kJ/kg}$$

The combustion equation can be written as

1 kg fuel $+ 3.41$ kg $O_2 = 1.26$ kg $H_2O + 3.15$ kg CO_2

The relation between $Q_{gr,p}$ and $Q_{gr,v}$ will be identical with that between $-\Delta H_0$ and $-\Delta U_0$, i.e. equation (15.11). Remembering that only gaseous reactants and products should be included, the equation becomes

$$Q_{gr,p} = 46\,890 - 8.314 \times 298 \left(\frac{3.15}{44} - \frac{3.41}{32} \right)$$

$$= 46\,890 + 87 = 46\,977 \text{ kJ/kg}$$

Finally, $Q_{net,p}$ is given by

$$Q_{net,p} = Q_{gr,p} - 1.26 \times 2442 = 46\,977 - 3077 = 43\,900 \text{ kJ/kg}$$

Note that the difference between $Q_{gr,v}$ and $Q_{net,v}$ is 6·2 per cent, whereas the two net calorific values differ by only about 0·2 per cent.

15.6 Efficiency of power plant and combustion processes

The cycle efficiency of a heat engine is calculated without reference to the source of heat. When considering the efficiency of the power plant as a whole, however, we are also concerned with the efficiency of the means of producing the heat, i.e. whether the fuel is in fact being completely burnt under the most advantageous conditions. The *overall thermal efficiency* of a power plant can be suitably defined as the ratio of the work produced to the latent energy in the fuel supplied. The question arises as to how this latent energy should be assessed.

Any power plant, considered as a whole, is an open system undergoing a steady-flow process—with air and fuel entering the system, and the products of combustion leaving it, at a steady rate. This is illustrated in Fig. 15.7. The fuel and air enters at atmospheric temperature, and the largest quantity of energy that can be transferred to the surroundings will be produced if (1) the fuel is burnt completely, (2) the products leave at atmospheric temperature, and (3) any water in the products is in the liquid phase. The energy transferred under these conditions is $-\Delta H_a$, where suffix *a* refers to atmospheric temperature. If this quantity is regarded as the effective latent energy in the fuel, the efficiency of the power plant is suitably defined by

$$\eta = \frac{W}{-\Delta H_a} \qquad (15.12)$$

From what has been said in the previous section it is evident that this is substantially the same as

$$\eta = \frac{W}{Q_{gr,p}} \qquad (15.13)$$

Now the temperature of the products is never reduced to atmospheric temperature in practice, for this would entail an infinitely large heat-exchange surface somewhere in the plant (e.g. in the economiser of a steam power plant). Moreover this condition

Fig. 15.7   Fig. 15.8

is not even approached, if for no other reason than the undesirability of allowing the water vapour to condense. Combustion products usually include a small quantity of sulphur dioxide which would dissolve in the water and form corrosive sulphuric acid. Considerations of this type have led many engineers to prefer to define thermal efficiency in terms of the net calorific value

$$\eta = \frac{W}{Q_{net,p}} \qquad (15.14)$$

It is clear that power plant efficiencies are process efficiencies and, like all such efficiencies, they are comparisons between an achieved result and some arbitrary standard: it is most important always to state clearly which standard, $Q_{gr,p}$ or $Q_{net,p}$, is being used. A more rational definition of power plant efficiency is discussed in section 17.5, from which it is apparent that of the two arbitrary definitions presented here, definition (15.14) is to be preferred.

It must be emphasised that the foregoing discussion applies equally to steam power plant and internal-combustion engines. For the former, Fig. 15.7 can be analysed further as in Fig. 15.8—part A of the system being regarded as the boiler furnace and part B as the working fluid which is taken through a thermodynamic cycle. Most of the heat rejected, Q_2, passes out in the condenser cooling water. When the combustion process itself can be isolated from the rest of the plant, as in a steam boiler or gas turbine combustion chamber, it is necessary to have some measure of its efficiency so that the performance of different combustion systems can be compared. Combustion efficiency is most generally defined as the percentage of the latent energy input which is effectively utilised in the desired manner. The detailed definition will depend upon the method of designating the latent energy input and the effectively used output.

In the case of a steam boiler, for example, the combustion is used to produce heat and so transform water into steam. The combustion efficiency can therefore be alternatively defined as

$$\eta = \frac{Q}{Q_{gr,p}} \quad \text{or} \quad \frac{Q}{Q_{net,p}} \tag{15.15}$$

where Q is the heat transferred to the water, i.e. the increase in enthalpy of the water per unit mass of fuel supplied.

To take a second example, the object of the adiabatic combustion in a gas turbine combustion chamber is to increase the *sensible* enthalpy of the fluid flowing through it—so that this can be used to produce work in the turbine. The efficiency may be defined, therefore as the ratio of the actual increase of sensible enthalpy per kilogram of fuel, to the increase obtained when each kilogram of fuel is burnt completely. The energy equation is

$$(H_{P2} - H_{R1}) = (H_{P2} - H_{P0}) + \Delta H_0 + (H_{R0} - H_{R1}) = 0$$

With complete combustion, the increase in sensible enthalpy is given by

$$(H_{P2} - H_{P0}) - (H_{R1} - H_{R0}) = -\Delta H_0$$

With incomplete combustion, the increase in sensible enthalpy will be

$$\{(H_{P2} - H_{P0}) - (H_{R1} - H_{R0})\}_{\text{actual}} = -\Delta H_0 - \Sigma m_i(-\Delta H_{0i})$$

The summation is of unburnt or partially burnt constituents: m_i is the mass of constituent i per unit mass of fuel and ΔH_{0i} is its enthalpy of combustion. The choice of reference temperature t_0 has little effect on the efficiency, and $-\Delta H_0$ can be replaced by the gross or net calorific value at constant pressure according to convention. The expression for combustion efficiency becomes

$$\eta = \frac{\{(H_{P2} - H_{P0}) - (H_{R1} - H_{R0})\}_{\text{actual}}}{Q_{gr,p} \text{ (or } Q_{net,p})} \tag{15.16}$$

The numerator is best determined from the results of an experimental analysis of the products. A second definition of efficiency frequently employed is the percentage of the actual fuel used which would have been sufficient to achieve the actual product temperature with complete combustion. Further discussion of these and other ways of expressing gas turbine combustion efficiencies can be found in Ref. (26).

Example 15.9 Find the efficiency of the combustion process of Example 15.5, on the basis of the net calorific value of the fuel and the percentage of complete combustion attained. The following data can be assumed:

$$Q_{net,p} \approx Q_{net,v} \text{ of the coal} = 24\ 490 \text{ kJ/kg}$$

$$-\Delta H_{25} \text{ of solid carbon} = 32\ 790 \text{ kJ/kg}$$

$$-\Delta H_{25} \text{ of carbon monoxide} = 10\ 110 \text{ kJ/kg}$$

The efficiency may be expressed by

$$\eta = \frac{Q_{net,p} - \Sigma m_i(-\Delta H_{25})_i}{Q_{net,p}}$$

From the solution of Example 15.5, the unburnt and partially burnt constituents in the products, per kg of coal burnt, are

0·17 kg of CO in the gaseous products

$0.4 \times 0.217 = 0.0868$ kg of C in the refuse

$$\Sigma m_i(-\Delta H_{25})_i = (0.17 \times 10\ 110) + (0.0868 \times 32\ 790) = 4565 \text{ kJ/kg}$$

$$\eta = \frac{24\ 490 - 4565}{24\ 490} = 0.814$$

15.7 Dissociation

It was pointed out in section 15.4 that the maximum temperature reached in an adiabatic combustion process is limited by the phenomenon of dissociation. Consider, for example, the reaction

$$CO + \tfrac{1}{2}O_2 \rightarrow CO_2$$

When it proceeds in the direction indicated by the arrow it is accompanied by a release of energy. The reaction can be made to proceed in the reverse direction, however, if sufficient energy is supplied to molecules of CO_2. Some of the CO_2 molecules in the combustion products do receive sufficient energy in collision for this to occur, i.e. some of the molecules undergo the reaction

$$CO_2 \rightarrow CO + \tfrac{1}{2}O_2$$

This possibility is symbolised by writing the equation as

$$CO + \tfrac{1}{2}O_2 \rightleftharpoons CO_2$$

The reversed reaction is accompanied by an absorption of energy and it is termed an *endothermic* reaction. It is found that at any particular temperature and pressure the proportions of CO_2, CO and O_2 adjust themselves until the two reactions proceed at the same rate, i.e. until the number of CO_2 molecules being formed is equal to the number dissociating. A state of stable chemical equilibrium is then said to exist. The state of equilibrium is not a static one, because the two reactions are going on continuously and simultaneously. It is only at high temperatures, above about 1500 K, that a significant proportion of the CO_2 molecules must dissociate to provide an equilibrium mixture.

The preceding remarks apply equally to H_2O molecules in combustion products. It is now possible to appreciate why the adiabatic combustion temperature is less than that predicted by the simple calculation of Example 15.7. Evidently the products contain an equilibrium mixture of CO_2, CO and O_2, and an equilibrium mixture of H_2O, H_2 and O_2. The presence of CO and H_2 indicates that not all the chemical energy in the fuel is released. The question arises as to how the states of chemical equilibria can be predicted. We shall proceed to show that a condition for equilibrium can be deduced from the Second Law of Thermodynamics. The first step is to visualise how a thermodynamically reversible chemical reaction can be achieved, and the argument suggested by van't Hoff in 1887 will be used here.

(i) *The van't Hoff equilibrium box*

Consider a stoichiometric reaction between CO and O_2, each initially at pressure p_1 and temperature T, to form CO_2 at the same pressure and temperature. This reaction can be carried out in many ways. In the gas calorimeter, for example, all the energy released is transferred as heat and the reaction is irreversible. When the reaction is made reversible, however, we shall find that some of the energy is transferred as work. Fig. 15.9 illustrates one conceivable steady-flow open system in which the reaction can proceed reversibly. The reaction itself occurs in the reaction chamber, or 'equilibrium box', which contains CO_2, CO and O_2 in such proportions that the reactions

$$CO + \tfrac{1}{2}O_2 \rightarrow CO_2 \quad \text{and} \quad CO_2 \rightarrow CO + \tfrac{1}{2}O_2$$

are proceeding simultaneously at equal rates. The box is maintained at a constant temperature T by surrounding it with a reservoir at T. Any heat exchanges will therefore take place reversibly. The total pressure in the box may have any arbitrary value p, which may be greater or smaller than p_1, determined by the total mass of gas present. The CO, O_2 and CO_2 enter or leave the box through semipermeable membranes, the mass flows at entry and exit being equal. If these mass transfers are to take place reversibly (see section 14.3), the pressure of each constituent outside the box must be equal to its partial pressure inside the box (i.e. p_{CO}, p_{O_2} or p_{CO_2}), and the temperature of each constituent outside the box must be equal to T. Since the pressure and temperature of the CO, O_2 and CO_2 are p_1 and T at the boundary, reversible isothermal compressors and expanders must be included in the system to maintain the respective pressures at p_{CO}, p_{O_2} and p_{CO_2} outside the semipermeable membranes.

Imagine that CO and O_2 are transported slowly into the reaction box at the steady molar flow rates of n and $n/2$ respectively, and that the reaction proceeds in the normal direction. The equilibrium mixture will remain unchanged if CO_2 is discharged at the steady molar flow rate of n. Only an infinitesimal change in the external conditions is needed to reverse the expanders and compressors and to cause the reaction to proceed in the opposite direction. All the quantities of heat and work transferred across the boundary then have the same magnitude but are reversed in sign. The whole isothermal steady-flow process can therefore be regarded as reversible.

Since the gases in the equilibrium box can be assumed to be perfect, the mole fractions are proportional to the partial pressures, equation (14.17). The relative proportions in the equilibrium mixture can therefore be expressed in terms of the partial pressures. What we shall proceed to show as a consequence of the Second Law is that the constituents must be in certain proportions defined by the equation

$$\frac{p_{CO_2}}{p_{CO}(p_{O_2})^{1/2}} = K$$

where K can be a function only of the temperature of the mixture T and not of the pressure p. K is called the *equilibrium constant* or *dissociation constant*.

[Drawn for the case where the net reaction proceeds according to $CO + \frac{1}{2}O_2 \rightarrow CO_2$, and $p_1 <$ any partial pressure]

Fig. 15.9   Van't Hoff equilibrium box

(ii) *The equilibrium constant*

The work done when a perfect gas undergoes a reversible, isothermal steady-flow process is given by equation (10.6). When the gas flows at the molar rate of n, equation (10.6) can be written as

$$W = nR_0T \ln\frac{p_1}{p_2}$$

The net work done per unit time by the system shown in Fig. 15.9 is therefore

$$W = nR_0T \left(\ln\frac{p_1}{p_{CO}} + \frac{1}{2}\ln\frac{p_1}{p_{O_2}} + \ln\frac{p_{CO_2}}{p_1} \right)$$

$$= nR_0T \left(\ln\frac{p_{CO_2}}{p_{CO}(p_{O_2})^{1/2}} + \ln p_1^{1/2} \right) \tag{15.17}$$

Now suppose that the pressure p in the reaction box is changed to p', but all other conditions remain unchanged. The new partial pressures will be p'_{CO}, p'_{O_2} and p'_{CO_2}, such that

$$p'_{CO} + p'_{O_2} + p'_{CO_2} = p'$$

and the net work done per unit time will be

$$W' = nR_0T \left(\ln \frac{p'_{CO_2}}{p'_{CO}(p'_{O_2})^{1/2}} + \ln p_1^{1/2} \right)$$

Either $W = W'$ or $W \neq W'$. If the latter, the system producing the lesser work can be reversed and the two systems can be coupled together. The combined system will then operate in a cycle and produce a net amount of work while exchanging heat with a single reservoir of uniform temperature T. This contradicts the Second Law and so W must equal W'. The same argument must apply to any reversible system operating with the same boundary conditions of p_1 and T, and the work W is therefore the maximum which can be obtained from this chemical reaction. When the process is irreversible the work done will be less—in the calorimeter process it is zero. Since $W = W'$, it follows that

$$\frac{p_{CO_2}}{p_{CO}(p_{O_2})^{1/2}} = \frac{p'_{CO_2}}{p'_{CO}(p'_{O_2})^{1/2}}$$

We can therefore write

$$K = \frac{p_{CO_2}}{p_{CO}(p_{O_2})^{1/2}} \tag{15.18}$$

where K is a quantity which is independent of the pressure p of the equilibrium mixture K may, however, be a function of the temperature of the mixture T, and is in fact found to be so by experiment.

The whole of the foregoing argument could have been given in general terms for any reaction, e.g.

a kmol of A $+ b$ kmol of B $\rightleftharpoons c$ kmol of C $+ d$ kmol of D

We should then have arrived at the general expression

$$K = \frac{(p_C)^c(p_D)^d}{(p_A)^a(p_B)^b}$$

For the reaction $H_2 + \tfrac{1}{2}O_2 \rightleftharpoons H_2O$ the equilibrium constant is evidently

$$K = \frac{p_{H_2O}}{p_{H_2}(p_{O_2})^{1/2}} \tag{15.19}$$

The value of K for any given reaction can be found by experiment at various temperatures. K is not dimensionless and the actual values depend upon the units of pressure used—usually the partial pressures are expressed in atmospheres. It is also possible to write the reaction equation in a variety of ways, e.g. the H_2O reaction could be written as

$$2H_2 + O_2 \rightleftharpoons 2H_2O$$

In which case

$$K = \frac{(p_{H_2O})^2}{(p_{H_2})^2(p_{O_2})}$$

and yet a different set of numbers would be obtained. Both the unit of pressure and form of chemical equation adopted must be specified when quoting values of equilibrium constants.

Remembering that the partial pressures are proportional to the mole fractions, it is evident that K is a measure of the amount of dissociation. In (15.18) or (15.19) a high value implies that the mixture contains a large proportion of the undissociated product.

Fig. 15.10   Equilibrium constants

CO_2 or H_2O. Fig. 15.10, showing the variation of K_{CO_2} and K_{H_2O} with temperature, indicates that the amount of dissociation increases with increase of temperature. Values of these and other dissociation constants are tabulated against temperature in Ref. (17).

The following example illustrates the method of calculating the equilibrium constant when the percentage dissociation has been obtained by a products analysis, and also how this equilibrium constant can then be used to predict the percentage dissociation in other circumstances.

Example 15.10 The products from the combustion of a stoichiometric mixture of CO and O_2 are at a pressure of 1 atm and a certain temperature. The products analysis shows that 35 per cent of each kmol of CO_2 is dissociated. Determine the equilibrium constant for this temperature, and thence find the percentage dissociation when the products are at the same temperature but compressed to 10 atmospheres.

The combustion equation for the reaction to products in equilibrium is

$$CO + \tfrac{1}{2}O_2 \rightarrow (1 - a)CO_2 + aCO + \frac{a}{2}O_2$$

where a is the fraction of CO_2 dissociated.

At 1 atm the products consist of

$$0 \cdot 65 \text{ kmol } CO_2 + 0 \cdot 35 \text{ kmol } CO + 0 \cdot 175 \text{ kmol } O_2$$

The total amount of substance of products is $1 \cdot 175$ kmol, and the partial pressures of the constituents in atmospheres are therefore

$$p_{CO_2} = \frac{n_{CO_2}}{n}p = \frac{0 \cdot 65}{1 \cdot 175} = 0 \cdot 553 \text{ atm}$$

$$p_{CO} = \frac{n_{CO}}{n}p = \frac{0 \cdot 35}{1 \cdot 175} = 0 \cdot 298 \text{ atm}$$

$$p_{O_2} = \frac{n_{O_2}}{n}p = \frac{0 \cdot 175}{1 \cdot 175} = 0 \cdot 149 \text{ atm}$$

Based on the equilibrium equation $CO + \tfrac{1}{2}O_2 \rightleftharpoons CO_2$,

$$K = \frac{p_{CO_2}}{p_{CO}(p_{O_2})^{1/2}} = \frac{0 \cdot 553}{0 \cdot 298(0 \cdot 149)^{1/2}} = 4 \cdot 81 \text{ (atm)}^{-1/2}$$

At any pressure p, the partial pressures will be

$$p_{CO_2} = \frac{1 - a}{1 + a/2}p, \quad p_{CO} = \frac{a}{1 + a/2}p, \quad p_{O_2} = \frac{a/2}{1 + a/2}p$$

And the equilibrium constant will be

$$K = \frac{\left(\dfrac{1 - a}{1 + a/2}\right)}{\left(\dfrac{a}{1 + a/2}\right)\left(\dfrac{a/2}{1 + a/2}\right)^{1/2}} \times \frac{p}{p \times p^{1/2}} \tag{15.20}$$

Since the temperature is unchanged, K will still equal $4 \cdot 81$ (atm)$^{-1/2}$. At 10 atm therefore,

$$4 \cdot 81 = \frac{(1 - a)(2 + a)^{1/2}}{a^{3/2}} \times \frac{1}{10^{1/2}}$$

Squaring and simplifying, we have

$$230 \cdot 4a^3 + 3a = 2$$

and the solution, by trial and error or graphical means, is

$$a = 0 \cdot 185$$

Dissociation $= 18 \cdot 5$ per cent of the CO_2.

The preceding example illustrates the important point that although the equilibrium constant is independent of the pressure of the reacting mixture, the actual fraction of the dissociated product usually varies with pressure. Inspection of equation (15.20) in the example shows that the term containing p only vanishes when the number of kilomoles of dissociating compound and of dissociated constituents are equal.

We may now see how a knowledge of the equilibrium constants enables the combustion equation to be established, and the adiabatic combustion temperature to be predicted, for the reaction of Example 15.7. The reactants are liquid heptane and air in stoichiometric proportions. Assuming that some of the CO_2 and H_2O molecules have dissociated, the combustion equation will be

$$C_7H_{16} + 11O_2 + \frac{79}{21}(11)N_2 \rightarrow aCO_2 + bH_2O + cCO + dO_2 + eH_2 + \frac{79}{21}(11)\,N_2$$

There are five unknowns: a, b, c, d and e. Three equations can be written down directly, by equating the number of atoms of carbon, hydrogen and oxygen on each side of the equation. Thus,

$$a + c = 7$$
$$2b + 2e = 16$$
$$2a + b + c + 2d = 22$$

The remaining equations can be formulated with the aid of the equilibrium constants, which establish the proportions in which CO_2, CO and O_2, and H_2O, H_2 and O_2, can exist in equilibrium. If n is the total number of kilomoles of products, given by

$$n = a + b + c + d + e + \frac{79}{21}(11)$$

the partial pressures are

$$p_{CO_2} = \frac{a}{n}p, \quad p_{CO} = \frac{c}{n}p, \quad \text{etc.}$$

The two equations are therefore

$$(CO + \tfrac{1}{2}O_2 \rightleftharpoons CO_2) \quad K_{CO_2} = \frac{a}{c(d/n)^{1/2}} \times \frac{1}{p^{1/2}}$$

$$(H_2 + \tfrac{1}{2}O_2 \rightleftharpoons H_2O) \quad K_{H_2O} = \frac{b}{e(d/n)^{1/2}} \times \frac{1}{p^{1/2}}$$

Since p appears in these equations, the pressure at which the reaction occurs must be known. (In Example 15.7 this was irrelevant because the enthalpies can be assumed to be independent of pressure). Assuming the pressure is specified, there is still the difficulty that the dissociation constants are functions of the unknown temperature. The method of solution is to assume a final temperature, obtain the appropriate values of K from tables, e.g. Ref. (17), and then solve the five simultaneous equations for a, b, c, d and e. Once the composition of the products is known, the final temperature can be computed from the energy equation. If this does not agree with the assumed temperature, a better approximation to the values of K can be obtained and the calculation repeated until the desired agreement is reached.

Since the fuel and air were initially at 15 °C, the energy equation for the reaction in Example 15.7 was

$$(H_{Pt} - H_{P25}) + \Delta H_{25} + (H_{R25} - H_{R15}) = 0$$

The enthalpy changes are best found for one kmol of liquid heptane supplied. Owing to dissociation the heptane is not burnt completely, and the term ΔH_{25} will in this case be the enthalpy of combustion of 1 kmol of liquid heptane less the enthalpy of combustion of c kmol of CO and e kmol of H_2. The numerical solution of such a dissociation problem is too lengthy to be given here, but the reader can find a variety of worked examples in Ref. (7).

When the fuel/air mixture is appreciably richer than stoichiometric, it will be found that the number of kilomoles of oxygen (d) in the products is negligible. A somewhat simpler treatment involving only one dissociation constant can then be used. The argument rests on the fact that the oxygen produced by the dissociation of CO_2 can be regarded as being wholly used in the combustion of H_2. The two reaction equations

$$CO_2 + \tfrac{1}{2}O_2 \rightleftharpoons CO_2 \quad \text{and} \quad H_2 + \tfrac{1}{2}O_2 \rightleftharpoons H_2O$$

can then be combined to yield the so-called *water-gas reaction*

$$CO_2 + H_2 \rightleftharpoons CO + H_2O$$

With the O_2 eliminated from the products, one fewer equation is required, and the water-gas dissociation constant is all that is needed for the solution of the simultaneous equations. Values of

$$K = \frac{(p_{H_2O})(p_{CO})}{(p_{H_2})(p_{CO_2})}$$

are also given in Ref. (17).

The foregoing method of solution is for an adiabatic steady-flow process at a known constant pressure, such as occurs in a ram-jet combustion chamber. (This type of process also occurs in gas turbine combustion chambers, but the temperatures are too low for dissociation to be significant.) In the cylinders of many reciprocating internal-combustion engines the reaction occurs at substantially constant volume and the final pressure is unknown. The following example deals with this case, using carbon monoxide as the fuel for simplicity.

Example 15.11   The analysis of the products obtained from the adiabatic combustion of a stoichiometric mixture of CO and air at constant volume showed that there were 0·808 kmol of CO_2, 0·192 kmol of CO and 0·096 kmol of O_2, per kmol of CO in the reactants. The initial pressure and temperature of the reactants were 1 atm and 60 °C, and the final temperature was found to be 2465 °C.

Check the accuracy of the results by an atom balance and by verifying that the equilibrium condition is satisfied. The equilibrium constant, based on the reaction $CO + \frac{1}{2}O_2 \rightleftharpoons CO_2$ and with pressures in atmospheres, is 8·73 $(atm)^{-1/2}$ at the temperature of 2465 °C.

The reactants for a stoichiometric mixture, per kmol of CO, are

$$CO + \tfrac{1}{2}O_2 + \frac{79}{21}(\tfrac{1}{2})\,N_2$$

The combustion equation for the reaction is therefore

$$CO + 0·5\,O_2 + 1·881\,N_2 \rightarrow 0·808\,CO_2 + 0·192\,CO + 0·096\,O_2 + 1·881\,N_2$$

Checking the atom balance we have

(C) $0·808 + 0·192 = 1$

(O) $1·616 + 0·192 + 0·192 = 2$

The relevant partial pressures are

$$p_{CO_2} = \frac{0·808}{n}\,p, \quad p_{CO} = \frac{0·192}{n}\,p, \quad p_{O_2} = \frac{0·096}{n}\,p$$

Since the reactants and products are assumed to be perfect gases, and the reaction occurs at constant volume,

$$\frac{n_1 R_0 T_1}{p_1} = \frac{n R_0 T}{p}$$

where suffix 1 refers to the initial conditions. The number of kilomoles of reactants are

$$n_1 = 1 + 0·5 + 1·881 = 3·381$$

and $p_1 = 1$ atm and $T_1 = 333$ K. Hence

$$\frac{p}{n} = \frac{T}{3·381 \times 333}$$

The equilibrium constant is

$$K = \frac{p_{CO_2}}{p_{CO}(p_{O_2})^{1/2}} = \frac{0·808}{0·192\left(\dfrac{0·096T}{3·381 \times 333}\right)^{1/2}}\,(atm)^{-1/2}$$

When T is 2738 K, the calculated K equals 8·71 $(atm)^{-1/2}$. This is in good agreement with the true value of 8·73 $(atm)^{-1/2}$.

(The check can be completed by showing that the energy equation is satisfied. For this adiabatic constant volume process, the energy equation will be

$$(U_{P2465} - U_{P25}) + \Delta U_{25} + (U_{R25} - U_{R60}) = 0$$

ΔU_{25} is the internal energy of combustion of 0·808 kmol of CO, per kmol of CO supplied in the reactants.)

The treatment of the combustion of hydrocarbon fuels in air given here requires some qualification. Firstly, at high temperatures nitrogen is not the inert gas we have hitherto supposed it to be; some of it combines with oxygen to form nitric oxide according to the equation $\frac{1}{2}O_2 + \frac{1}{2}N_2 \rightleftharpoons NO$. The combination is endothermic and causes a reduction in products temperature. Secondly, some of the molecules of water vapour dissociate not only into hydrogen and oxygen, but also into hydrogen and hydroxyl ($\frac{1}{2}H_2 + OH \rightleftharpoons H_2O$). And, finally, some of the molecules of oxygen, nitrogen and hydrogen dissociate into their respective atoms. The dissociation constants for all these reactions are known, and their effect on the products temperature can be taken into account in more accurate calculations.

It must be emphasised that dissociation does not affect the heats of combustion or calorific values, which are determined at about atmospheric temperature. In general, if the products of combustion of a hydrocarbon fuel-air mixture are cooled below 1500 K before leaving the system, the chemical energy liberated is not much affected by dissociation. As the products of combustion are cooled, the dissociated products recombine and the combustion is effectively complete below 1500 K.* The efficiency of the combustion process itself is therefore not affected by any dissociation that may occur due to a high temperature reached at some intermediate stage. The efficiency of a power plant in which the combustion process occurs may be affected, however, because the heat transferred to the working fluid is now transferred at a lower average temperature—the maximum temperature reached being reduced by the dissociation. In other words, the quantity of energy liberated is the same as if there were no dissociation, but its effectiveness for producing work is diminished. For steam or gas turbine plants this point is of no importance owing to the use of relatively low maximum temperatures imposed by the metallurgical limit. The discussions in sections 17.4 and 18.4 (ii), however, indicate that dissociation is an important factor limiting the efficiency of reciprocating internal-combustion engines and rocket motors.

15.8 Tabulations of thermodynamic reaction data

It is worth emphasising that although this chapter has been concerned with the particular class of chemical reaction known as combustion, most of the theory presented applies equally to chemical reactions in general. The enthalpy of combustion,

---

*When the products are cooled very rapidly, however, as when a sample is withdrawn through a water-cooled sampling tube, the reaction rates may be reduced to such an extent that very little recombination takes place.

for example, is then referred to as the *enthalpy of reaction*. In this wider context, it is impracticable to tabulate the enthalpies of all reactions which a chemical compound may undergo with other substances. Instead it is the enthalpy of formation ΔH_{f0} which is tabulated and, as will be shown, it is a simple matter to calculate the enthalpy of reaction ΔH_0 in any particular case from the values of ΔH_{f0} of the substances taking part in the reaction.

The *enthalpy of formation* is defined as the increase in enthalpy when a compound is formed from its constituent elements in their natural forms and in a standard state: it is usually expressed in energy units per mol of compound, and referred to the case where each of the reacting elements is at 1 atm and 25 °C and the product is at 1 atm and 25 °C. The qualification 'natural form' implies that the enthalpy of, say, gaseous hydrogen H_2 is put equal to zero, and not that of the dissociated monatomic gas H. Or again, some elements may exist in several 'natural' forms and only one can be used as a datum state—e.g. for carbon it is graphite and not diamond which is used for this purpose.

The enthalpy of formation of three compounds is listed below;

$$C(\text{graphite}) + 2H_2 \rightarrow CH_4;\ (\Delta H_{f25})_{CH_4} = -74\,870\text{ kJ/kmol of } CH_4$$
$$C(\text{graphite}) + O_2 \rightarrow CO_2;\ (\Delta H_{f25})_{CO_2} = -393\,520\text{ kJ/kmol of } CO_2$$
$$H_2 + \tfrac{1}{2}O_2 \rightarrow H_2O(\text{vap});\quad (\Delta H_{f25})_{H_2O} = -241\,830\text{ kJ/kmol of } H_2O$$

Since the change in enthalpy is independent of the process by which the end state is reached, we can view a reaction in two stages—a breaking up of the reactants into elements followed by a recombination to form the products. Thus to find the enthalpy of reaction of methane and oxygen we can write

$$CH_4 + 2O_2 \rightarrow C + 2H_2 + 2O_2 \rightarrow CO_2 + 2H_2O(\text{vap})$$

and hence

$$\Delta H_0 = (\Delta H_f)_{CO_2} + 2(\Delta H_f)_{H_2O} - (\Delta H_f)_{CH_4} - 2(\Delta H_f)_{O_2}$$
$$= (-393\,520) + 2(-241\,830) - (-74\,870) - 0$$
$$= -802\,310\text{ kJ/kmol of } CH_4$$

We have seen in section 7.9 (iii)* that data for enthalpy of reaction ΔH_0, change in Gibbs function ΔG_0, and dissociation constant K, are interchangeable. Thus from equations (7.72), (7.73) and (7.75)

$$-\Delta G_0 = -\Delta H_0 + T_0(S_{P0} - S_{R0}) \tag{15.21}$$

$$\ln K = -\frac{\Delta G_0}{R_0 T_0} \tag{15.22}$$

By virtue of the Third Law [see section 7.9 (i)], which enables absolute entropies to be determined for the reactants and products relative to a common datum at absolute

---

*The remainder of section 15.8 should be studied only after Chapter 7, and therefore on a second reading of the book.

zero of temperature, the entropy term in equation (15.21) can be calculated. The *Gibbs function of formation* ΔG_{f0}* can then be defined in an analogous manner to the value of ΔH_{f0}; and the values of ΔG_0 for any reaction can be found by adding the formation values as for ΔH_0. Consequently it follows from equation (15.22) that it is also possible to define logarithms of the *equilibrium constant of formation* $\ln K_f$, from which the logarithm of the equilibrium constant K for any reaction can be calculated from a similar summation procedure.† A short compilation of various chemical thermodynamic data from Refs. (36) and (37) is given in Fig. 15.11: it contains values of ΔH_{f0}, ΔG_{f0}, $\log_{10}K_f$, and also of the molar heat C_p and the absolute molar entropy S_0, all in the standard state of 1 atm and 25 °C.

	M $\left[\dfrac{\text{kg}}{\text{kmol}}\right]$	At 298·15 K and 1 atm				
		ΔH_{f0} $\left[\dfrac{\text{kJ}}{\text{kmol}}\right]$	ΔG_{f0} $\left[\dfrac{\text{kJ}}{\text{kmol}}\right]$	$\log_{10}K_f$ (K_f in atm units)	C_{p0} $\left[\dfrac{\text{kJ}}{\text{kmol K}}\right]$	S_0 $\left[\dfrac{\text{kJ}}{\text{kmol K}}\right]$
O_2 (gas) . .	31·999	0	0	0	29·37	205·03
O (gas) . .	15·9995	249 190	231 770	− 40·604	21·91	160·95
H_2 (gas) . .	2·016	0	0	0	28·84	130·58
H_2O (liq) .	18·015	−285 820	−237 190	41·554	75·30	70·01
H_2O (vap) .	18·015	−241 830	−228 590	40·048	33·58	188·72
C (graphite) .	12·011	0	0	0	8·53	5·69
C (diamond) .	12·011	1 900	2 870	− 0·502	2·44	6·06
C (gas) . .	12·011	714 990	669 550	−117·302	20·84	157·99
CO (gas) . .	28·011	−110 530	−137 160	24·029	29·14	197·54
CO_2 (gas) . .	44·010	−393 520	−394 390	69·095	37·13	213·68
CH_4 (gas) .	16·043	− 74 870	− 50 810	8·902	35·64	186·10

Fig. 15.11 A selection of chemical thermodynamic data

Both the enthalpy and equilibrium constant for perfect gases are independent of pressure, and in practice the data for ΔH_{f0} and $\log_{10}K_f$ can be assumed constant for real gases and vapours over wide ranges of pressure. Similarly these values are very nearly independent of pressure for liquids and solids. The Gibbs function, however, involves an entropy term which varies appreciably with pressure for gaseous substances (although not for liquids and solids), and this variation must be taken into account when any gaseous reactants or products are not at the standard pressure of 1 atm.

---

*Note that this is often referred to as the 'free energy of formation' and given the symbol ΔF_{f0}: it should not be confused with the Helmholtz function for which F is used in Chapter 7.

†Because values of $\ln K$ (or $\log_{10}K$) can be found from additions of the appropriate values of $\ln K_f$, it follows that values of K can be found by multiplication of the K_f values. This should also be clear from the definition of the equilibrium constant in section 15.7, since by multiplication of suitable equilibrium constants and cancellation of partial pressure terms any required equilibrium constant can be found.

A particular difficulty appears with regard to substances which cannot exist at the standard state. Thus H_2O vapour cannot exist at 1 atm at 25 °C, and H_2O in the standard state can exist only as a compressed liquid. The tables make use of a fictitious vapour at the standard state, obtained by assuming that the substance behaves as a perfect gas between the saturated vapour state at 25 °C (0·031 66 bar in the case of H_2O) and the standard state. This is permissible because we are only interested in using the tabulated data to calculate *changes* in H or G. As an example, let us consider the differences between the tabulated liquid and vapour values for water. For the enthalpy of formation, omitting the suffix 25 for brevity:

$(\Delta H_f)_{\text{vap}}$ at 1 atm $-(\Delta H_f)_{\text{liq}}$ at 1 atm

$= [(\Delta H_f)_{\text{vap}}$ at 1 atm $-(\Delta H_f)_{\text{vap}}$ at 0·031 66 bar] $+$

$\quad [(\Delta H_f)_{\text{vap}}$ at 0·031 66 bar $-(\Delta H_f)_{\text{liq}}$ at 0·031 66 bar] $+$

$\quad [(\Delta H_f)_{\text{liq}}$ at 0·031 66 bar $-(\Delta H_f)_{\text{liq}}$ at 1 atm]

$= 0 + H_{fg} + 0 = 43\,990 \text{ kJ/kmol of } H_2O$

The first and third terms are zero because the enthalpy of vapour and of liquid is taken as independent of pressure; the middle term represents the latent heat of vaporisation per kmol of H_2O. Considering the Gibbs function of formation, again at 25 °C,

$(\Delta G_f)_{\text{vap}}$ at 1 atm $-(\Delta G_f)_{\text{liq}}$ at 1 atm

$= [(\Delta G_f)_{\text{vap}}$ at 1 atm $-(\Delta G_f)_{\text{vap}}$ at 0·03166 bar] $+$

$\quad [(\Delta G_f)_{\text{vap}}$ at 0·03166 bar $-(\Delta G_f)_{\text{liq}}$ at 0·03166 bar] $+$

$\quad [(\Delta G_f)_{\text{liq}}$ at 0·03166 bar $-(\Delta G_f)_{\text{liq}}$ at 1 atm]

$= 8\cdot3144 \times 298\cdot15 \ln \dfrac{1\cdot01325}{0\cdot03166} + 0 + 0 = 8590 \text{ kJ/kmol of } H_2O$

The first term represents extrapolation to the fictitious vapour state using (15.21) and assuming that the vapour behaves as a perfect gas over that range. The second term is zero because the Gibbs function remains constant in an isothermal change of phase [see sections 7.7 and 7.9 (i)], and the third term is zero because the Gibbs function of a liquid is taken as independent of pressure.

Finally, considering the entropy at 25 °C we can write

$(S_0)_{\text{vap}}$ at 1 atm $-(S_0)_{\text{liq}}$ at 1 atm

$\quad = [(S_0)_{\text{vap}}$ at 1 atm $-(S_0)_{\text{vap}}$ at 0·031 66 bar]

$\quad + [(S_0)_{\text{vap}}$ at 0·031 66 bar $-(S_0)_{\text{liq}}$ at 0·031 66 bar]

$\quad + [(S_0)_{\text{liq}}$ at 0·031 66 bar $-(S_0)_{\text{liq}}$ at 1 atm]

$\quad = -8\cdot3144 \ln \dfrac{1\cdot013\,25}{0\cdot031\,16} + \dfrac{43\,990}{298\cdot15} + 0$

$\quad = -28\cdot82 + 147\cdot54 = 118\cdot72 \text{ kJ/kg K}$

Or, since $G = H - TS$ and the temperature is constant at T_0, the change in entropy can also be found from

$$\frac{\Delta H}{T_0} - \frac{\Delta G}{T_0} = \frac{43\,990 - 8590}{298 \cdot 15} = 118 \cdot 73 \text{ kJ/kg K}$$

We conclude this brief explanation of the way in which thermodynamic reaction data are presented, with an example illustrating its use to calculate the values of ΔH, ΔG and K for a reaction at a specified temperature and pressure which differs from the standard state.

Example 15.12 Calculate ΔH, ΔG and $\log_{10} K$ at 25 °C for the reaction

$$CO_2 + H_2 \rightarrow CO + H_2O \text{ (vap)}$$

when the *total* pressure of the reactants and products is 1 atm each. Use the data of Fig. 15.11 and check them for consistency.

Also calculate the value of $\log_{10} K$ at $T = 400$ K, using (a) equation (15.22), and (b) equation (7.77). Assume in each case that the molar heats C_p are constant over the temperature range considered.

The enthalpy of reaction ΔH_0 when each constituent is at 1 atm is

$$\Delta H_0 = (-110\,530) + (-241\,830) - (-393\,520) - 0 = +41\,160 \text{ kJ}$$

This will also be equal to ΔH for the specified pressure of each reactant and product (in this case $\frac{1}{2}$ atm each), because the enthalpy can be taken as independent of pressure.

The change in Gibbs function ΔG_0 is given by

$$\Delta G_0 = (-137\,160) + (-228\,590) - (-394\,390) - 0 = +28\,640 \text{ kJ}$$

To find the change of Gibbs function for the reactants and products each at $\frac{1}{2}$ atm, the Gibbs function of each term will have to be modified by $nR_0T_0 \ln(p_0/p)$, where $p_0 = 1$ atm and $p = \frac{1}{2}$ atm. It is obvious that these modifications (with appropriate signs) will cancel in this case and $\Delta G = \Delta G_0$: there will be a net effect only when the reaction is such that the number of kilomoles of gaseous products differ from that of the gaseous reactants.

The equilibrium constant is

$$\log_{10} K = 24 \cdot 029 + 40 \cdot 048 - 69 \cdot 095 - 0 = -5 \cdot 018$$

To check consistency, let us first apply equation (15.21).

$$-\Delta H_0 = -\Delta G_0 - T_0(S_{P0} - S_{R0})$$

$$= -28\,640 - 298 \cdot 15(1 \times 197 \cdot 54 + 1 \times 188 \cdot 72 - 1 \times 213 \cdot 68 - 1 \times 130 \cdot 58)$$

$$= -28\,640 - 298 \cdot 15 \times 42 \cdot 00 = -41\,160 \text{ kJ}$$

Also, from equation (15.22),

$$\log_{10} K = - \frac{1}{2 \cdot 302\ 59} \frac{\Delta G_0}{R_0 T_0}$$

$$= - \frac{1}{2 \cdot 302\ 59} \frac{28\ 640}{8 \cdot 3144 \times 298 \cdot 15} = -5 \cdot 018$$

a) To calculate the value of $\log_{10} K_T$ at $T = 400$ K, we will first write equations for ΔH_T and $(S_{PT} - S_{RT})$ at T, taking the values of C_p from Fig. 15.11. From equation (15.10),

$$\Delta H_T = (1 \times 29 \cdot 14 + 1 \times 33 \cdot 58)(T - 298 \cdot 15) + 41\ 160$$

$$+ (1 \times 37 \cdot 13 + 1 \times 28 \cdot 84)(298 \cdot 15 - T)$$

$$= 41\ 160 - 3 \cdot 25(T - 298 \cdot 15) = 42\ 130 - 3 \cdot 25T \text{ kJ}$$

and, remembering that the pressure terms cancel in this case,

$$(S_{PT} - S_{RT}) = \sum_P n C_p \ln \frac{T}{T_0} + (S_{P0} - S_{R0}) + \sum_R n C_p \ln \frac{T_0}{T}$$

$$= 42 \cdot 00 - 3 \cdot 25 \ln \frac{T}{298 \cdot 15}$$

At $T = 400$ K,

$$\Delta G_T = \Delta H_T - T(S_{PT} - S_{RT})$$

$$= (42\ 130 - 3 \cdot 25 \times 400) - 400 \left(42 \cdot 00 - 3 \cdot 25 \ln \frac{400}{298 \cdot 15} \right)$$

$$= 24\ 410 \text{ kJ}$$

and from (15.22)

$$\log_{10} K = - \frac{1}{2 \cdot 302\ 59} \frac{24\ 410}{8 \cdot 3144 \times 400} = -3 \cdot 188$$

(b) Alternatively, using equation (7.77), and treating ΔH as a variable between T_0 and T,

$$\int_{T_0}^{T} d \ln K = - \frac{1}{R_0} \int_{T_0}^{T} \Delta H \, d \left(\frac{1}{T} \right)$$

and hence, using the intermediate result for ΔH_T in (a),

$$\ln K_{400} - \ln K_{298} = - \frac{1}{R_0} \int_{T_0}^{T} \left(42\ 130 - \frac{3 \cdot 25}{1/T} \right) d \left(\frac{1}{T} \right)$$

$$= - \frac{1}{R_0} \left[\frac{42\ 130}{T} - 3 \cdot 25 \ln \frac{1}{T} \right]_{T_0}^{T}$$

$$= - \frac{1}{R_0} \left(\frac{42\ 130}{400} - \frac{42\ 130}{298 \cdot 15} - 3 \cdot 25 \ln \frac{298 \cdot 15}{400} \right) = 4 \cdot 213$$

Thus, using $\log_{10} K = -5 \cdot 018$ at 25 °C from the first part of the example,

$$\log_{10} K_{400} = \frac{4 \cdot 213}{2 \cdot 30259} - 5 \cdot 018 = -3 \cdot 188$$

which agrees with the result obtained in (a).

15.9 Limitations of the thermodynamic analysis

In the previous chapter it was emphasised that the thermodynamic analysis of the processes occurring in condensers and air-conditioning plant is limited to the prediction of final states and energy transfers from a knowledge of the initial states and certain operating conditions. A more detailed analysis of the way in which heat is transferred must be undertaken before an actual plant can be designed.

The thermodynamic analysis of combustion processes, given in this chapter, is limited in a similar way—it does not enable us to specify the conditions under which the process can actually be carried out in practice. For example, it cannot be used to predict the air/fuel ratio and initial temperature necessary for self-ignition to occur. Nor has the simple thermodynamic treatment anything to say about the rate at which the process can be carried out: for example, it cannot predict flame speeds, or the conditions under which a stable flame is possible, or the delay periods experienced in internal-combustion engines.

A great deal of empirical data has been amassed over the years to explain the detailed phenomena encountered in practical combustion systems such as coal and oil fired furnaces, gas burners, reciprocating engines, gas turbines, ram-jets and rocket motors. The concepts necessary for a synthesis of this mass of data are now being formulated with the growth of the fundamental sciences of combustion kinetics and combustion wave theory. Quantitatively there is still a wide gap to be bridged before these sciences can be applied directly so that new combustion systems can be designed on the drawing board, but even now they can be used to aid consistent thinking and to reduce the amount of *ad hoc* development which is still required before a new design can be made to work efficiently. Ref. (30) provides an excellent nonmathematical introduction to combustion kinetics and combustion wave theory.

Part III   Work Transfer

Introduction

So far in this book we have been concerned mainly with relations between fluid properties and the quantities of work and heat which accompany changes of state. Before it is possible to design plant which will effect the desired changes of state, it is necessary to analyse the mechanism of work and heat transfer in some detail: the former mode of energy transfer is discussed here and the latter in Part IV. This division is easily justified in such cases as steam plant, refrigeration plant and air-conditioning plant, because the work and heat transfers are carried out in separate components. It is not so easily justified when considering reciprocating air compressors or internal-combustion engines. To cover these latter examples it may be said that Part III deals with those systems whose main purpose is the transfer of work and for which exchanges of heat with the surroundings are only incidental.

Fundamentally, there are two types of machine used to perform the process of work transfer: (a) positive displacement machines, and (b) non-positive displacement machines. The former comprise all reciprocating expanders and compressors and also some types of rotary compressor such as the Roots blower and the vane pump. The features common to class (a) are: (1) these machines possess members (e.g. pistons and valves) which ensure positive admission and delivery of the working fluid and prevent any undesired reversal of flow within the machine; (2) their operation is intermittent and they subject the fluid to non-flow processes; (3) the work is transferred by virtue of a hydrostatic pressure force exerted on a moving boundary. In non-positive displacement machines, which comprise axial-flow and radial-flow turbines and compressors, there are no means for preventing a reversal of flow in the machine; the processes are steady-flow processes; and work is transferred by virtue of the change in momentum of a stream of fluid flowing at a high speed over blades or vanes attached to a rotor.

In all practical power plant, work is done during successive compressions and expansions of the working fluid. To obtain a net amount of work, it is essential to change the state of the working fluid between the compression and expansion processes. So far we have mainly considered plant in which this change of state is brought about by external heat addition, e.g. steam power plant and closed-cycle gas turbine plant. The one exception in Part II has been the open-cycle gas turbine wherein the change of state is effected by a chemical reaction. In Part III we shall describe the reciprocating internal-combustion engine, where the compression, change of state by combustion, and subsequent expansion, are all performed in a single component—the engine cylinder. Reciprocating internal-combustion engines fall within the class of positive displacement machines, as does the newer rotary type known as the Wankel engine which is also described briefly.

The scheme of Part III may now be outlined. Chapter 16 deals with positive displacement compressors and expanders, with the reciprocating air compressor and steam engine discussed as important examples of each type. The following chapter describes the processes in positive-displacement internal-combustion engines. The knowledge of non-flow processes already gained from a study of Parts I and II is adequate for the purposes of these two chapters. For a study of non-positive displacement machines, however, the treatment of steady-flow processes given so far is insufficiently detailed: we need to know something about the nature of the flow *within* the open system. Chapter 18 analyses the simplest type of flow—one-dimensional steady flow—and the results are then applied in Chapter 19 to the processes which occur in non-positive displacement machines.

When dealing with stationary power plant, the useful net work is always delivered by the engine as a torque in a rotating shaft. When the power plant is used for propelling a vehicle, this need not necessarily be the case. Admittedly in rail, road and ship propulsion the power is usually delivered via a shaft, but in the more recent developments of aircraft propulsion the force required to move an aircraft against atmospheric drag or gravitational pull is derived without the intermediary of shaft-work. The analysis of the behaviour of jet power plant is basically a problem of one-dimensional flow, and as such it is treated in Chapter 18.

Finally, it should be realised that the major fraction of the total mechanical work produced by the world's power plant is utilised in the production of electrical power, i.e. it is used to drive electric generators. If the primary source of energy—whether chemical or nuclear—can be converted directly into electricity, both a higher efficiency and an overall simplification of plant might be attained. Part III closes with a chapter describing briefly the main devices which are being developed for 'direct conversion'. The excuse for including this topic in a Part entitled Work Transfer is that in principle electrical and mechanical work are mutually convertible without loss and in this sense are equivalent.

16 Reciprocating Expanders and Compressors

The function of a compressor is to admit fluid from a low pressure region, compress it, and deliver it to a high pressure region. An expander carries out the reverse processes with the object of producing work.

Although this chapter is mainly concerned with reciprocating machines, of which the air compressor and steam engine are considered as typical examples, the last section deals with rotary positive-displacement compressors. These are more akin to their reciprocating equivalent than they are to the non-positive displacement compressors described in Chapter 19, even though the latter are rotary machines. The essential feature common to all positive-displacement machines is that the fluid is prevented by a solid boundary from flowing back in the direction of the pressure gradient. All the compressors described here are therefore capable of producing large pressure ratios.

As already pointed out, reciprocating machines can be regarded as open systems being steadily supplied with working fluid and transferring work and heat between the fluid and its surroundings at a uniform rate. Nevertheless the internal processes are intermittent, and this fact imposes an important limitation on reciprocating machines—the comparatively low rate at which machines of reasonable size can handle the working fluid.

16.1 Work transfer in reversible reciprocating machines

A reciprocating machine consists essentially of a cylinder and piston, with the piston connected to a rotating shaft by a connecting rod and crank as shown in Fig. 16.1.

Fig. 16.1 The reciprocating expander or compressor

Valves permit the fluid to flow into and out of the cylinder at appropriate times. The total piston travel is called the *stroke*, and the volume displaced is called the *swept volume*. There is always some minimum volume, the *clearance volume*, when the piston is at the end of the stroke. This is necessary to accommodate the valves and avoid contact of piston and cylinder head.

The succession of events can best be explained with the aid of a diagram (Fig. 16.2a) showing how the pressure in the cylinder varies during one revolution of the crank-shaft, i.e. during one complete *machine cycle*. The change in state of the fluid as it passes through the machine will be depicted on a *p-v* diagram (Fig. 16.2b). We shall refer explicitly to the operation of an expander, and bear in mind the fact that in a

Fig. 16.2 *p-V* and *p-v* diagrams for an ideal expander

compressor the machine cycle is traversed in the reverse direction. The following analysis refers to an idealised machine in which the processes are reversible, the state of the fluid at inlet and outlet remains constant with respect to time, and the kinetic energy at inlet and outlet is negligible.

a–b The inlet valve is open, and a mass of fluid Δm_f in state 1 enters the cylinder without change of state. It mixes reversibly with a mass Δm_c already present in the clearance volume in the same state 1.

b–c Both valves are closed and the mass $(\Delta m_f + \Delta m_c)$ changes its state during the expansion to pressure p_2.

c–d The exhaust valve is open and a mass Δm_f is ejected from the cylinder without further change of state.

d–a Both valves are closed and a mass Δm_c is compressed to the original state 1.

It is important to note that although Fig. 16.2a represents the events during a machine cycle, it does not represent a thermodynamic cycle. Admittedly the mass Δm_c is carried through a thermodynamic cycle, although it consists merely of the traversing and retraversing of a single path of states as shown on the *p-v* diagram of Fig. 16.2b.

But the mass Δm_f is not carried through a cycle at all, and simply has its state changed from state 1 to state 2.

Before the work done can be estimated, the processes of expansion and compression between states 1 and 2 must be defined. Since the speed of flow through the machine is not very great, these processes cannot be assumed to be adiabatic. A better approximation is to assume that they are polytropics, i.e. of the form $pV^n = $ constant. An expression for the work done can be derived in two ways—either by considering the forces on the piston during a complete machine cycle, or by applying the steady-flow energy equation between inlet and outlet of the machine. The same assumption of constant inlet and outlet conditions has to be made in both cases. That is, it must be assumed that there is sufficient intake and receiver volume for the intermittent working of the machine to make a negligible difference to the fluid properties at inlet and outlet.

(i) *Machine cycle analysis*

The net work done per machine cycle is the algebraic sum of the work transfers in the four processes, each of which can be found from $\int pA \, dL$ or $\int p \, dV$. A is the area of the piston and L the distance moved. Referring to Fig. 16.2,

$$\oint dW = \int_a^b p \, dV + \int_b^c p \, dV + \int_c^d p \, dV + \int_d^a p \, dV = \oint p \, dV$$

$\oint dW$ is evidently equal to the area of the p-V diagram. Carrying out the integration, making use of equation (3.12), we have

$$\oint dW = p_1(V_b - V_a) + \frac{(p_2 V_c - p_1 V_b)}{1 - n} + p_2(V_d - V_c) + \frac{(p_1 V_a - p_2 V_d)}{1 - n}$$

$$= \frac{n}{1 - n} \{p_1(V_a - V_b) + p_2(V_c - V_d)\}$$

The work done can be expressed in terms of the fluid properties at inlet and outlet by noting that

$$V_a = \Delta m_c v_1, \quad V_b = (\Delta m_f + \Delta m_c)v_1$$
$$V_c = (\Delta m_f + \Delta m_c)v_2, \quad V_d = \Delta m_c v_2$$

We then have

$$\oint dW = \Delta m_f \frac{n}{1 - n}(p_2 v_2 - p_1 v_1) \tag{16.1}$$

The rate of work done is obtained by multiplying the net work done during the machine cycle by the number of cycles N per unit time, e.g. by the rev/min. If the mass flow per unit time ($N \Delta m_f$) is denoted by m_f, the rate of work done is

$$W = m_f \frac{n}{1 - n}(p_2 v_2 - p_1 v_1) \tag{16.2}$$

Equation (16.2) can be recast in terms of the pressure ratio as follows

$$W = m_f p_1 v_1 \frac{n}{1-n} \left(\frac{p_2 v_2}{p_1 v_1} - 1 \right)$$

and since $v_2/v_1 = (p_1/p_2)^{1/n}$,

$$W = m_f p_1 v_1 \frac{n}{1-n} \left\{ \left(\frac{p_2}{p_1} \right)^{(n-1)/n} - 1 \right\} \qquad (16.3)$$

Identical expressions are obtained for the case where the machine cycle is carried out in the reverse direction as in a compressor; p_2 is then greater than p_1 and W becomes a negative quantity.

The exact path of the expansion and compression processes, i.e. the value of the polytropic index n, depends upon the heat flow through the cylinder walls. Ideally, heat is only transferred during the processes b–c and d–a, because processes a–b and c–d are merely mass transfers to and from the cylinder without change of state. The heat transferred during the compression of Δm_c along d–a is equal and opposite to the heat transferred during its expansion from b to c. The net heat transfer per machine cycle is therefore that associated with the expansion of Δm_f along b–c. Since the system is closed during this process the appropriate energy equation is

$$\oint dQ = \Delta m_f (u_2 - u_1) + \Delta m_f \int_1^2 p \, dv$$

$$= \Delta m_f (u_2 - u_1) + \Delta m_f \frac{1}{1-n} (p_2 v_2 - p_1 v_1)$$

Since $u = h - pv$, this can also be written as

$$\oint dQ = \Delta m_f (h_2 - h_1) + \Delta m_f \frac{n}{1-n} (p_2 v_2 - p_1 v_1)$$

The rate of heat transfer per unit time is therefore

$$Q = m_f (h_2 - h_1) + m_f \frac{n}{1-n} (p_2 v_2 - p_1 v_1) \qquad (16.4)$$

(ii) *Steady-flow analysis*

The foregoing expressions for the work and heat transfers can also be obtained by applying the steady-flow energy equation, without the necessity for such a detailed analysis of the events which take place within the cylinder. The energy equation, neglecting the kinetic energy terms, is

$$Q = m_f (h_2 - h_1) + W$$

Since we are considering a reversible process, we may also write

$$Q = m_f \int_1^2 T \, ds$$

The integral can be evaluated by making use of equation (6.13), which relates the entropy and other thermodynamic properties during any infinitesimal change of state, viz.

$$T \, ds = dh - v \, dp$$

Thus

$$Q = m_f \int_1^2 (dh - v \, dp) = m_f(h_2 - h_1) - m_f \int_1^2 v \, dp$$

By direct comparison with the steady-flow energy equation it is evident that

$$W = -m_f \int_1^2 v \, dp \qquad (16.5)$$

Making the further assumption that the process is polytropic, as before, we can perform the integration and obtain

$$W = m_f \frac{n}{1 - n} (p_2 v_2 - p_1 v_1)$$

and hence also

$$Q = m_f(h_2 - h_1) + m_f \frac{n}{1 - n} (p_2 v_2 - p_1 v_1)$$

which are identical with equations (16.2) and (16.4).

There are two important points to be noted from the analysis of the reversible reciprocating machine.

(a) If the machine is an expander the work done increases as the value of n decreases, while for a compressor the amount of negative work decreases as n decreases. There is no need to evaluate equation (16.2) to show this—it is evident from the p-v diagrams of Fig. 16.3. Equation (16.2) is represented by ($m_f \times$ shaded area), i.e. by $-m_f \int_1^2 v \, dp$, and the negative slope of a polytropic curve decreases with decrease of n (section 3.4).

Fig. 16.3

(b) Equation (16.1) indicates that the clearance volume has no effect on the work done *per unit mass flow*. The mass Δm_c in the clearance volume is expanded and recompressed along the same state path and no net work is done on or by it. The clearance volume, however, affects the capacity of the machine, i.e. the mass Δm_f of fluid which a cylinder of given size can handle per machine cycle. The greater the clearance volume per unit swept volume, the smaller the amount of fluid that can be handled by the machine. This will be shown formally for an air compressor in section 16.3.

16.2 The indicator diagram

There is one important respect in which the actual p-V diagram of all real reciprocating machines differs from the ideal diagram of a reversible machine. Due to the fact that

(a) Expander (b) Compressor

Fig. 16.4

valves take a finite time to open and close, the corners of the diagram are rounded and ill defined. When a valve port is only partly open, throttling occurs and there is a considerable pressure drop in the direction of flow. Even when the port is fully open, some pressure drop is inevitable due to fluid friction. Thus the real p-V diagrams for an expander and compressor appear as in Fig. 16.4a and b. It must be remembered too, that in the real machine the expansion and compression processes will not necessarily be exactly polytropic.

Since the real processes are not reversible, the actual work and heat transfers cannot be individually expressed in terms of the fluid properties at inlet and outlet as in equations (16.2) and (16.4). Nevertheless a good approximation to the actual work done can still be obtained from $\oint(\text{force on piston} \times dL)$ or $\oint p\,dV$, provided the variation of average pressure in the cylinder is plotted during the machine cycle. Strictly speaking, it is the pressure on the piston face which should be measured, but this presents too difficult a problem of measurement for ordinary purposes. [The error can become significant when considerable turbulence is present as in high-speed compression-ignition engines—see section 17.3 (ii).]

The real p-V diagram is obtained by fitting an indicator to the machine. Fig. 16.5

illustrates a simple *mechanical indicator* suitable for machines of low rotational speed.*
The main part of the indicator is itself a small cylinder and piston which can be put
into communication with the cylinder of the machine. The volume of the indicator
cylinder must be small compared with the clearance volume of the machine cylinder,
otherwise the diagram will be altered appreciably when the indicator cock is opened.
The indicator piston is spring loaded so that its movement is directly proportional
to the pressure, and this movement is magnified by a link mechanism connecting the
piston and a pencil. The spring is calibrated, the *spring constant k* being expressed as
the number of units of pressure required to move the pencil point vertically

Fig. 16.5   The mechanical indicator

through unit distance. The pencil point moves along the surface of a drum
parallel to its axis, while the drum is rotated by a cord connected, via a reducing
linkage, to the piston of the machine. Thus while the pencil moves along the
drum a distance proportional to the pressure, the drum surface moves past the
pencil a distance proportional to the engine piston stroke and in phase with it. In
this way the p-V diagram can be recorded on an *indicator card* wrapped round
the drum. When the card is unwrapped the area of the diagram can be measured
with a planimeter, this area being proportional to the work done per machine
cycle. The constant of proportionality is not, in practice, evaluated, and the work
done is usually deduced by making use of the concept of mean effective pressure
already introduced in section 12.7.

Consider the p-V diagram of Fig. 16.6a. The work done per machine cycle can be
expressed by

$$\oint dW = \int_{V_a}^{V_b} (p_1 - p_2)\, dV = \int_0^L (p_1 - p_2) A\, dL$$

---

*Descriptions of other indicators, e.g. electronic and optical types, can be found in the *Proceedings of the
Institution of Mechanical Engineers* and in the *Transactions of the Society of Automotive Engineers*. Such
indicators are more suitable for high-speed machines because the inertia of their moving parts is much less
than that of the mechanical indicator described here.

where A is the area of the piston and L the stroke. This can be written as

$$\oint dW = p_m A L \qquad (16.6)$$

where

$$p_m = \frac{1}{L} \int_0^L (p_1 - p_2)\, dL$$

Thus p_m is the height of a rectangle of the same area and length as the p-V diagram. p_m is known as the *indicated mean effective pressure*, and it can be regarded as that constant pressure which, if it acted on the piston over one stroke, would do the same amount of work as is done in one machine cycle. Since the actual indicator card is drawn to a reduced scale, Fig. 16.6b, it is first necessary to find its mean height a/l,

Fig. 16.6 Finding the mean effective pressure

say in cm, and then deduce p_m in bar by multiplying the mean height by the spring constant k in bar/cm, i.e.

$$\frac{p_m}{[\text{bar}]} = \frac{k}{[\text{bar/cm}]} \frac{a/[\text{cm}^2]}{l/[\text{cm}]} \qquad (16.7)$$

The work done per machine cycle can then be found from (16.6). If N is the rotational speed in rev/s, A the piston area in m^2 and L the stroke in m, the numerical equation for the rate of work done becomes

$$\frac{W}{[\text{kW}]} = 10^2 \frac{p_m}{[\text{bar}]} \frac{A}{[\text{m}^2]} \frac{L}{[\text{m}]} \frac{N}{[\text{rev/s}]}$$

The work so found is known as the *indicated power* (i.p.). The power of machines is often still expressed in horse-power units, and it is then referred to as the indicated horse-power (i.h.p.). The British and metric horse-power units are related to the kilowatt by the following equation, to four significant figures.

$$1 \text{ kW} = \frac{1}{0 \cdot 7457} \text{ British h.p.} = \frac{1}{0 \cdot 7355} \text{ metric h.p.}$$

The actual power produced at the shaft of an expander will be less than the i.p. because some work is dissipated by friction between piston and cylinder and in the bearings. The *shaft power* (s.p.) is measured by some type of dynamometer,* of which the friction brake is a common example, and consequently it is often referred to as the *brake power* (b.p.). The ratio

$$\eta_m = \frac{\text{b.p.}}{\text{i.p.}} \tag{16.9}$$

is called the *mechanical efficiency* of the expander. If the machine is a compressor, equation (16.8) gives the rate of work done *on* the fluid by the piston. In fact more work than this must be supplied to the shaft to overcome mechanical friction losses, and the mechanical efficiency is therefore defined in this case as

$$\eta_m = \frac{\text{i.p.}}{\text{b.p.}} \tag{16.10}$$

For reasons similar to those given in section 12.7, a machine cycle yielding a high indicated mean effective pressure is likely to provide the basis for a compressor or expander having a high mechanical efficiency.

In this section we have concentrated on the essential differences which exist between all real and ideal reciprocating machines. There are other differences which are peculiar to particular types of machine and these will be brought out in the following sections. For brevity we shall consider one example of a compressor—the air compressor, and one example of an expander—the steam engine.

16.3 Reciprocating air compressors

One special feature of a compressor is the use of spring-loaded, non-return valves which open and close automatically (Fig. 16.7). This considerable mechanical simplification is made possible by the fact that the fluid is forced in the direction opposing the pressure gradient. In an expander it is necessary to use mechanically operated valves, timed to open and close by some linkage from the crankshaft. Automatic spring-loaded valves tend to be less positive in their action and lead to greater throttling losses than mechanically-operated valves, but the mechanical simplification is considered to be worth the small amount of additional compression work entailed. A

---

*A description of the more common types of dynamometer can be found in textbooks on 'Mechanics of Machines'.

typical *p-V* diagram is shown in Fig. 16.7. The wavy lines during the suction and delivery processes are due to valve bounce which arises because initially a rather large pressure difference is required across the valves to overcome their inertia. Apart from these peculiarities the actual indicator diagram follows the ideal *p-V* diagram closely, and for the remainder of this section only the ideal diagram will be considered.

Fig. 16.7   Self-acting valves for compressor and their effect on the actual indicator diagram

When air is the working fluid, the equation of state for a perfect gas enables equations (16.2), (16.3) and (16.4) to be put in the form:

$$W = m_f \frac{n}{1-n} R(T_2 - T_1) \tag{16.11}$$

$$W = m_f \frac{n}{1-n} RT_1 \left\{ \left(\frac{p_2}{p_1}\right)^{(n-1)/n} - 1 \right\} \tag{16.12}$$

$$Q = m_f c_v (T_2 - T_1) + m_f \frac{n}{1-n} R(T_2 - T_1)$$

Since $R = c_v(\gamma - 1)$, the last equation reduces to

$$Q = m_f \left\{ \gamma c_v + \frac{n}{1-n} c_v(\gamma - 1) \right\} (T_2 - T_1)$$

$$= m_f \frac{\gamma - n}{1-n} c_v (T_2 - T_1) \tag{16.13}$$

In section 16.1 it was pointed out that the work done *on* the air decreases as the value of *n* decreases. For this reason the compressor cylinder is either water cooled, or finned and air cooled. Even with good cooling the value of *n* is usually between

1·2 and 1·3, whereas the compression work is a minimum when $n = 1$, i.e. for an iso-thermal compression. (If $n < 1$ the air is cooled below the inlet temperature, implying the use of refrigeration on which work would have to be expended.) These considerations give rise to the practice of quoting the performance of a compressor in terms of an *isothermal efficiency* defined by

$$\eta_{\text{iso}} = \frac{\text{isothermal work}}{\text{actual indicated work}} \tag{16.14}$$

If the mechanical friction losses are included, we have the *overall isothermal efficiency* defined as the ratio of the isothermal work to the actual brake work.

The reversible isothermal work cannot be obtained by putting $n = 1$ in (16.11), since this would yield the indeterminate solution 0/0. The anomaly arises from the fact that $\int dv/v$ is a special case of $\int dv/v^n$, for which the solution is $\ln v$. A repetition of the derivation of equation (16.2) for the special case $n = 1$ can be shown to yield

$$W_{\text{iso}} = -m_f p_1 v_1 \ln \frac{p_2}{p_1} = -m_f R T_1 \ln \frac{p_2}{p_1} \tag{16.15}$$

This is the quantity used as the numerator in the expression for the isothermal efficiency (16.14).

When the air is compressed non-isothermally, i.e. when $1 < n < \gamma$, the extra work expended is not necessarily wasted. If the high pressure air were to be used directly in an expander with the same polytropic index as the compressor, all the extra work could ideally be recovered. There are two main reasons why recovery is not feasible in practical applications. First, for n to lie between 1 and γ during an expansion heat has to be supplied *to* the air. But this heat cannot be drawn from the cooler surroundings, and some means would have to be provided for transferring the heat rejected during the compression directly to the expander in a reversible manner. Secondly, the compressed air is normally used some distance away from the reciprocating compressor and it cools to atmospheric temperature in the delivery pipe. In fact the air is sometimes cooled deliberately immediately after compression, to condense the water vapour always present in atmospheric air and permit it to be drained off. This cooling has the incidental advantage of reducing the air viscosity and volume flow and hence the friction losses in the delivery pipe.

One other point noted in section 16.1 was the importance of the clearance volume in so far as it limits the mass flow which can be passed by a machine of given cylinder volume and speed. This consideration gives rise to another quantity used to assess the performance of a compressor—the *volumetric efficiency*. The volumetric efficiency is defined as the ratio of the actual volume induced per cycle to the swept volume. In terms of easily measured quantities it is given by

$$\eta_{\text{vol}} = \frac{m_f v_1}{N V_s}$$

The quantity $m_f v_1$ is known as the *free air delivery*; it is the rate of volume flow measured at inlet (e.g. atmospheric) pressure and temperature.

An expression for the volumetric efficiency of a reversible compressor can be deduced as follows. With reference to Fig. 16.7,

$$\eta_{\text{vol}} = \frac{V_b - V_a}{V_s} = \frac{V_s + V_d - V_a}{V_s}$$

$$= 1 + \frac{V_d}{V_s} - \frac{V_d}{V_s} \times \frac{V_a}{V_d} = 1 - \frac{V_d}{V_s}\left\{\left(\frac{p_2}{p_1}\right)^{1/n} - 1\right\} \qquad (16.16)$$

Evidently with a fixed value of V_d/V_s, called the *clearance ratio*, the volumetric efficiency decreases with increase of pressure ratio. This is also obvious from Fig. 16.8a. At

Fig. 16.8   The effect of multi-stage compression on volumetric efficiency

some pressure ratio the flow will be zero, but even before this extreme value is reached it is more economical to split the compression into two or more stages, rather than simply increase the size of cylinder to obtain the desired flow.

The *p-V* diagram for a reversible two-stage machine is shown in Fig. 16.8b. Air is compressed to some intermediate pressure p_i in the low-pressure cylinder and is then transferred to the high-pressure cylinder for final compression to p_2. The intermediate pressure p_i is determined by the relative values of the swept volumes in the two stages and their respective clearance ratios (see Example 16.1). The volumetric efficiency is usually expressed in terms of the first stage swept volume V_s, i.e. $(V_b - V_a)/V_s$. If the whole compression were accomplished in the low-pressure cylinder, the dotted curve in Fig. 16.8b indicates that the volumetric efficiency would be reduced to $(V_b - V_a')/V_s$.

An improvement in volumetric efficiency is not the only benefit to be obtained by compounding the compression. Equation (16.12) indicates that the work required decreases as the inlet temperature T_1 decreases. If an *intercooler* is fitted between the stages, cooling the air leaving the first stage to T_1 before it enters the second stage, the work required to drive the second stage is reduced. Fig. 16.9 illustrates this—the shaded area representing the saving in work. Intercooling brings the compression nearer to the ideal of isothermal compression. With an infinite number of stages this ideal would be reached, but in practice more than three stages are not used unless unusually high pressure ratios are required.

The actual saving in work depends on the choice of intermediate pressure. It is easy to show that in the case of reversible polytropic compression the total work required is a minimum when the work is divided equally between the stages. This applies to any number of stages, but here it will only be proved for the two-stage

Fig. 16.9   The effect of intercooling on the work of a two-stage compressor

machine. With complete intercooling to T_1 at constant pressure, the total work required in the two stages is

$$W = W_{1i} + W_{i2}$$

$$= m_f \frac{n}{1-n} RT_1 \left\{ \left(\frac{p_i}{p_1} \right)^{(n-1)/n} - 1 \right\} + m_f \frac{n}{1-n} RT_1 \left\{ \left(\frac{p_2}{p_i} \right)^{(n-1)/n} - 1 \right\}$$

Differentiating with respect to p_i and equating to zero,

$$\frac{n-1}{n} \left(\frac{1}{p_1} \right)^{(n-1)/n} \left(\frac{1}{p_i} \right)^{1/n} - \frac{n-1}{n} p_2^{(n-1)/n} p_i^{(1-2n)/n} = 0$$

or

$$p_i = \sqrt{(p_2 p_1)} \tag{16.17}$$

It then follows that

$$p_i/p_1 = p_2/p_i, \quad \text{and} \quad W_{1i} = W_{i2} \tag{16.18}$$

357

Finally, it is worth noting two mechanical advantages of compounding which help to offset the additional complication and cost. (a) Only the small high-pressure cylinder need be designed to withstand the delivery pressure. (b) A multi-cylinder arrangement provides a less difficult balancing problem and entails the use of a smaller flywheel.

In the whole of the foregoing analysis of reciprocating machines it has been implied that the working fluid is only introduced to one side of the piston. The work capacity of the machine can be very nearly doubled if both sides of the piston are used; the machine is then said to be *double-acting*. The chief difference in construction between a single- and a double-acting machine is that the latter requires the addition of a piston rod and crosshead, as depicted in Fig. 16.10, because it is necessary to seal the

Fig. 16.10   The double-acting compressor

'open' end of the cylinder with a gland. The piston performs two complete machine cycles per revolution. To determine the indicated power of a double-acting machine, it is necessary to take indicator diagrams from each end of the cylinder, and evaluate each diagram separately. It must be remembered that the effective piston area [A in equation (16.8)] is not the same in each case—the cross-sectional area of the piston rod must be subtracted from the piston area when evaluating the indicated power associated with the crosshead side of the piston.

Example 16.1   A single-acting two-stage compressor with complete intercooling delivers 6 kg/min of air at 16 bar. Assuming an intake state of 1 bar and 15 °C, and that the compression and expansion processes are reversible and polytropic with $n = 1.3$, calculate the power required, the isothermal efficiency and the free air delivery. Also calculate the net heat transferred in each cylinder and in the intercooler.

If the clearance ratios for the low- and high-pressure cylinders are 0.04 and 0.06 respectively, calculate the swept and clearance volumes for each cylinder. The speed is 420 rev/min. (See Fig. 16.11.)

The pressure ratio per stage is

$$r_p = \frac{p_i}{p_1} = \frac{p_2}{p_i} = \sqrt{\left(\frac{16}{1}\right)} = 4$$

The work done in the two stages is

$$W = W_1 + W_2 = 2m_f \frac{n}{1-n} RT_1(r_p^{(n-1)/n} - 1)$$

$$= -2 \times \frac{6}{60} \times \frac{1\cdot3}{0\cdot3} \times 0\cdot287 \times 288(4^{0\cdot3/1\cdot3} - 1) = -2 \times 13\cdot50 = -27\cdot00 \text{ kW}$$

The isothermal work is

$$W_{iso} = -m_f RT_1 \ln\frac{p_2}{p_1}$$

$$= -\frac{6}{60} \times 0\cdot287 \times 288 \ln\frac{16}{1} = -22\cdot92 \text{ kW}$$

and the isothermal efficiency is

$$\eta_{iso} = \frac{W_{iso}}{W} = \frac{22\cdot92}{27\cdot00} = 84\cdot9 \text{ per cent}$$

The free air delivery is

$$V = m_f \frac{RT_1}{p_1} = 6 \times \frac{0\cdot287 \times 288}{100 \times 1} = 4\cdot96 \text{ m}^3/\text{min}$$

The temperature at the exit of each stage is

$$T_c = T_g = T_1 r_p^{(n-1)/n} = 288 \times 4^{0\cdot3/1\cdot3} = 397 \text{ K}$$

Hence the net heat rejected from each cylinder is found from the energy equation

$$Q = m_f c_p(T_c - T_1) + W_1 = m_f c_p(T_g - T_1) + W_2$$

$$= \frac{6}{60} \times 1\cdot005(397 - 288) - 13\cdot50 = -2\cdot55 \text{ kW}$$

and from the intercooler

$$Q = m_f c_p(T_1 - T_c) = \frac{6}{60} \times 1\cdot005(288 - 397) = -10\cdot95 \text{ kW}$$

The volumetric efficiency for each stage is found from (16.16) as

$$\eta_{vol,1} = 1 - 0\cdot04(4^{1/1\cdot3} - 1) = 0\cdot924$$
$$\eta_{vol,2} = 1 - 0\cdot06(4^{1/1\cdot3} - 1) = 0\cdot887$$

Hence for the first stage the swept volume is

$$V_b - V_d = \frac{\text{free air delivery}}{\text{speed} \times \eta_{vol,1}} = \frac{4\cdot96}{420 \times 0\cdot924} = 0\cdot012\ 78 \text{ m}^3$$

and the clearance volume is

$$V_d = 0\cdot04(V_b - V_d) = 0\cdot000\ 51 \text{ m}^3$$

Fig. 16.11

For the second stage the swept volume is

$$V_f - V_h = \frac{1}{r_p} \frac{\text{free air delivery}}{\text{speed} \times \eta_{\text{vol},2}} = \frac{1}{4} \times \frac{4 \cdot 96}{420 \times 0 \cdot 887} = 0 \cdot 003 \ 33 \ \text{m}^3$$

and the clearance volume is

$$V_h = 0 \cdot 06(V_f - V_h) = 0 \cdot 000 \ 20 \ \text{m}^3$$

It must be noted that points d and e need not coincide on the indicator diagram because they represent different states and different clearance masses Δm_{c1} and Δm_{c2}. In this particular case

$$V_e = V_h r_p^{1/n} = 0 \cdot 000 \ 20 \times 4^{1/1 \cdot 3} = 0 \cdot 000 \ 58 \ \text{m}^3$$

16.4 The steam engine

The reciprocating steam engine is a particular type of expander designed to produce work from steam expanding from a high to a low pressure. It is used as an alternative to a turbine in small steam plant (up to a few hundred kilowatts) where a turbine would be very inefficient, and where the inherent intermittent working of a reciprocating machine, which only allows moderate rates of flow and rotational speeds, is no disadvantage. (The reasons for the inefficiency of small turbines will become apparent after a reading of Chapter 19.) The range of expansion for the engine is governed by considerations of the thermodynamic cycle, and it will be remembered from Chapter 11 that a very wide range of pressure is required if a steam cycle is to be efficient. This requirement leads to multi-stage expansion in successive cylinders, but only the single-cylinder steam engine is described here in any detail.

There are two features peculiar to the steam engine which result in its ideal indicator diagram being different from the ideal diagram of an expander shown in Fig. 16.2a. The first lies in the fact that the temperature of the working fluid is always much greater than the surroundings. Thus, although the work done increases as the index of expansion n decreases, the best that can be hoped for in practice is that the expansion shall be isentropic. An index of expansion less than the isentropic index would imply heat reception from the surroundings. The second peculiarity arises because with steam at low pressures there is a very rapid increase of specific volume as the pressure is decreased isentropically, i.e. $-(dv/dp)_s$ is very large. With the pressure ranges employed in practice, this implies that a very long stroke would be required if the steam were to be expanded reversibly down to condenser pressure inside the engine cylinder. Such a long stroke is undesirable not only for structural reasons, but also because it would reduce the mean effective pressure of the machine cycle and so increase the proportion of the indicated power expended in overcoming mechanical friction. That complete expansion entails a decrease in mean effective pressure should be evident from the p-V diagram of Fig. 16.12, the mean effective pressure of the additional area 433' being so small.

Fig. 16.12 shows a sketch of a double-acting steam engine fitted with a simple slide valve which is actuated by an eccentric on the crankshaft. As with all expanders, mechanically operated valves are necessary, and this feature makes possible the opening and closing of ports at any desired point in the machine cycle. The events on one side of the piston are as follows, those on the other side being similar but displaced in time by half a revolution of the crankshaft.

Fig. 16.12   A double-acting steam engine cylinder with an ideal *p-V* diagram

1–2 *Admission:* When the piston is at outer dead centre the inlet port is opened, and a mass Δm_f of high pressure steam is admitted by the time the inlet port is closed at the point of *cut-off*. The steam admitted mixes with a mass Δm_c which remained in the clearance volume after completion of the previous cycle.

2–3 *Expansion:* The steam expands isentropically until the exhaust port is opened at the end of the stroke.

3–4 *Blow-down:* The pressure at the end of the expansion stroke p_3 is higher than the condenser pressure p_4. When the exhaust port opens some of the steam 'blows down' into the condenser until the pressure in the cylinder is p_4.

4–5 *Exhaust:* On its return stroke the piston displaces the steam left in the cylinder after the blow-down until the exhaust port closes at point 5.

5–1 *Compression or 'Cushioning':* The steam Δm_c left in the cylinder is compressed isentropically during the remainder of the stroke. Point 5 may be so chosen that the final pressure is equal to p_1, when Δm_c will mix reversibly with the next charge Δm_f.

The state path of the steam in the cylinder is depicted in Fig. 16.13. Admission 1–2 occurs ideally with no change in state, and the ensuing expansion follows an isentropic path 1, 2–3 up to the blow-down. Although the blow-down is an irreversible expansion process, it may be imagined to consist of two parts, one reversible and the other irreversible. First the steam in the cylinder expands isentropically, displacing some steam into the condenser: the expelled steam acquires some kinetic energy in the process. The state of the steam remaining in the cylinder is represented by point 4. Secondly, the kinetic energy of the expelled steam is dissipated irreversibly in the condenser; this part of the process is not shown on the *h-s* diagram. During the exhaust process 4–5 the state of the steam in the cylinder is unchanged, though the displaced

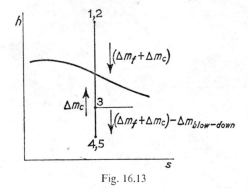

Fig. 16.13

steam will mix with steam in the condenser irreversibly. Finally, the mass Δm_c in state 4,5 is returned isentropically to state 1,2.

Due to the blow-down, as well as for other reasons elaborated below, the expansion of the mass Δm_f passing through the cylinder is not isentropic. The overall enthalpy drop $\Delta m_f(h_1 - h_6)$ is less than the isentropic drop $\Delta m_f(h_1 - h_5)$, if h_6 denotes the enthalpy of the exhaust steam in the condenser space before it has rejected any heat to the cooling water. The ratio between the actual and isentropic enthalpy drops is the *isentropic efficiency* of the engine. Example 16.2 shows how the effect of blow-down on the isentropic efficiency can be estimated.

Example 16.2 A steam engine cylinder has a swept volume of $0.03\ \text{m}^3$ and a clearance volume of $0.001\ \text{m}^3$. Cut-off is at 0.4 of the stroke. Saturated steam is admitted at 7 bar and expands isentropically to the blow-down. The mass which expands *inside* the cylinder during the blow-down may be assumed to do so isentropically. Cushioning is timed so that isentropic compression brings the clearance steam to the admission state. Calculate the work done per machine cycle and the isentropic efficiency when the engine exhausts at atmospheric pressure.

Using the notation of Fig. 16.12,

$$V_1 = 0.001\ \text{m}^3, \quad V_3 = V_4 = 0.031\ \text{m}^3$$
$$V_2 = V_1 + 0.4(V_3 - V_1) = 0.013\ \text{m}^3$$

During admission 1–2 the state is

$$p_1 = p_2 = 7 \text{ bar}, \quad v_1 = v_2 = 0.2728 \text{ m}^3/\text{kg}$$
$$u_1 = u_2 = 2573 \text{ kJ/kg}, \quad h_1 = h_2 = 2764 \text{ kJ/kg}$$
$$s_1 = s_2 = 6.709 \text{ kJ/kg °K}$$

During exhaust 4–5 the pressure is 1.013 bar, and the dryness fraction is found from $s_1 = s_5 = s_4$ to be 0.893. Then

$$v_4 = v_5 = 1.494 \text{ m}^3/\text{kg}$$
$$u_4 = u_5 = 2283 \text{ kJ/kg}, \quad h_4 = h_5 = 2434 \text{ kJ/kg}$$

At the beginning of the blow down, state 3, the specific volume is

$$v_3 = v_2 \left(\frac{V_3}{V_2}\right) = 0.2728 \times \frac{0.031}{0.013} = 0.6505 \text{ m}^3/\text{kg}$$

By trial and error from tables, or more quickly from the h–s chart, we find that in state 3, defined by $v_3 = 0.6505 \text{ m}^3/\text{kg}$ and $s_3 = s_2 = 6.709 \text{ kJ/kg K}$, $p_3 = 2.6 \text{ bar}$ and $x_3 = 0.939$. Hence $u_3 = 2417 \text{ kJ/kg}$.

The masses of steam in the cylinder at the points 1 to 5 are

$$m_1 = m_5 = \frac{V_1}{v_1} = 0.0037 \text{ kg}$$

$$m_2 = m_3 = \frac{V_2}{v_2} = \frac{V_3}{v_3} = 0.0477 \text{ kg}$$

$$m_4 = \frac{V_4}{v_4} = 0.0207 \text{ kg}$$

Cushioning must begin when

$$V_5 = m_5 v_5 = 0.0055 \text{ m}^3$$

The total work done per machine cycle can now be found from five non-flow processes, i.e

$W_{12} = p_1(V_2 - V_1) = 100 \times 7 \times 0.012 =$	$+8.40 \text{ kJ}$
$W_{23} = -m_2(u_3 - u_2) = 0.0477 \times 156 =$	$+7.44 \text{ kJ}$
$W_{34} =$	$0 \quad \text{kJ}$
$W_{45} = p_4(V_5 - V_4) = -100 \times 1.013 \times 0.0255 =$	-2.58 kJ
$W_{51} = -m_5(u_1 - u_5) = -0.0037 \times 290 =$	-1.07 kJ
$W =$	$+12.19 \text{ kJ}$

The isentropic work can be found in two ways. Assuming that full expansion is achieved by lengthening the stroke to 3′, corresponding to the exhaust state 4, 5

$$V_3' = m_3 v_4 = 0.0713 \text{ m}^3$$

Hence the total work is

$W_{12} =$	$+8.40 \text{ kJ}$
$W_{23'} = -m(u_3' - u_2) = 0.0477 \times 290 =$	$+13.83 \text{ kJ}$
$W_{3'5} = p_3'(V_5 - V_3') = -100 \times 1.013 \times 0.0658 =$	-6.67 kJ
$W_{51} =$	-1.07 kJ
$W_{\text{isen}} =$	$+14.49 \text{ kJ}$

We arrive at the same answer more easily using the steady-flow analysis. The flow steam is

$$m_f = m_2 - m_1 = 0.0477 - 0.0037 = 0.0440 \text{ kg}$$

and

$$W_{\text{isen}} = -m_f(h_3' - h_1) = 0.0440 \times 330 = +14.52 \text{ kJ}$$

The isentropic efficiency is

$$\eta_{\text{isen}} = \frac{W}{W_{\text{isen}}} = \frac{12.19}{14.52} = 0.84$$

Note that the mean effective pressure for the complete reversible expansion is

$$(p_m)_{\text{isen}} = \frac{W_{\text{isen}}}{V_3' - V_1} = \frac{14.52}{100 \times 0.0703} = 2.07 \text{ bar}$$

and for the irreversible expansion,

$$p_m = \frac{W}{V_4 - V_1} = \frac{12.19}{100 \times 0.030} = 4.06 \text{ bar}$$

Thus in practice, when mechanical friction is taken into account, the irreversible expansion would probably yield a greater brake power.

It has already been mentioned that the ports may be opened and closed at any desired point in the cycle. The point of cut-off and the point where cushioning begins require special comment. If the cut-off is too early in the expansion stroke the work

Fig. 16.14   The effect of varying (a) cut-off and (b) cushioning on the *p-V* diagram

done per cycle will be unduly small, and if too late, the additional work will not be commensurate with the additional amount of steam admitted (see Fig. 16.14a). The work done per machine cycle can also be increased by delaying the onset of cushioning, Fig. 16.14b, but again this will be at the expense of an increase in steam consumption. This is because the pressure of the clearance steam after cushioning is then below the steam main pressure, and when the inlet port is opened steam rushes in to raise the cylinder pressure in an irreversible manner.

The indicator diagram taken from an actual engine differs from the ideal diagram
of Fig. 16.12 for a number of reasons. There are the usual differences already mentioned in section 16.3 associated with the finite time required for opening and closing
valve ports and the pressure drops incurred by fluid friction as the steam enters and
leaves the cylinder. There is another important difference associated with the fact
that it is not possible to eliminate heat transfers between the steam and the cylinder.
The steam undergoes a considerable change in temperature during its passage through
the machine but, due to the resistance to heat flow between fluid and cylinder, the
cylinder temperature fluctuates between much narrower limits—this is indicated in
Fig. 16.15. Evidently the temperature difference between the steam and cylinder
varies in such a way that heat is rejected by the steam during admission and early
expansion, while heat is received during the latter part of the expansion and during the

Fig. 16.15

exhaust process. Even though the cylinder be well lagged, or steam jacketed, and the
heat rejected to the surrounding atmosphere is small, the cylinder itself acts alternately
as a sink and source of heat.

The effect of the heat exchanges is not very marked if superheated steam is being
used, but it becomes very pronounced if the steam is only just dry at entry. Heat
rejection during admission is then manifested by condensation of steam on the cylinder
walls. Once condensation occurs there is a sudden and very large increase in the rate
of heat transfer,* and hence the condensation process is accelerated. Heat transfer
from the cylinder during the latter part of the expansion and exhaust stroke re-
evaporates some of the condensate, but too late for the resulting vapour to do much
work. The main effect of the re-evaporation is an increase in exhaust pressure due
to the increased specific volume of the steam flowing through the exhaust port. The
net result of all these heat exhanges is an increase in steam consumption and decrease
of work output.

Both the loss of work associated with the failure to expand down to condenser
pressure in the cylinder, and that due to the heat transfers, become more pronounced

*A condensing vapour exchanges heat with a surface at a rate which may be several hundred times larger
than that for a dry vapour or gas—see Fig. 22.25.

as the pressure range is increased. More efficient operation can then be achieved by compounding, i.e. by carrying out the expansion in two or more stages. The large volume of low-pressure steam can be handled in a large diameter cylinder without the need for an excessively long stroke, and at the same time the range of temperature variation in each cylinder is only a fraction of the total temperature change of the steam. Sometimes, to keep the low-pressure cylinder within reasonable dimensions, the low-pressure steam is split between two cylinders in parallel. The expansion may be completed down to the condenser pressure if a turbine is added as a final stage. Even if the plant is of relatively low power output, a turbine can be quite efficient if it is called upon to deal with the steam at the low-pressure end where the volume flow is large. The combination of a reciprocating engine and turbine has been used frequently for medium-size marine plant.

The power output of a steam engine can be controlled in one of two ways. From what has been said about the choice of the point of cut-off (Fig. 16.14a), it will be

Fig. 16.16   Cut-off and throttle governing

clear that one possibility is the use of a special valve gear which will permit the timing of the cut-off to be varied; this is called *cut-off governing*. The second method, *throttle governing*, consists of throttling the steam in a regulating valve before admission to the steam chest, thereby reducing the inlet pressure and effective enthalpy drop available from the expansion. The effect of both methods is illustrated in Fig. 16.16 for the simple case where the clearance volume is zero. In the cut-off governed engine the reduction in work is obtained by a reduction in the mass of steam admitted and the *ideal* enthalpy drop per unit mass is unchanged. In the throttle governed engine the work is reduced by an irreversible reduction of steam chest pressure, which results in an appreciable reduction in available enthalpy drop per unit mass, and the mass of steam admitted is only slightly reduced by virtue of the fact that the specific volume of the steam is increased. The former is clearly the more efficient method of governing, and is always preferred for such applications as steam locomotives where the engine is not speed governed. The heavy valve gear required for varying the point of cut-off is unsuitable for use with the normal type of speed governor, and where constant speed is required (e.g. for pumping plant) the ease with which a simple throttle valve can be operated by a governor outweighs considerations of fuel economy. Reference to Fig. 16.16a will indicate another reason why cut-off governing is used for traction engines. By operating the engine non-expansively a rectangular diagram is obtained. This provides the particularly large torques required for starting purposes, and the disproportionate increase in steam consumption is of no significance for such a short duration.

16.5 Rotary positive-displacement compressors

In the type of compressor discussed earlier, the component which displaces the fluid from the low- to the high-pressure region performs a reciprocating motion. Here we

Fig. 16.17   The Roots blower

shall discuss compressors in which the displacement is performed by a rotating component. The rotary machine can operate at high speeds and can therefore handle larger rates of flow than its reciprocating counterpart—or is smaller for a given flow. The processes are for this reason more nearly adiabatic than in reciprocating compressors, and the indicated isothermal efficiency will be lower. To some extent this is offset by the fact that the mechanical efficiency is usually higher than for reciprocating machines, thus giving similar *overall* isothermal efficiencies for both classes of compressor. Rotary compressors are particularly suitable as superchargers for internal-combustion engines and as vacuum pumps, and they are often arranged in stages.

There are several distinct designs of rotary compressor, and one of the simplest and best known is the *Roots blower* shown in Fig. 16.17. Its essential parts are two rotors, each rotor having two or more lobes with profiles formed by complete epicycloids and hypocycloids. This geometric form ensures that at all angular positions the high-pressure space is sealed from the low-pressure space. One rotor is coupled to the source of power, the other being driven from it by gears external to the casing (not through the lobes). There is always a slight clearance between the meshing lobes to avoid wear at the sealing surfaces.

The mode of action is as follows. In the position shown, a volume V_s of air, at atmospheric pressure say, is being trapped between the left-hand rotor and the casing.

Fig. 16.18

Assuming adiabatic conditions, this trapped volume does not change its state until the space V_s is opened to the high-pressure region. At that instant some high-pressure air will rush back from the receiver and mix irreversibly with the air in the blower until the pressure is equalised. The air is then displaced into the receiver. This happens four times per complete revolution and the free air delivery of the compressor is thus $4V_s$ per revolution. The air flow into the receiver is intermittent, even though the rotors revolve with uniform speed.

To see how the work done in this irreversible flow process may be calculated, let us consider the reciprocating compressor of Fig. 16.18 with no clearance and a swept volume of V_s. After drawing in air at atmospheric pressure p_1, the inlet valve is closed and the delivery valve opened immediately afterwards with the piston still at inner dead centre. Back-flow from the receiver will equalise the pressure in the receiver and cylinder by an irreversible process. We shall consider the receiver to be of infinite volume compared with the volume V_s, i.e. that the equalised pressure will be p_2, which is the steady pressure of the air leaving at section 2. This compressor performs the same process as a Roots blower with an infinite receiver.* The work required per four revolutions of the reciprocating machine, equivalent to that required per revolution of the Roots blower, is

$$W = -(p_2 - p_1)4V_s \qquad (16.19)$$

Note that this is greater than the work required to drive a reversible adiabatic compressor.

Fig. 16.19a shows a *vane pump*, whose mode of action differs slightly from that of the Roots blower. Volumes V_a, V_b, V_c, V_d, are trapped between the vanes and casing,

(a)   (b)

Fig. 16.19   The rotary vane compressor

the vanes sliding in slots in the eccentrically placed rotor. Consider the volume V_a of air which has just been trapped at atmospheric pressure p_1. When it reaches the position of volume V_d, which is on the verge of being uncovered, it will have been reduced in size due to the asymmetrical placing of the inlet and outlet ports; thus $p_d > p_1$ assuming adiabatic conditions. In practical applications the pressure p_d is normally less than the delivery pressure p_2, and further compression is obtained by an irreversible back-flow of air from the receiver as in the Roots blower. An idealised equivalent reciprocating compressor of swept volume V_s equal to V_a is shown in Fig. 16.19b. The initial part of the compression from p_1 to p_d is reversible, and after opening the delivery valve at d, further compression to p_2 is achieved irreversibly.

*The analysis for the case of a finite receiver is complicated, and for this the reader is referred to Ref. (17).

The total work per N revolutions, being equivalent to the work required per revolution by a vane pump with N vanes, is

$$W = N \frac{\gamma}{1 - \gamma} p_1 V_a \left\{ \left(\frac{p_d}{p_1} \right)^{(\gamma - 1)/\gamma} - 1 \right\} - N(p_2 - p_d) V_d \qquad (16.20)$$

The vane compressor is said to have internal compression due to the reduction in volume from V_a to V_d, a feature which is absent in the Roots blower. By comparing the p-V diagrams in Figs. 16.18 and 16.19, it is evident that internal compression reduces the work required. The *Lysholm compressor*, which is similar to the Roots blower but having helical lobes, possesses a high degree of internal compression. It is described fully in Ref. (9).

17 Reciprocating Internal-Combustion Engines

There are two principal classes of reciprocating internal-combustion (or I.C.) engine. They are distinguished mainly by whether the combustion is initiated by a spark, or spontaneously by virtue of the rise in temperature during the compression process. Petrol and gas engines are in the former class and they will be referred to here as

Fig. 17.1   The S.I. engine

spark-ignition (or S.I.) engines. The term compression-ignition (or C.I.) engine will be used to refer to the second class, although the terms Diesel and oil engine are also common.

After describing the mode of action of some typical S.I. and C.I. engines, suitable criteria of performance are discussed, and the factors which limit the performance of practical I.C. engines are outlined. This is followed by a comparison of the real processes of an I.C. engine with the equivalent air-standard cycle. The chapter ends with a brief discussion of the limits that are imposed by the laws of thermodynamics on the conversion of chemical energy into work.

17.1 The working of I.C. engines

Both S.I. and C.I. engines can be designed so that one complete cycle of events in the cylinder is completed during either four or two strokes of the piston. The main events occurring in a *four-stroke* petrol S.I. engine are illustrated in Fig. 17.1.

1–2 *Induction stroke:* A mixture of air and petrol vapour is drawn into the cylinder, the pressure of the mixture being a little below atmospheric due to friction in the induction pipe. The inlet valve closes at the end of the stroke.

Fig. 17.2

2–3 *Compression stroke:* With both inlet and exhaust valves closed, the air-fuel mixture is compressed. Just before the piston reaches outer dead centre, ignition is effected by an electric spark. Combustion is not instantaneous, but occupies a finite period of time. Nevertheless much of it occurs at nearly constant volume because the piston is moving relatively slowly near dead centre.

3–4 *Expansion or working stroke:* Combustion is nominally completed at the beginning of the expansion stroke, and the products expand until the exhaust valve opens just before the end of the stroke. As the valve opens, the gases blow down the exhaust duct until the pressure in the cylinder has fallen to approximately atmospheric pressure.

4–1 *Exhaust stroke:* The products which have not escaped from the cylinder during the blow-down are displaced by the piston. To provide the pressure difference necessary to overcome friction in the exhaust duct, the pressure in the cylinder is slightly above atmospheric. At the end of the stroke there will be some residual gases in the clearance volume, which will dilute the next charge drawn into the cylinder.

The events in a four-stroke C.I. engine, illustrated in Fig. 17.2, differ from those in the S.I. engine in the following manner.

1–2 *Induction stroke:* Air alone is admitted to the cylinder.

2–3 *Compression stroke:* The air is compressed, and towards the end of the stroke liquid fuel is sprayed into the cylinder. The temperature of the air at the end of compression is sufficiently high for the droplets of fuel to vaporise and ignite as they enter the cylinder.

3–4 *Expansion or working stroke:* Since the fuel is sprayed into the cylinder at a controlled rate, the pressure may remain fairly constant during combustion. In modern high-speed engines, however, there is often a rise in pressure during the initial stages of combustion. At some point in the expansion stroke the combustion is nominally complete and the pressure then falls steadily until the exhaust valve is opened and blow-down occurs. This is followed by the exhaust stroke as in the S.I. engine.

When the *two-stroke* cycle is employed, the induction and exhaust strokes are eliminated. The piston stroke must be longer, because part of the compression and

Fig. 17.3   The two-stroke engine

expansion strokes are used for the processes of exhaust and induction. Fig. 17.3 shows a section of a common type of two-stroke engine. Apart from the self-acting spring-loaded valve on the crank case, no valves are required, and the piston itself is used to open and close the inlet and exhaust ports. In Fig. 17.3a the piston is shown near the end of the compression stroke. The upward motion has decompressed the crank-case and air has been admitted through the self-acting valve to the crank-case. During the expansion stroke the air in the crank-case is compressed, and near the end of this stroke the exhaust port is uncovered to allow the hot gases to blow down the exhaust duct. Further movement of the piston uncovers the inlet port, and compressed air from the crank-case flows into the cylinder. The exhaust and inlet ports are open simultaneously for a short period so that the incoming air can assist in clearing the cylinder of combustion products. The machine cycle is finally completed by compression of the air trapped in the cylinder. The two-stroke engine can be built as a S.I. engine, with petrol vapour and air being admitted to the crankcase, or as a C.I. engine.

A full discussion of the relative merits of two-stroke and four-stroke engines would be out of place here, but the following points are worth noting.

(a) The two-stroke engine does not require any mechanically operated valves. This implies not only considerable mechanical simplification, but also that the engine can be run in either direction. The two-stroke C.I. engine has established itself in marine practice primarily because it eliminates the need for a reversing gear.

(b) For the same rotational speed, the two-stroke engine has twice the number of working strokes. The power output from an engine of given bulk and weight is therefore greater for a two-stroke engine, although for a number of reasons it is not twice as great. The greater frequency of working strokes in the two-stroke engine also implies a more uniform torque and a smaller flywheel.

(c) The two-stroke engine tends to have a higher fuel consumption than the four-stroke engine. Little time is available for induction and exhaust in the two-stroke engine, so that the process of clearing the combustion products from the cylinder and recharging it is less complete. Moreover, less time is available for the fresh charge of air to cool the cylinder, and overheating may be a more serious problem. In two-stroke S.I. engines some fuel may pass through the cylinder without combustion because of the necessary overlap of the periods when the inlet and exhaust ports are open.

A description of the general working of I.C. engines would not be complete without the mention of a rotary variant, a recent design of which is known as the *Wankel engine*. It is akin to the two-stroke engine in that it requires no valves. The diagrammatic sketch in Fig. 17.4 shows the principles of operation. The triangular rotor A, having internal teeth meshing with a central gear fixed to the engine casing, performs a planetary motion around this gear. An eccentric B, integral with the power output shaft, revolves in a recess of rotor A at three times its speed. The shape of the outer casing is such that the apexes of the triangular rotor seal off three distinct spaces 1, 2 and 3. The shape of the curved surfaces of rotor A is unimportant, except in so far as it determines the maximum and minimum volumes of the sealed spaces, and therefore the compression ratio. The curved surfaces are recessed as shown, mainly to avoid 'throttling' of the flame spreading from the spark plug through the combustion space.

The succession of events is not difficult to follow. Fig. 17.4a shows the situation where a charge is being induced into space 1, the preceding charge is fully compressed in space 2, and the combustion products are being exhausted from space 3. Fig. 17.4b represents the situation a little later with the new charge occupying the maximum volume and about to be compressed in space 1, combustion gases expanding in space 2 to provide the 'working stroke', and space 3 being reduced in volume and thus exhausting the products through the outlet port.

The rotary engine has one important feature in common with its reciprocating counterpart; it is a positive-displacement machine and the processes occur in a cyclic

Fig. 17.4 The Wankel engine

intermittent manner. We therefore consider it here for the same reason that the Roots blower and vane pump were considered in the chapter on reciprocating compressors. Its obvious advantages *vis-à-vis* the reciprocating engine are that it is capable of handling fairly large mass flows for its size and thus has a larger power to weight ratio, and also that its rotor can be perfectly balanced by the addition of appropriate masses.

Three disadvantages of the rotary design are apparent. Firstly, because the processes occur in a sequence of positions around the casing, the surfaces near the spark plug are intermittently cooled by the *compressed* charge whereas the reciprocating engine cylinder is intermittently cooled by the colder *uncompressed* charge. Furthermore, the casing surface adjacent to space 3 is permanently exposed to the expanding burnt gases. This implies the need for good local cooling arrangements. Although the problem is eased by the high surface/volume ratio of the combustion space, it does of course result in high heat losses. Secondly, the long narrow shape of the combustion space is not good from the point of view of using high compression ratios, as will become clear after the discussion of detonation in section 17.3 (i). Thirdly—and this is the reason why the many attempts to develop a rotary engine over the years have proved unsuccessful—there is an inherently difficult sealing problem at the apexes and flat side faces of the triangular rotor. It is too early to say to what extent, with further development, the rotary engine is likely to displace the reciprocating engine from the numerous fields of application where the latter is well established. Certainly a C.I. version of the rotary design, possible in principle, will also have to be developed before the rotary engine can hope to capture a major part of the market. Further details of the Wankel engine can be found in Ref. (22).

17.2 Criteria of performances for I.C. engines

It was explained in section 15.6 that the overall efficiency of a power plant is usually expressed as

$$\eta = \frac{W}{Q_{net,p}}$$

where W is the net output per unit mass of fuel supplied and $Q_{net,p}$ is the net calorific value of the fuel at constant pressure. The net output of an engine is called the *brake power* (b.p.) and for this reason the overall efficiency is termed the *brake thermal efficiency* η_b. The latter is defined by

$$\eta_b = \frac{b.p.}{m_f \times Q_{net,p}} \tag{17.1}$$

The *specific fuel consumption* is frequently used as an alternative criterion of performance. It is defined as the rate of fuel consumption per unit of brake power, i.e.

$$s.f.c. = \frac{m_f}{b.p.} \tag{17.2}$$

Evidently, for any given fuel the s.f.c. is inversely proportional to η_b. Most common hydrocarbon fuels have very similar calorific values, and the s.f.c. can therefore be used when comparing the economy of engines using different fuels.

In a reciprocating engine an appreciable part of the loss is due to mechanical friction, and it is informative if the thermodynamic and mechanical friction losses are separated.

The overall efficiency is therefore analysed as the product of the *indicated thermal efficiency* η_i and the *mechanical efficiency* η_m. Thus

$$\eta_b = \left(\frac{\text{i.p.}}{m_f \times Q_{\text{net},p}}\right)\left(\frac{\text{b.p.}}{\text{i.p.}}\right) = \eta_i \times \eta_m \qquad (17.3)$$

The *indicated power* (i.p.) is the actual rate of work done by the working fluid on the piston, and the difference between the i.p. and the b.p. is the power absorbed by mechanical friction (piston, bearings, etc.). As its name implies, the i.p. can be determined from an *indicator diagram*. The procedure has been explained in section 16.3. The only point of difference here is that with a four-stroke engine the area of the small anticlockwise loop (Fig. 17.5) must be subtracted from the area of the main diagram when determining the *indicated mean effective pressure* (i.m.e.p.). The negative work of the small loop is termed the *pumping loss*—it is the result of viscous friction in the

Fig. 17.5

induction and exhaust strokes. The physical equation equivalent to (16.8) for the i.p. is

$$\text{i.p.} = p_m LAN \qquad (17.4)$$

Here N is the number of machine cycles per unit time, viz. half the rotational speed for a four-stroke engine, and the rotational speed for a two-stroke engine.

Fig. 17.6

When an engine is run at constant speed, the variation of η_b and η_m with load appears as in Fig. 17.6. At constant speed the power expended on mechanical friction—termed the friction power (f.p.)—is approximately constant. Therefore, as the b.p. is reduced the mechanical friction loss exercises an increasing influence on the efficiencies. At zero b.p. all the fuel is consumed in overcoming friction, and both η_b and η_m are zero. With an engine of good mechanical design, η_m should be between 80 and 90 per cent at full load. η_b depends upon the type of engine, but at full load it is usually 25–35 per cent for S.I. engines and 30–40 per cent for C.I. engines.

When discussing air-standard cycles for reciprocating engines it was suggested that the mean effective pressure provides a useful guide to the relative size of an engine. In practice it is usual to use the concept of *brake mean effective pressure* (b.m.e.p.) for this purpose. The b.m.e.p., which may be denoted by p_{mb}, is defined by an equivalent equation to (17.4), i.e.

$$\text{b.p.} = p_{mb} LAN \qquad (17.5)$$

377

The b.m.e.p. may be regarded as that part of the i.m.e.p. which is imagined to contribute to the b.p., the remainder being used to overcome friction. An increase in b.m.e.p. implies that the cylinder volume (*LA*) can be smaller for a given output. As an illustration, the values of b.m.e.p. used in S.I. car engines have steadily increased through the years from 4·6 bar (when the R.A.C. rating was introduced) to approximately 10 bar. The corresponding b.p. obtained per litre of cylinder swept volume has increased from about 10 to 30 kW. This considerable reduction of specific engine size is not due solely to the increase in b.m.e.p. but is partly due to a simultaneous increase in permissible engine speed. In the following section we shall discuss the more important factors which limit the efficiency and b.m.e.p. of S.I. and C.I. engines.

17.3 Factors limiting the performance of I.C. engines

When discussing gas turbine power plant it was pointed out that the performance was limited by the maximum gas temperature that can be permitted in the engine.

Fig. 17.7

This 'metallurgical limit' is imposed because the turbine is continuously exposed to high temperatures under conditions of severe stress. Provided that proper attention is paid to the cooling of the cylinder (particularly the cylinder head) and exhaust valves, no such limitation is encountered in reciprocating engines where the working stresses are much lower. The reason for this is indicated in Fig. 17.7 which shows qualitatively the temperature variation of the gas in an engine cylinder. The peak gas temperature may be as high as 2800 K, but this temperature is reached due to intermittent combustion and not due to a transfer of heat through the cylinder wall. The metal temperature therefore fluctuates between much narrower limits. Moreover, due to cylinder cooling, and the insulating properties of a gas boundary layer and a film of lubricating oil which are adhering to the wall, the mean metal temperature is well below the mean gas temperature.

There are several factors of a different nature limiting the performance of reciprocating engines and, as they differ for S.I. and C.I. engines, each type of engine will be considered separately.

(i) *The S.I. Engine*

The analysis of the air-standard Otto cycle showed that the cycle efficiency improves with increase of compression ratio. This is also true of the brake thermal efficiency of S.I. engines. There is an upper limit to the compression ratio that can be employed, however, which has its origin in the fact that the fluid compressed in the cylinder is a mixture of fuel vapour and air. The temperature of the combustible mixture increases on compression, and if the compression ratio is too high it is possible for self-ignition to occur before spark-ignition. The onset of combustion at too early a point in the compression stroke would lead to a serious loss of power and must be avoided.

Even a compression ratio sufficiently low to avoid *pre-ignition* may be too high for efficient operation. This is due to the phenomenon of *detonation*. Although a considerable effort has been made to determine the exact nature of detonation in an engine

Fig. 17.8   Combustion in a S.I. engine

cylinder, it is still a controversial topic. There are undoubtedly several phenomena involved and such terms as 'knocking' and 'pinking' are often used in an attempt to distinguish between them. Here we can only give a very brief and over-simplified picture, with the aid of Fig. 17.8.

Under normal conditions combustion begins near the spark plug after a short delay following the formation of the spark, and the flame spreads through the mixture with a rapid but finite velocity. The energy released by the spark is only small, and it sets off a relatively slow flameless reaction which occupies the *delay period*. Further energy is released at an accelerating rate by the reaction, until a proper flame front develops. (The delay period is very short—approximately 0·002 second—but it may correspond to an appreciable movement of the crank in a high-speed engine. For example, at 3000 rev/min the crank turns through 36 degrees in 0·002 second. This is the reason why the point of ignition is always well in advance of outer dead centre.)

As the flame front spreads in a uniform manner across the combustion space, it compresses the unburnt part of the mixture before it. The temperature of the unburnt portion is raised by intense radiation from the flame as well as by the compression. If the initial pressure and temperature are too high, the unburnt portion appears to reach a critical condition resulting in self-ignition at numerous points in the unburnt 'end-gas'. A violent pressure rise is then produced which may result in a knocking noise. This phenomenon of self-ignition in the 'end-gas', though occurring after the spark, is akin to the phenomenon of pre-ignition. It should not be confused with detonation proper.

Strictly speaking, the term detonation refers to phenomena associated with the sudden acceleration of a flame front. As the flame front progresses at a moderate rate through the mixture, a critical condition is sometimes reached at which a violent pressure wave, i.e. a shock wave, is formed. The violence of the shock wave so accelerates the rate of reaction that the flame front keeps pace with the shock wave and they traverse the cylinder together at a very high velocity. This rapidly moving front is termed a *detonation wave*, and it is reflected and re-reflected by the cylinder walls; it results in a noise—though usually of a higher frequency than that produced by self-ignition in the 'end-gas'.*

The main objection to detonation of any kind is that it causes excessive turbulence which removes most of the insulating gas boundary layer. It therefore increases the rate of heat transfer between gas and cylinder wall and may result in serious overheating of the engine. Even if this be avoided, continuous detonation may cause the formation of local hot spots which eventually lead to pre-ignition. Excessive turbulence also results in a reduction of power output. From the mechanical point of view, detonation may cause deterioration of the bearing surfaces in a lightly built engine. Lastly, the appearance of carbon in the exhaust gases when detonation occurs indicates that it results in incomplete combustion. For all these reasons a compression ratio sufficiently high to cause detonation must be avoided.

Some fuels permit the use of a higher compression ratio than others,† and there are compounds ('dopes') which can be added to petrols to suppress the tendency to detonate. It is difficult to devise a measure of the detonating tendency of a fuel. One method is to compare the performance of a fuel with that of a heptane-octane mixture in a special variable compression-ratio engine. n-heptane (C_7H_{16}) is a fuel liable to give detonation at low compression ratios, while iso-octane (C_8H_{18}) detonates at much higher ratios: different mixtures of the two provide a continuous *octane number* scale from 0 to 100 per cent in octane content. The compression ratio of the engine is raised until the fuel under test detonates, and the highest percentage of octane in a heptane-octane mixture that produces detonation at the same compression ratio is called the octane number of the fuel. The need for a test engine of agreed design arises

---

*A discussion of various theories can be found in Refs. (10) and (16).

†Fuel oils have a marked tendency to detonate and therefore cannot be used in S.I. engines. Apart from this, oils do not evaporate as readily as petrols and homogeneous combustible mixtures are not easily obtained.

from the fact that the size and shape of the combustion space and the location of the spark plug are also important factors governing detonation, since they determine the length of time during which the 'end gas' is exposed to the advancing flame front. For the same reason, in an engine of given size a large number of small cylinders is preferable to a small number of large cylinders. Over the last fifty years, developments in 'doped' fuels and combustion chamber design [Ref. (12)] have succeeded in raising the permissible compression ratio of ordinary car engines from about 4·5 to 9.

With the compression ratio fixed, the other factor limiting the performance is the rise in pressure obtainable during combustion. The i.m.e.p. increases with increase in peak pressure, and an increase in i.m.e.p. is desirable because it implies that the engine will be smaller for a given power output and that a relatively smaller share of the i.p. will be absorbed in overcoming mechanical friction. The peak pressure is fixed by the amount of fuel that can be burnt and hence by the amount of oxygen available. Theoretically it will be a maximum if a stoichiometric mixture is used. Stoichiometric air-fuel ratios by weight are nearly identical for all common liquid fuels, i.e. about 14·5:1. In practice, maximum brake thermal efficiency is achieved with a slightly weak mixture giving complete combustion of the fuel, and maximum power with a slightly rich mixture giving complete combustion of the oxygen (Fig. 17.9).

Fig. 17.9 Effect of mixture strength on engine performance

The variation of efficiency with load has already been mentioned in the previous section, and the method of regulating the power output deserves some mention here. At first sight it would seem that the work output could be reduced by weakening the mixture and thereby reducing the rise in pressure during combustion. Unfortunately, reliable ignition by a spark is only possible over a fairly narrow range of air-fuel ratio—18:1 being about the weakest limit. It follows that variation of the air-fuel ratio is quite unsuitable as a method of governing S.I. engines. Instead, approximately stoichiometric ratios are used at all loads, and regulation is achieved by reducing the mass of mixture induced in each working cycle. This method is referred to as *throttle* or *quantity governing*. Throttling results in sub-atmospheric pressure in the cylinder during the induction stroke, and an increase in the 'pumping loss' (Fig. 17.10). Since the exhaust process always occurs at atmospheric pressure, the dilution of the fresh charge with residual products is more serious at part load. In fact, at very low loads, mixtures richer than stoichiometric must be supplied to ensure satisfactory ignition. Evidently throttle governing is an inefficient method, and the part load efficiency of S.I. engines is poor for this reason.

The normal method of achieving the appropriate mixture of air and fuel is by means of a *carburettor*, which consists of a venturi throat in the induction manifold, and one or more fuel jets through which petrol can be introduced into the air stream at this section. The pressure depression at the throat, whose magnitude depends on the air flow, causes petrol to be sucked from the jets into the air stream. From what has been said previously it is clear that the normal requirement from a carburettor is an approximately stoichiometric mixture at most speeds and throttle settings in order to work near the point of maximum efficiency. But in addition a rich mixture is required (a) at full throttle to attain maximum power, and (b) at very low throttle settings when there is marked dilution by residual products. Carburettors usually satisfy these requirements only moderately well, and moreover it is found that they produce non-homogeneous mixtures of air and fuel and uneven distribution to the various cylinders. Fitting a multi-cylinder engine with several carburettors may improve this situation, but such an arrangement requires frequent adjustment. Solutions giving

Fig. 17.10 Effect of throttle governing on the indicator diagram

more positive control have been sought, using injection pumps and nozzles capable of metering the correct amounts of petrol—preferably into individual cylinders. *Direct petrol injection* has been used successfully on several engines but, because of the high cost of injection equipment compared with that of carburettors, it has been restricted to the more expensive car engines.

One way of overcoming our apparent inability to govern an S.I. engine by weakening the mixture, is by charging the cylinder with a controlled non-uniform mixture. Part of the cylinder is filled with a layer of ignitable mixture, while the remainder is filled with air. In this way ignition can be achieved with an average charge well below stoichiometric ratio. This method of filling a cylinder with a non-homogeneous air-fuel mixture is called *stratified charging*. Although engines have been successfully governed in this way under laboratory conditions, no commercial engine using this principle has yet been developed. A successful engine of this type would provide a major break-through in improving the part-load, and therefore overall, economy of S.I. engines.

(ii) *The C.I. engine*

Since the fuel is introduced into the cylinder of a C.I. engine only when combustion is required, i.e. towards the end of the compression stroke, pre-ignition cannot occur. Moreover, since the fuel is injected at a controlled rate, the simultaneous combustion of the whole quantity of fuel cannot occur, and the problem of detonation as in S.I. engines does not arise. It follows that the compression ratio can be much higher in C.I. engines, and in fact there is a lower limit of about 12:1 below which compression-ignition of common fuel oils is not possible.

In a C.I. engine most of the combustion occurs at substantially constant pressure, and the maximum pressure reached in the cylinder is largely determined by the compression ratio employed. The upper limit of compression ratio is therefore fixed by the strength of the cylinder, the bearings, and other parts whose stresses are determined by the peak pressure forces. Evidently the designer must reach a compromise between high efficiency and low weight and cost. From the analysis of the air-standard Diesel cycle, summarised in Fig. 12.16, it is evident that compression ratios in excess of 20:1 give diminishing returns, and this is in fact about the maximum value used in practical engines. A comparison of the air-standard efficiencies of the Diesel and Otto cycles (Fig. 12.16), when made at appropriate compression ratios, suggests that the C.I. engine will be the more efficient. This is borne out in practice, and in general C.I. engines are more efficient than S.I. engines.

We must next consider a factor limiting the i.m.e.p. of C.I. engines. It is evident from the *p-V* diagram that the mean effective pressure increases with increase in the change of volume which occurs during the combustion at constant pressure (Fig. 17.2). This change in volume is a maximum when all the fuel that can be burnt in the oxygen available is injected into the cylinder. At first sight it appears that this implies the use of a stoichiometric air-fuel ratio ($\approx 14.5:1$) as in S.I. engines. To see why this is not so, it is necessary to outline briefly the way in which combustion is accomplished in C.I. engines.

Although the fuel is injected as a deeply penetrating spray of fine droplets, the fact that it is in liquid form means that the delay period between injection and initiation of combustion is appreciably longer than in S.I. engines. Each droplet must first partially evaporate until sufficient vapour diffuses into the surrounding air to form a layer having an air-fuel ratio which is combustible. This vaporisation cools the droplet and surrounding layer, and time is required for the necessary temperature to be re-established. Combustion then begins simultaneously at many points in the combustion space, and there is an initial rapid and uncontrolled rise of pressure. As the cylinder fills with combustion products, the rate of reaction becomes more controlled and the pressure remains sensibly constant. If the delay period is too long, there is time for a large fraction of the charge to enter the cylinder, and a rather violent initial combustion may occur resulting in rough running and the so-called *Diesel knock*.

Diesel knock is thus seen to be a function of the fuel, and it also depends on droplet size, the rate at which droplets are introduced into the cylinder, and the heat transferred between droplets and their environment. If the latter parameters are kept substantially constant, by using a standard engine as is done when measuring octane numbers, the Diesel knock tendency can be measured by matching the fuel with a two-component fuel-oil mixture. For this purpose cetane ($C_{16}H_{34}$) and alpha-methyl-naphthalene ($C_{11}H_{10}$) are used to establish the so called *cetane number scale*. It is found, not surprisingly, that a high octane number fuel implies a low cetane number and vice versa, and this is one of the principal reasons why petrols are unsuitable for C.I. engines.

Apart from reducing the delay period to avoid Diesel knock another difficulty, particularly in the later stages of injection, is to ensure that all the fuel molecules collide with oxygen molecules to complete the process of combustion. To assist the mixing process, some turbulence or swirl is deliberately stimulated by careful design of the inlet port and combustion chamber. The turbulence must not be excessive, otherwise the initial uncontrolled rise of pressure may be too violent, and the work in the expansion stroke may be reduced unduly. What is required, therefore, is a controlled movement of the bulk of the air, and not small-scale random turbulence. Ideally the spray of fuel should be carried with the air through one complete sector between successive sprays by the time combustion is completed (Fig. 17.11a). Unfortunately this approach requires several small fuel jets (which may become choked), and very high injection pressures to achieve good fuel 'atomisation'. Another way of solving this problem (Fig. 17.11b) is to have some kind of separate *swirl chamber*

Fig. 17.11

into which the fuel is sprayed in a relatively coarse spray. Break-up of the fuel and mixing with oxygen is achieved as a result of the intense vortex motion produced as the air is forced into the chamber during the compression stroke. In spite of much ingenuity expended on combustion chamber design [Ref. (14)], it is not possible to burn all the fuel completely if stoichiometric ratios are used, and the minimum air-fuel ratio, corresponding to full load, usually lies between 18:1 and 25:1. Consequently the engine must be larger than would be the case were it to be capable of burning stoichiometric ratios. The swirl-chamber design, while allowing the use of air-fuel ratios down to 18:1, also results in higher heat losses during combustion and a lower efficiency than direct injection.

When comparing the size and weight of C.I. and S.I. engines, three main factors must be born in mind: utilisation of air, cylinder peak pressure and thermal efficiency. The C.I. engine, being incapable of using stoichiometric ratios and working with high peak pressures, tends to be larger and heavier than an equivalent S.I. engine although the former has a higher thermal efficiency.

Turning to the question of governing, it is found that very weak mixtures can be ignited and burnt in a C.I. engine, so that it is possible to reduce the power output by reducing the fuel supply. Reduction of the fuel supply is equivalent to a reduction of the cut-off ratio r_c of the air-standard Diesel cycle, and Fig. 12.16 shows that this improves the cycle efficiency. In practice, although an improvement in the indicated thermal efficiency may be achieved at part load, the fall in mechanical efficiency more than outweighs this effect and the brake thermal efficiency always falls off. Nevertheless the reduction in efficiency with decrease of load is not so marked as in S.I. engines. Governing the power output by weakening the mixture is usually referred to as *quality governing*.

So far in our discussion of both S.I. and C.I. engines we have considered factors limiting the work done per complete cycle*—but the actual power output will also depend on the rotational speed. It would seem that, other things being equal, the power output of an engine can be raised by increasing the speed—at least up to the limit imposed by purely mechanical considerations of stress and piston sliding velocities. In practice, however, it is found that the maximum indicated work per engine cycle varies considerably with speed, and beyond a certain speed the i.p. may even fall with further increase in speed. Typical curves are shown in Fig. 17.12. The reduction in indicated work per cycle at high speeds is chiefly due to a reduction in the mass of the charge induced during each engine cycle. Theoretically, an unsupercharged engine should induce a mass equivalent to the swept volume at ambient pressure and temperature. As a result of fluid friction and expansion of

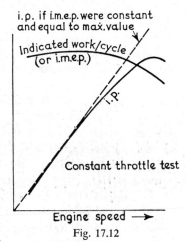

Fig. 17.12

the charge in the hot inlet manifold, considerably less is taken in at high engine speeds when gas velocities are high and the manifold is very hot. (This is only a rough generalisation, and whether or not the mass induced falls with increase in speed depends also upon the valve timing and the design of the inlet and exhaust manifolds.)

The 'breathing' capacity of an engine is expressed in terms of a *volumetric efficiency* defined as

$$\frac{\left[\begin{array}{c}\text{Volume of induced charge per induction stroke}\\ \text{reckoned at a reference pressure and temperature}\end{array}\right]}{\text{Swept volume of piston}} \qquad (17.6)$$

The reference pressure and temperature should logically be taken as the ambient values prevailing when the volumetric efficiency is determined, but sometimes they are assumed to be at some arbitrary level such as 0 °C and 1 atm. When comparing

---

*The work done per engine cycle is, of course, proportional to the mean engine torque, i.e. it equals $4\pi \times$ torque for the four-stroke engine and $2\pi \times$ torque for the two-stroke engine.

volumetric efficiencies care should be taken to ensure that they are based on the same definition.

Fig. 17.12 shows that the indicated work per engine cycle also falls at low speeds. This is due partly to increased leakage past valves and piston for which more time is available at low speed. It is also due to reduced turbulence and the increased cooling time available, which make the combustion process less efficient. At very low speeds the combustion tends to be erratic, and below a certain speed the engine will not run at all.

Other things being equal, the total power developed in an engine cylinder depends upon the mass of the charge induced, and from the earlier remarks it might be supposed that this is limited by the prevailing ambient pressure and temperature of the air. In S.I. engines the induction temperature is lowered slightly by the evaporative cooling effect of liquid fuels. This is the reason for the superiority of alcohol over petrol as a fuel—its latent heat is about three times that of petrol and the improved volumetric efficiency is sufficient to outweigh the disadvantage of the calorific value being only about half that of petrol. Sometimes a liquid of high latent heat, such as methanol, is injected into the induction manifold to lower the temperature still further; this device has been used in some aircraft engines for short periods of thrust boosting. Finally by utilising resonance effects in the inlet and exhaust manifolds, a certain amount of pressure charging can be achieved at selected speeds.

A much greater increase in charge density can be obtained, however, by precompressing the charge before induction with some type of *supercharger*. A Root blower or a centrifugal compressor is commonly used for this purpose. The supercharger may be gear driven from the crankshaft, or its power may be derived from a turbine driven by the exhaust gases.

Although the net increase in power obtained by supercharging can be very considerable, it has little effect on the brake thermal efficiency. The reason is that the fuel supply must be increased in proportion to the increase in air charge in order to maintain the required air-fuel ratio. Owing to its greater mechanical complexity, a supercharged engine is not always preferred to its larger and heavier unsupercharged equivalent. Nevertheless, supercharging has become established practice in two important fields. Firstly, a reciprocating aircraft engine must be supercharged at high altitudes, as otherwise the power falls off due to reduced atmospheric pressure. Secondly, two-stroke C.I. engines are usually supercharged because there is the added advantage that the cylinders can be more effectively cleared—or *scavenged*—of combustion products.*

Enough has been said to indicate the main factors which influence the design of an engine. In practice the design must be a compromise between thermodynamic and mechanical considerations which are closely interlinked, and for detailed discussions the reader is referred to specialised texts on the subject such as Refs. (4), (21) and (27). A good discussion of combustion in S.I. and C.I. engines is presented in Ref. (19).

*A useful survey of supercharging practice is given in Ref. (15).

17.4 Comparison of real I.C. engine processes with air-standard cycles

That an analogy exists between the processes in real I.C. engines and the equivalent air-standard Otto, Diesel or mixed cycles has been suggested in Chapter 12, but further explanation is required. The analogy rests on the fact that the p-v diagram for the air-standard cycle is similar in shape to the p-V diagram obtained from the engine, i.e. that the m.e.p. of the cycle is approximately equal to the i.m.e.p. of the engine. This implies that the indicated work W_i produced by an engine inducing a mass m of air is approximately equal to mW, where W is the net work per unit mass of air taken through the air-standard cycle. The energy of reaction made available for work, by burning a mass m_f of fuel is approximately $m_f \times Q_{net,p}$. Ideally this is equal to the heat input $mc_v(T_3 - T_2)$ in the Otto cycle (see Fig. 12.14) or $mc_p(T_3 - T_2)$ in the Diesel cycle (see Fig. 12.15). It follows that the indicated thermal efficiency of the engine is approximately equal to the air-standard efficiency; e.g. for the S.I. engine,

$$\eta_i = \frac{W_i}{m_f \times Q_{net,p}} = \frac{mW}{mc_v(T_3 - T_2)} = \text{Otto cycle efficiency}$$

In practice this analogy is far less close than the comparable analogy between the processes in the open-cycle gas turbine and the air-standard Joule cycle. The reasons for the poorer analogy in the case of the reciprocating engine are as follows.

(1) Since the processes are carried out intermittently in a single component they are not so clearly defined, e.g. the corners of the p-V diagram are rounded owing to the time taken for valves to open and close, and combustion does not occur precisely at constant volume (or constant pressure as the case may be).
(2) The temperature changes involved in the expansion and compression are much larger than in gas turbines, and the consequent variation of specific heats is appreciable.
(3) The air-fuel ratios used in reciprocating engines are much smaller, so that the properties of the working fluid are markedly different from those of air.
(4) The maximum temperature reached at the end of the combustion process is much higher than in gas turbines, and dissociation plays an important role. This implies that the calorific value of the fuel is used much less effectively for the production of work.
(5) Owing to the use of cylinder cooling and the relatively low mass flow per unit surface area in a reciprocating machine, the compression and expansion processes cannot be so nearly described as adiabatic.

Nothing further need be said about item (1), but the remaining points require amplification. This will be done by considering the compression, combustion and expansion processes in turn.

(a) *Compression process:* The ratio of specific heats γ decreases with increase of temperature, and its mean value for the temperature range normally encountered in the compression process is appreciably less than 1·4 even if the gas is air. The mixing of the incoming charge with residual products of combustion results in a lower value for γ for the mixture, because the products contain CO_2 and H_2O which individually have a relatively low value of γ. In S.I. engines the gas compressed is an air-fuel mixture, and this too has a lower value of γ than air. Finally, heat transfers from the cylinder to the gas in the early stages of the compression, and from the gas to the cylinder in the later stages, mean that the compression does not follow an isentropic law. The net result of all these effects is that the pressure and temperature at the end of the compression are appreciably less than the values predicted by air-standard calculations (see Example 17.1).

(b) *Combustion process:* Due to the high temperature reached in the combustion process there is an appreciable proportion of dissociated products at the nominal end of combustion. The result is that the full calorific value of the fuel is not released and the actual temperature rise is less than would be predicted if dissociation were ignored. The calculation of the adiabatic temperature rise in a chemical reaction was discussed in sections 15.4 and 15.7.

Fig. 17.13

(c) *Expansion process:* As the temperature during the expansion falls, some of the dissociated molecules combine, and strictly speaking the combustion process still continues during the expansion. By the end of the expansion the temperature has fallen to a level where the proportion of dissociated products is very small, and the fuel can then be regarded as completely burnt. Evidently the unused part of the calorific value is recovered to some extent, although the energy recovered in this way is not so effective for the production of work. This is because the increased pressures resulting from the completion of the reaction are available only over part of the stroke. The efficiency with which the calorific value is converted into work is therefore decreased by the dissociation occurring in the early stages of combustion. The expansion is also markedly different from the isentropic expansion of air because the hot combustion products have a low value of γ and the expansion is far from adiabatic. An appreciable quantity of energy is rejected as heat, particularly during the period of intense chemical activity when radiation from the burning gases plays an important role.

The effect of all these factors on the p-V diagram of an S.I. engine is shown in Fig. 17.13. The net result is a considerable reduction in both mean effective pressure and efficiency. It is possible to analyse the processes in I.C. engines taking the more important of these factors into account and thermodynamic charts enabling this to be done have been produced by Hottel et al., Ref. (3). The air-standard cycle is a simple model adequate only for predicting the effect of compression ratio on the performance of an engine. The more complex models can predict the effect of other parameters such as the air/fuel ratio. Nevertheless, such calculations are more useful in showing the relative importance of the various 'losses' than for the prediction of an actual engine's performance. This is because other variables, such as the design of the induction system, or those mentioned in the first of the five points listed earlier, also play an important part in determining the performance.

Example 17.1   A charge enters a S.I. engine at 330 K and 1 bar, and is compressed isentropically through a compression ratio of 7:1. Estimate the temperature and pressure at the end of compression, taking the charge to be (a) pure air with constant specific heats, (b) a stoichiometric mixture of air and octane (C_8H_{18}) with variable specific heats. (Neglect the effect of dilution with residual products of combustion.)

(a) With air, the temperature and pressure at the end of compression are given by

$$T_2 = T_1 r_v^{\gamma-1} = 330 \times 7^{0.4} = 719 \text{ K}$$

$$p_2 = p_1 \frac{v_1 T_2}{v_2 T_1} = 1 \times 7 \times \frac{719}{330} = 15.25 \text{ bar}$$

(b) The combustion equation for octane is

$$C_8H_{18} + 12.5\,O_2 + 3.76 \times 12.5\,N_2 \rightarrow 8\,CO_2 + 9\,H_2O + 3.76 \times 12.5\,N_2$$

Hence the air-fuel ratio by mass is

$$\frac{(12.5 \times 32) + (3.76 \times 12.5 \times 28)}{114} = 15.05$$

The value of γ for the mixture will vary with temperature. Assuming that T_2 found in (a) is approximately correct, a mean value of γ for the compression can be found from the properties of the mixture at $(330+719)/2 \approx 525$ K. The value of the gas constant R is

$$R = \left(\frac{15.05}{16.05} \times 0.287\right) + \left(\frac{1}{16.05} \times \frac{8.314}{114}\right) = 0.274 \text{ kJ/kg K}$$

At 525 K, c_p for air is 1.035 and for octane 2.691 kJ/kg K. Hence for the mixture

$$c_p = \left(\frac{15.05}{16.05} \times 1.035\right) + \left(\frac{1}{16.05} \times 2.691\right) = 1.138 \text{ kJ/kg K}$$

$$\gamma = \frac{c_p}{c_p - R} = \frac{1.138}{0.864} = 1.317$$

The temperature and pressure after compression are therefore

$$T_2 = 330 \times 7^{0.317} = 612 \text{ K}$$

$$p_2 = 1 \times 7 \times \frac{612}{330} = 13.0 \text{ bar}$$

A better estimate can now be obtained by taking the properties of the constituents at the mean temperature of $(330+612)/2 = 471$ K, when for the mixture $c_p = 1.115$ kJ/kg K and $\gamma = 1\cdot326$. T_2 is then found to be 622 K and p_2 13·2 bar. Evidently a further approximation is not necessary.

17.5 Maximum work of a chemical reaction

It has been seen that there are practical limits to the work output which can be obtained from the combustion of a fuel in an internal-combustion engine in which the reaction is rapid and highly irreversible. Of rather more academic interest is the question what is the maximum possible work that can be derived from any particular chemical reaction. Since the power is derived from a non-cyclic process while heat is exchanged with a single reservoir (i.e. the surrounding atmosphere), and not from a cycle working between two reservoirs, the Carnot efficiency cannot be a guide in answering this question.

In effect the question posed has been answered in section 15.7, but the following paragraphs discuss this problem more explicitly. For an efficient conversion of chemical energy into work it is essential for the reaction to be carried out reversibly. A hypothetical reversible I.C. plant has already been described in section 15.7 (i)—the van't Hoff equilibrium box. The particular van't Hoff box described therein performs the reaction $CO + \frac{1}{2}O_2 \rightarrow CO_2$ reversibly while exchanging heat with one reservoir. In doing so it produces the maximum possible amount of work, and hence it rejects the least possible amount of heat. The reaction proceeds isothermally, indicating that a high temperature at some point in a combustion process is not essential for maximum work output. The reason why the efficiency of a real I.C. engine increases with increase of compression ratio, and hence of peak temperature, is therefore not obvious. It lies in the fact that in the hypothetical reversible engine the chemical energy is transformed directly into work, which is not the case in a conventional I.C. engine. The rapid combustion occurring in the latter results in the chemical energy being first of all transformed into random molecular energy (i.e. sensible internal energy) as manifested by the rise in temperature accompanying the combustion process. Apart from the chemical change of some of the components, this irreversible combustion has the same effect of increasing the sensible internal energy of the gases (mainly nitrogen) as an external heat addition at the appropriate temperatures. Once the intermediate transformation into sensible internal energy is accepted as inevitable, the analogy between the real process and the equivalent air-standard cycle has shown that the practical I.C. engine efficiency will improve with increase of peak temperature (i.e. compression ratio). It should not be forgotten, however, that a reversible isothermal reaction would result in a much higher proportion of the chemical energy being transformed into work.

A numerical example may be helpful. The expression for the maximum work which can be done in any reversible process by the reactants CO and O_2, exchanging heat with a single reservoir at temperature T, was found in section 15.7 (ii) to be

$$W_{\max} = R_0 T \left\{ \ln \frac{p_{CO_2}}{p_{CO}(p_{O_2})^{1/2}} + \ln (p_1)^{1/2} \right\} \text{ per kmol of CO}$$

The reactants are assumed to enter in stoichiometric proportions at pressure p_1 and temperature T, with the product CO_2 leaving at the same pressure and temperature. With all pressures in atmospheres, and p_1 taken as one atmosphere, the second term in the bracket is zero. For a temperature $T = 298$ K, the equilibrium constant is found from tables to be

$$K = \frac{p_{CO_2}}{p_{CO}(p_{O_2})^{1/2}} = 10^{45 \cdot 07} \text{ (atm)}^{-1/2}$$

Hence

$$W_{max} = 8 \cdot 314 \times 298 \ln (10)^{45 \cdot 07} = 257\ 100 \text{ kJ/kmol CO}$$

Since the enthalpy of reaction, ΔH_{25}, is $-283\ 000$ kJ/kmol CO, it is evident that W_{max} is very nearly equal to $(-\Delta H_{25})$. The heat transferred to the surroundings can be found from the steady-flow energy equation

$$Q = \Delta H_{25} + W_{max}$$
$$= -283\ 000 + 257\ 100 = -25\ 900 \text{ kJ/kmol CO}$$

The work W_{max} need not necessarily be less than the negative of the enthalpy of reaction. If a similar calculation is carried out for the reversible combustion of a stoichiometric mixture of C and O_2, it can be shown that W_{max} is greater than $(-\Delta H_{25})$; in this case heat is received from the surroundings during the process. The definition of efficiency that is normally used for I.C. plant, whether $W/(-\Delta H_0)$ or $W/Q_{net,p}$, is therefore not logically sound because theoretically it can be greater than unity. A process efficiency should be defined as a ratio X/X_{max}, where X and X_{max} are the actual and maximum possible values of the physical quantities concerned in the process. A more rational definition would be W/W_{max}.

It is found that for hydrogen and the usual hydrocarbon fuels the net calorific value $Q_{net,p}$ is numerically close to W_{max}, while the gross calorific value $Q_{gr,p}$ differs from W_{max} by a much wider margin. Hence

$$\frac{W}{W_{max}} \approx \frac{W}{Q_{net,p}}$$

and this provides the principal argument in favour of defining thermal efficiencies by equation (15.14) rather than (15.13). It is interesting to note that the value of W_{max}, unlike that of ΔH_0 or calorific value, does not depend on the phase of H_2O in the products or of the fuel, and this is proved formally in section 7.9 (iii).

One final comment: we have seen that there is no fundamental thermodynamic reason why the efficiency of an I.C. engine should not be 100 per cent, and therefore present-day utilisation of our fuel resources is far removed from the ultimate standard of perfection. Because an appreciable fraction of the work produced by the world's power plant is used for the production of electricity, an improvement in fuel utilisation would undoubtedly follow if we could convert chemical (and nuclear) energy directly into electrical energy in a reversible reaction. Although there are some chemical reactions which allow chemical → electrical energy conversions with a high degree of reversibility, as in the lead-acid accumulator and the hydrogen-oxygen 'fuel-cell' (see section 20.4), no equally efficient device has yet been developed for common fuels and air. Any such device would almost certainly not permit high *rates* of energy transformation: the advantages of irreversible combustion at high temperature is that very high power outputs can be produced by relatively small power plant.

18 One-Dimensional Steady Flow and Jet Propulsion

The steady-flow energy equation was deduced in Chapter 4 as a consequence of the First Law of Thermodynamics. The equation relates the heat and work transfers across the boundary of an open system, to the mechanical and thermodynamic properties of the fluid at the inlet and outlet where these properties can be assumed constant across the cross-section of flow. No assumption had to be made about whether the flow was frictionless or not, and the derivation did not require any knowledge of the flow *within* the open system. We have seen that this equation enables a useful range of problems to be solved. For example, when considering adiabatic flow in a nozzle, a knowledge of the initial enthalpy, initial velocity and final enthalpy enables the final velocity to be calculated. If the adiabatic flow is also assumed to be frictionless, the fact that the entropy is constant—a consequence of the Second Law of Thermodynamics—enables the final thermodynamic state to be predicted. The principle of Conservation of Mass—in the form of the continuity equation—then enables the inlet and outlet areas to be determined for any given mass flow.

The foregoing analysis, however, has nothing to say about how the cross-sectional area of the nozzle must change between inlet and outlet for the results to be physically possible. Before this kind of information can be obtained from the analysis, some knowledge of the flow within the open system is evidently necessary. This extension of the analysis is undertaken in the science of *fluid dynamics*.

With the advent of high-speed flight, turbo-jet engines, ram-jets and rockets, fluid dynamics has been enormously extended in recent years—particularly the branch dealing with compressible flow known as *gas dynamics*. Gas dynamics is now studied as a subject in its own right, and can no longer be adequately covered by a chapter in a book of this kind. All that will be attempted here is the consideration of a few special cases with the object of emphasising the point at which a simple thermodynamic analysis must be reinforced by a more complete treatment. Such a treatment (a) must involve some assumption about the flow within the open system, and (b) may involve *explicitly* Newton's Second Law of Motion. A case where item (a) is necessary and sufficient will be considered first before any attempt is made to explain what is meant by item (b).

18.1 Isentropic flow in a duct of varying area

When a fluid flows through a duct with no work or heat transfer, the only factors which can cause a change in fluid properties are (i) a change in flow area and (ii) frictional forces. In this section we shall assume that frictional forces are absent. The effect of variation in the cross-sectional area of the duct may be determined if some assumption is made about the nature of the flow.

The simplest type of flow conceivable is *one-dimensional* steady flow. With reference to Fig. 18.1, the flow is said to be one-dimensional if the following conditions are fulfilled:

(a) Changes in area and curvature of the axis are gradual.

(b) Thermodynamic and mechanical properties are uniform across planes normal to the axis of the duct.

No real flow is truly one-dimensional, but provided that there are no sudden changes in direction or area, and that average values of the properties at any cross-section

Fig. 18.1

are used in the analysis, the one-dimensional treatment yields results which are sufficiently accurate for many purposes.

When the flow is one-dimensional the properties are uniform over any cross-section of the flow within the open system—not merely at inlet and outlet. It is therefore possible to apply the energy equation to the flow between any two planes 1 and 2, and so, for given inlet conditions, to determine the properties at any section along the duct. Consider the case of isentropic flow in a nozzle (or diffuser). Given p_1 and T_1 (or x_1 if the fluid is a wet vapour), the thermodynamic properties at various values of p_2 can be found using the equation of state (or tables of properties) together with the fact that s_2 equals s_1. Given also the initial velocity C_1, corresponding values of C_2 can be found from the energy equation

$$h_2 + \frac{C_2^2}{2} = h_1 + \frac{C_1^2}{2}$$

Finally, given the inlet cross-sectional area A_1, corresponding values of A_2 can be found from the continuity equation

$$\frac{A_2 C_2}{v_2} = \frac{A_1 C_1}{v_1}$$

The variation of v_2, C_2 and A_2 with p_2 can then be represented by curves of the form shown in Fig. 18.2. Although the precise shape of the curves depends upon the equation of state of the fluid, all gases and vapours exhibit the same general behaviour under these conditions. One point of interest is the very gradual change in specific volume v at low velocities. For some analyses of low-speed gas flow it is possible to simplify the calculations without much loss of accuracy by assuming that the specific volume is constant. This realm of fluid dynamics is called *incompressible flow* and the treatment is then identical for liquids and gases.

Fig. 18.2 can be interpreted as applying to either a nozzle or diffuser. Moving from left to right on the diagram the pressure p_2 is falling and the duct acts as nozzle. Although both the specific volume and velocity increase continuously during the expansion, the flow area decreases to a minimum and then increases. A glance at the continuity equation shows that this will happen if the rate of increase of specific volume is at first less than, and finally greater than, the rate of increase of velocity. Such behaviour is always exhibited by gases and vapours. The pressure ratio at which the minimum area is reached in called the *critical pressure ratio*; at this point the mass flow per unit area is a maximum.

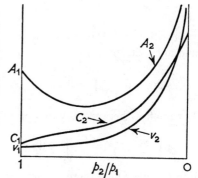

Fig. 18.2   Variation of v, C add A with pressure along a duct

i) Derivation of the basic equations

It is more informative if the foregoing step-by-step method of calculation is replaced by the analytical approach which follows. First it is necessary to arrive at equations expressing the First and Second Laws, and the principle of conservation of mass (continuity), in differential form. Applying the steady-flow energy equation between any two planes an infinitesimal distance apart we arrive at the differential energy equation

$$dQ = dh + d\left(\frac{C^2}{2}\right) + dW \tag{18.1}$$

When applied to the simple case of adiabatic flow in a nozzle or diffuser, (18.1) reduces to

$$dh + d\left(\frac{C^2}{2}\right) = 0 \tag{18.2}$$

Applying the Second Law to the flow between planes an infinitesimal distance apart, we have the result that for a reversible process

$$dQ = T\,ds$$

We also have the general relation between properties, equation (6.13), which states that

$$T\,ds = dh - v\,dp$$

395

For an isentropic process therefore,

$$\mathrm{d}h = v\,\mathrm{d}p \tag{18.3}$$

The differential form of the continuity equation is most easily obtained by logarithmic differentiation. Thus we have

$$\log A + \log C - \log v = \text{constant}$$

and differentiating,

$$\frac{\mathrm{d}A}{A} + \frac{\mathrm{d}C}{C} - \frac{\mathrm{d}v}{v} = 0 \tag{18.4}$$

Equations (18.2) to (18.4) can be used to find expressions for the changes in velocity and area with pressure in an isentropic process. Thus, combining (18.2) and (18.3),

$$v\,\mathrm{d}p + \mathrm{d}\left(\frac{C^2}{2}\right) = 0 \tag{18.5}$$

For the flow between any two planes 1 and 2 we therefore have

$$\tfrac{1}{2}(C_2{}^2 - C_1{}^2) = -\int_1^2 v\,\mathrm{d}p \tag{18.6}$$

For any given fluid there is a definite relation between p and v in an isentropic process, and when this is known in algebraic form the R.H.S. of the equation can be evaluated.

To find the change in area required to accommodate the flow, equation (18.5) can be substituted in (18.4) to yield

$$\frac{\mathrm{d}A}{A} = \frac{v}{C^2}\,\mathrm{d}p + \frac{\mathrm{d}v}{v}$$

$$= v\,\mathrm{d}p\left(\frac{1}{C^2} + \frac{1}{v^2}\frac{\mathrm{d}v}{\mathrm{d}p}\right) \tag{18.7}$$

It is shown in texts on fluid dynamics that the velocity of propagation of a small pressure wave through a fluid (i.e. a sound wave) is given by

$$a^2 = -v^2(\partial p/\partial v)_s \tag{18.8}$$

The bracketed term is the rate of change of pressure with specific volume at constant entropy.* Since equation (18.7) refers to isentropic flow, (18.8) may be used to substitute for $\mathrm{d}v/\mathrm{d}p$ to give

$$\frac{\mathrm{d}A}{A} = v\,\mathrm{d}p\left(\frac{1}{C^2} - \frac{1}{a^2}\right) \tag{18.9}$$

where a is the local velocity of sound in the fluid at the point in the flow where the pressure is p and the specific volume is v.

*It is conventional to use the density ρ in place of v in fluid dynamics; the expression is therefore usually written as $a^2 = (\partial p/\partial \rho)_s$.

Equation (18.9) may be interpreted directly without integration. Considering accelerated flow, decelerated flow, and constant velocity flow in turn, the most important consequences are as follows:

Accelerated flow (nozzle)
  dp must be negative so that:
  (1) When $C < a$, dA is negative, i.e. the duct must converge.
  (2) When $C > a$, dA is positive, i.e. the duct must diverge.

Decelerated flow (diffuser)
  dp must be positive so that:
  (1) When $C < a$, dA is positive.
  (2) When $C > a$, dA is negative.

Constant velocity flow
  dp is zero so that:
  (1) Whether $C \gtrless a$, dA is zero.

It follows immediately that if the flow is to be continuously accelerated from a subsonic to a supersonic velocity the duct must have a throat between the two regions, at which dA is zero and the velocity C is equal to the local sonic velocity a—see Fig. 18.3a. The same remarks apply to a flow continuously decelerating from a supersonic

Fig. 18.3   Flow in a duct of varying area

to a subsonic velocity as depicted in Fig. 18.3b. When there is not a continuous rise (or fall) of pressure, the presence of a throat does not necessarily imply that the velocity there is sonic: Fig. 18.3c illustrates the case of a venturi where a subsonic flow is accelerated and then decelerated without sonic velocity ever being reached. The condition at the throat is that of constant velocity flow with dp zero. One final point is worth noting from the foregoing analysis. It is that *the critical pressure ratio is that pressure ratio which will accelerate the flow to a velocity equal to the local velocity of sound in the fluid*, because this is the point in the flow where the area is a minimum and the mass flow per unit area a maximum.

Fig. 18.4

All these deductions have been made without reference to any particular fluid, and they apply equally to liquids and gases. The reason why it is never necessary in practice to use convergent-divergent nozzles or diffusers for liquids, is that the velocity of sound in a liquid is about 1500 m/s. Such a velocity is quite outside the range of practicable velocities of flow in view of the friction which would be involved.

The analysis can be carried further if we restrict our attention to certain types of fluid for which there is a simple algebraic relation between p and v in an isentropic process; equations (18.5) and (18.7) may then be integrated analytically. For a perfect gas we have the isentropic relation

$$pv^\gamma = \text{constant}$$

Although no such simple equation can be formulated for the isentropic expansion or compression of a vapour, it was pointed out in section 9.4 that a good approximation is provided by the same type of polytropic law, i.e.

$$pv^n = \text{constant}$$

If the vapour is steam, n is approximately 1·135 when most of the process occurs in the wet region* and 1·3 when mostly in the superheat region. To provide a uniform solution for gases and vapours, the symbol n will be used for the index of isentropic expansion. The resulting equations, however, will only be exact for perfect gases, i.e. for the case where $n = \gamma$.

To avoid obscuring the essentials by lengthy equations, we shall only consider the simple case of *a continuous expansion from an initial state where the velocity may be considered negligible*. Fig. 18.4 shows the nomenclature employed. Using the polytropic relation between p and v, and putting C_1 equal to zero, equation (18.6) gives the velocity at any plane 2 as

$$\tfrac{1}{2}C_2{}^2 = \frac{n}{1 - n}(p_2 v_2 - p_1 v_1)$$

*This statement will be qualified later in section 18.2.

and hence

$$C_2 = \left[\frac{2n}{1-n} p_1 v_1 \left\{\left(\frac{p_2}{p_1}\right)^{(n-1)/n} - 1\right\}\right]^{1/2} \tag{18.10}$$

Note that, for any given initial state, there is a maximum possible velocity which is achieved when the fluid is expanded to zero pressure. This maximum velocity is

$$C_{\max} = \left(\frac{2n}{n-1} p_1 v_1\right)^{1/2} \tag{18.11}$$

The curve representing C_2 in Fig. 18.2 does not, therefore, proceed to infinity. The required area at any plane 2 can be found by integrating (18.7), but it is more easily obtained by computing C_2 and v_2 and using the continuity equation directly. Thus, if m is the mass flow, we have

$$m = \frac{A_2 C_2}{v_2} \quad \text{or} \quad \frac{A_2}{m} = \frac{v_2}{C_2} = \frac{v_1}{C_2(p_2/p_1)^{1/n}} \tag{18.12}$$

An expression for the mass flow per unit area can be found by taking the reciprocal of (18.12) and substituting for C_2 from (18.10). This yields

$$\frac{m}{A_2} = \left[\frac{2n}{1-n} \frac{p_1}{v_1} \left\{\left(\frac{p_2}{p_1}\right)^{(n+1)/n} - \left(\frac{p_2}{p_1}\right)^{2/n}\right\}\right]^{1/2} \tag{18.13}$$

The mass flow per unit area is a maximum at the throat, and the pressure at the throat can therefore be found by differentiating (18.13) with respect to p_2 and equating to zero. The result is that m/A_2 is a maximum when

$$\frac{p_2}{p_1} = \left(\frac{2}{n+1}\right)^{n/(n-1)}$$

In this equation p_2 is the throat pressure p_t, and it is also referred to as the critical pressure p_c. The critical pressure ratio is therefore

$$\frac{p_t}{p_1} = \frac{p_c}{p_1} = \left(\frac{2}{n+1}\right)^{n/(n-1)} \tag{18.14}$$

The following tables gives the critical pressure ratio for four values of n:

n	1·135	1·300	1·400	1·667
p_c/p_1	0·577	0·546	0·528	0·487

The value of the maximum mass flow per unit area can be found by substituting (18.14) in (18.13) giving

$$\frac{m}{A_t} = \left\{n \left(\frac{2}{n+1}\right)^{(n+1)/(n-1)} \frac{p_1}{v_1}\right\}^{1/2} \tag{18.15}$$

The general result, that the velocity at the throat will be sonic, has already been deduced from equation (18.9); it is easy to check this for the special case of fluids which expand isentropically according to the law $pv^n = $ constant. Thus substituting (18.14) in (18.10), the throat velocity is

$$C_t = \left(\frac{2n}{1+n} p_1 v_1\right)^{1/2}$$

Using $p_1 v_1^n = p_t v_t^n$ in conjunction with (18.14), this can be transformed to

$$C_t = (np_t v_t)^{1/2} \tag{18.16}$$

Sonic velocity at the throat can be found from (18.8), since by differentiating the isentropic relation $pv^n = $ constant, we have

$$\left(\frac{\partial p}{\partial v}\right)_s = -\frac{np}{v}$$

and hence when $p = p_t$ and $v = v_t$,

$$a_t = (np_t v_t)^{1/2}$$

The required identity between C_t and the local sonic velocity is thereby established.

It must be emphasised once again that this analysis is for a *continuous* expansion. The pressure p_t and velocity C_t, given by the foregoing equations, are the minimum and maximum values respectively which can be achieved at the throat with the given initial conditions p_1 and v_1. Higher values of p_t, and lower values of C_t, are possible in a venturi where the pressure increases after the throat.

When the fluid is a perfect gas, we may substitute $n = \gamma$ and $pv = RT$ in the previous equations to obtain

$$C_2 = \left[\frac{2\gamma}{1-\gamma} RT_1 \left\{\left(\frac{p_2}{p_1}\right)^{(\gamma-1)/\gamma} - 1\right\}\right]^{1/2} \tag{18.17}$$

$$\frac{m}{A_2} = p_1 \left[\frac{2\gamma}{1-\gamma} \frac{1}{RT_1} \left\{\left(\frac{p_2}{p_1}\right)^{(\gamma+1)/\gamma} - \left(\frac{p_2}{p_1}\right)^{2/\gamma}\right\}\right]^{1/2} \tag{18.18}$$

$$\frac{p_t}{p_1} = \frac{p_c}{p_1} = \left(\frac{2}{\gamma+1}\right)^{\gamma/(\gamma-1)} \tag{18.19}$$

$$C_t = a_t = (\gamma RT_t)^{1/2} \tag{18.20}$$

$$\frac{m}{A_t} = p_1 \left\{\frac{\gamma}{RT_1} \left(\frac{2}{\gamma+1}\right)^{(\gamma+1)/(\gamma-1)}\right\}^{1/2} \tag{18.21}$$

A simple expression for the ratio T_t/T_1 can also be formulated in this instance since

$$\frac{T_t}{T_1} = \left(\frac{p_t}{p_1}\right)^{(\gamma-1)/\gamma} = \frac{2}{\gamma+1} \tag{18.22}$$

Equations (18.17) to (18.22) can be deduced more easily and directly by using the energy equation in the form

$$c_p T_1 = c_p T_2 + \tfrac{1}{2} C_2^2$$

but this is left as an exercise for the student.

Much of the preceding analysis has only been necessary to obtain an expression for the critical pressure ratio, equation (18.14). Most calculations on the design of nozzles are easily made from first principles once (18.14) has been established. The following example illustrates this.

Example 18.1 Determine the throat area, exit area and exit velocity for a steam nozzle to pass a mass flow of 0·2 kg/s when the inlet conditions are 10 bar and 250 °C and the final pressure is 2 bar. Assume that the expansion is isentropic and that the inlet velocity is negligible.

Fig. 18.5 illustrates the process on an *h–s* diagram. Using $n = 1·3$ when finding the critical pressure ratio, we have from (18.14)

$$\frac{p_t}{p_1} = \left(\frac{2}{1·3 + 1} \right)^{1·3/0·3} = 0·546$$

and hence

$$p_t = 0·546 \times 10 = 5·46 \text{ bar}$$

Since the expansion is isentropic,

$$s_t = s_1 = 6·926 \text{ kJ/kg K}$$

Fig. 18.5

The steam is still superheated at the throat because s_t is greater than the saturation value 6·794 kJ/kg K at the throat pressure of 5·46 bar. By a double interpolation in the superheat table, of which only the second step is given here, the enthalpy and specific volume at the throat are found to be

$$h_t = 2753 + (2854 - 2753) \frac{6·926 - 6·794}{7·018 - 6·794} \text{ (at 5·46 bar)}$$

$$= 2813 \text{ kJ/kg}$$

$$v_t = 0·3476 + (0·3916 - 0·3476) \frac{6·926 - 6·794}{7·018 - 6·794} \text{ (at 5·46 bar)}$$

$$= 0·3735 \text{ m}^3/\text{kg}$$

Now from the energy equation with $C_1 = 0$,

$$C_t = \{2(h_1 - h_t)\}^{1/2}$$

$$= \{2 \times 10^3 (2944 - 2813)\}^{1/2} = 511·9 \text{ m/s}$$

401

From the continuity equation the required throat area is

$$A_t = \frac{mv_t}{C_t} = \frac{0 \cdot 2 \times 0 \cdot 3735}{511 \cdot 9} \times 10^4 = 1 \cdot 46 \text{ cm}^2$$

The dryness fraction at exit may be found from

$$s_2 = s_1 = s_{f2} + x_2 s_{fg2}$$

$$x_2 = \frac{6 \cdot 926 - 1 \cdot 530}{5 \cdot 597} = 0 \cdot 964$$

Therefore

$$h_2 = 505 + (0 \cdot 964 \times 2202) = 2628 \text{ kJ/kg}$$
$$v_2 = 0 \cdot 964 \times 0 \cdot 8856 = 0 \cdot 8537 \text{ m}^3/\text{kg}$$

The exit velocity is therefore

$$C_2 = \{2 \times 10^3 (2944 - 2628)\}^{1/2} = 795 \cdot 0 \text{ m/s}$$

and the exit area required is

$$A_2 = \frac{0 \cdot 2 \times 0 \cdot 8537}{795 \cdot 0} \times 10^4 = 2 \cdot 15 \text{ cm}^2$$

[Note that the approximate expression $pv^{1 \cdot 3}$ = constant has only been used to determine the critical pressure ratio, and hence indirectly the throat area. The throat and exit velocities have been calculated from the correct isentropic enthalpy drop and not from $\int v \, dp$. Note also that such calculations are greatly simplified if values of the enthalpy drop and specific volume can be read directly from an h–s chart.]

(ii) *Phenomena in nozzles operating off the design pressure ratio*

This section has so far been devoted to establishing the area changes necessary for a continuous isentropic expansion from a given initial state to a given final pressure. It is also of interest to know what happens to the flow in a nozzle of given dimensions when the downstream pressure differs from the value for which the nozzle is designed. In this respect the behaviour of convergent and convergent-divergent nozzles is different, and the two types must be considered separately.

(a) *Convergent nozzle:* Consider the experimental set-up shown in Fig. 18.6. The inlet conditions are maintained constant and the downstream pressure p_e is gradually reduced from the inlet pressure p_1 by opening the valve.

When $p_e = p_1$ there is no pressure drop and no flow: curve (i). As p_e is reduced, the axial pressure distribution takes the form of curve (ii), and the mass flow increases as shown in Fig. 18.6b. The pressure p_2 in the plane of the nozzle exit is always equal to p_e. This state of affairs continues until p_e/p_1 is reduced to the critical pressure ratio p_c/p_1, corresponding to curve (iii). At this operating condition the nozzle passes the maximum mass flow, given by equation (18.15), and $p_2 = p_e = p_c$.

Further reduction of p_e cannot produce any change in the flow within the nozzle because we have seen that p_2/p_1 *cannot be less than the critical pressure ratio unless there is a throat upstream of plane* 2. Consequently, for all values of p_e less than p_c, the pressure p_2 in the plane of the exit remains constant at the critical pressure p_c. There is therefore no further change in specific volume or velocity at plane 2 and no change in mass flow. The mass flow remains at the maximum value, as shown in Fig. 18.6b, and the nozzle is said to be *choked*. On leaving plane 2 the fluid undergoes an unrestrained and irreversible expansion to p_e, as illustrated by curve (iv), and the

Fig. 18.6   Flow in a convergent nozzle with varying back pressure

flow is no longer amenable to simple one-dimensional treatment. What happens is that the pressure oscillates around p_e, as shown in Fig. 18.6c, and the supersonic velocity achieved in the initial divergence of the stream is rapidly reduced by viscous friction.

(b) *Convergent-divergent nozzle:* Fig. 18.7 shows a similar set-up with a convergent-divergent nozzle. When operating at the design pressure ratio, appropriate to the throat and exit areas of the nozzle, the flow accelerates continuously from a subsonic to a supersonic velocity. The axial pressure and velocity distributions are then represented by curves (iv) in Figs. 18.7a and b.

Now consider what happens as p_e is slowly reduced from p_1 to the design exit pressure. At first the duct acts as a subsonic nozzle followed by a subsonic diffuser, i.e. as a venturi, and the pressure distribution appears as curve (i). The mass flow increases during this range of p_e until the conditions are such that the throat pressure p_t is equal to the critical pressure p_c: curve (ii). The mass flow is then a maximum and the nozzle is choked. Further reduction of p_e cannot alter the conditions at the throat and the mass flow thereafter remains constant.

If p_e is reduced below the value giving rise to curve (ii), the process of acceleration followed by diffusion can no longer continue in the same manner as before. This is because once the stream has acquired a supersonic velocity, a *converging* duct is necessary for an isentropic diffusion process. Evidently the normal type of continuous isentropic flow process can no longer exist under these conditions. Mathematically speaking, the equations expressing the constancy of energy, mass, and entropy, cannot be satisfied simultaneously in the given duct. The energy and continuity equations are fundamental and must be satisfied—it is the assumption of isentropic

Fig. 18.7 Flow in a convergent-divergent nozzle with varying back pressure

flow which must be discarded. What happens is that a discontinuity—known as a *shock wave*—is formed in the flow. A shock wave involves a rise of pressure, an increase of entropy, and a deceleration from a supersonic to a subsonic velocity.* These events occur suddenly in a very short distance, and the resulting subsonic stream is then diffused normally in the remaining length of the divergent duct to satisfy the condition that p_2 equals p_e. The pressure distribution then corresponds to curve (iii).

As p_e is reduced towards the design value, the shock wave moves down the divergent part towards the exit. When it reaches the exit, the shock wave changes its character and becomes inclined to the direction of flow, and the downstream velocity can then remain supersonic. Finally, if p_e is reduced below the design value, the pressure p_2 remains constant at the design value and an uncontrolled expansion to p_e occurs outside the nozzle in a manner similar to that already described for the convergent nozzle.

*A one-dimensional treatment can still be used to analyse flow involving some types of shock wave, but for this the student must turn to texts on fluid dynamics, e.g. Ref. (5).

18.2 Metastable flow of vapours

Before leaving the subject of flow in nozzles, it is necessary to describe a phenomenon peculiar to the rapid expansion of a vapour. The ideal case of isentropic expansion of a superheated vapour to a state in the wet region is depicted on the T-s diagram in Fig. 18.8a. At the point in the expansion where the pressure is p_2 a change of phase should begin to occur. At this point the random kinetic energy of the molecules has fallen to a level which is insufficient to overcome the attractive forces of the molecules and some of the slower moving molecules coalesce to form tiny droplets of condensate. Now this process, although rapid, does not have time to occur in a nozzle where the flow velocity is very great. The achievement of equilibrium between the liquid and vapour phases is therefore delayed, and the vapour continues to expand in a dry state.

If the constant pressure line p_3 in the superheat region is extended into the wet region, the final state could be represented by the point 3 in Fig. 18.8b. Between

Fig. 18.8   Metastable states

pressures p_2 and p_3 the vapour is said to be *supersaturated* or *supercooled*. The latter term owes its origin to the fact that at any pressure between p_2 and p_3 the temperature of the vapour is always less than the saturation temperature corresponding to that pressure. The path of the process between 2 and 3 is shown dotted in Fig. 18.8b because these states are not states of stable equilibrium—they are called *metastable* states. The type of equilibrium established is certainly not stable because a disturbance, such as the introduction of a measuring instrument, can bring about condensation and a restoration of the correct state of equilibrium between liquid and vapour. Neither can the state be regarded as unstable, because this term is only applied to states for which any infinitesimal disturbance will cause a major change of state. An infinitesimal disturbance, such as the formation of a tiny droplet, is insufficient to bring about general condensation.

When designing steam nozzles, it may be assumed that the supersaturated vapour expands according to the laws of the superheated vapour. As steam tables do not normally extend the stable superheat range into the metastable supersaturated region, extrapolation of property values is most easily carried out by reference to approximate property relations for superheated steam. The simplest equation which can be used for this purpose is equation (9.8), which led to the following two useful relations for an isentropic process:

$$pv^{1 \cdot 3} = \text{constant and } p/T^{1 \cdot 3/0 \cdot 3} = \text{constant.}$$

These equations enable the final state and isentropic enthalpy drop to be predicted. Thus in practice, due to the phenomenon of supersaturation, the value of $1 \cdot 3$ should be used for n even if the steam is only just superheated initially and most of the expansion occurs in the 'wet region'.

The following example shows that the mass flow through a nozzle is increased by supersaturation. In early experiments on nozzle flow it was noticed that the measured mass flow was often greater than the value estimated on the assumption of a normal expansion with phase equilibrium, and it was this fact that drew the attention of engineers to the phenomenon of supersaturation.

Example 18.2 Dry saturated steam at $2 \cdot 8$ bar is expanded in a simple convergent nozzle to a pressure of $1 \cdot 7$ bar. The throat area is 3 cm^2 and the inlet velocity is negligible. Estimate the exit velocity and mass flow (a) if phase equilibrium is assumed throughout the expansion and (b) if the steam is assumed to be supersaturated.

(a) Assuming isentropic expansion, the dryness fraction at the exit is given by

$$s_1 = s_2 = s_{f2} + x_2 s_{fg2}$$
$$7 \cdot 015 = 1 \cdot 475 + x_2 5 \cdot 707, \quad \text{and hence} \quad x_2 = 0 \cdot 971$$

Hence

$$h_1 - h_2 = 2722 - \{483 + (0 \cdot 971 \times 2216)\} = 87 \text{ kJ/kg}$$
$$C_2 = \{2(h_1 - h_2)\}^{1/2} = (2 \times 10^3 \times 87)^{1/2} = 417 \text{ m/s}$$
$$v_2 = x_2 v_{g2} = 0 \cdot 971 \times 1 \cdot 031 = 1 \cdot 001 \text{ m}^3/\text{kg}$$
$$m = \frac{A_2 C_2}{v_2} = \frac{3 \times 417}{10^4 \times 1 \cdot 001} = 0 \cdot 125 \text{ kg/s}$$

(b) Assuming the supersaturated expansion to conform to $pv^{1 \cdot 3} = \text{constant}$, and using (18.10), we have

$$C_2 = \left[\frac{2n}{1 - n} p_1 v_1 \left\{ \left(\frac{p_2}{p_1} \right)^{(n-1)/n} - 1 \right\} \right]^{1/2}$$

v_1 is $0 \cdot 6462$ m^3/kg from tables, so that

$$C_2 = \left[\frac{2 \times 1 \cdot 3}{0 \cdot 3} 10^5 \times 2 \cdot 8 \times 0 \cdot 6462 \left\{ 1 - \left(\frac{1 \cdot 7}{2 \cdot 8} \right)^{0 \cdot 3/1 \cdot 3} \right\} \right]^{1/2} = 413 \text{ m/s}$$
$$v_2 = v_1 \left(\frac{p_1}{p_2} \right)^{1/n} = 0 \cdot 6462 \left(\frac{2 \cdot 8}{1 \cdot 7} \right)^{1/1 \cdot 3} = 0 \cdot 949 \text{ m}^3/\text{kg}$$
$$m = \frac{3 \times 413}{10^4 \times 0 \cdot 949} = 0 \cdot 131 \text{ kg/s}$$

[Note that part (b) may be solved using equation (9.11) to find $(h_2 - h_1)$ and hence C_2, and (9.10) to find v_2.]

Provided that the vapour remains supersaturated up to the outlet of the nozzle, the method used in the previous example is adequate because, in the absence of friction, the process within the nozzle can be considered reversible and therefore isentropic. The restoration of equilibrium is an irreversible process, however, and if this occurs at some point within the nozzle the prediction of the final state is difficult.

The supersaturated condition cannot continue indefinitely—the delay in condensation leads to a build-up of molecular cohesive forces which finally results in sudden condensation at many points in the vapour. Much work has been done to determine the minimum pressure to which steam may be expanded before condensation occurs, and a description of some of these experiments is given in Refs. (7) and (8). It has been found that even if the steam is only just superheated at the beginning of the expansion, the condensation pressure is less than the critical pressure. Thus the irreversible restoration of stable equilibrium always occurs *after* the throat of a steam nozzle, and an estimate of the mass flow for a given throat area can always be made in the manner indicated by Example 18.2b.

The difficulty arises in predicting the conditions at the outlet of convergent-divergent nozzles of large pressure drop. Experiments have shown that the condensation occurs suddenly with an increase of both entropy and pressure. Fig. 18.9 illustrates a frictionless expansion of this kind on the *h-s* diagram. When designing nozzles in practice, it is always necessary to introduce a 'coefficient' or 'efficiency' to allow for the increase of entropy due to friction, and this factor, being experimentally determined, will automatically take

Fig. 18.9 Restoration of equilibrium

into account the increase in entropy due to the condensation process. The prediction of the final state by such means will be made clearer in the next section when the effects of friction are discussed.

18.3 The momentum equation

It is now appropriate to explain what was meant in the introduction to this chapter by the statement that it is sometimes necessary to invoke *explicitly* Newton's Second Law of Motion. It may seem surprising that we have been able to solve any problem concerned with motion without reference to this fundamental principle. The fact is, however, that so far we have been concerned with problems for which the Second Law of Motion is expressed *implicitly* in the energy equation. This point may be appreciated if we think for a moment of how the energy equation is deduced for a purely mechanical system. The starting point is Newton's Second Law, that the force acting on a body is equal to the rate of change of momentum in the direction of the force, i.e.

$$F = \frac{\mathrm{d}(mC)}{\mathrm{d}t} \tag{18.23}$$

Equation (18.23) is the simplest form of *momentum equation*.

If a body of constant mass m is moved a distance dx by the force F we can write

$$F \, dx = m \left(\frac{dC}{dt} \right) dx$$

And since $dx = C \, dt$ this becomes

$$F \, dx = mC \, dC$$

which on integration yields

$$\int F \, dx = \frac{m}{2} (C_2{}^2 - C_1{}^2)$$

The L.H.S. can be integrated directly if there exists a function of x, which we may denote by P, such that $dP/dx = -F$. We then have

$$\int_1^2 -dP = \frac{m}{2} (C_2{}^2 - C_1{}^2)$$

or

$$P_1 + \tfrac{1}{2}mC_1{}^2 = P_2 + \tfrac{1}{2}mC_2{}^2 \tag{18.24}$$

In these circumstances there is a quantity $(P + mC^2/2)$ which remains constant during the motion—it is called the energy of the body. $mC^2/2$ is the kinetic energy and P the potential energy. If the force is simply that due to gravity, $F = \text{constant} = -mg$, and the change in potential energy is

$$P_2 - P_1 = mg(x_2 - x_1)$$

Or again, if the force is the restoring force of an extended spring, $F = -kx$ where k is the spring stiffness, and

$$P_2 - P_1 = \frac{k}{2} (x_2{}^2 - x_1{}^2)$$

The integrated equation, (18.24), which is an equation of mechanical energy, is equivalent to the equation of momentum and the two are interchangeable. It is often more convenient to use the energy equation because it relates scalar quantities; the momentum equation, on the other hand, relates vector quantities.

The simple energy equation for a purely mechanical system is only another form of the momentum equation because it has been deducted from it without further reference to empirical data. But this energy equation applies only to frictionless systems. Friction forces are not simple functions of position (i.e. we cannot write $dP/dx = -F$), and they involve the transformation of mechanical forms of energy into other quite different forms (e.g. internal energy) with which classical mechanics is not concerned. A more comprehensive energy equation can be formulated only when the additional generalisation of the First Law of Thermodynamics has been introduced. Then the energy equation rests upon its own empirical ground and applies to any system whether frictionless or not and whether adiabatic or not; it is no longer simply another form of the momentum equation. We should not be surprised, however, if the energy and momentum equations appear to be identical when friction forces are assumed to be absent and the process is adiabatic. Conversely, when these conditions are *not* fulfilled, we can usefully employ both the general energy equation *and* the momentum equation, because they are no longer identical but express different fundamental principles.

It is now necessary to apply Newton's Second Law of Motion to fluid flow. The deduction of the most general form of this momentum equation is left to texts on fluid dynamics; we shall restrict our attention to one-dimensional steady flow. Heat may be transferred to the fluid provided that it changes the fluid properties uniformly across the cross-section of flow.

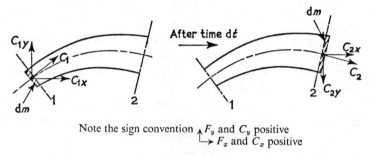

Note the sign convention F_y and C_y positive
F_x and C_x positive

Fig. 18.10

Consider the open system between planes 1 and 2 in Fig. 18.10. In time dt the mass between 1 and 2 plus the mass equal to dm moves to the right. Newton's law states that the vector sum ΣF of all the external forces acting on the mass is equal to the rate of change of momentum of the mass. Assuming there are no components of the vector quantities of force, velocity and momentum in a direction perpendicular to the plane of the paper, and resolving in the x and y directions, we obtain

$$\Sigma F_x = \frac{1}{dt}\{(M_{12x} + C_{2x}\,dm) - (M_{12x} + C_{1x}\,dm)\}$$

$$\Sigma F_y = \frac{1}{dt}\{(M_{12y} + C_{2y}\,dm) - (M_{12y} + C_{1y}\,dm)\}$$

M_{12x} and M_{12y} are the vector components of the total momentum of the fluid within the boundary of the open system, and for steady flow they do not change with time. Moreover, dm/dt is the rate of mass flow which we usually denote simply by m. We therefore have as the momentum equation

$$\Sigma F_x = m(C_{2x} - C_{1x}) \quad \text{and} \quad \Sigma F_y = m(C_{2y} - C_{1y}) \tag{18.25}$$

409

From the continuity equation, m is given by

$$m = \frac{A_1 C_1}{v_1} = \frac{A_2 C_2}{v_2}$$

For our purpose in this chapter it is sufficient to consider only a straight duct, as in Fig. 18.11. The resultant of all forces in the y-direction is then zero, so that the momentum equation reduces to

$$\Sigma F = m(C_2 - C_1) \tag{18.26}$$

The term ΣF includes all externally impressed forces on the mass of fluid between planes 1 and 2 which affect the change of momentum. It may include (1) gravitational forces; (2) hydrostatic forces, due to p_1 and p_2 acting on the areas A_1 and A_2, and due to the pressure exerted by the wall on the fluid; and (3) frictional forces due to the shear stress at the wall.* We may ignore gravitational forces in any application we shall consider in this book, just as we have already done when applying the energy equation. Equation (18.26) will now be applied to three simple cases in turn: (i) isentropic flow in a duct of varying area, (ii) heat addition to frictionless flow in a duct of constant area, and (iii) irreversible adiabatic flow in a duct of varying area.

Fig. 18.11

(i) *Isentropic flow in a duct of varying area*

The differential form of equation (18.26) may be obtained by applying it to an element between planes 1 and 2 an infinitesimal distance apart. The properties are p, v, C, etc., at plane 1 and $(p + dp)$, $(v + dv)$, $(C + dC)$, etc., at plane 2. With isentropic flow there are no frictional shear stresses so that the only forces affecting the flow are the pressure forces shown in Fig. 18.11. These are the pressure forces pA and $(p + dp)(A + dA)$ at each end of the element, and the reaction on the fluid from the hydrostatic pressure along the wall. The average force normal to the surface of the wall is $(p + dp/2)\, dA_s$, where dA_s is the surface area. It is only the component of this force in the axial direction which contributes to the change in momentum, and this is

$$\left(p + \frac{dp}{2}\right) dA_s \sin \phi = \left(p + \frac{dp}{2}\right) dA$$

Equation (18.26) in differential form is therefore

$$pA - (p + dp)(A + dA) + \left(p + \frac{dp}{2}\right) dA = m\, dC$$

---

*The only logical way of allowing for friction in a one-dimensional treatment is to assume that all the friction is concentrated at the wall. In the real flow, friction between the layers of fluid results in a non-uniform distribution of properties at right angles to the flow and the flow cannot, strictly speaking, be visualised as 'one-dimensional' at all.

Neglecting second-order terms, and remembering that m is equal to AC/v, this reduces to

$$A \, dp + \left(\frac{AC}{v}\right) dC = 0 \tag{18.27}$$

or alternatively,

$$v \, dp + C \, dC = 0 \tag{18.28}$$

(18.28) is the same as the energy equation applied to isentropic flow, i.e. equation (18.5). Thus, as expected for adiabatic frictionless flow, the momentum and energy equations are identical. This is the reason why the momentum equation did not appear explicitly in the analysis given in section 18.1.

(ii) *Heat addition to frictionless flow in a duct of constant area*

The case now to be considered approximates to the process in the combustion chamber of a turbo-jet engine.* Since the flow is not adiabatic, the energy and momentum equations will not be identical, and it is easy to show that the flow cannot be completely described without using both equations.

When the fluid is a perfect gas the energy equation is

$$Q = c_p(T_2 - T_1) + \tfrac{1}{2}(C_2{}^2 - C_1{}^2)$$

Owing to the increase of temperature, and therefore of specific volume, the velocity will increase in a duct of constant area, i.e. $C_2 > C_1$. Without knowing the change of pressure, this increase of velocity cannot be predicted, and the energy equation is evidently not sufficient to enable Q to be calculated for a given temperature rise or vice versa.

The differential form of the momentum equation for frictionless flow has already been deduced as equation (18.27), i.e.

$$A \, dp + \frac{AC}{v} dC = 0$$

In this case both A and (AC/v) are constant, so that the equation can be integrated directly to give

$$(p_2 - p_1) = -\frac{C_1}{v_1}(C_2 - C_1) \tag{18.29}$$

This equation, together with the energy equation, equation of state, and continuity equation, provide four equations which can be solved for the four unknowns. That is, if Q and the initial conditions are known, p_2, T_2, v_2 and C_2 can be predicted; or when T_2 and the initial conditions are given, Q, p_2, v_2 and C_2 can be found.

---

*Although there is a considerable amount of friction in an actual chamber, arising from the use of baffles to stabilise the flame, it is usual to take this into account separately by assuming that all the friction occurs prior to the combustion which is here simulated by an addition of heat.

The calculation is too lengthy to be elaborated here, but the foregoing outline suffices to show that the momentum equation is essential for the solution of this type of problem. Furthermore, it also shows clearly that the heat addition to a fluid in a constant area duct cannot be achieved without a drop of pressure. It is apparent from the momentum equation that $p_2 = p_1$ only if $C_2 = C_1$. To achieve this it would be necessary to use a divergent duct, or a duct of such large cross-sectional area that changes in velocity are negligible.

(iii) *Irreversible adiabatic flow in a duct of varying area*

When the effect of friction is considered, the frictional force due to the shear stress along the wall of the duct must be included in the momentum equation. With reference to Fig. 18.12, and denoting the shear stress by τ, the force due to friction is $\tau \, dA_s$. The component in the axial direction opposing motion is therefore

$$\tau \, dA_s \cos \phi = \tau \left(\frac{dA}{\sin \phi} \right) \cos \phi$$

Fig. 18.12

Including this force in equation (18.27), we have

$$-A \, dp - \tau \, dA \cot \phi = \frac{AC}{v} \, dC$$

or

$$v \, dp + C \, dC + \frac{\tau v}{A} \cot \phi \, dA = 0 \qquad (18.30)$$

(18.30) is evidently different from the energy equation for adiabatic flow, which is

$$dh + C \, dC = 0$$

The two equations become identical when $\tau = 0$ because then the process also becomes isentropic and dh is equal to $v \, dp$.

τ can be determined by experiment, but it is found to be a function of the velocity and turbulence of the fluid and the surface roughness of the duct (see section 22.1). It is therefore not a constant and varies along the duct. If the variation of τ along the duct is known, equation (18.30) can be used in conjunction with the energy equation, equation of state, and continuity equation, to predict the state of the fluid at any point in the duct. Although such a procedure is sometimes illuminating [see Ref. (18)], it is very laborious. For practical purposes, e.g. when designing nozzles, it is usual to take account of friction by introducing empirical coefficients which can be used directly in conjunction with calculations based on the assumption of isentropic flow. Three such coefficients are in common use in nozzle design and they are defined as follows.

When the primary interest is in the exit velocity produced by the nozzle it is usual to use either a *velocity coefficient* k_c, or a *nozzle efficiency* η_N. The former, k_c, is defined as the ratio of the actual exit velocity to that calculated for the same pressure ratio on the assumption of isentropic flow, i.e.

$$k_c = \frac{C_2}{(C_2)_{\text{isen}}}$$

The nozzle efficiency is the ratio of the actual to isentropic enthalpy drop, i.e.

$$\eta_N = \frac{h_1 - h_2}{(h_1 - h_2)_{\text{isen}}}$$

If $C_1 \simeq 0$, as we have been assuming throughout our discussion of nozzles, it is evident that

$$\eta_N = k_c{}^2$$

When the primary interest lies in the mass flow passed by the nozzle, a *coefficient of discharge* is used, defined by

$$k_d = \frac{m}{m_{\text{isen}}}$$

Nozzles with straight axes usually have efficiencies ranging from 94 to 99 per cent, whereas those with curved axes, as used in turbines, have efficiencies ranging from

Fig. 18.13

90 to 95 per cent. The efficiency depends largely upon the profile of the nozzle. According to a one-dimensional isentropic treatment, there is nothing to choose between the two nozzles shown in Fig. 18.13. In practice, the shorter nozzle, having the smaller surface area swept by the fluid, incurs less friction loss provided the included angle of the divergent part does not exceed 30 degrees.

The use of a coefficient of discharge and a nozzle efficiency in the design of a convergent-divergent nozzle is illustrated by the following example.

Example 18.3   A nozzle is required to pass an air flow of 1·5 kg/s. The inlet conditions are zero velocity, pressure 3·5 bar and temperature 425 °C; the air is to be expanded to 1·4 bar. Determine the throat area required if the coefficient of discharge is assumed to be 0·98.

Also calculate the exit velocity and exit area if the nozzle efficiency is assumed to be 95 per cent.

The critical pressure ratio under isentropic conditions is

$$\frac{p_c}{p_1} = \left(\frac{2}{\gamma + 1}\right)^{\gamma/(\gamma-1)} = \left(\frac{2}{2\cdot4}\right)^{1\cdot4/0\cdot4} = 0\cdot528$$

Since the overall pressure ratio is 0·4, a convergent-divergent nozzle will be required for complete expansion within the nozzle. The conditions at the throat with isentropic flow will be

$$p_t = p_c = 0·528 \times 3·5 = 1·848 \text{ bar}$$

$$T_t = \frac{2}{\gamma + 1} T_1 = \frac{2 \times 698}{2·4} = 581·7 \text{ K}$$

$$v_t = \frac{RT_t}{p_t} = \frac{0·287 \times 581·7}{100 \times 1·848} = 0·903 \text{ m}^3/\text{kg}$$

The corresponding velocity at the throat may be found directly from the energy equation instead of using (18.17), i.e.

$$\tfrac{1}{2}C_t^2 = c_p(T_1 - T_t)$$

Hence

$$C_t = \{2 \times 10^3 \times 1·005(698 - 581·7)\}^{1/2} = 483·5 \text{ m/s}$$

The mass flow under isentropic conditions would be (1·5/0·98) kg/s. The throat area required is therefore

$$A_t = \frac{mv_t}{C_t} = \frac{1·5 \times 0·903}{0·98 \times 483·5} \times 10^4 = 28·6 \text{ cm}^2$$

If the expansion were isentropic, the final temperature would be

$$T_2' = T_1 \left(\frac{p_2}{p_1}\right)^{(\gamma-1)/\gamma} = 698 \left(\frac{1·4}{3·5}\right)^{0·4/1·4} = 537 \text{ K}$$

Assuming a nozzle efficiency of 0·95, the actual final temperature will be given by

$$0·95 = \frac{T_1 - T_2}{T_1 - T_2'} = \frac{698 - T_2}{698 - 537}$$

from which $T_2 = 545$ K. The exit velocity will therefore be

$$C_2 = \{2 \times 10^3 \times 1·005(698 - 545)\}^{1/2} = 554·6 \text{ m/s}$$

Also,

$$v_2 = \frac{RT_2}{p_2} = \frac{0·287 \times 545}{100 \times 1·4} = 1·117 \text{ m}^3/\text{kg}$$

and hence the exit area required is

$$A_2 = \frac{mv_2}{C_2} = \frac{1·5 \times 1·117}{554·6} \times 10^4 = 30·2 \text{ cm}^2$$

Summary: The fundamental principles which may be used in the solution of problems of one-dimensional steady flow are:

(1) First Law of Thermodynamics,
(2) Second Law of Thermodynamics,
(3) Conservation of Mass,
(4) Newton's Second Law of Motion.

In addition we have

(5) Equation of State, or its equivalent in tables of properties.

Only four equations are necessary to determine the flow. When the flow is reversible and adiabatic we may make direct use of a consequence of the Second Law of Thermodynamics, i.e. that the entropy is constant. Under these conditions, separate consideration of Newton's Law of Motion is superfluous because the momentum and energy equations are identical. When the flow is either irreversible or non-adiabatic we cannot make direct use of the consequences of the Second Law, but we can then use Newton's Law of Motion.

We have only carried out analyses for very simple cases—and not all of these have been carried to completion owing to the length of the calculation. It is not surprising that more sophisticated methods are used in the proper study of gas dynamics. For example, in order to obtain the simple expressions of nozzle flow, (18.10) to (18.16), we had to assume a negligible inlet velocity. In fact equally simple expressions can be obtained when the inlet velocity is finite once the concepts of *total* (or *stagnation*) temperature and pressure are introduced. The velocity term is then taken into account implicitly rather than explicitly. Where velocity terms must occur explicitly, much simplification of the equations is possible if the velocity is expressed as a fraction of the local sonic velocity—this ratio is known as the *Mach number*. The use of such concepts is essential if manageable equations are to be obtained when dealing simultaneously with changes of area, heat addition and friction [see Ref. (11)]. Gas dynamic concepts have not been introduced here because they would have obscured the fundamental links between thermodynamics and gas dynamics which it has been our object to present.

18.4 Jet propulsion engines

As a final example of the use of the momentum equation we shall discuss briefly the nature of *propulsive work* produced by aircraft power plant. When an aircraft is flying horizontally at a constant speed C_A, the drag of the aircraft must be balanced by an equal and opposite thrust F_N produced by the power plant. The rate of work done is then equal to the product $F_N C_A$, and this is known as the propulsive work (strictly 'power'). Any estimation of this quantity must involve an application of the momentum equation since we are concerned directly with forces.

Propulsion engines fall into two distinct classes: those which make use of atmospheric air as the main propulsive fluid, and those which carry their own propulsive fluid. The latter are known as rocket motors and will be considered separately. For a full treatment of jet propulsion see Ref. (23).

(i) *Propulsion engines using atmospheric air*

All propulsion engines consist of one or more of the three components depicted in Fig. 18.14, viz. a diffuser, source of energy, and propelling nozzle. No diffuser or nozzle is used in the aircraft piston engine, and the energy source consists of a piston engine producing mechanical work which is used to drive a propeller. All three components are used in a *turbo-jet* engine, and in this case the energy source comprises a compressor, combustion chamber and turbine. No net work is produced by the turbine, but some of the energy available in the gases leaving the turbine is used in the propelling nozzle to produce a high-speed jet. At very high aircraft speeds sufficient pressure rise is obtained in the diffuser alone and there is no need for a compressor and therefore for the turbine which drives it. This leads to the idea of the *ram-jet* engine, consisting simply of a diffuser, combustion chamber and propelling nozzle.

Fig. 18.14 Schematic representation of a propulsion engine

Due to the direct dependence of pressure rise upon aircraft speed, the ram-jet propelled vehicle must be given an initial velocity before the engine can function. Moreover, pressures sufficiently high for good efficiency can only be obtained at supersonic speeds, when the pressure rise across shock waves at the inlet can be utilised.

In all these forms of power plant the thrust is produced by accelerating a mass of air and increasing its momentum in a rearward direction. The net reaction upon the aircraft is the useful thrust F_N. An expression for F_N can be obtained by considering the momentum equation. In what follows it will be assumed that the mass flow of propulsive fluid is constant between inlet and outlet of the engine, i.e. that the mass of fuel added is negligible. The exhaust gases leaving a piston engine provide only a minute part of the total propulsive fluid and the assumption is obviously permissible in this instance. It is also valid for a turbo-jet engine because the fuel/air ratios used are small, but it is not such a good approximation in the case of a ram-jet engine where the fuel/air ratios are approximately stoichiometric.

The momentum equation (18.26), states that

$$\Sigma F = m(C_2 - C_1)$$

When deriving this equation we were considering flow through a stationary duct; the velocities relative to such a duct are absolute velocities (i.e. relative to the earth) and consequently they have been denoted by the symbol C. We are here considering a duct possessing motion relative to the earth, but if the duct is moving with a *uniform* velocity the equation is still valid.* The velocities in the equation can be regarded as either absolute velocities or velocities relative to the duct. Velocities relative to a moving duct will always be denoted by V. The momentum equation can therefore also be written as

$$\Sigma F = m(V_2 - V_1) \tag{18.31}$$

and it is this form which we shall use in what follows. Suffixes 1 and 2 will be used to denote properties at the inlet and outlet, referring once again to Fig. 18.14. We shall denote by R' the resultant of all the forces exerted *by the duct* upon the fluid passing through it. These comprise the hydrostatic pressure forces normal to the wall, and the friction forces along the wall (where 'wall' refers to the internal surface of the duct and the external surface of any body attached to the duct and immersed in the stream). From the discussion of ΣF in the previous section it will be evident that ΣF is equal to R' plus the net pressure force $(p_1A_1 - p_2A_2)$. The momentum equation can therefore be written as

$$R' + (p_1A_1 - p_2A_2) = m(V_2 - V_1)$$

From Newton's Third Law of Motion, the net force exerted by the fluid *on the duct*, R say, must be equal and opposite to R', i.e. $R = -R'$. So far we have always considered forces to be positive if they act in the direction for which the fluid velocity is taken as positive. When discussing propulsive forces, it is more convenient *to change the sign convention and treat the forces as positive in the direction of motion of the aircraft*. If we denote by F_T the force exerted by the fluid on the duct which produces a thrust in the direction of motion of the aircraft, then $F_T = -R = R'$. The momentum equation can therefore be written as

$$F_T + (p_1A_1 - p_2A_2) = m(V_2 - V_1)$$

and hence

$$F_T = (p_2A_2 + mV_2) - (p_1A_1 + mV_1)\dagger$$

Although F_T is the total force exerted by the propulsive fluid upon the duct, it is not the net force exerted by the duct upon the aircraft. There are pressure forces acting on the external surface of the duct which are not balanced unless A_2 is equal to A_1. The external pressure is the atmospheric pressure p_a, and the net external force assisting the motion of the duct is therefore $p_a(A_1 - A_2)$. The net thrust F_N which the power plant exerts upon the aircraft is found by adding this external force to F_T.

---

*It is shown in books on mechanics that any dynamical equation based on Newton's Laws of Motion, and which is known to hold in one frame of reference, must also hold in any other frame of reference that moves with a uniform velocity relative to the first. Any such frame in which a Newtonian equation is valid is termed an 'inertial frame'.

†In gas dynamics the quantity $(pA + mV)$ at any plane in the flow is known as the *stream thrust* at that plane.

There will be other external forces due to friction on the outside of the duct, but these are normally included in the drag of the aircraft as a whole. It follows that the net thrust is given by

$$F_N = (p_2 A_2 + mV_2) - (p_1 A_1 + mV_1) + p_a(A_1 - A_2) \qquad (18.32)$$

Under certain conditions (18.32) can be reduced to a simpler form. When the aircraft is flying at subsonic speeds, p_1 is equal to p_a, and V_1 is equal in magnitude to the aircraft speed C_A. (These equalities do not hold at supersonic speeds because of the effect of shock waves at the inlet.) With these modifications, and writing V_J (jet velocity relative to the aircraft) for V_2, equation (18.32) appears as

$$F_N = m(V_J - C_A) + A_2(p_2 - p_a) \qquad (18.33)$$

The first term on the R.H.S. of (18.33) is called the *momentum thrust* and the second the *pressure thrust*. The pressure thrust is in some cases zero. For example, when the propulsive jet is produced by a propeller the downstream pressure is atmospheric, i.e. $p_2 = p_a$. Or again, the propelling nozzle of a turbo-jet engine is a simple converging nozzle: p_2 will therefore be equal to p_a when the pressure ratio (outlet/inlet) is greater than the critical pressure ratio [section 18.1 (ii)]. In such circumstances the thrust is simply

$$F_N = m(V_J - C_A) \qquad (18.34)$$

From the simple expression (18.34), it is obvious that the same thrust can be achieved by using either an engine producing a high-velocity jet of low mass flow (turbo-jet), or one which produces a low-velocity jet of high mass flow (propeller). The choice can be shown to depend upon the aircraft speed, as follows. The energy supplied by the fuel is used in two ways, (a) in producing the propulsive work $F_N C_A$, and (b) in producing a jet having an absolute kinetic energy of $m(V_J - C_A)^2/2$. The latter represents a waste of energy, although it cannot be made zero without at the same time reducing the thrust to zero. If the wasted kinetic energy in the jet is to be small, however, $(V_J - C_A)$ must not be too great, i.e. we require a high-speed jet for a fast aircraft and a low-speed jet for a slow aircraft.

Such considerations as these have led to the development of the range of power plant shown diagrammatically in Fig. 18.15. Proceeding from (a) to (e), the propulsive jet decreases in mass flow and increases in velocity. (a) represents the piston engine and propeller. (b) is a propeller turbine engine, where the low-velocity jet from the propeller is combined with the high-velocity jet from the turbine engine. In (c), the by-pass engine, the propeller is replaced by a fan giving a jet of smaller mass flow but of higher velocity. (d) is the simple turbo-jet and (e) the ram-jet.

(ii) *Rocket motors*

There is one remaining type of propulsion unit to consider, which is in a class by itself—namely the *rocket motor*. It is distinguished from the other means of propulsion by the fact that it makes no use of the atmosphere but carries its own oxidant. For this reason it is the only power plant suitable for use at very high altitudes or in outer space.

The rocket motor consists essentially of a combustion chamber and propelling nozzle. For short durations, such as those involved in temporary thrust boosting or

(a) Piston engine   (b) Propeller turbine engine

(c) By-pass engine

(d) Turbo-jet engine

(e) Ram-jet engine

Fig. 18.15   Propulsion engines

starting a ram-jet propelled vehicle, the most economical rocket is obtained by using solid fuel. For longer durations a liquid fuel (e.g. kerosene) and liquid oxidant (e.g. nitric acid or hydrogen peroxide) are always preferred. In what follows we shall adopt rocket practice and use the term fuel to include the oxidant: calorific values, for example, are always quoted per unit mass of reacting material which the rocket must carry. The fuel is either fed into the combustion chamber by mechanical pumps, or forced in by compressed nitrogen carried in a separate container as indicated in

Fig. 18.16. Since it is possible to operate a rocket combustion chamber at relatively high pressures—20 bar or more—very high jet velocities and large thrusts can be obtained.

To make full use of the large pressure drop across the propelling nozzle, a convergent-divergent nozzle is always employed. It will be operating in the choked condition and the mass flow passed by the nozzle (equal to the mass flow of fuel) is therefore given by equation (18.21), viz.

$$m = A_t p_1 \left\{ \frac{\gamma}{RT_1} \left(\frac{2}{\gamma + 1} \right)^{(\gamma+1)/(\gamma-1)} \right\}^{1/2}$$

This is assuming isentropic expansion and that the combustion gases behave as a perfect gas. The values of γ and the combustion temperature T_1 are more or less fixed by the properties of the reactants employed, i.e. by the calorific value and specific heats. A rough estimate of T_1 can be obtained by treating the combustion as a constant pressure, adiabatic process. Due account must be taken of dissociation, however, because approximately stoichiometric mixtures are used and the temperatures are very high—in the region of 2500 °C. This type of calculation has been described in section 15.7. One of the unknowns is the extent to which the reaction still continues during the expansion in the nozzle. The calculations are further complicated by the fact that the combustion chamber and nozzle must be cooled—usually by the liquid fuel prior to injection—and the process is therefore not strictly adiabatic.

With the combustion temperature more or less fixed, it is evident from the equation that the mass flow per unit throat area is a function only of combustion chamber pressure—or, conversely, that the combustion pressure is fixed by the rate at which fuel is fed into the chamber. The maximum permissible pressure is determined by the high-temperature strength of the materials used in the construction of the chamber.

Fig. 18.16
Rocket motor

The main factors affecting the performance of the rocket motor are apparent from the expression for the thrust produced. The working fluid has no initial momentum relative to the rocket, and V_1, p_1 and A_1 are zero in the momentum equation (18.32). The net thrust is therefore

$$F_N = mV_2 + A_2(p_2 - p_a)$$

If the nozzle is designed for complete expansion from the combustion pressure to p_a, the pressure thrust is zero and the thrust is simply

$$F_N = mV_2$$

The exit velocity V_2 is given by equation (18.17), and with complete expansion the pressure ratio can be denoted by p_a/p_1 so that,

$$V_2 = \left[\frac{2\gamma}{1 - \gamma} RT_1 \left\{ \left(\frac{p_a}{p_1} \right)^{(\gamma-1)/\gamma} - 1 \right\} \right]^{1/2}$$

Combining this equation with the mass flow equation we have the thrust per unit throat area as

$$\frac{F_N}{A_t} = p_1\gamma \left[\frac{2}{\gamma - 1} \left(\frac{2}{\gamma + 1} \right)^{(\gamma+1)/(\gamma-1)} \left\{ 1 - \left(\frac{p_a}{p_1} \right)^{(\gamma-1)/\gamma} \right\} \right]^{1/2} \tag{18.35}$$

(18.35) shows that for a given nozzle (A_t) and given reactants (γ), the thrust at any altitude (p_a) is a function only of the combustion pressure p_1, i.e. of the rate of fuel supply. The thrust is not dependent upon the combustion temperature, except in so far as γ depends slightly upon this quantity.

The thrust itself is not, however, the most important criterion of the performance of a rocket. Such a large proportion of the total weight of a rocket is comprised of fuel, and fuel containers, that it is the thrust per unit flow of fuel which really matters. This quantity, called the *specific impulse I*, can be found by combining the mass flow and thrust equations to give

$$I = \frac{F_N}{m} = \left[RT_1 \frac{2\gamma}{\gamma - 1} \left\{ 1 - \left(\frac{p_a}{p_1} \right)^{(\gamma-1)/\gamma} \right\} \right]^{1/2} \tag{18.36}$$

From this expression it may be seen that a high combustion temperature is necessary for good performance. Remembering that the gas constant for the expanding products is equal to R_0/M, where R_0 is the universal gas constant and M is the average molecular weight, it is also apparent that it is desirable to choose a fuel which yields products of low molecular weight. It is in fact a high value of the ratio T_1/M which is important when considering the choice of fuel. In general, fuels having a high hydrogen content prove to be the best, enabling high specific impulses to be obtained without the use of excessively high combustion temperatures.

In arriving at equations (18.35) and (18.36) we supposed that the gases are expanded to ambient pressure in the nozzle, and therefore that the pressure thrust is zero. When operating at high altitudes this is not practicable, because complete expansion would require a nozzle having a very large exit to throat area ratio (and in space it would be infinite). This in turn implies a long nozzle with considerable friction loss. Theoretically, with isentropic expansion, the thrust can be shown to be a maximum with complete expansion: the reduction in momentum thrust as a result of using a smaller area ratio is never entirely compensated by the additional pressure thrust due to p_2 being greater than p_a. This is not necessarily true when friction is taken into account because the theoretical jet velocity, and hence momentum thrust, are not achieved. A considerable departure from the optimum nozzle area ratio can in fact be permitted without much loss of thrust. Equations (18.35) and (18.36) may be expanded to include the pressure thrust term, but it is easier to work from first principles as indicated in the following example.

Example 18.4 A rocket nozzle has an (exit area/throat area) ratio of 3:1. Assuming that the expansion is isentropic, estimate the thrust per unit throat area and the specific impulse, when the combustion temperature is 3000 K, the combustion pressure is 20 bar, and atmospheric pressure is 1 bar. Assume that the gases have a molar mass of 33.5 kg/kmol and that γ is 1·2.

The mass flow per unit throat area from equation (18.21) is

$$\frac{m}{A_t} = p_1 \left\{ \frac{\gamma}{RT_1} \left(\frac{2}{\gamma+1} \right)^{(\gamma+1)/(\gamma-1)} \right\}^{1/2}$$

Now

$$R = \frac{R_0}{M} = \frac{8 \cdot 314 \left[\dfrac{kJ}{kmol\,K} \right]}{33 \cdot 5 \left[\dfrac{kg}{kmol} \right]} = 0 \cdot 248 \left[\frac{kJ}{kg\,K} \right]$$

and hence

$$\frac{m}{A_t} = 20[\text{bar}] \times 10^5 \left[\frac{N/m^2}{bar} \right] \times$$

$$\left\{ \frac{1 \cdot 2}{0 \cdot 248 \left[\dfrac{kJ}{kg\,K} \right] \times 10^3 \left[\dfrac{J\,or\,N\,m}{kJ} \right] \times 3000\,[K]} \left(\frac{2}{2 \cdot 2} \right)^{2 \cdot 2/0 \cdot 2} \right\}^{1/2}$$

$$= 1504 \left[\frac{N\,kg}{m^5} \times \frac{kg\,m/s^2}{N} \right]^{1/2} = 1504 \left[\frac{kg}{s\,m^2} \right]$$

The pressure in the plane of the exit p_2 will be greater than atmospheric pressure and its value is determined by the area ratio. A simple equation for p_2 in terms of known quantities cannot be formulated, but the following graphical method may be adopted

$$\frac{m}{A_t} = \frac{A_2 V_2}{A_t v_2} = \frac{3 V_2}{v_2}$$

(N.B. Here V is relative velocity and v specific volume.) V_2 may be found directly from the energy equation which, with V_1 equal to zero, yields

$$V_2 = \{2c_p(T_1 - T_2)\}^{1/2}$$

Since $c_p = \gamma R/(\gamma - 1) = 1 \cdot 488$ kJ/kg K,

$$V_2 = \left\{ 2 \times 1488 \left[\frac{J}{kg\,K} \right] (3000[K] - T_2) \right\}^{1/2} = 54 \cdot 55 \left[\frac{m}{s\,K^{1/2}} \right] (3000[K] - T_2)^{1/2}$$

Also, v_2 can be found in terms of T_2 from

$$v_2 = \frac{RT_2}{p_2} = \frac{RT_2 (T_1/T_2)^{\gamma/(\gamma-1)}}{p_1}$$

$$= \frac{248 \left[\dfrac{J}{kg\,K} \right] T_2 \left(\dfrac{3000[K]}{T_2} \right)^{1 \cdot 2/0 \cdot 2}}{20 \times 10^5 \left[\dfrac{N}{m^2} \right]} = 0 \cdot 000\,124 \left[\frac{m^3}{kg\,K} \right] T_2 \left(\frac{3000[K]}{T_2} \right)^6$$

Substituting for V_2 and v_2 in the expression for the mass flow per unit throat area (m/A_t), we have

$$1504\left[\frac{\text{kg}}{\text{s}\,\text{m}^2}\right] = \frac{3\times54\cdot55\left[\frac{\text{m}}{\text{s}\,\text{K}^{1/2}}\right](3000[\text{K}]-T_2)^{1/2}}{0\cdot000\,124\left[\frac{\text{m}^3}{\text{kg}\,\text{K}}\right]T_2\left(\frac{3000[\text{K}]}{T_2}\right)^6}$$

$$0\cdot001\,14\frac{T_2}{[\text{K}]}\left\{\frac{3000}{T_2/[\text{K}]}\right\}^6 = \left\{3000-\frac{T_2}{[\text{K}]}\right\}^{1/2}$$

Evaluating each side of the equation for a series of values of $T_2/[\text{K}]$ and plotting the results, we arrive at the solution $T_2 = 1906$ K. Hence

$$V_2 = 54\cdot55\left[\frac{\text{m}}{\text{s}\,\text{K}^{1/2}}\right](3000[\text{K}]-1906[\text{K}])^{1/2} = 1804\left[\frac{\text{m}}{\text{s}}\right]$$

$$p_2 = \frac{20[\text{bar}]}{(3000/1906)^6} = 1\cdot315\,[\text{bar}]$$

The thrust per unit throat area is therefore

$$\frac{F_N}{A_t} = \frac{m}{A_t}V_2 + \frac{A_2}{A_t}(p_2-p_a)$$

$$= 1504\left[\frac{\text{kg}}{\text{s}\,\text{m}^2}\right]\times1804\left[\frac{\text{m}}{\text{s}}\right]+3\times10^5(1\cdot315-1\cdot0)\left[\frac{\text{N}}{\text{m}^2}\right]$$

$$= (2\cdot713+0\cdot095)\times10^6\left[\frac{\text{N}}{\text{m}^2}\right] = 2\cdot808\times10^6[\text{N/m}^2]$$

The specific impulse is then given by

$$I = \frac{F_N/A_t}{m/A_t} = \frac{2808[\text{kN/m}^2]}{1504[\text{kg/s}\,\text{m}^2]} = 1\cdot87\,\text{kN per kg/s}$$

19   Rotary Expanders and Compressors

This chapter describes the way in which work is transferred in rotary expanders and compressors. The former are always referred to as turbines. In a turbine the working fluid enters at a high pressure and acquires increased kinetic energy as it expands to a lower pressure in a ring of fixed nozzles. The stream of fluid then undergoes a change of momentum as it flows through passages between blades attached to the rotor, and the component in a direction tangential to the circle of rotation produces the output torque at the shaft.

This series of events is reversed in the rotary compressor. Input torque from some external source imparts a change of momentum to the working fluid passing between the rotor blades. Having acquired an increased velocity, the fluid then slows down with an accompanying rise of pressure while flowing through a ring of fixed diffusers.

There are two main types of turbine and rotary compressor, distinguished by whether the flow through the machine is in the axial or radial direction. Accordingly, this chapter may be divided into four main sections, dealing respectively with the axial-flow turbine, axial-flow compressor, radial-flow turbine, and radial-flow compressor. In all these machines the fluid undergoes a continuous steady-flow process and the speed of flow is very high. For this reason comparatively small machines can handle large mass flows and large work transfers, and the processes can be assumed to be adiabatic.

It is worth emphasising here that although the Roots blower and vane pump are rotary compressors, they are both positive-displacement machines and as such have been described in Chapter 16. In a positive-displacement compressor the fluid is prevented from flowing back in the direction of the pressure gradient by solid surfaces. No such positive control of the flow exists in the compressors considered in this chapter, where the compression is produced by the processes of acceleration and subsequent diffusion.

19.1 Momentum principles applied to flow through a rotor

Before dealing with the four types of rotary machine, it is helpful to consider the general case of steady flow through passages in a rotor when the absolute velocity of the fluid has components in the axial, radial, and tangential directions. The axial and radial components will be denoted by suffixes a and r, and the tangential component, commonly called the *whirl velocity*, by suffix w. For simplicity we shall assume that

424

the flow is 'one-dimensional', i.e. that the fluid enters the rotor with uniform properties through an area A_1, at a mean radius r_1, and leaves with uniform properties through an area A_2, at a mean radius r_2. The passages joining the inlet and outlet areas may be curved, and the area change (i.e. whether the passages are convergent or divergent) is governed by the velocity and pressure changes which the fluid is required to undergo in the rotor. With the aid of Fig. 19.1 we shall consider the effect *on the rotor* of a steady rate of mass flow m, i.e. the axial thrust and torque which the fluid exerts on the rotor.

(i) *Axial thrust on the rotor*

We have seen in section 18.3, that the vector sum of all the external forces acting on the fluid in a passage is equal to the rate of change of momentum between the inlet

Fig. 19.1

and outlet sections: equation (18.25). Applying this result to the axial components of the forces and velocities we have

$$\Sigma F_a = m(C_{2a} - C_{1a})$$

Forces are always taken as positive in the direction in which the velocity is treated as positive. The vector sum ΣF_a may be split into two parts: (a) the sum R'_a of the pressure forces and friction shear forces which the passage wall exerts on the fluid, and (b) the net hydrostatic pressure force $(p_1 A_1 - p_2 A_2)_a$ which the fluid exerts over the inlet and outlet areas in the axial direction.* The equation then becomes

$$R'_a + (p_1 A_1 - p_2 A_2)_a = m(C_{2a} - C_{1a})$$

*A word of warning is necessary about the general validity of the momentum equation which we are now applying to a curved passage in a rotor. We have always tacitly assumed, and will continue to do so, that any shear forces acting along the planes 1 and 2 at inlet and outlet are negligible. Although there are no such forces across planes at right angles to the flow in the straight ducts considered in Chapter 18, they will be present at the inlet and outlet of a curved passage in a rotor when there is a component of velocity along the plane. In such cases our steady-flow energy equation is also not strictly true because, since the rotor is moving, these shear forces will result in energy transfers which would have to be included in the work term W.

Finally, since the vector sum R_a of the pressure and friction forces which the fluid exerts *on the passage wall* must be equal to $-R'_a$, the axial force on the rotor is given by

$$R_a = m(C_{1a} - C_{2a}) + (p_1A_1 - p_2A_2)_a \qquad (19.1)$$

No work is done by R_a because the rotor does not move in the axial direction; it is merely a thrust which must be taken up by a suitable thrust bearing.

(ii) *Torque exerted on the rotor*

Since the fluid may not be flowing along a passage of constant radius, it is not convenient to deduce the torque directly from the tangential (or whirl) forces exerted by the fluid on the passage. An easier approach is provided by the use of the concept of *angular momentum*, i.e. moment of momentum. The principle of angular momentum is commonly stated in mechanics as follows: *The rate of change of angular momentum of a system of particles about a fixed point is equal to the sum of the moments of the external forces about that point.* The sum of these moments is the torque applied to the system undergoing the change of momentum. To apply the principle of angular momentum to the flow through a rotor, the usual argument adopted for dealing with open systems can be used.

Consider a system consisting of the mass of fluid in the rotor passage together with an elemental mass δm which is about to enter the rotor through A_1. In time δt this elemental mass enters, and an equal mass δm leaves through A_2. For steady flow, the angular momentum of the fluid within the passage does not change with time. The whole change in angular momentum during time δt is therefore equal to the difference between the angular momentum of the mass δm at inlet and outlet, viz.

$$\delta m(r_2C_{2w} - r_1C_{1w})$$

and the rate of change of angular momentum becomes

$$\frac{\delta m}{\delta t}(r_2C_{2w} - r_1C_{1w})$$

This must be equal to the moment about the axis of all the external forces applied by the rotor to the fluid, and this is the torque we may denote by T'. In the limit, $\delta m/\delta t$ is the rate of mass flow m, so that

$$T' = m(r_2C_{2w} - r_1C_{1w})$$

Here again the forces which the rotor exerts on the fluid, resulting in the torque T', may be both pressure forces and friction forces. From Newton's Third Law it follows that the torque T which the fluid exerts on the rotor is equal to $-T'$, and hence

$$T = m(r_1C_{1w} - r_2C_{2w}) \qquad (19.2)$$

The rate of work done when the rotor has the uniform angular velocity ω is given by

$$W = T\omega = m\omega(r_1C_{1w} - r_2C_{2w})$$

If U_1 and U_2 are the rotor velocities at the entry and exit sections this reduces to

$$W = m(U_1C_{1w} - U_2C_{2w}) \qquad (19.3)$$

When $r_1 = r_2 = r$, as in an axial-flow machine, the torque is simply

$$T = m(C_{1w} - C_{2w})r$$

In this simple case the torque can be represented by a force

$$F_w = m(C_{1w} - C_{2w})$$

acting at the radius r. The work done is then

$$W = mU(C_{1w} - C_{2w}) \qquad (19.4)$$

where U is the rotor blade velocity (ωr) of the axial-flow machine.

Little need be said about the radial component of the fluid velocity. Normally the fluid enters and leaves the rotor through annuli, so that the flow is symmetrically

Fig. 19.2   Axial-flow turbine stage

disposed about the rotor axis and a change in the radial component of velocity has no net effect upon the rotor. If the flow is not symmetrical about the axis of rotation a net radial force may have to be taken up by the shaft bearings, but no work is done by this force.

427

19.2 Axial-flow turbines

An axial-flow turbine consists of one or more stages, each stage comprising one annulus of fixed nozzles and one annulus of moving blades as in Fig. 19.2a. Usually the total pressure drop across the stage is divided between the nozzles and blades as illustrated in Fig. 19.2b. The division is usually expressed, not in terms of pressure drops, but in terms of the corresponding enthalpy drops. The criterion used is the *degree of reaction* Λ defined as

$$\Lambda = \frac{\text{enthalpy drop in rotor blades}}{\text{enthalpy drop in stage}} = \frac{h_1 - h_2}{h_0 - h_2} \tag{19.5}$$

The two most common values of Λ used are 0 and 0·5 (or 0 and 50 per cent). When the reaction is zero, all the stage pressure drop occurs in the nozzles and the stage is

Fig. 19.3   Velocity triangles for turbine stage

called an *impulse stage*. In a 50 per cent reaction design the pressure ratios across the nozzles and blades are approximately equal. This will be referred to here simply as a *reaction stage*, since we shall not be considering the use of other values of Λ.

The mode of action of the turbine can best be studied by following the path of the fluid through a single stage at the mean radius of the annulus, as in Fig. 19.3a. The fluid enters the nozzles with velocity C_0 at pressure p_0 and is expanded to pressure p_1. It leaves the nozzles with a velocity C_1, in a direction making an angle α_1 with the tangential direction, i.e. the plane of rotation. C_1 must satisfy the energy equation

$$\tfrac{1}{2}(C_1^2 - C_0^2) = h_0 - h_1 \tag{19.6}$$

The velocity of the fluid relative to the moving blades can be found by subtracting vectorially the blade speed U. This is easily accomplished by drawing the inlet velocity triangle. To avoid a multiplicity of indices, relative velocities will be denoted by V

428

and the relative velocity at inlet to the blades is therefore V_1. V_1 makes an angle β_1 with the tangential direction, and if the fluid is to flow smoothly into the blade passages without undue disturbance, the inlet angle of the blades must be made approximately equal to β_1. If the outlet angle of the rotor blades is β_2 the direction of the relative velocity at outlet, V_2, will also be approximately β_2. Applying the energy equation to the flow *relative to the rotor blades* it follows that

$$\tfrac{1}{2}(V_2^2 - V_1^2) = h_1 - h_2 \tag{19.7}$$

since no work is done *relative to the blades*, i.e. in the frame of reference of the blades. V_2 must satisfy this equation. Vectorial addition of the blade speed then gives the absolute velocity C_2 and the direction α_2. Usually, if this is one stage of a multi-stage turbine, C_2 will be made equal to C_0 and α_2 equal to α_0; the fluid can then pass on to another similar stage.

Equation (19.4) gives the work done, i.e.

$$W = mU(C_{1w} - C_{2w}) = mU\Delta C_w$$

Note that the whirl component of the absolute outlet velocity, C_{2w}, is negative in the case depicted in Fig. 19.3. It is usually convenient to combine the inlet and outlet velocity triangles to give the velocity diagram of Fig. 19.3b. It is then easy to see that the algebraic difference of C_{1w} and C_{2w} is equal to the algebraic difference of the whirl components of the relative velocities at inlet and outlet. Thus,

$$W = mU\Delta C_w = mU\Delta V_w* \tag{19.8}$$

The axial thrust can be obtained from equation (19.1). For an axial flow stage the areas A_1 and A_2 are normal to the axial direction and consequently (19.1) becomes

$$R_a = m(C_{1a} - C_{2a}) + (p_1 A_1 - p_2 A_2) \tag{19.9}$$

Evidently the term $m(C_{1a} - C_{2a})$ is negative for the particular case shown in Fig. 19.3. It must be pointed out that equation (19.9) does not necessarily give the total thrust exerted on the shaft bearings. In addition to R_a, there may be a net pressure thrust acting on the solid rotor disk, of area A_d say, equal to $A_d(p_1 - p_2)$. In impulse turbines p_2 is equal to p_1, and if $A_1 = A_2$ as in Fig. 19.2, the thrust due to the pressure terms is zero. Thus only small thrust bearings are required. To reduce the thrust in multi-stage reaction turbines, the fluid is sometimes introduced at the mid-point of the casing and allowed to expand in opposite directions; the machine is then called a *double-flow turbine*.

As an alternative to equation (19.8), an expression for the work done can be found in terms of the fluid velocities by eliminating the enthalpy term from the energy

*The interchangeability of ΔC_w and ΔV_w is to be expected because, as we have already pointed out in section 18.4 (i), the momentum equation upon which (19.8) is based applies in any inertial frame of reference and we are here treating the rotating blade passages as if they are moving linearly with a uniform absolute velocity U.

equation. Thus, applying the energy equation between planes 1 and 2, Fig. 19.2a, we have

$$m(h_2 - h_1) + \frac{m}{2}(C_2{}^2 - C_1{}^2) + W = 0$$

Absolute velocities are used here (i.e. velocities relative to the turbine casing) so that the work term must be included. No work is done in the nozzles and W is therefore the work done in the stage. Using (19.7) the enthalpy term can be eliminated to give

$$W = \frac{m}{2}\{(C_1{}^2 - C_2{}^2) + (V_2{}^2 - V_1{}^2)\} \tag{19.10}$$

(19.10) can be shown to be identical with (19.8) by considering the geometry of the velocity triangles. Thus,

$$C_1{}^2 = V_1{}^2 + U^2 + 2V_1 U \cos \beta_1$$
$$C_2{}^2 = V_2{}^2 + U^2 - 2V_2 U \cos \beta_2$$

Subtracting,

$$C_1{}^2 - C_2{}^2 = V_1{}^2 - V_2{}^2 + 2U(V_1 \cos \beta_1 + V_2 \cos \beta_2)$$

But

$$\Delta V_w = V_1 \cos \beta_1 + V_2 \cos \beta_2$$

Fig. 19.4 Effect of increasing stage enthalpy drop

and therefore

$$W = mU\Delta V_w = \frac{m}{2}\{(C_1{}^2 - C_2{}^2) + (V_2{}^2 - V_1{}^2)\}$$

The agreement between the two derivations is not surprising because the energy equation contains implicitly the laws of motion (see section 18.3). Once the thermodynamic properties are eliminated the energy equation should yield the same result as an equation derived from purely mechanical considerations.

So far we have concentrated on what happens in an individual stage and we must now briefly consider the turbine as a whole. The chief aim in the design of a turbine is to utilise the available isentropic enthalpy drop with the least possible friction loss and in the smallest number of stages. It should be clear from equation (19.8) that the work done per stage increases with the blade speed U, but consideration of the centrifugal stresses in the rotating parts leads to a maximum permissible blade speed. With the blade speed fixed, an attempt to use a small number of stages (and hence a large enthalpy drop per stage) may mean that a smaller fraction of the enthalpy drop is converted into work and that a larger fraction remains as kinetic energy in the stream leaving the rotor blades. This should be clear from Fig. 19.4. Furthermore, high fluid velocities lead to large friction losses. A useful criterion which indicates the effectiveness of energy abstraction in a stage is the *diagram efficiency* η_d (sometimes called the *utilisation factor*). This is defined as

$$\eta_d = \frac{\text{work calculated from velocity diagram}}{\text{energy available to rotor blades}} \tag{19.11}$$

The denominator is the kinetic energy at inlet to the rotor blades, $mC_1^2/2$, plus the energy $m(V_2^2 - V_1^2)/2$ made available by any expansion in the rotor. From equation (19.10) it is apparent that the difference between the denominator and numerator is the rejected kinetic energy $mC_2^2/2$.

It must be emphasised that although a low diagram efficiency indicates the probability of large friction losses, $(1 - \eta_d)$ does not itself necessarily represent a loss. This is because the kinetic energy leaving one stage can be used in the following stage. The total work done by an expansion in a multi-stage turbine can be found by summing the individual stage outputs; i.e. per unit mass flow we have:

$$
\begin{aligned}
W_{01} &= (h_0 - h_1) + \tfrac{1}{2}(C_0^2 - C_1^2) \\
W_{12} &= (h_1 - h_2) + \tfrac{1}{2}(C_1^2 - C_2^2) \\
&\cdots \\
W_{n-1,n} &= (h_{n-1} - h_n) + \tfrac{1}{2}(C_{n-1}^2 - C_n^2) \\
\hline
\Sigma W &= (h_0 - h_n) + \tfrac{1}{2}(C_0^2 - C_n^2)
\end{aligned}
$$

Evidently the interstage velocities are irrelevant *except* in so far as they affect the friction losses.

Even the kinetic energy at outlet from the last stage, or *leaving loss*, can be utilised to some extent by fitting a diffuser (usually in the form of a volute) to the turbine exhaust. The effect of the diffuser is to increase the pressure drop across the turbine as illustrated in Fig. 19.5. In the figure the diffusion process is depicted as being isentropic, but in practice friction only permits part of the leaving loss to be recovered in this way.

The various relations derived in the foregoing paragraphs will now be applied to particular stage designs.

Fig. 19.5   Effect of diffuser at turbine exit

(i) *Impulse stage*

When all the enthalpy drop occurs in the nozzles there can be no change in the relative velocity of the fluid passing through the rotor blade passages. In the absence of friction there will also be no change in pressure or specific volume across the rotor blades, and the continuity equation then requires the area at right angles to the direction of flow to be the same at inlet and outlet. With reference to Fig. 19.6, these areas are $l_1 np \sin \beta_1$ at inlet and $l_2 np \sin \beta_2$ at outlet: n is the number of blades in the row, p is the pitch, and l is the height of the blade. The simplest method of obtaining equal areas is to maintain the blade height constant and make β_2 equal to β_1. If this is done the velocity diagram for the impulse stage* appears as in Fig. 19.6. In practice, friction will occur in the passages, and if the blade passages have equal inlet and outlet areas there will be a small pressure and enthalpy drop

*In steam turbine practice it is customary to refer to a single-stage impulse turbine as a de Laval turbine, and a turbine consisting of a number of such stages in series as a Rateau turbine.

resulting in a slight acceleration of the fluid relative to the blades.* In the following analysis friction in the rotor blade passages will be neglected.

Expressions for the work done by the stage and the diagram efficiency can be found as follows. With reference to Fig. 19.6,

$$\Delta V_w = V_1 \cos \beta_1 + V_2 \cos \beta_2$$

For the case where $V_2 = V_1$ and $\beta_2 = \beta_1$ this reduces to

$$\Delta V_w = 2V_1 \cos \beta_1$$

But

$$V_1 \cos \beta_1 = C_1 \cos \alpha_1 - U$$

Fig. 19.6 Impulse stage

and hence

$$W = mU\Delta V_w = 2mU(C_1 \cos \alpha_1 - U) \tag{19.12}$$

With an impulse stage the only energy available to the rotor blades is $mC_1^2/2$. The diagram efficiency is therefore

$$\eta_d = \frac{2mU(C_1 \cos \alpha_1 - U)}{mC_1^2/2} = 4\frac{U}{C_1}\left(\cos \alpha_1 - \frac{U}{C_1}\right) \tag{19.13}$$

The ratio U/C_1 is called the *blade speed ratio*.

Other things being equal, equation (19.12) indicates that the work output increases with reduction in nozzle outlet angle α_1. A reduction in α_1, however, implies a reduced axial velocity ($C_1 \sin \alpha_1$) and hence an increase in annulus area if the same mass flow is to be passed through the stage. The nozzle area over which the fluid flows is therefore increased, and this in turn implies an increase in nozzle friction loss. Considerations of this kind lead to the use of nozzle outlet angles of between 15 and 30 degrees. If α_1 is assumed to be fixed, equation (19.13) can be differentiated with respect to the blade speed ratio, and the result can be equated to zero to find the value of U/C_1 which yields a maximum value of η_d. Thus,

$$\frac{d(\eta_d)}{d(U/C_1)} = 4 \cos \alpha_1 - 8\frac{U}{C_1}$$

*This effect of friction is not obvious, but all that can be done here is to refer the reader to the well-known Fanno lines of fluid mechanics.

which equals zero when

$$\frac{U}{C_1} = \frac{\cos \alpha_1}{2} \qquad (19.14)$$

It will be seen by inspection of the velocity diagram that this optimum blade speed ratio corresponds to the case where the outlet velocity is axial, i.e. when the outlet kinetic energy is a minimum. Substituting (19.14) in (19.13) we have

$$(\eta_d)_{\text{max}} = \cos^2 \alpha_1 \qquad (19.15)$$

Similarly, the work at optimum blade speed ratio becomes

$$W = 2mU^2 \qquad (19.16)$$

The variation of η_d with U/C_1 is shown in Fig. 19.7.

The maximum permissible blade speed that can be used depends upon the temperature of the working fluid, the strength of the heat-resisting alloys available, and the life required of the turbine; blade speeds of about 250 m/s are common, but speeds as high as 370 m/s are sometimes used. When the rotational speed at which the turbine is required to operate is low, the blade speed is often chosen well below the permissible limit. It is then determined by the need to use a well-proportioned annulus; too high a blade speed at low rotational speed implies an annulus of large mean diameter and small height (for a given mass flow and therefore fixed annulus area). This might involve the use of unduly short blades, particularly at the high pressure end of the turbine where the specific volume of the fluid is small. Irrespective of how the

Fig. 19.7

blade speed is determined, once it is fixed equation (19.14) can be used to determine C_1, and hence the stage enthalpy drop or number of stages for a given overall enthalpy drop. This procedure is illustrated by the following example.

Example 19.1 A multi-stage gas turbine is to be designed with impulse stages, and is to operate with an inlet pressure and temperature of 6 bar and 900 K, and an outlet pressure of 1 bar. The isentropic efficiency of the turbine is likely to be 85 per cent. All the stages are to have a nozzle outlet angle of 15 degrees, equal inlet and outlet blade angles, a mean blade speed of 250 m/s and equal inlet and outlet gas velocities. Estimate the number of stages required. Assume $c_p = 1 \cdot 15$ kJ/kg K and $\gamma = 1 \cdot 333$.

The final temperature, assuming isentropic expansion, is

$$T' = \frac{900}{(6/1)^{0 \cdot 333/1 \cdot 333}} = 575 \text{ K}$$

Hence the actual overall temperature drop is

$$0 \cdot 85(900 - 575) = 276 \cdot 3 \text{ K}$$

The energy equation for the flow through any nozzle row is

$$\tfrac{1}{2}(C_1{}^2 - C_0{}^2) = c_p(T_0 - T_1)$$

If the optimum blade speed ratio is to be used, C_1 is given by

$$C_1 = \frac{2U}{\cos \alpha_1} = \frac{2 \times 250}{0 \cdot 966} = 517 \cdot 6 \text{ m/s}$$

C_0 is to be equal to C_2, and with $\beta_2 = \beta_1$ we have

$$C_2 = C_1 \sin \alpha_1 = 517 \cdot 6 \times 0 \cdot 259 = 134 \cdot 1 \text{ m/s}$$

Hence

$$T_0 - T_1 = \frac{517 \cdot 6^2 - 134 \cdot 1^2}{2 \times 10^3 \times 1 \cdot 15} = 108 \cdot 7 \text{ K}$$

Since there is no expansion in the rotor blades, $(T_0 - T_1)$ will also be the stage temperature drop, and therefore the number of stages becomes

$$\frac{276 \cdot 3}{108 \cdot 7} = 2 \cdot 54$$

Evidently three stages are required. The blade speed ratio used will be greater than the optimum, and the required expansion will be obtained with slightly lower gas velocities.

(ii) *Velocity-compounded impulse stage*

Before considering reaction designs it is necessary to mention an important modification to the simple impulse stage. Sometimes it is desirable to carry out a large enthalpy drop in one stage, in which case the fluid velocities are very high and, due to the blade speed limitation, the U/C_1 ratio would be low and the work output consequently a low fraction of the available enthalpy drop. The work output can be increased by *velocity compounding*, even though still using a low blade speed ratio. The method is illustrated in Fig. 19.8, which shows a *two-row velocity-compounded stage.**

The flow through the nozzles and first moving row is similar to the flow through a simple impulse stage, except that the ratio U/C_1 is much smaller. The fluid leaves the first row of rotor blades with a high velocity C_2 at an angle α_2 and enters a row of fixed guide blades having an inlet angle α_2. No further expansion takes place and these blades merely turn the stream into the direction required for entry to a second row of rotor blades, i.e. C_3 equals C_2. The guide blades are often made symmetrical, with the outlet angle α_3 equal to the inlet angle α_2, and the blade height will then be the same at inlet and outlet to give equality of areas. The geometry of the velocity triangles then fixes the magnitude V_3 and direction β_3 of the relative velocity at inlet to the second moving row. If these blades are also symmetrical, β_4 is equal to β_3 and V_4 is equal to V_3. The velocity diagrams are then determined, and the work done is

$$W = mU(\Delta V_{w1} + \Delta V_{w2}) \tag{19.17}$$

where suffixes 1 and 2 refer to the first and second rows of moving blades.

*A velocity-compounded stage is known as a Curtis stage.

With the foregoing assumptions about symmetrical moving and guide blades, and unchanged relative velocities in the moving and guide blade passages, it is easy to show that for a given value of α_1 the diagram efficiency is a maximum when

$$\frac{U}{C_1} = \frac{\cos \alpha_1}{4} \tag{19.18}$$

The leaving velocity C_4 is then axial. It can also be shown that with this optimum value of U/C_1 the diagram efficiency is the same as for the simple impulse stage, i.e.

$$(\eta_a)_{\max} = \cos^2 \alpha_1$$

but that the work done becomes

$$W = 6mU^2 + 2mU^2 = 8mU^2 \tag{19.19}$$

Thus, for any given blade speed, a two-row stage can utilise four times the enthalpy drop of a simple impulse stage without any reduction in diagram efficiency. The

Fig. 19.8   Velocity-compounded stage

velocities are of course high in the nozzles and first rotor row, and friction losses may be rather large. One important use of a velocity-compounded stage is mentioned in section 19.2 (iv).

Example 19.2   The nozzles of a two-row velocity-compounded stage have outlet angles of 22 degrees and utilise an isentropic enthalpy drop of 200 kJ per kg of steam. All moving and guide blades are symmetrical, and the mean blade speed is 150 m/s. Assuming an isentropic efficiency for the nozzles of 90 per cent, find all the blade angles and calculate the specific power output produced by the stage. The velocity at inlet to the stage can be neglected.

435

Since the inlet velocity may be assumed zero, the outlet velocity from the nozzles is given by

$$\tfrac{1}{2}C_1^2 = \eta_N \Delta h'$$
$$C_1 = (2 \times 10^3 \times 0{\cdot}90 \times 200)^{1/2} = 600 \text{ m/s}$$

The velocity diagram for the first row can now be drawn, as in Fig. 19.9, from which

$$\beta_1 = \beta_2 = 29{\cdot}0°, \quad \alpha_2 = 41{\cdot}3°, \quad \Delta V_{w1} = 812 \text{ m/s}$$

With $C_3 = C_2$ and $\alpha_3 = \alpha_2$, the velocity diagram for the second row can also be drawn, from which

$$\beta_3 = \beta_4 = 64{\cdot}8°, \quad \Delta V_{w2} = 211 \text{ m/s}$$

The work output per kg/s of steam is therefore

$$W = U(\Delta V_{w1} + \Delta V_{w2}) = \frac{150(812 + 211)}{10^3} = 153{\cdot}5 \text{ kW}$$

Note that even with the comparatively low blade speed used in the foregoing example, the relative velocity at entry to the first row of rotor blades is very high.

Fig. 19.9

If too near the local velocity of sound in the fluid, losses will be incurred by the formation of shock waves in the blade passages. For this reason the two-row impulse stage is not suitable for use with high blade speeds—and for this reason also, three-row stages, which have an even lower optimum U/C_1 ratio, are seldom used. It is worth noting that sonic velocity, given by $[-v^2(\partial p/\partial v)_s]^{1/2}$ in general or $(\gamma RT)^{1/2}$ for a perfect gas [see section 18.1 (i)], is higher at the high-pressure end of an expansion. It follows that higher fluid velocities relative to the blade are permissible in the first few stages of a multi-stage turbine and therefore that this might be where a two-row impulse stage can be used without incurring losses from shock waves. We shall see in section 19.2 (iv) that there are in fact good reasons for choosing this type of stage for the high-pressure end of a turbine.

(iii) *Reaction stage*

Before considering the special case of 50 per cent reaction, it is convenient to derive a general expression for the degree of reaction in terms of the main variables. This is

436

most easily done when the velocities at inlet and outlet of the stage are equal and when the axial velocity remains constant throughout the stage. Both these conditions are commonly used in reaction designs because they enable the same blade shapes to be used in successive stages.

In general the degree of reaction is

$$\Lambda = \frac{h_1 - h_2}{h_0 - h_2}$$

Since the velocities are to be the same at inlet and outlet of the stage, the energy equation, per unit mass flow, reduces to

$$W = h_0 - h_2$$

Therefore, from equation (19.8), and using the nomenclature of Fig. 19.3,

$$h_0 - h_2 = U\Delta V_w = UC_a(\cot \beta_1 + \cot \beta_2)$$

Also, applying the energy equation to the flow relative to the rotor blades,

$$h_1 - h_2 = \tfrac{1}{2}(V_2{}^2 - V_1{}^2)$$
$$= \tfrac{1}{2}C_a{}^2(\mathrm{cosec}^2\, \beta_2 - \mathrm{cosec}^2\, \beta_1)$$
$$= \tfrac{1}{2}C_a{}^2(\cot^2 \beta_2 - \cot^2 \beta_1)$$

It follows that

$$\Lambda = \frac{C_a}{2U}(\cot \beta_2 - \cot \beta_1) \tag{19.20}$$

Putting $\Lambda = 0$ in (19.20) yields the impulse stage with $\beta_1 = \beta_2$, and putting $\Lambda = 0\cdot5$ we have the reaction stage with

$$\frac{U}{C_a} = \cot \beta_2 - \cot \beta_1 \tag{19.21}$$

From the geometry of the velocity triangles we also have the equation

$$U = C_a(\cot \alpha_1 - \cot \beta_1) = C_a(\cot \beta_2 - \cot \alpha_2)$$

Comparing this with (19.21) we get

$$\alpha_1 = \beta_2 \quad \text{and} \quad \alpha_2 = \beta_1$$

Also, from the initial assumptions that $C_2 = C_0$ and $C_a = C_0 \sin \alpha_0 = C_2 \sin \alpha_2$, it follows that

$$\alpha_2 = \alpha_0$$

Evidently, for a 50 per cent reaction design the nozzles and blades have the same shape, and the velocity diagram is symmetrical as shown in Fig. 19.10a. (Since the blades also act as nozzles in a reaction design, the term nozzle is sometimes discarded and the stage is said to consist of a row of *stator blades* followed by a row of *rotor blades*.)

The optimum blade speed ratio for the reaction stage can be found as follows. From the symmetry of Fig. 19.10a,

$$\Delta V_w = 2C_1 \cos \alpha_1 - U$$

and therefore, per unit mass flow,

$$W = U\Delta V_w = U(2C_1 \cos \alpha_1 - U) \tag{19.22}$$

Since there is an enthalpy drop in the rotor blades, the energy available to the rotor blades per unit mass flow is

$$\frac{C_1^2}{2} + \frac{(V_2^2 - V_1^2)}{2}$$

And since $V_2 = C_1$ this becomes

$$C_1^2 - \tfrac{1}{2}V_1^2$$

Fig. 19.10   Velocity diagram for reaction stage

Furthermore

$$V_1^2 = C_1^2 + U^2 - 2C_1 U \cos \alpha_1$$

and therefore the energy available is

$$\tfrac{1}{2}(C_1^2 - U^2 + 2C_1 U \cos \alpha_1)$$

It follows that

$$\eta_d = \frac{2U(2C_1 \cos \alpha_1 - U)}{(C_1^2 - U^2 + 2C_1 U \cos \alpha_1)}$$

$$= \frac{2\dfrac{U}{C_1}\left(2\cos \alpha_1 - \dfrac{U}{C_1}\right)}{1 - \left(\dfrac{U}{C_1}\right)^2 + 2\left(\dfrac{U}{C_1}\right)\cos \alpha_1} \tag{19.23}$$

Differentiating (19.23) with respect to U/C_1 and equating to zero we have the optimum blade speed ratio given by

$$\frac{U}{C_1} = \cos \alpha_1 \tag{19.24}$$

When the optimum blade speed ratio is used the velocity diagram appears as in Fig. 19.10b.

Substituting (19.24) in (19.22) the work done for a rate of flow m becomes

$$W = mU^2 \tag{19.25}$$

and substituting (19.24) in (19.23) yields

$$(\eta_d)_{max} = \frac{2 \cos^2 \alpha_1}{1 + \cos^2 \alpha_1} \tag{19.26}$$

Fig. 19.11 shows the variation of diagram efficiency with U/C_1 for the three types of stage described. The significance of these curves will be discussed in the next sub-section.

(iv) *Multi-stage turbines and internal losses*

The overall enthalpy drop across a turbine, particularly in steam turbine plant, is usually great enough to require the use of a large number of stages in series. As many as fifty stages may be required. Assuming that optimum blade speed ratios are used in each case, the work done per stage for the three types considered has been shown to be

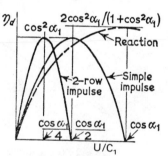

two-row impulse: $8mU^2$
simple impulse: $2mU^2$
reaction: mU^2

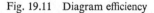

Fig. 19.11   Diagram efficiency

It might therefore be supposed that it is advantageous to use a succession of two-row impulse stages, since this would result in the least number of blade rows for a given overall enthalpy drop. However, it is also important to utilise the given isentropic enthalpy drop with the minimum loss, and when losses are taken into account the order of merit is not so obvious.

The losses in a turbine may be divided conveniently into two groups. There will be *external* losses due to bearing friction and the power required to drive auxiliaries. These losses will be similar for all types of turbine and will not be considered here. The second and major group, the *internal losses*, may be dealt with under two heads. (a) *Blading friction losses:* The fluid friction in the nozzle and rotor blade passages results in the actual enthalpy drop being less than the isentropic enthalpy drop. These friction losses are proportional to the square of the average relative velocity through the blading, and the surface area of blading swept by the fluid; they also depend on the nature of the flow. In general, for the same surface area and average relative velocity, the friction incurred by flow in a curved passage is less when the flow is accelerating, as in reaction blading, than it is when the velocity is substantially constant as in impulse blading. The reason for this is as follows. Most of the friction occurs in a boundary layer where there is a steep velocity gradient and it is much less in a laminar boundary layer than in a turbulent boundary layer.* The boundary layer on a blade is normally turbulent over the concave surface, but on the convex surface it is initially laminar and only becomes turbulent at some distance from the leading edge. The effect of having a falling pressure gradient in the direction of flow is to cause the desirable laminar condition to prevail over a greater part of the blade surface.

A comparison of the velocity diagrams for the three types of stage (Fig. 19.12) shows that the average velocity through the nozzle and rotor blading is less for the

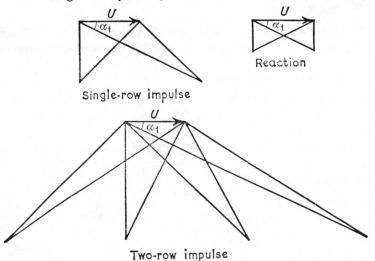

Fig. 19.12   Comparative velocity diagrams—optimum U/C_1 ratios, same values of
U and α_1

reaction stage than for either of the others. This fact, coupled with the beneficial pressure gradient across the reaction rotor blades, is sufficient to outweigh the effect of the greater surface area incurred by the larger number of stages in a reaction design. The net result is that friction losses are least in a multi-stage reaction turbine.

(b) *Leakage losses:* Turbine rotor blading is carried either on a series of disks or a drum, as shown in Figs. 19.13a and b. Because clearance is needed between the moving and stationary parts, some fluid passes through the turbine without doing its full complement of work on the blading. Losses incurred in this way are called leakage losses, and since they involve a form of throttling, they contribute to the irreversible increase of entropy which occurs during the expansion.

*The concept of a boundary layer is described in section 22.1.

The leakage flow area between the nozzle diaphragm and rotor shaft in (a) is obviously much less than that between the nozzle tips and rotor drum in (b). The disk construction is therefore always used for impulse stages, because the pressure difference across a nozzle row is comparatively large. On the other hand, the drum construction is usually preferred for reaction stages where the pressure drop across a nozzle row is not so great, because this type of construction does avoid the inevitable friction loss incurred by any disk spinning in a fluid.

The leakage past the tips of the rotor blades will clearly be greater in a reaction stage, owing to the pressure drop which is absent in any impulse design. In general, therefore, the total leakage losses are rather greater in a reaction design than in an impulse design. This is particularly true at the high pressure end of a turbine where,

Fig. 19.13   Disk and drum construction

owing to the high density of fluid, the annulus area required is small and the tip clearance is a relatively large proportion of the blade height.

Summing up, we have seen that there are two main sources of loss to consider:
  (i) Blading friction losses which are least for a reaction design.
  (ii) Leakage losses which are least for an impulse design.
A glance at the expansion on a *p-v* diagram, Fig. 19.14, will emphasise the fact that when the work is divided equally between the stages (i.e. when approximately equal stage pressure *ratios* are used), the pressure *differences* causing leakage are greatest at the high-pressure end of a turbine. Furthermore, these pressure differences are particularly large in turbines of high pressure ratio (inlet/outlet). As a general rule, the leakage losses predominate over the friction losses at the high-pressure end of high pressure-ratio turbines (e.g. most steam turbines), whereas at the low-pressure end, or throughout low pressure-ratio turbines (e.g. gas turbines), the friction losses predominate.

Taken together, the points made in the previous paragraph explain why reaction blading is invariably used for gas turbines, and why steam turbines normally have impulse stages at the high-pressure end. The choice of stages in steam turbines varies widely with the range of steam conditions employed and the relative importance of

441

capital cost and turbine efficiency. (Generally speaking turbine efficiency is the primary consideration because, apart from the saving in fuel, a lower steam consumption implies a smaller boiler and condenser which together account for the largest fraction of the capital cost of the whole plant.) In high pressure-ratio steam turbines it is usual to use at least two turbines in series: a high-pressure turbine with impulse stages followed by a low-pressure reaction turbine. The first impulse stage is commonly a two-row velocity-compounded design, because with the pressures and superheats used in modern plant it is important to achieve a large enthalpy drop in the first row of nozzles. Not only does this avoid exposing the turbine casing and rotor to the extreme steam conditions, but it also appreciably increases the volume flow before the steam reaches the first row of rotor blades. Without this increase in volume the blades might otherwise be too small for good efficiency. Even with an initial two-row

Fig. 19.14

impulse stage, it is sometimes necessary to use *partial admission* to avoid too small a blade height: the required flow area is obtained with a reasonable annulus height by blanking off sectors of the row of nozzles. It is of interest to note that the power output of steam turbines is commonly controlled by varying the inlet area in this way.

After this brief discussion of the sources of loss in a turbine, it can now be explained why small turbines (i.e. turbines of low mass flow and small work output) have poorer efficiencies than large turbines. When the blades are very small both the fluid friction loss and leakage loss become a large proportion of the work done: the former because the boundary layer occupies a relatively large portion of the passage, and the latter the because clearances between stationary and moving parts cannot be reduced proportionately.

We opened this sub-section with a comparison of the three types of stage on the basis of number of stages required, and went on to show that, when losses are considered, reaction designs are usually preferred in spite of the relatively large number of stages they involve. The position as regards number of stages is not quite as poor as our simple analysis has indicated however, because the optimum blade speed ratio need not necessarily be used. Fig. 19.11 indicates that the curve of diagram efficiency for a reaction stage is fairly flat near the optimum U/C_1 ratio, and it is possible to use a lower value of U/C_1 (i.e. larger enthalpy drop per stage) without an appreciable loss in efficiency. It must be remembered however, that the diagram efficiency is only a rough guide to the probable susceptibility of a stage to friction losses, and the true optimum blade speed ratio for maximum *isentropic* efficiency ($\Delta h/\Delta h_{\text{isen}}$) is not necessarily the nominal optimum value we have been deducing. The true optimum can only be found from a more complete analysis which makes use of friction data obtained from tests on rows of blades in 'cascade tunnels'.* Although this cannot be entered into here, we shall consider qualitatively the likely effect of using higher and lower values of the U/C_1 ratio in a reaction stage.

*The type of analysis used for gas turbines can be found in Ref. (13).

Fig. 19.15 shows the velocity diagrams for blade speed ratios less than, equal to, and greater than, the nominal optimum value. The blade speed U and nozzle angle α_1 have been assumed the same in each case. It will be seen that the use of a low U/C_1 ratio involves:

(1) higher fluid velocities,
(2) decrease in surface area swept by the fluid.

The surface area is reduced both because of the smaller number of stages required, and because the higher axial velocity implies that a smaller annulus area is required to

$U/C_1 < \cos \alpha_1$ $U/C_1 = \cos \alpha_1$ $U/C_1 > \cos \alpha_1$

Fig. 19.15   Reaction stage with low and high values of U/C_1 (same U and α_1)

pass the given mass flow. Item (2) partially offsets (1), but since the friction is proportional to the square of the fluid velocity item (1) predominates and there is a decrease in isentropic efficiency of the stage.

The use of a high U/C_1 ratio involves:

(1) lower fluid velocities,
(2) increase in surface area due to the greater number of stages and lower axial velocity.

In this case there may even be a slight rise in isentropic efficiency as the U/C_1 ratio is increased from the nominal optimum value, and the eventual fall in efficiency is more gradual. Fig. 19.16 shows a typical curve of isentropic efficiency versus U/C_1 ratio. It follows that a rather greater sacrifice in efficiency may be incurred by attempting to use a smaller number of stages (i.e. lower U/C_1 ratio), than might be supposed from simple consideration of the variation in diagram efficiency.

To close this discussion of losses, one form of loss peculiar to steam turbines must be mentioned. It was suggested in section 11.3 that it is undesirable for much of the expansion to occur in the wet region. It seems that liquid in the steam flow appears in three places, namely as a film running along the casing, as a fine mist suspended in the steam, and

Fig. 19.16   Isentropic efficiency of reaction stage

as a film on the surface of the nozzle blades (a liquid film does not form on the rotor blades because any moisture is flung off by centrifugal effects). The film on the nozzle blades detaches itself in relatively large droplets at velocities well below the mean velocity of the vapour, although sufficiently high to cause erosion damage at the leading edges of the rotor blades, as indicated by the dotted arrows in Fig. 19.17. The turbine efficiency is also reduced. An annular collecting slot in the casing is often used to reduce erosion, but clearly this does not remove the moisture which does the damage. More recently, slots in the trailing edges of hollow nozzle blades through which the liquid film is sucked away have been tried on an experimental scale, see Ref. (20). Provided that the liquid from the trailing edge can be removed, the 0·88 lower limit for dryness fraction suggested in section 11.3 might not represent the absolute limit it has been thought to be.

(v) *Stage efficiency, overall efficiency, reheat factor and polytropic efficiency*

Both the isentropic efficiency of a stage, and the overall isentropic efficiency of a turbine, can be defined in various ways depending on whether it is desired to take account of the whole, or a particular part, of the losses incurred. We have seen that the work done on the blading is less than can be expected from the isentropic enthalpy drop because of blading friction loss and leakage loss. Both these forms of loss increase the entropy of the expanding fluid and lead to a reduced actual enthalpy drop. Before the work done on the blading reaches the shaft, some of it will be dissipated in overcoming the disk friction loss when disk construction is employed. This loss also contributes to the increase in entropy and reduction in enthalpy drop. All these internal losses affect the state of the fluid flow-ing through the stage or turbine. Efficiencies defined in terms of the internal work delivered *to* the shaft are called *internal efficiencies*. Finally, the work delivered *by* the shaft, or *shaft work*, is even smaller by reason of the external losses (e.g. bearing friction) which do not affect the state of the fluid: efficiencies based upon the shaft work may be called *external efficiencies*. We shall here be concerned solely with the internal efficiency of a stage or turbine.

Fig. 19.17   Effect of a wet vapour

The internal work done during the expansion of unit mass flow from state 1 to state 2 is

$$W = (h_1 - h_2) + \tfrac{1}{2}(C_1{}^2 - C_2{}^2)$$

In a multi-stage turbine the interstage velocities are equal, or nearly so, and for our purposes we can simply write

$$W = (h_1 - h_2) = \Delta h$$

The internal work is a maximum when the expansion between the given pressures is isentropic, i.e.

$$W_{\text{max}} = (h_1 - h_2') = \Delta h'$$

Defining the efficiency always as W/W_{max}, we arrive at the following definitions of the isentropic efficiency of a stage, η_s, and overall isentropic efficiency of a turbine, η_o:

$$\eta_s = \frac{\Delta h_s}{\Delta h'_s} \quad \text{and} \quad \eta_o = \frac{\Delta h_o}{\Delta h'_o}$$

The following argument may be used to show that η_o is always greater than η_s.

For simplicity, we shall consider a four-stage turbine with each stage having the same efficiency η_s. The expansion is shown on an h-s diagram in Fig. 19.18. The dotted curve indicates the average state of the fluid during the actual expansion: in steam turbine practice it is called the *line of condition*. The total work Δh_o can be expressed either in terms of the overall isentropic enthalpy drop as

$$\eta_o \Delta h'_o$$

or in terms of the isentropic stage enthalpy drops as

$$\eta_s(\Delta h'_{s1} + \Delta h'_{s2} + \Delta h'_{s3} + \Delta h'_{s4}) = \eta_s \Sigma \Delta h'_s$$

Equating these expressions we have

$$\eta_o = \eta_s \frac{\Sigma \Delta h'_s}{\Delta h'_o} = \eta_s \mathscr{R} \qquad (19.27)$$

Fig. 19.18

$\Sigma \Delta h'_s$, called the *cumulative enthalpy drop*, is evidently greater than $\Delta h'_o$, because the vertical height between any pair of constant pressure lines increases with increase of entropy. The ratio $\mathscr{R}$ of the cumulative enthalpy drop to the isentropic enthalpy drop is called the *reheat factor*, and it is always greater than unity.

The physical interpretation of this result is that the internal losses in any stage (except the last) are partially recoverable in subsequent stages due to the 'reheating' effect of friction. It follows that it is more important to use high-efficiency stages at the low-pressure end of a turbine than at the high-pressure end. It should also be evident from the foregoing analysis that the difference between η_o and η_s increases as the overall pressure-ratio of the turbine is increased. For this reason high pressure-ratio turbines tend to be more efficient than low pressure-ratio turbines.

Example 19.3   Twelve successive stages of a reaction steam turbine have blades with effective inlet and outlet angles of 80° and 20° respectively. The mean diameter of the blade rows is 1·2 m and the speed of rotation is 3000 rev/min. If the axial velocity is constant throughout the stages, estimate the enthalpy drop per stage.

The steam inlet conditions are 10 bar and 250 °C, and the outlet pressure is 0·2 bar. Assuming a reheat factor of 1·04, estimate the stage efficiency (defined without reference to the kinetic energy term).

Determine the approximate blade height at the outlet of the sixth stage, if the twelve stages are to develop 7500 kW. Assume the line of condition to be straight on the h-s diagram.

The mean blade speed is

$$U = \pi dN = \frac{\pi \times 1\cdot2 \times 3000}{60} = 188\cdot5 \text{ m/s}$$

The velocity diagram can now be drawn to scale as in Fig. 19.19a, or the required velocities can be calculated as follows.

$$\frac{V_1}{\sin 20} = \frac{U}{\sin 60} \quad \text{hence} \quad V_1 = 74\cdot4 \text{ m/s}$$

From symmetry,

$$\Delta V_w = U + 2V_1 \cos 80 = 214\cdot3 \text{ m/s}$$

Therefore

$$W = U\Delta V_w = \frac{188\cdot5 \times 214\cdot3}{10^3} = 40\cdot4 \text{ kJ/kg}$$

Since there is no change of kinetic energy between inlet and outlet, the enthalpy drop per stage must also be 40·4 kJ/kg.

The actual enthalpy drop for the 12 stages is $12 \times 40\cdot4 = 484\cdot8$ kJ/kg. Using the *h–s* chart for steam, as illustrated in Fig. 19.19b, the initial enthalpy is found to be 2944 kJ/kg,

Fig. 19.19

and after isentropic expansion to 0·2 bar the enthalpy is 2281 kJ/kg; i.e. an isentropic drop of 663 kJ/kg. The overall efficiency is therefore

$$\eta_o = \frac{484\cdot8}{663} = 0\cdot731$$

If the reheat factor is 1·04, the stage efficiency must be

$$\eta_s = \frac{0\cdot731}{1\cdot04} = 0\cdot703$$

The assumed line of condition can be located on the *h–s* diagram since the actual final enthalpy is $2944 - 484\cdot8 = 2459$ kJ/kg. After the sixth stage half of the actual enthalpy drop has occurred, i.e. the enthalpy is $2944 - 242\cdot4 = 2702$ kJ/kg. At this point the steam is still superheated and the pressure is 1·75 bar. By interpolation in the tables at this pressure and $h = 2702$ kJ/kg the specific volume is found to be $v = 1\cdot02$ m³/kg. The power output is 7500 kW (or kJ/s), so that the mass flow of steam must be

$$\frac{7500}{484\cdot8} = 15\cdot47 \text{ kg/s}$$

The axial velocity, from the velocity diagram, is

$$V_{1a} = V_1 \sin 80 = 73\cdot27 \text{ m/s}$$

and therefore the annulus area required is

$$A = \frac{mv}{V_{1a}} = \frac{15\cdot47 \times 1\cdot02}{73\cdot27} = 0\cdot215 \text{ m}^2$$

If the mean diameter is d and the blade height l,

$$A = \pi dl$$

and hence

$$l = \frac{0\cdot215}{\pi \times 1\cdot2} = 0\cdot0570 \text{ m} = 5\cdot70 \text{ cm}$$

We have seen that the overall isentropic efficiency of a turbine increases as the overall pressure ratio (p_1/p_2) increases. When carrying out a series of cycle calculations to determine the optimum pressure ratio, as in section 12.2 for example, it is clearly misleading to assume a constant value for the overall turbine efficiency η_o over the range of pressure ratio. For such purposes the *small-stage* or *polytropic efficiency* η_∞ is a more useful concept. It can be arrived at by taking the argument leading to equation (19.27) a step further. There we assumed the expansion to be carried out in several stages each having the isentropic efficiency η_s. The polytropic efficiency is identical with η_s when the number of stages is infinite, i.e. it is the isentropic efficiency of an infinitesimal step in the expansion which is assumed constant throughout the whole expansion. Thus it is defined by

$$\eta_\infty = \frac{dh}{dh'} = \text{constant} \tag{19.28}$$

where $-dh$ and $-dh'$ are respectively the actual and isentropic enthalpy drops over the pressure drop $-dp$. That the relation between η_∞ and η_o is simple when the fluid is a perfect gas can be shown as follows. For an isentropic expansion we have $T/p^{(\gamma-1)/\gamma} = \text{constant}$, which on differentiation becomes

$$\frac{dT'}{T} = \frac{\gamma-1}{\gamma}\frac{dp}{p}$$

Combining this with (19.28), and remembering that for a perfect gas $dh = c_p\, dT$ and $dh' = c_p\, dT'$, we get

$$\frac{dT}{T} = \eta_\infty \frac{\gamma-1}{\gamma}\frac{dp}{p}$$

and on integration, η_∞ being constant by definition,

$$\frac{T_2}{T_1} = \left(\frac{p_2}{p_1}\right)^{\eta_\infty(\gamma-1)/\gamma} \tag{19.29}$$

Finally, from the definition of η_0, we have

$$\eta_0 = \frac{T_1 - T_2}{T_1 - T_2'} = \frac{1 - (T_2/T_1)}{1 - (T_2'/T_1)} = \frac{1 - (p_2/p_1)^{\eta_\infty(\gamma-1)/\gamma}}{1 - (p_2/p_1)^{(\gamma-1)/\gamma}} \tag{19.30}$$

This equation enables the variation of η_0 with pressure ratio to be determined for any given value of η_∞.

Note that if we write

$$\eta_\infty(\gamma - 1)/\gamma = (n - 1)/n$$

equation (19.29) is the relation between p and T for a polytropic process undergone by a perfect gas. Thus the definition of η_∞ implies the assumption that this non-isentropic process is polytropic, and hence the origin of the term 'polytropic efficiency'.

Fig. 19.20 Axial-flow compressor stage

Note also that the concepts of polytropic efficiency and reheat factor are alternative ways of accounting for the 'reheating' effect of friction—the former has been found the more useful for gas turbine calculations and the latter for steam turbine calculations.

19.3 Axial-flow compressors

We have already noted in the introduction to this chapter that from the point of view of the exchange of energy between the fluid and the rotor, the compressor can be regarded as a reversed turbine. An axial compressor stage is therefore always thought of as a row of rotor blades followed by a row of stator blades, Fig. 19.20a. The velocity triangles are usually depicted as in Fig. 19.20b, it being conventional to denote the flow directions by angles measured from the rotor axis and not from the plane of the

448

disk as in turbine design. The processes undergone by the fluid as it passes through the stage at the mean diameter of the annulus are as follows.

The fluid approaches the rotor blades with absolute velocity C_0 at an angle α_0. The relative velocity at inlet V_0 is obtained by vectorial subtraction of the blade speed U. If V_0 is found to be in the direction β_0, the rotor blade inlet angle must also be approximately equal to β_0 for the fluid to flow over the blade without interference. The rotor blade passage is made divergent and the flow is diffused, the fluid leaving with a reduced relative velocity V_1 at an angle β_1. Vectorial addition of the blade speed gives the absolute velocity C_1 at outlet from the rotor and its direction α_1; C_1 is larger than C_0 due to the work done on the fluid by the rotor. The inlet angle of the stator blade is made approximately equal to α_1, and the fluid is diffused in the stator passage. The velocity is reduced to C_2, and the stream leaves in the direction α_2. If the fluid

Fig. 19.21

is to leave the stage in a condition suitable for entry to another similar stage, C_2 and α_2 must be made equal to C_0 and α_0.

The work done in the stage, given by equation (19.4) but with the notation of Fig. 19.20, is

$$W = mU(C_{0w} - C_{1w})$$

It is evident from the velocity triangles that this is a negative quantity, in agreement with our sign convention for work. If the two velocity triangles are combined as in Fig. 19.21a it is convenient to write the equation as

$$W = -mU\Delta C_w \qquad\qquad (19.31)$$

where ΔC_w is $(C_{1w} - C_{0w})$.

In the stage just described, some increase of enthalpy, and therefore of pressure, is obtained in both rotor and stator rows. The concept of degree of reaction is also useful here, and for compressors it is defined as the ratio of the enthalpy rise in the rotor to the enthalpy rise in the stage, i.e.

$$\Lambda = \frac{h_1 - h_0}{h_2 - h_0}$$

The losses associated with fluid friction and blade tip clearance are found to be a minimum when the pressure rise is approximately equally divided between the rotor

449

and stator, and therefore a 50 per cent reaction design is normally used. By a similar analysis to that employed in section 19.2 (iii), it can be shown that with equal inlet and outlet velocities, and a constant axial velocity through the stage, 50 per cent reaction implies that the rotor and stator blades are identical in shape, i.e.

$$\beta_0 = \alpha_1 \quad \text{and} \quad \beta_1 = \alpha_0$$

The velocity diagram is then symmetrical as in Fig. 19.21b. This is the only type of stage design that will be considered here.

When a value for the isentropic efficiency is available, the stage pressure-ratio can be deduced as follows. Since axial compressors are used exclusively for compressing gases (normally air), the perfect gas relations may be introduced. With equal inlet and outlet velocities, the steady-flow energy equation applied to the whole stage is simply

$$W = -mc_p(T_2 - T_0)$$

and hence

$$c_p(T_2 - T_0) = U\Delta C_w \tag{19.32}$$

If the isentropic efficiency of the stage, η_s, is defined as the ratio of the isentropic temperature rise $(T_2' - T_0)$ to the actual temperature rise $(T_2 - T_0)$, we have

$$(T_2' - T_0) = \eta_s(T_2 - T_0)$$

$$\frac{T_2'}{T_0} = 1 + \eta_s\frac{(T_2 - T_0)}{T_0}$$

Consequently,

$$\frac{p_2}{p_0} = \left(\frac{T_2'}{T_0}\right)^{\gamma/(\gamma-1)} = \left\{1 + \eta_s\frac{(T_2 - T_0)}{T_0}\right\}^{\gamma/(\gamma-1)} \tag{19.33}$$

Substituting equation (19.32) in (19.33) we get finally

Fig. 19.22

$$\frac{p_2}{p_0} = \left(1 + \eta_s\frac{U\Delta C_w}{c_pT_0}\right)^{\gamma/(\gamma-1)} \tag{19.34}$$

Equation (19.33) may also be used to obtain the overall pressure-ratio of a multi-stage compressor if the overall isentropic efficiency is substituted for η_s and the suffixes 0 and 2 refer to the inlet and outlet of the whole compressor. The overall temperature rise is the sum of the stage temperature rises, and if the same work is done in each of N stages it is simply $N\Delta T_s$. By using a similar argument to that used when deducing the reheat factor for turbines, in conjunction with Fig. 19.22, it is easy to show that the overall isentropic efficiency of a compressor is *less* than the stage efficiency, and that this difference increases as the pressure ratio of a compressor is increased by adding more stages. The concept of polytropic efficiency is useful here also: for a compression it is defined as

$$\eta_\infty = \frac{\mathrm{d}h'}{\mathrm{d}h}$$

For a perfect gas, by repeating the argument used in deriving equation (19.29), we have the actual outlet temperature given by

$$\frac{T_2}{T_1} = \left(\frac{p_2}{p_1}\right)^{(\gamma-1)/\gamma\eta_\infty}$$

We must now discuss briefly the factors which limit the pressure rise that can be obtained efficiently in one stage. Before doing so it is necessary to emphasise the essential difference between a compressor and turbine, viz. that the flow in the rotor

Turbine blades   Compressor blades
Fig. 19.23

and stator passages of a compressor is diffusing, whereas it is accelerating in a turbine (Fig. 19.23). It is always more difficult to arrange for an efficient deceleration of flow than for an efficient acceleration. In a diffusion process there is a natural tendency for the fluid to break away from the wall of the diverging passage, reverse its direction, and flow back in the direction of the pressure gradient. This will occur if the rate of divergence of the passage is too great, and the tendency is particularly marked when the passage is curved. The difference between the nature of the flow in a compressor and turbine has a number of important consequences. Firstly, the blade design is much more critical in a compressor. Whereas turbine blade profiles can be constructed quite satisfactorily of circular arcs and straight lines, compressor blades must be of aerodynamic shape, i.e. of aerofoil section. A less obvious consequence is that the work done in a compressor stage cannot actually be estimated from equation (19.31) as it stands. The result of trying to induce a flow in a direction opposing the pressure gradient is that the axial velocity distribution across the annulus settles down after the first few stages into the peaky profile shown in Fig. 19.24. The effect of this is that

451

the one-dimensional treatment of this chapter is not nearly so adequate for compressors as it is for turbines, and the work done on unit mass of fluid depends to a greater extent than in a turbine on the radius at which the fluid is flowing through the annulus. In order to obtain an estimate of the average work done from one-dimensional considerations it has been found necessary to multiply equation (19.31) by an empirical coefficient, called the *work done factor*. A typical value, except for the first few stages where it is nearer unity, is 0·86.

The work done factor expresses a limitation upon the work that can be put into a stage. It should not be confused with the isentropic efficiency which indicates how much of this actual work is usefully employed in raising the pressure and how much is dissipated by friction. The other most important factor limiting the work which can be put into a stage is the permissible rate of divergence of the blade passage. Fig. 19.23 shows that this will be a function of the angle through which the fluid is

Fig. 19.24 Axial velocity distributions

deflected, i.e. $(\beta_0 - \beta_1)$ for the rotor and $(\alpha_1 - \alpha_2)$ for the stator. For the 50 per cent reaction design we are considering, equation (19.31) can be expressed as

$$W = -mUC_a(\tan \beta_0 - \tan \beta_1)$$

In this form it is clear that a limitation on $(\beta_0 - \beta_1)$ is reflected in a limitation on the work that can be put into a stage, and hence on the pressure ratio that can be obtained. With the blade speed limited by stressing considerations, and the axial velocity C_a limited by the need to keep the gas velocities relative to the blading below the sonic value and so avoid losses due to the formation of shock waves, any attempt to obtain a stage pressure-ratio much in excess of 1·2 is accompanied by a serious drop in isentropic efficiency.

We are now in a position to explain an important difference between the operating characteristics of positive displacement and non-positive displacement compressors. Assuming the compressor to be running at a constant rotational speed, the delivery pressure of a positive displacement compressor is increased steadily by throttling the outlet and reducing the mass flow. This process can be continued until the mass flow is zero. With the non-positive displacement rotary compressor there is only a restricted range of mass flow, at any given speed, within which the compressor can operate. If the mass flow is reduced beyond a certain point, the directions of the velocities relative to the blading are so widely different from the blade angles that the flow breaks down completely. The compressor is then said to *surge*. Even within the restricted range of operation the efficiency varies considerably, and it can only be maintained at a high value by varying the rotational speed and mass flow simultaneously in such a way that the velocity triangles maintain the same shape.

19.4 Radial-flow turbines

The axial-flow turbine is nearly always preferred to its radial-flow counterpart. There are, however, a few special applications for which the radial-flow turbine is sometimes used and it deserves a brief mention here. There are two possibilities to be considered —outward flow and inward flow.

(i) *Outward-flow turbines*

When the pressure ratio is large, as with steam turbines, it is essential to use outward radial flow rather than inward radial flow. This is because the increase in volume

Fig. 19.25   Rotor for Ljungström turbine

as the expansion proceeds can be partly accommodated by the increase in flow area associated with increase in diameter.

The Ljungström steam turbine is the only commercial outward-flow turbine. Briefly, it consists of concentric rows of blades, similar in shape to those of an axial-flow reaction turbine, attached to the opposing faces of two rotor disks, Fig. 19.25. The axial cross-section of the rotor exhibits a flow path of a convergent-divergent character. At first the increase in flow area with diameter is rather greater than is required to accommodate the increase in volume, but later the reverse applies and the blades must be increased in length. The disks rotate in opposite directions, each shaft driving a generator. Thus each row of blades produces a share of the work output, and the work output per pair of rows is thereby larger than in an axial-flow machine where half the rows are stationary. (There is no theoretical reason why the 'stator' rows should not be allowed to rotate in an axial-flow turbine but the mechanical difficulties are rather greater.) The velocity triangles for any row are similar to those for an axial-flow reaction stage, the main difference being that the blade velocity is greater at outlet than at inlet. The work done on any blade row is found from the rate of change of angular momentum in the usual way, using equation (19.3).

The Ljungström turbine has the advantage that it is very short in length, and this reduces the difficulties due to differential expansion on starting. It can in fact be started from cold in a matter of minutes, and it is this feature which has made it popular as a power plant for 'stand-by' or 'peak-load' generator sets.

(ii) *Inward-flow turbines*

When the turbine pressure ratio is comparatively small, as in gas turbine plant, it is possible to obtain the required increase in flow area with an inward-flow arrangement such as that shown in Fig. 19.26. This type of turbine is similar to the Francis hydraulic turbine. The gas enters the turbine via a volute and flows through a row of fixed

Fig. 19.26   Inward radial-flow turbine

nozzles situated around the periphery of the rotor. The high velocity stream then flows inward between vanes attached to the side of the rotor, and the passages are so shaped that the flow finally exhausts in an axial direction. Straight radial vanes are generally used, partly for ease of manufacture, but mainly because centrifugal forces would produce undesirable bending stresses if curved vanes were used.

With radial vanes, the velocity of the gas relative to the rotor at entry must be in the radial direction if the gas is to enter the passages smoothly. The inlet velocity triangle therefore appears as in Fig. 19.27. For a given rotor tip speed U_1, nozzle angle α_1, and absolute velocity C_1, the relative velocity will be V_1. The whirl component of the absolute velocity, C_{1w}, is equal to U_1. A cross-section through the 'eye' of the rotor, at the mean radius of the exit annulus, is also shown in Fig. 19.27. If the outlet velocity C_2 is to be truly axial, the outlet velocity triangle shows that the relative velocity V_2 must make an angle α_2 with the plane of rotation. The rotor vanes must therefore be bent into this direction at the exit. Since there is no angular momentum at the outlet, equation (19.3) becomes simply

$$W = mU_1C_{1w}$$

And since C_{1w} is equal to U_1, this reduces to

$$W = mU_1{}^2 \tag{19.35}$$

A simple expression for the pressure ratio required to produce this work can be deduced when:

(a) the fluid is a perfect gas,
(b) the velocities at inlet and outlet of the turbine are equal,
(c) a value is assumed for the overall isentropic efficiency of the turbine, η_0.

The steady-flow energy equation applied to the whole turbine, with inlet conditions denoted by suffix 0, is then

$$W = mc_p(T_0 - T_2) = mc_p\eta_0(T_0 - T_2')$$

Section at mean radius
of exit r_2

Fig. 19.27

where T_2' is the temperature after an isentropic expansion to the final pressure. This equation can be written as

$$W = mc_p\eta_0 T_0 \left\{ 1 - \left(\frac{p_2}{p_0}\right)^{(\gamma-1)/\gamma} \right\}$$

Equating this to (19.35) we have

$$\frac{p_2}{p_0} = \left(1 - \frac{U_1{}^2}{\eta_0 c_p T_0}\right)^{\gamma/(\gamma-1)} \tag{19.36}$$

The permissible rotor tip speed depends upon the gas temperature; it is not likely to be much greater than 460 m/s when the gas inlet temperature is 1200 K. Using these values of U_1 and T_0, an isentropic efficiency of 85 per cent, and typical values of c_p and γ for combustion gases of 1·15 kJ/kg K and 1·333, the pressure ratio becomes

$$\frac{p_2}{p_0} = \left(1 - \frac{460^2}{10^3 \times 1\cdot15 \times 0\cdot85 \times 1200}\right)^{1\cdot333/0\cdot333} = \frac{1}{2\cdot22}$$

455

It is apparent that if inlet/outlet pressure ratios of much more than 2 are required, more than one stage is necessary. Unfortunately the ducting arrangements necessary for compounding radial turbines involve a rather tortuous flow path, as illustrated in Fig. 19.28. The additional friction loss associated with this 'does not arise in axial-flow machines where the multiplication of stages presents no problem. For this reason the axial-flow turbine is always preferred when the pressure ratio is such that more than one radial stage would be required. Even when only low pressure-ratio turbines of the two types are compared, the axial machine normally has the higher efficiency. A radial design only becomes competitive for small turbines of low mass flow because then, as was pointed out in section 19.2 (iv), the axial-flow turbine becomes very inefficient.

Fig. 19.28

19.5 Radial-flow compressors

Although the inward radial-flow turbine has so far been little used, its opposite number, the centrifugal compressor, has been used extensively both in small sizes as superchargers and in large sizes as compressors for gas turbine plant. Centrifugal compressors have lower isentropic efficiencies than axial compressors, but other compensating advantages have proved to be important for some applications.

In appearance the centrifugal compressor is very similar to the radial turbine shown in Fig. 19.26. The rotor, or *impeller*, is driven in the opposite direction, so increasing

Section at mean radius
of inlet r_0

Fig. 19.29   Velocity triangles for centrifugal compressor

the angular momentum of the fluid (normally air). There is also a pressure rise due to centrifugal effects. The fluid leaves the impeller with a high velocity and flows through fixed diffuser passages in which the remaining pressure rise is obtained.

The velocity triangles at the inlet and outlet of the impeller are shown in Fig. 19.29, for the normal case where the absolute inlet velocity is axial. Ideally the fluid should be ejected from the periphery of the impeller at such a velocity that the whirl component C_{1w} is equal to the impeller tip speed U_1. Owing to the inertia of the air trapped

456

between the impeller vanes, C_{1w} is always less than U_1, the difference depending upon the number of vanes. The work done, from equation (19.3), is

$$W = -mU_1 C_{1w}$$

Introducing the overall isentropic efficiency, and assuming equal inlet and outlet velocities for the compressor as a whole, the energy equation for unit mass becomes

$$W = -\frac{c_p}{\eta_o}(T_2' - T_0)$$

Hence the pressure ratio is given by

$$\frac{p_2}{p_0} = \left(\frac{T_2'}{T_0}\right)^{\gamma/(\gamma-1)} = \left(1 + \frac{T_2' - T_0}{T_0}\right)^{\gamma/(\gamma-1)}$$

$$= \left(1 + \frac{\eta_o U_1 C_{1w}}{c_p T_0}\right)^{\gamma/(\gamma-1)} \tag{19.37}$$

For an air compressor, the maximum permissible peripheral speed would be about 460 m/s because, although not subject to high temperatures, the impeller is normally made of light alloy. With the usual number of impeller vanes C_{1w} is about $0.9\,U_1$. Using these figures, together with $c_p = 1.005$ kJ/kg K, $\gamma = 1.4$, $T_0 = 288$ K, and $\eta_o = 0.8$, the pressure ratio becomes

$$\frac{p_2}{p_0} = \left(1 + \frac{0.8 \times 0.9 \times 460^2}{10^3 \times 1.005 \times 288}\right)^{1.4/0.4} = 4.39$$

Impellers made from titanium permit higher tip speeds to be used and pressure ratios of between 6 and 7 are then possible. As with radial turbines, the losses associated with compounding usually mean that the axial-flow machine is preferred when more than one radial stage would be required. Centrifugal compressors producing pressure ratios much greater than 6 are therefore not used when high efficiency and low weight are important requirements. When size and weight are unimportant, a high efficiency can be obtained from multistage radial compressors if intercoolers are fitted between the stages.

This chapter provides the briefest possible introduction to the study of rotary expanders and compressors. A more detailed introduction, but one which still employs an integrated approach to the subject of turbo-machinery as a whole, is provided by Ref. (6). For more practical details the reader must turn to specialised books on each type of machine, such as Ref. (2) for steam turbines, and volumes X, XI and XII of Ref. (1) for components of gas turbine plant.

20 Direct Conversion

Often the production of mechanical work from a source of energy is merely an intermediate step in the production of electricity. It might be expected that if this intermediate step were eliminated the conversion of chemical or nuclear energy into electrical energy could be made with smaller and lighter equipment or perhaps with greater efficiency. Devices for *direct conversion* which are receiving most attention at the present time can be classified as follows:

1. Thermionic converters.
2. Thermoelectric converters.
3. Photovoltaic generators.
4. Magnetohydrodynamic generators.
5. Fuel cells.

It is much too early to say which of these has the greater potential for development or what are the applications for which each type is most suited. Broadly, the first three are likely to be suitable as sources of auxiliary power, the fourth for large-scale power generation, and the fifth for medium power outputs such as are required for road vehicles.

The theories on which the design of these systems are based are likewise at an early stage of development. An understanding of them demands a knowledge of thermodynamics, electromagnetics, fluid mechanics and solid-state physics which is beyond the scope of this book.* All we shall attempt here is a brief description of the main features of each type of device, with an indication of the points at which thermodynamics plays a part in the theory. It is hoped that this introduction to the topic will stimulate interest and encourage the engineering student to pay greater attention to supporting science courses. Such is the pace of change that subjects of study which are now peripheral may well become central in his lifetime.

The performance of four of the devices listed above is limited by thermodynamic considerations as will be apparent later. Thermionic and thermoelectric converters are heat engines, even if unconventional forms, in which electrons acting as the 'working fluid' can be said to be taken through a cycle.

---

*Ref. (24) is recommended as the first attempt to present a coherent introduction to the subject of direct conversion in a single volume, as opposed to books comprising collections of papers by specialists.

458

The maximum efficiency of such converters is therefore limited to the Carnot value $(T_1 - T_2)/T_1$, based on the temperatures of the source and sink between which they are operating. On the other hand, magnetohydrodynamic generators and fuel cells are essentially open systems, drawing on steady supplies of reactants at ambient conditions (p_0, T_0) which ideally could leave as products at ambient conditions. Their work output is thus limited to the maximum work of reaction, or change in Gibbs function ΔG_0, as explained in sections 17.5 and 7.9(iii). Photovoltaic generators are the exception, because here electromagnetic radiation is converted directly to electricity without the intermediate step of a conversion to heat or a chemical reaction.

It is worth pointing out that the laws of thermodynamics, and the energy equation $Q = \Delta U + W$, apply to more complex systems than those with which this book is primarily concerned. For example, they apply equally to systems involving electric and magnetic effects. The concept of work has to be widened, however, to embrace other than mere mechanical work, and more than two properties may have to be specified to fix the state of the system. Consider for example a system comprising an electrolytic cell, with an external circuit through which flows a quantity of electricity or charge $\delta \mathscr{Z}$ impelled by a potential difference $\mathscr{V}$. If the external circuit consists of a simple resistance, the energy flow $\mathscr{V} \delta \mathscr{Z}$ will be dissipated as ohmic heating: this being analogous to the situation where mechanical work done by a system is dissipated by friction. But in principle it is possible to pass $\delta \mathscr{Z}$ through a perfect electric motor (i.e. one with no copper, iron, and windage losses) which will do mechanical work equal to $\mathscr{V} \delta \mathscr{Z}$. Any mode of energy transfer *which can in principle be converted into mechanical work without loss*, can be included in the work term of the energy equation. W then becomes a symbol for a more generalised concept of thermodynamic work. If the operation of the electrolytic cell involves the liberation of gases, there may be work transfer due to a change in volume also. Thus for a reversible cell the energy equation could be written as:

$$dQ = dU + p\,dV + \mathscr{V}\,d\mathscr{Z}$$

It would be found that in this case the internal energy U is a function of three properties, e.g. p, T and $\mathscr{V}$. Before turning to books on direct conversion such as Ref. (24), the student will find it helpful to obtain this kind of background material from Ref. (8) of Part I.

20.1 Thermionic converters

Thermionic devices make use of the fact that when a metal plate is heated some of the free electrons are emitted from the surface. The distribution of energy among the free electrons, and the binding forces which exist at the surface, are such that only the faster moving electrons are emitted. The number emitted increases when the general energy level is raised by increasing the temperature of the metal.

The simplest form of thermionic device is the vacuum diode shown in Fig. 20.1a, in which two opposing plates are maintained at different temperatures, T_1 and T_2.

Since the current density (A/m^2) emitted from the surface is a function of temperature, there will be a net flow of electrons from the high temperature electrode (cathode) to the low temperature electrode (anode), returning to the cathode via an external circuit. The retarding potential barrier which the electrons leaving the cathode must overcome is illustrated in Fig. 20.1b as $(\delta + \phi_2 + \mathscr{V} - \phi_1)$. ϕ_1 and ϕ_2 are the potential barriers for electrons within the material of the cathode and anode respectively: in solid-state physics ϕ is referred to as the *work-function* of the material and its value varies greatly with the state of the metal surface. δ is the retarding potential due to the cloud of electrons in the space between the electrodes and is called the *space-charge barrier*. $\mathscr{V}$ is the output voltage. Using electromagnetic theory to calculate the number of electrons per unit area which have sufficient energy to overcome the barrier, a relation may be deduced for the net current density, $\mathscr{J}$, from cathode to anode in terms of ϕ, δ, $\mathscr{V}$ and the temperatures T_1 and T_2. The net power output is then $\mathscr{J}\mathscr{V}$ per unit area of emissive surface.

It is possible to apply the steady flow energy equation to an open system containing the cathode, with inlet and outlet planes across the conductor leading to the cathode

Fig. 20.1 A diode converter

and the space between the electrodes. The rate of heat flow, q_1, required to maintain unit area of cathode at T_1, is equal to the algebraic sum of the energy of the electrons flowing across the gap and in the conducting lead, the thermal radiation from the plate, and the conduction of heat in the lead. An expression can thereby be obtained for q_1, and hence finally for the efficiency of conversion $\mathscr{J}\mathscr{V}/q_1$.

With present day materials, reasonable efficiencies can be achieved only if the space-charge barrier, δ, is kept to a minimum. And with a vacuum diode this can be done only by maintaining a very small gap between the electrodes (<0.025 mm), which is difficult at the high temperatures required ($T_1 > 1600$ K). Another approach is to introduce an ionising agent into the gap to neutralise the space-charge. It is expected that, with further development, power densities of up to 300 kW/m² will be achieved with an efficiency of about 15 per cent.

If we view the system with its external leads and load as a whole, we can see that the electrons pass through a cycle of states while heat is supplied and rejected at the cathode and anode respectively. The maximum efficiency is therefore limited to $(T_1 - T_2)/T_1$. Having no highly stressed parts, very high temperatures can be used.

460

But the inevitably high rate of irreversible heat transfer by radiation to the anode prohibits the use of large temperature *differences* and the efficiency obtainable cannot be very high. Thermionic converters are likely to be used either as a source of auxiliary power in space applications where small size and low weight are important, or as 'topping' devices in conjunction with conventional power plant to make use of the hitherto unused temperature drop between available combustion temperatures and permissible boiler operating temperatures. In the latter case, the electrons would be the high temperature working fluid in a binary cycle. Experiments are proceeding on these lines in the nuclear power field, by incorporating the converter into nuclear fuel elements as shown in Fig. 20.2. The tubular sheath around the fuel rod is the cathode and the coolant, which is the heat source for the conventional power plant, flows along the outside of the tubular anode.

Fig. 20.2 Possible nuclear application

20.2 Thermoelectric converters

When the junctions of a loop of two wires of dissimilar metal are maintained at different temperatures an electric current is produced in the circuit: this is the net result of the Peltier and Seebeck effects. The open-circuit potential difference produced, per unit temperature difference between the junctions, is called the *Seebeck coefficient* or *thermoelectric power* α [see Fig. 20.3a]. With pairs of ordinary metals the value of α is only of the order of a few $\mu V/K$; adequate for use as thermocouples but much too low to be of interest for direct conversion. The development of semi-conductors, however, which have thermoelectric powers in the region of mV/K has now made thermoelectric converters a practical possibility. Useful power outputs can be obtained by connecting a large number of pairs in series.

The Seebeck effect in metals is due to the diffusion of electrons along the temperature gradient from the hot to the cold junction. If this is greater in one metal than the other there will be a net flow of electrons around the circuit. In semi-conductors the rate of diffusion is much greater, and furthermore the effect in each limb of the pair can be *additive* for the following reason. In some types of semi-conductor the charge carriers are electrons as in ordinary metals—these are referred to as n-type materials. But in others, p-type materials, the charge carriers are positive and can be regarded as 'holes' vacated by electrons. The current produced in a pair of n- and p-type materials is in effect equal to the sum of the rates of diffusion of negative and positive charge carriers.

A high thermoelectric power α is not the only criterion of importance in the choice of material for a thermoelectric converter. Fig. 20.3b shows a typical arrangement with hot junctions at T_1 and cold junctions at T_2. It should be clear that a low thermal conductivity, k, is necessary to keep the heat flow by conduction across the elements

461

to a minimum, and a low electrical resistivity, ρ, is required to reduce the internal ohmic ($\mathscr{I}^2\mathscr{R}$) losses. Semi-conductors are also good in these respects, because they are essentially materials of low thermal and electrical conductivity with the latter alone improved by the addition of impurities. α, k and ρ are temperature dependent and, when arriving at what is called the *figure of merit*, M, for a pair of n- and p-type materials, it has been found convenient to include the temperature range which it is intended to employ. The figure of merit commonly used is

$$M = \frac{\alpha^2}{k\rho} \times \frac{(T_1 + T_2)}{2} \tag{20.1}$$

The thermoelectric converter is like the thermionic variety in that it is a heat engine with a flow of charge carriers as the working fluid. The maximum possible efficiency is therefore $(T_1 - T_2)/T_1$. This could be achieved only if the figure of merit were infinite, i.e. if there were no irreversibilities due to thermal conduction or ohmic loss between

Fig. 20.3 Thermoelectric converter

the hot and cold junctions. A thermodynamic analysis can be used to show that the practical ideal efficiency of conversion for a given value of M is given by

$$\eta = \frac{T_1 - T_2}{T_1} \times \frac{(1 + M)^{\frac{1}{2}} - 1}{(1 + M)^{\frac{1}{2}} + (T_2/T_1)} \tag{20.2}$$

A not very rigorous derivation of this equation can be found in Ref. (24). Since we are here dealing with essentially irreversible processes, the relatively new science of irreversible thermodynamics is necessary for a rigorous treatment [see the chapter by Hatsopoulos and Keenan in Ref. (25)]. With present-day materials, M is about unity and T_1 and T_2 might be about 600 °C and 200 °C respectively. The corresponding ideal efficiency from equation (20.2) is thus about 10 per cent. Various losses not taken into account in deriving (20.2), such as those due to thermal contact resistance between the elements and the heat source and sink, result in actual efficiencies being nearer 5 per cent.

Unless new semi-conductor materials can be developed which are capable of operating at high temperatures, thermoelectric converters are likely to be used as 'bottomers' rather than 'toppers' in binary cycles. They are especially suitable for waste-heat recovery: perhaps using the exhaust gases of gas turbines as a source of heat, or the heat generated by a radioactive source as in the American 'Snap III' generator designed for space applications. Thermoelectric generators using paraffin heating have been used in the U.S.S.R. as a power source for radio receivers in rural areas. Thermoelectric devices can also be run in reverse—by supplying a current from an external source they act as refrigerators. The coefficient of performance is quite low, about 0·6, but the simplicity of the device makes it useful for some applications.

20.3 Photovoltaic generators

Electricity can be produced directly by exposing p-n junctions to electromagnetic radiation. The conversion is due to quantum interactions between individual photons of radiation and the electronic structure of the cell material, which take place when the photon energy ($= hc/\lambda$) exceeds a certain threshold value (h is Planck's constant, c the velocity of light and λ the wavelength). The two most important types of energy source are (a) radioactive materials, which yield radiations of very short wavelength ($\approx 10^{-12}$ m), and (b) sunlight whose wavelength ranges around 10^{-6} m. Photovoltaic cells designed to use sunlight have received the most attention.

Photovoltaic effects in a solid were first observed in 1877 by Adams and Day who worked with selenium. The work led to the development of photographic exposure meters, but they convert no more than 0·8 per cent of the incident radiation into electrical energy. With the introduction of n- and p-type semiconductors in the 1950s, conversion efficiencies of about 6 per cent were obtained and it was possible to think of using photovoltaic cells as a power source. Efficiencies of 15 per cent are now obtained with silicon p-n junctions.

Energy is produced by photons of radiation, creating free electrons and holes. To achieve a high efficiency it is necessary to produce electron-hole pairs very close to the p-n interface. Those far from the junction recombine without contributing to the cell's output. Furthermore, the internal resistance must be small. For these reasons cells consist of a very thin layer ($\approx 10^{-3}$ mm thick) of n-type material on a wafer of p-type material ($\approx 0·5$ mm thick). Unwanted surface recombination of electron-hole pairs is minimised by chemically etching the surface of the cell to leave it in a highly polished state, and reflection losses are reduced by using anti-reflection coatings. By considering the various losses inherent in the collecting and conversion processes, it can be shown by solid-state theory as in Ref. (24) that the theoretical maximum efficiency of a silicon cell is about 20 per cent, and with other possible materials such as gallium arsenide or cadmium telluride it is about 25 per cent. At present the actual efficiency of cells made of the theoretically better materials is much lower than that of silicon cells.

Because the source of energy is free, the factor limiting the use of photovoltaic cells as a power source is not so much the conversion efficiency as the high cost of production. So far this has limited their use to space applications, such as communication satellites, where the cost of the cells is a small proportion of the total cost of the system. If the cost can be reduced to a commercial level, then the efficiency will become important because it determines the number of solar cells required for a given power output. The power produced per unit surface area of cell might be increased by concentrating the radiation with lenses or reflectors. Unfortunately the conversion efficiency of present materials is reduced as the temperature of the cell increases so that the net effect is small. Cooling the cell is no answer: the result would simply be a thermoelectric system. Another possibility is the construction of multi-layer cells of different materials, matching their electronic characteristics to the energy spectra of the radiation in such a way that a greater proportion of the wavelength range can be used. Efficiencies of over 30 per cent might then be possible.

20.4 Magnetohydrodynamic generators

Strictly speaking, it is doubtful whether the magnetohydrodynamic (mhd) generator can be called a direct conversion device: certainly there is no direct conversion of heat

Fig. 20.4 Magneto hydrodynamic duct vis-à-vis the conventional generator

into electrical energy as in thermionic and thermoelectric converters. A gas at high pressure and temperature is still required as in conventional plant, but instead of using this via a turbine to drive metallic conductors through a magnetic field as in a conventional generator, the gas itself acts as a moving conductor (see Fig. 20.4). At very high temperatures a gas is ionised and the degree of ionisation can be enhanced by 'seeding' the gas with a small quantity of another substance such as caesium or potassium. Such ionised gases are referred to as plasmas. This conducting fluid is accelerated in a nozzle and passed at high velocity through a rectangular duct across which a magnetic field is applied at right angles to the direction of flow. An e.m.f. is thereby generated in the fluid, which can be utilised by placing electrodes along the sides of the duct.

The flow in the duct is governed by the momentum and energy equations which in this case must include electrical force and energy terms. A simple derivation of these equations for one-dimensional, frictionless, adiabatic flow is given in Ref. (24), where they are shown to be:

$$\text{momentum equation: } v\frac{dp}{dx} + C\frac{dC}{dx} = -v\mathscr{J}\mathscr{B}$$

$$\text{energy equation: } \frac{C}{v}\frac{d}{dx}(h + \tfrac{1}{2}C^2) = -\frac{\mathscr{J}\mathscr{V}}{b}$$

Here $\mathscr{B}$ is the magnetic flux density, $\mathscr{V}$ the potential difference across the electrodes, b the distance between electrodes, and $\mathscr{J}$ the current density given by

$$\mathscr{J} = \sigma\left(\frac{\mathscr{V}}{b} - C\mathscr{B}\right)$$

σ is the electrical conductivity of the fluid. From these equations it is possible to deduce such useful performance criteria as the rate at which enthalpy is converted into electrical energy, the ratio of actual to isentropic enthalpy drop, and the power produced per unit volume of duct. It can be shown that the latter, viz. the power density, is proportional to $\sigma C^2\mathscr{B}^2$. The velocity of flow C is much higher than the speed at which a solid conductor can be rotated in a conventional generator and this partially offsets the inherent disadvantage that σ (even for a seeded gas at 3000 K) is only about 100 (ohm-m)$^{-1}$ compared with 10^7 (ohm-m)$^{-1}$ for solid copper.

The overall efficiency will depend on the way in which the mhd duct is incorporated in the complete plant. Because sufficient ionisation can only be achieved at temperatures of 2500 °C or greater, the mhd generator is essentially a high temperature device. Furthermore, the temperature must still be very high at outlet from the duct, so that to obtain a reasonable overall efficiency it must be used as a 'topper' in conjunction with a conventional cycle. A typical open-cycle scheme is shown in Fig. 20.5: it would be possible to use a completely closed cycle if the combustion chamber was replaced by a nuclear reactor but the temperatures required for adequate ionisation are well beyond those at present possible in reactors. Considering the open-cycle plant shown, the mhd portion to the left of the dotted line A–A is an open system for which the efficiency might be expressed as

$$\frac{\text{net electrical work—compressor work}}{\Delta H_0}$$

The net electrical output is the gross output less the power supplied to the electromagnets producing the field $\mathscr{B}$, and ΔH_0 is the enthalpy of combustion. (Those who have studied section 7.9 (iii) will appreciate that the denominator should strictly be the change in Gibbs function ΔG_0.) For the plant as a whole, the numerator becomes

Fig. 20.5 Possible mhd generator plant

the net electrical output, including that from the steam cycle, minus the feed pump work.

The major sources of loss are fluid friction, heat loss through the walls of the duct, and internal electrical resistance. These losses can be made a small fraction of the output only by using large ducts and high flow rates. It appears that the mhd generator is unlikely to be used for powers below the megawatt range. This is a case where little useful experimental work can be done on a small scale and thus development costs are very high.

20.5 Fuel cells

A fuel cell is a practical method of achieving a near reversible chemical reaction in which there is a conversion of chemical energy into electrical energy without even the intermediate step of conversion into random molecular energy (i.e. sensible internal energy). Thus, whereas the thermionic and thermoelectric forms of direct converter are essentially heat engines and subject to the Carnot limitation on efficiency, the fuel cell is not. Theoretically, the maximum work of a chemical reaction discussed in section 17.5 (and section 7.9) could be made available as electrical energy. The development of such a device, if capable of operating with hydrocarbon fuels or cheap derivatives such as carbon monoxide, would be a great step forward in the efficient utilisation of fuel resources.

Fig. 20.6 Hydrogen-oxygen fuel cell

The fuel cell, Fig. 20.6, is an electrochemical device consisting basically of an electrolyte and two non-consumable electrodes (often porous plates through which the fuel and oxidant diffuses). The electrodes play a catalytic role in facilitating the ionising reactions. In a simple type of hydrogen-oxygen fuel cell the reactions might be written as

Anode: $H_2 \rightarrow 2H^+ + 2e^-$

Cathode: $2e^- + 2H^+ + \frac{1}{2}O_2 \rightarrow H_2O$

The open-circuit emf can be related to a function of the enthalpy and entropy of the reactants and products (in fact to the change in the Gibbs function ΔG_0), and the effect of temperature and pressure on the emf can be determined by chemical thermodynamic analysis [see Ref. (26)]. The ideal maximum work for the reaction is not made available in practice because of internal losses. These include inhibiting effects due to absorption of molecules on the surface of the electrodes, uneven concentration of the electrolyte when current is flowing, changes in the conductivity of the electrolyte near the electrodes, and ohmic heating ($\mathscr{I}^2\mathscr{R}$ loss) due to the resistance of the electrolyte. Depending on the fuel and oxidant employed, the emf produced is between 0·5 and 1·5 volts and any required output voltage can be obtained by connecting cells in series as in the ordinary electric battery. Current densities of up to 400 A/m^2 have been achieved.

Until recently, research in this field has been done by only a few individuals and mostly on the hydrogen-oxygen cell. The picture has been transformed since 1959 and now a large number of laboratories and research teams are actively engaged in this development. The result has been not only a marked increase in performance— in power density and efficiency—but also such a bewildering variety of types of cell that classification is difficult. Attempts have been made to classify them on the basis of operating temperature or pressure, and type of electrolyte, fuel, oxidant, or electrode, but it is too early for a neat picture to have emerged. There is little doubt that fuel cells will find a place for special military and space applications, but whether they will come into more general use depends on finding means of using cheap fuel and extending their working life.

Part IV Heat Transfer

Introduction

Heat is energy in transition under the motive force of a temperature difference, and the study of heat transfer deals with the rate at which such energy is transferred. When discussing thermodynamic processes, time is never considered a limiting factor, and ideal processes can be conceived during which heat is transferred by virtue of an infinitesimally small temperature difference. In actual processes some of the available temperature drop, which might be used for the production of work, must be sacrificed to ensure that the required quantity of heat is transferred in a reasonable time across a surface of reasonable size. Two types of problem are frequently of interest; the first, met in the design of boilers, heat exchangers, etc., is concerned with the promotion of the required rate of heat transfer with the minimum possible surface area and temperature difference; the other is concerned with the prevention of heat transfer, i.e. with thermal insulation.

Three modes of heat transfer may be distinguished, but the fact that a temperature difference is necessary is common to all. One mode of heat transfer is that of *conduction*, discussed in Chapter 21, in which energy is transferred on a molecular scale with no movement of macroscopic portions of matter relative to one another. Fourier's law forms the basis of all calculations of heat transfer by conduction.

A second mode of heat transfer occurs when temperature differences exist between a fluid and a solid boundary. Here the redistribution of energy is partly due to conduction and partly due to the motion of the fluid itself. Such motion can be due to a pump, as in the case of a fluid passing through the tubes of a heat exchanger, and this mode of heat transfer is called *forced convection*. On the other hand, the motion may be entirely due to density gradients in the fluid, caused by the temperature gradients—as when a stove heats the air in a room—and this is called *free* or *natural convection*. Convection is far more difficult to analyse than conduction because it is a combined problem of heat flow and fluid flow. The rate of heat transfer is influenced by all the fluid properties that can affect both heat and fluid flow, such as velocity, thermal conductivity, viscosity, density, thermal expansion, etc. In finding an exact solution to any particular problem, it is essential to satisfy the equation of motion (Newton's Second Law), the equation of energy, and the equation of continuity of flow, in addition to Fourier's law. The differential equations governing convection are complicated and exact solutions can be given in only a few simple cases. Some approximate analytical methods are given in Chapter 22, together with a description of an empirical approach using dimensional analysis.

The third mode of heat transfer is *radiation,* and this mode does not depend upon the existence of an intervening medium. All matter at temperatures above absolute zero emits energy in the form of electromagnetic waves. Gases at low pressure emit waves of particular frequencies only, which are the natural frequencies of vibration of the gas molecules. In gases at high pressure, or liquids and solids, the molecular vibrations are more damped, and the emission of energy is over wide frequency bands. It is impossible here to go into the mechanism of radiation, and from the engineer's point of view it is usually sufficient to know the total quantities of energy emitted and absorbed by matter at various temperatures. The calculation of radiation heat transfer, discussed in Chapter 23, is based mainly on the laws of Stefan-Boltzmann, Kirchhoff and Lambert.

Often all three forms of heat transfer are involved simultaneously. It is then usual to calculate the rate of heat transfer by each mode separately, adding the separate effects to provide an estimate of the total rate of heat transfer. In this way a complex problem can be resolved, and the relative importance of the various modes of heat transfer can be assessed. A few simple examples are given in Chapter 24. The final result may not be exact, however, because a certain amount of interaction usually exists between the different modes.

In the study of heat transfer it is usual to talk about the flow of heat, and of heat being contained in a material, as if heat were something tangible. Although in thermodynamics we permit ourselves to talk loosely about the flow of heat, we are always careful to distinguish between the energy in transition (heat), and the energy residing in the material (internal energy). This distinction is necessary because the change in internal energy can be due to either work or heat crossing the boundary of a system. Normally this distinction is unnecessary in heat transfer because we are considering processes during which the change in internal energy is due solely to a flow of heat, and we are not concerned with transformations of one form of energy into another. In heat transfer problems involving flowing fluids, some of the kinetic energy is transformed into random molecular energy by viscous forces, and greater care is needed in the use of the term heat. Nevertheless, for low velocities of flow the quantities of kinetic energy involved are usually small compared with the rate at which heat is transferred to or from the fluid, so that again no harm is done by the loose usage of the term. In high speed flow, however, when the dissipation of kinetic energy into random molecular energy may be considerable, the thermodynamic view of the concept of heat must be retained.

21   Conduction

When temperature differences are present in any matter, heat flows from the hot to the cold regions until the temperatures are equalised. This occurs even when the movement of macroscopic portions of matter is prevented as in solids. The actual mechanism is complicated, but a brief explanation is as follows.

In solids the conduction of heat is partly due to the impact of adjacent molecules vibrating about their mean positions, and partly due to internal radiation. When the solid is a metal, there are also large numbers of mobile electrons which can easily move through the matter, passing from one atom to another, and they contribute to the redistribution of energy in the metal. Indeed the contribution of the mobile electrons predominates in metals, thus explaining the relation which is found to exist between the thermal and electrical conductivity of such materials.*

In liquids, single molecules may make large excursions with frequent collisions, even when large-scale motion (convection) of the fluid is completely suppressed, and such movement assists the conduction of heat. Although the mobility of individual molecules is even greater in gases, the conductivity is appreciably lower than that of liquids and solids due to the relatively long 'mean free path' in gases and hence the comparatively infrequent impact between the molecules.

The detailed analysis of the mechanism of conduction is not discussed in this chapter. The first sections deal with Fourier's law of conduction, and its application to one-dimensional steady-flow problems. The equation for three-dimensional heat flow is then derived, and in the course of introducing relaxation methods, a two-dimensional steady-flow problem is solved. Finally, one-dimensional unsteady (transient) problems are considered with the aid of a numerical method.

For a brief discussion of the status of Fourier's law, vis-à-vis the laws of thermodynamics, the reader may turn to the end of section 6.7.

21.1 Fourier's law of heat conduction

Fourier's law of conduction is based on the empirical observation of one-dimensional steady heat flow through a solid. *One-dimensional* flow implies that the temperature is uniform over surfaces perpendicular to the direction x of heat conduction (Fig. 21.1a),

---

*A fairly easy but comprehensive modern explanation is given by J. D. Eshelby in Ref. (35), p. 267.

and such surfaces are called *isothermal* surfaces. *Steady flow* implies that the temperature at any point does not vary with time. It follows from considerations of continuity, that in one-dimensional steady flow the rate of heat flow through successive surfaces is constant.

Consider a plane layer of thickness dx, with one face maintained at a temperature t and the other at $(t + dt)$. The rate of heat flow Q is found to be proportional to the

Fig. 21.1   One-dimensional steady heat conduction through a plane-parallel wall

area of flow A and to the temperature difference dt across the layer, and inversely proportional to the thickness dx. This is Fourier's law and it can be expressed by the equation

$$Q = -kA \left(\frac{dt}{dx} \right) \tag{21.1}$$

The constant of proportionality k is called the *thermal conductivity* of the material, and the negative sign indicates that the heat flow is positive in the direction of temperature fall.

It follows immediately from equation (21.1) that for steady heat flow (Q independent of x) the temperature gradient will be constant if k is constant, i.e. the temperature profile will be linear. Equation (21.1) can then be integrated between the limits x_1 and x_2, to enable the heat flow to be expressed in terms of the surface temperatures t_1 and t_2 by

$$Q = -kA \frac{t_2 - t_1}{x_{12}} \tag{21.2}$$

When Q is in kW (or kJ/s), A in m^2, and the temperature gradient (dt/dx) in K/m, the units of k become kW/m K. Fig. 21.2 indicates that more convenient numbers are obtained if k is expressed in W/m K, and this smaller unit will often be used.

Fig. 21.2 gives some representative values of thermal conductivity for various materials, and it will be seen that the thermal conductivity varies with temperature. However, if the temperature difference $(t_2 - t_1)$ between the outside faces of the slab in Fig. 21.1a is small, a mean value of k may be chosen and assumed constant. When

$T/[K]$	250	300	400	500	600	800
Aluminium . . .	208	202	209	222	234	277
Copper . . .	393	386	377	372	367	357
Steel (0·5 %C) . .	57	55	52	48	45	38
Glass . . .		0·8–1·1				
Building brick . .		0·35–0·7				
Concrete . . .		0·9–1·4				
Asbestos . .		0·163	0·194	0·206		
Mercury, liquid . .	7·5	8·1	9·4	10·7	12·8	13·7
Sodium, liquid . .	135 (sol.)	135 (sol.)	86	80	74	6·3
Water (sat.) . . .	2·22 (ice)	0·614	0·687	0·642		
Steam (sat.) . .		0·0184	0·0275	0·0432		
Steam (low p) . .		0·0188	0·0266	0·0357	0·0463	0·0708
Hydrogen . .	0·156	0·182	0·228	0·272	0·315	0·402
Air . . .	0·0223	0·0262	0·0337	0·0404	0·0466	0·0577
Carbon dioxide . .	0·0129	0·0166	0·0246	0·0335	0·0431	0·0560

Fig. 21.2   Thermal conductivity $k/[W/m\ K]$

the temperature difference across the slab is large, due account must be taken of the variation of thermal conductivity with temperature. Typical temperature profiles for variable thermal conductivities are shown in Fig. 21.1b. Unless otherwise stated, k will be assumed constant, but the method of calculating the heat transfer when k is a function of temperature is illustrated in the following example.

Example 21.1   The conductivity of a plane-parallel wall is not constant, but can be expressed with sufficient accuracy as a linear function of temperature, $k = at + b$. Show that equation (21.2) can be used to calculate the rate of heat flow, if the arithmetic mean of the conductivities k_1 and k_2, at the temperatures t_1 and t_2 respectively, is substituted for k.

Considering unit area of wall, and writing $q = Q/A$, equation (21.1) becomes

$$q = -k\left(\frac{dt}{dx}\right) = -(at + b)\left(\frac{dt}{dx}\right)$$

For one-dimensional steady flow q is independent of x and hence, by integration between x_1 and x_2 (Fig. 21.1a), we have

$$q(x_2 - x_1) = -\frac{a}{2}(t_2^2 - t_1^2) - b(t_2 - t_1)$$

or

$$q = -\left\{\frac{a}{2}(t_2 + t_1) + b\right\}\frac{t_2 - t_1}{x_{12}}$$

But the mean conductivity is

$$\frac{k_2 + k_1}{2} = \frac{(at_2 + b) + (at_1 + b)}{2} = \left\{\frac{a}{2}(t_2 + t_1) + b\right\}$$

and hence equation (21.2) can be applied in the manner suggested.

That there is some similarity between the flow of heat and that of electricity becomes apparent when equation (21.2) is compared with Ohm's law, $\mathcal{I} = \mathcal{V}/\mathcal{R}$. The temperature difference, responsible for the heat conduction, plays an analogous part to the potential difference, or voltage $\mathcal{V}$, in the conduction of electricity. The current $\mathcal{I}$ can then be compared to the heat flow per unit area q, and the resistance $\mathcal{R}$ to the value of x_{12}/k (see Fig. 21.1a). The practical value of this *electrical analogy* will become clear in the next section.

21.2 One-dimensional steady conduction through a composite wall

A wall built up of three different materials is shown in Fig. 21.3. The surface temperatures of the wall are t_1 and t_4, and the temperatures of the interfaces are t_2 and t_3. For steady flow through the wall, the heat flow through successive slabs must be the same for reasons of continuity, and hence

$$q = -k_{12}\frac{t_2 - t_1}{x_{12}} = -k_{23}\frac{t_3 - t_2}{x_{23}} = -k_{34}\frac{t_4 - t_3}{x_{34}} \tag{21.3}$$

It follows from (21.3) that

$$-(t_2 - t_1) - (t_3 - t_2) - (t_4 - t_3) = -(t_4 - t_1) = q\left(\frac{x_{12}}{k_{12}} + \frac{x_{23}}{k_{23}} + \frac{x_{34}}{k_{34}}\right) \tag{21.4}$$

For the complete wall as a whole we can write

$$q = -U(t_4 - t_1) \tag{21.5}$$

where U is an *overall heat transfer coefficient* for the wall, defined by (21.5). U is normally expressed in $kW/m^2\ K$. It follows from (21.4) and (21.5), that

$$\frac{1}{U} = \frac{x_{12}}{k_{12}} + \frac{x_{23}}{k_{23}} + \frac{x_{34}}{k_{34}}$$

or in general for any number of layers

$$\frac{1}{U} = \Sigma\frac{\Delta x}{k} \tag{21.6}$$

$1/U$ is the overall resistance to heat flow, and it is equal to the sum of the individual resistances x_{12}/k_{12}, etc., in *series*. This might be expected from the electrical analogy illustrated in Fig. 21.3. The approach by way of equations (21.5) and (21.6) makes it possible to find the heat flow through a composite wall in terms of the surface temperatures, without calculating the interface temperatures. We have assumed, however, that there is no contact resistance between the layers; poor contact can lead to a marked reduction in the rate of heat transfer because of the very low conductivity of air.

Frequently it is required to calculate the heat flow through a wall separating two fluids of known temperature, when the surface temperatures of the wall are unknown.

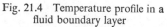

Fig. 21.3 Heat conduction through a   Fig. 21.4 Temperature profile in a
composite wall   fluid boundary layer

A typical temperature distribution in a fluid near a wall is shown in Fig. 21.4. Most of the temperature drop occurs very near the wall in a relatively stagnant boundary layer which adheres to the wall. Some distance from the wall the heat transfer is assisted by convection, and the temperature drop becomes more gradual until it vanishes completely. The heat flow in the fluid boundary layer could theoretically be calculated from Fourier's law

$$q = -k_f \left(\frac{\mathrm{d}t}{\mathrm{d}x}\right)_w \tag{21.7}$$

where $(\mathrm{d}t/\mathrm{d}x)_w$ is the temperature gradient *in the fluid* immediately adjacent to the wall, and k_f is the thermal conductivity of the fluid. But $(\mathrm{d}t/\mathrm{d}x)_w$ is a quantity which cannot easily be measured in practice and it is more common to express the heat flow from the wall to a fluid by an equation of the form

$$q = -h_c(t_s - t_w) \tag{21.8}$$

where h_c is a *convection heat transfer coefficient* or *film coefficient*, and t_s is the fluid

475

temperature some distance from the wall. To make the sign convention and general appearance of equation (21.8) consistent with that of Fourier's equation (21.7), we have introduced the negative sign in (21.8). The temperature change $(t_s - t_w)$ in the x-direction must always be written in the correct order to make q positive in the direction of temperature fall. Comparing (21.7) and (21.8), it is evidently possible to think of a stagnant fluid layer of thickness $\delta = k_f/h_c$, which has the same resistance to heat flow as the actual fluid. *The heat flow calculated from (21.8) includes the combined effects of conduction and convection in the fluid, and the two cannot be separated.* The value of h_c is difficult to calculate, and methods of evaluating it will be considered in Chapter 22. It is a function of numerous variables, such as the fluid viscosity and

Fig. 21.5   Heat exchange between two fluids through a composite wall

density, and the fluid velocity relative to the wall—it is *not* a physical constant valid for a particular fluid under all conditions of convective heat transfer. For the time being it will merely be assumed that h_c can be found.

In addition to heat transfer by convection from the wall surface to its surroundings, the wall may also lose or gain heat by radiation. It is again convenient to express this heat transfer by the equation

$$q = -h_r(t_s - t_w) \tag{21.9}$$

where h_r is a *radiation heat transfer coefficient*. In this case t_s is the temperature of the surface with which the wall is exchanging heat by radiation. h_r is a function of several variables, such as the temperatures, emissivities and absorptivities of the radiating surfaces. The evaluation of h_r is considered in Chapter 23, and here it is merely assumed that its value can be found.

Let us now consider a composite wall which separates two fluids at temperatures t_a and t_b (Fig. 21.5). It is assumed that the surfaces with which the wall is exchanging heat by radiation have the same temperature as the intervening fluid. The total heat flow to the wall is given by

$$q = -h_{ca}(t_1 - t_a) - h_{ra}(t_1 - t_a)$$

$$= -(h_{ca} + h_{ra})(t_1 - t_a) \tag{21.10}$$

Similarly the heat flow from the wall is

$$q = -(h_{cb} + h_{rb})(t_b - t_4) \tag{21.11}$$

An overall heat transfer coefficient U can now be defined by

$$q = -U(t_b - t_a) \tag{21.12}$$

It then follows from (21.3), (21.10), (21.11) and (21.12), that

$$\frac{1}{U} = \frac{1}{h_{ca} + h_{ra}} + \frac{x_{12}}{k_{12}} + \frac{x_{23}}{k_{23}} + \frac{x_{34}}{k_{34}} + \frac{1}{h_{cb} + h_{rb}}$$

or in general

$$\frac{1}{U} = \sum \frac{1}{h_c + h_r} + \sum \frac{\Delta x}{k} \tag{21.13}$$

Again considering the electrical analogy, the overall resistance to heat flow $1/U$ is equal to the sum of the individual resistances $1/(h_{ca} + h_{ra})$, x_{12}/k_{12}, etc. Here $1/h_{ca}$ and $1/h_{ra}$ are two resistances in *parallel*.

21.3 Radial steady conduction through the wall of a tube

When the inner and outer surfaces of a long tube are each at a uniform temperature, heat flows radially through the tube wall. Due to symmetry, any cylindrical surface concentric with the axis of the tube is an isothermal surface, and the direction of heat flow is normal to such a surface. To calculate the steady rate of conduction through the wall, it will be found convenient to consider unit length of tube rather than unit surface area. Continuity considerations indicate that the radial heat flow per unit length of the tube through successive layers must be constant if the flow is steady. Thus, since the area of successive layers increases with radius, the temperature gradient must decrease with radius. The temperature distribution will therefore appear as in Fig. 21.6.

Consider any elemental cylindrical tube of thickness dr. The area of flow per unit length of tube is $2\pi r$, and the temperature gradient normal to this tube is (dt/dr). The heat flow is, according to Fourier's law,

$$Q = -k2\pi r \left(\frac{dt}{dr}\right)$$

Fig. 21.6 Radial heat flow through a thick tube

Hence, since Q is independent of r, we obtain by integration

$$Q \ln \frac{r_2}{r_1} = -2\pi k(t_2 - t_1)$$

and thus the heat flow *per unit length of tube* is

$$Q = -\frac{2\pi k(t_2 - t_1)}{\ln (r_2/r_1)} \qquad (21.14)$$

Fig. 21.7 Radial flow through the wall of a composite tube

Considering a compound tube of n concentric layers, with fluid flowing through it and around it (Fig. 21.7), it is possible to derive an expression for the overall heat transfer coefficient *per unit length of tube*, U', analogous to equation (21.13) for the plane parallel wall. If t_a and t_b are the temperatures of the fluids inside and outside the wall respectively, the rate of heat flow per unit length of tube is

$$Q = -U'(t_b - t_a)$$

And using the same argument as in the previous section for deriving (21.13), it may be shown that U' is given by the expression

$$\frac{1}{U'} = \frac{1}{2\pi r_0 h_a} + \sum_{i=1}^{n} \frac{\ln (r_i/r_{i-1})}{2\pi k_i} + \frac{1}{2\pi r_n h_b} \qquad (21.15)$$

h_a and h_b are the respective total surface heat transfer coefficients, i.e. $(h_{ca} + h_{ra})$ and $(h_{cb} + h_{rb})$.*

In the case of a single thin tube of thickness Δx, for which the inner and outer radii are nearly equal, (21.15) reduces to

$$\frac{1}{U'} = \frac{1}{2\pi r h_a} + \frac{\Delta x}{2\pi r k} + \frac{1}{2\pi r h_b}$$

Hence the overall heat transfer coefficient *per unit area* of tube wall is

$$\frac{1}{U} = \frac{1}{h_a} + \frac{\Delta x}{k} + \frac{1}{h_b}$$

which is identical with the expression for a flat plate.

21.4 The differential equation of three-dimensional steady conduction

In the previous sections only one-dimensional steady conduction has been considered, i.e. when the temperature only varies in the direction of heat flow and is uniform over any plane surface normal to the direction of flow. To calculate the steady heat flow through a conductor of any arbitrary shape, it will be shown that the problem reduces to one of finding the temperature distribution in the conductor, and once this is known the problem is effectively solved. Consider a solid body having a plan view of the shape shown in Fig. 21.8. To get a picture of the temperature distribution, points of equal temperature can be joined to form *isothermal surfaces* (Fig. 21.8 showing these only in two dimensions). The picture of the temperature distribution is called the *temperature field*.

Here we shall only consider materials which are homogeneous and isotropic; homogeneous, because their density, specific heat, and thermal conductivity are the same everywhere in the temperature field; isotropic, because the thermal conductivity at any point is the same in all directions. For such materials the flow of heat is in the direction of maximum temperature gradient, i.e. at right angles to the isothermal surfaces. The *lines of heat flow*, which therefore cut the isothermal surfaces orthogonally, are also shown in Fig. 21.8.

*It will be apparent after reading Chapter 23 that, unless the internal fluid is radiating, h_{ra} must be zero.

Considering any elemental area dA, the heat flow through this element is given by Fourier's law as

$$dQ = -k \, dA \left(\frac{\partial t}{\partial n}\right) \tag{21.16}$$

where $(\partial t/\partial n)$ is the temperature gradient *normal* to dA. The following procedure may be adopted for calculating the total heat flow through the conductor once the temperature distribution is known. An isothermal surface is selected, divided into elemental

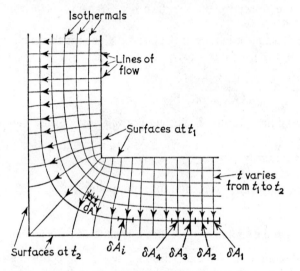

Fig. 21.8 Heat flow in the corner of a wall

areas δA_1, δA_2, etc., and the normal gradient found for each element. (Any isothermal surface may be chosen for steady flow, because the total rate of heat flow is the same for all such surfaces.) Thus the heat flow through δA_1 is

$$\delta Q_1 = -k \, \delta A_1 \left(\frac{\partial t}{\partial n}\right)_1$$

where $(\partial t/\partial n)_1$ is the temperature gradient normal to the isothermal surface at δA_1. The total heat flow through the material can be found by integrating this equation over the whole area A of the isothermal surface, and hence

$$Q = -\sum_{i=1}^{n} k \delta A_i \left(\frac{\partial t}{\partial n}\right)_i \tag{21.17}$$

The main problem lies in the prior determination of the temperature distribution in the field, and for this it is necessary to solve the differential equation that governs the conduction of heat. Consider any volume element (dx dy dz) of the conducting

480

material with its edges parallel to the coordinates (Fig. 21.9). The basic idea in establishing the differential equation is to satisfy the continuity of flow in any region of the temperature field; thus, when thinking of the volume element (dx dy dz), the heat inflow must be equal to the heat outflow, otherwise the temperature of the element will change with time and the flow will not be steady. The rate of heat inflow through the left-hand face (dy dz), i.e. in the x-direction, is

$$dQ_{1x} = -k(dy\ dz)\left(\frac{\partial t}{\partial x}\right)$$

and the outflow from the right-hand face (dy dz) is

$$dQ_{2x} = -k(dy\ dz)\frac{\partial}{\partial x}\left\{t + \left(\frac{\partial t}{\partial x}\right)dx\right\}$$

$$= -k(dy\ dz)\left\{\left(\frac{\partial t}{\partial x}\right) + \left(\frac{\partial^2 t}{\partial x^2}\right)dx\right\}$$

Fig. 21.9 The derivation of the differential equation of conduction in Cartesian coordinates

Similar equations can be written down for the y and z directions. The net rate of heat inflow into the element is thus

$$dQ = (dQ_{1x} - dQ_{2x}) + (dQ_{1y} - dQ_{2y}) + (dQ_{1z} - dQ_{2z})$$

$$= k(dx\ dy\ dz)\left(\frac{\partial^2 t}{\partial x^2} + \frac{\partial^2 t}{\partial y^2} + \frac{\partial^2 t}{\partial z^2}\right) \qquad (21.18)$$

For steady flow $dQ = 0$, and hence

$$\frac{\partial^2 t}{\partial x^2} + \frac{\partial^2 t}{\partial y^2} + \frac{\partial^2 t}{\partial z^2} = 0 \qquad (21.19)$$

There may be heat sources inside the conducting material as the result, for example, of nuclear fission, or of a flow of electric current and the ohmic heating of the material. If q_g is the heat generated per unit volume and per unit time, the total heat generated in the volume element (dx dy dz) per unit time is

$$dQ_g = q_g(dx\ dy\ dz)$$

For steady flow $dQ = -dQ_g$, and hence

$$k(dx\ dy\ dz)\left(\frac{\partial^2 t}{\partial x^2} + \frac{\partial^2 t}{\partial y^2} + \frac{\partial^2 t}{\partial z^2}\right) = -q_g(dx\ dy\ dz)$$

and finally,

$$\frac{\partial^2 t}{\partial x^2} + \frac{\partial^2 t}{\partial y^2} + \frac{\partial^2 t}{\partial z^2} + \frac{q_g}{k} = 0 \qquad (21.20)$$

q_g may be a function of position, i.e. $q_g = f(x, y, z)$, but for steady flow it cannot be a function of time.

Analytical solutions of equation (21.20) are only possible for relatively simple shapes of conductors. The following example of a simple one-dimensional problem with internal heat generation illustrates how the boundary conditions are applied to the solution of the differential equation.

Example 21.2 An electric current of 34 000 A flows along a flat steel plate 1·25 cm thick and 10 cm wide. The temperature of one surface of the plate is 80 °C and that of the other is 95 °C. Find the temperature distribution across the plate, and the value and position of the maximum temperature. Also calculate the total amount of heat generated per metre length of plate and the flow of heat from each surface of the plate.

The end effects along the short sides of the plate may be neglected, i.e. it may be assumed that no heat flows across these surfaces. It may also be assumed that the ohmic heating is generated uniformly across the section. The resistivity of the steel is $\rho = 12 \times 10^{-6}$ ohm cm, and its thermal conductivity is $k = 54$ W/m K.

This is essentially a one-dimensional problem with uniform heat generation throughout the conductor. The governing differential equation is

$$\frac{\partial^2 t}{\partial x^2} + \frac{q_g}{k} = 0$$

The heat generated per unit volume is ρi^2, where ρ is the resistivity, and i the current density.

Fig. 21.10

$$i = \frac{34\ 000}{1\cdot25 \times 10} = 2720 \text{ A/cm}^2$$

and hence

$$q_g = 12 \times 10^{-6} \times 2720^2 = 88\cdot78 \text{ W/cm}^3 = 88\ 780 \text{ kW/m}^3$$

The differential equation therefore becomes

$$\frac{\partial^2 t}{\partial x^2} + \frac{88\ 780}{0\cdot054} = 0 \quad \text{or} \quad \frac{\partial^2 t}{\partial x^2} + 1\ 644\ 000 = 0$$

Integrating this equation, we obtain

$$t = -822\ 000x^2 + Ax + B$$

where A and B are constants to be determined from the boundary conditions.

Choosing the origin of the x coordinate at the 95 °C surface of the plate (Fig. 21.10), we can write for $x = 0$

$$95 = B$$

and for $x = 0\cdot0125$ m

$$80 = -822\ 000(0\cdot0125)^2 + A(0\cdot0125) + B$$

Hence the equation for the temperature variation with x becomes

$$t = -822\,000x^2 + 9075x + 95$$

The temperature distribution is therefore parabolic. Differentiating with respect to x and equating to zero, we find the position of the maximum temperature, i.e.

$$\frac{\mathrm{d}t}{\mathrm{d}x} = -1\,644\,000x + 9075 = 0$$

and

$$x = 0 \cdot 00552 \text{ m} = 0 \cdot 552 \text{ cm}$$

The value of the maximum temperature is

$$t_{max} = -822\,000(0 \cdot 00552)^2 + 9075(0 \cdot 00552) + 95 = 120\,^\circ\text{C}$$

The heat flow *from* the plate across the 95 °C surface *per metre length of bar* is

$$Q_1 = kA \left(\frac{\mathrm{d}t}{\mathrm{d}x}\right)_{x=0} = 0 \cdot 054 \times \frac{10 \times 1}{100} \times 9075 = 49 \cdot 0 \text{ kW or kJ/s}$$

The heat flow *from* the plate across the 80 °C surface is

$$Q_2 = -kA \left(\frac{\mathrm{d}t}{\mathrm{d}x}\right)_{x=0 \cdot 0125} = -0 \cdot 054 \times \frac{10 \times 1}{100}(-11\,480) = 62 \cdot 0 \text{ kW or kJ/s}$$

The sum of Q_1 and Q_2 which is equal to 111 kW must be equal to the total heat generated per unit length of plate by ohmic heating, i.e. q_g (volume/metre length). Thus

$$Q_g = 88\,780(1 \times 0 \cdot 0125 \times 0 \cdot 1) = 111 \text{ kW}$$

In the next section a numerical method of solution is introduced which can be applied to complicated shapes, and this method is applied to a two-dimensional example. A good survey of mathematical methods used for the solution of conduction problems is given in Ref. (10).

One other approach is via the electrical analogy. It is essentially an experimental approach, but it is easier to arrange for a flow of electricity than for a flow of heat, and to measure potential differences and currents rather than temperature differences and heat flows. The method makes use of the fact that the steady flow of electricity in a three-dimensional conductor is governed by the equation

$$\frac{\partial^2 \mathcal{V}}{\partial x^2} + \frac{\partial^2 \mathcal{V}}{\partial y^2} + \frac{\partial^2 \mathcal{V}}{\partial z^2} = 0$$

where $\mathcal{V}$ is the potential distribution $\mathcal{V} = f(x, y, z)$ throughout the conductor. This equation is clearly similar to equation (21.19). The current $\delta \mathcal{I}_1$ flowing through any elemental area δA_1 in a direction normal to δA_1 is

$$\delta . \mathcal{I}_1 = -\frac{1}{\rho} \delta A_1 \left(\frac{\partial \mathcal{V}}{\partial n}\right)_1$$

where ρ is the resistivity of the material. Summing over the whole of an isopotential surface the total current is given by

$$\mathscr{I} = - \sum_{i=1}^{n} \frac{1}{\rho} \, \delta A_i \left(\frac{\partial \mathscr{V}}{\partial n} \right)_i$$

which is similar to equation (21.17). Thus potentials can be related to temperatures and currents to heat flows.

Thinking of the two-dimensional corner of Fig. 21.8, a simple electrical analogue can be produced from a sheet of metallised paper of uniform and relatively high resistivity, cut to the appropriate shape, with copper strips placed along the boundaries. The low resistivity of the copper strips ensures that any voltages $\mathscr{V}_1$ and $\mathscr{V}_2$ applied to them will be uniform and analogous to the isothermal boundaries t_1 and t_2. The potential at any point in the field can be measured by a simple probe and voltmeter, and the corresponding temperature found from

$$\frac{t}{t_1 - t_2} = \frac{\mathscr{V}}{\mathscr{V}_1 - \mathscr{V}_2}$$

Once the temperature distribution has been obtained, the heat flow can be calculated using equation (21.17). More sophisticated analogues, capable of handling three-dimensional problems, can be built using an electrolytic solution in a tank with conducting sides of appropriate shape. For further information the reader can turn to the last chapter of Ref. (29), and for a detailed treatment to Ref. (40).

21.5 Numerical solution of two-dimensional steady conduction

It has been explained in the previous section that the solution of a multi-dimensional conduction problem consists essentially of finding the temperature distribution in the conducting material. This involves the solution of the differential equation (21.20), which may be extremely difficult except for a few simple shapes.* To solve this equation it is necessary to know the boundary conditions of the temperature field, e.g. the temperature distribution at the edge of the region considered (see Fig. 21.8), and the distribution of the internal heat sources if any. It can be proved [Ref. (9)] that, for any given shape and boundary conditions, there is only one solution to (21.20), i.e. there is only one possible temperature distribution.

This section describes an approximate numerical method for solving two-dimensional conduction problems. For such problems equation (21.20) reduces to

$$\frac{\partial^2 t}{\partial x^2} + \frac{\partial^2 t}{\partial y^2} + \frac{q_g}{k} = 0 \qquad (21.21)$$

The object of the numerical method is to find values of the temperature in the conducting material at a limited number of points. It is most convenient to cover the region

*For analytical methods see Refs. (3) and (19).

considered by a square net, and to find the temperature at the mesh points (Fig. 21.11). Whereas a general analytic solution gives the values of the temperature everywhere in the field as a function $t = \phi(x, y)$, the numerical method merely finds numerical values of the temperature at the mesh points selected. Temperatures at other points can then be found by some method of interpolation.

It is possible to view the temperature field as a temperature surface or a relief map in the three-dimensional Cartesian coordinates (x, y, t); the variation of t in the x and y directions may be visualised by taking sections through the surface in these

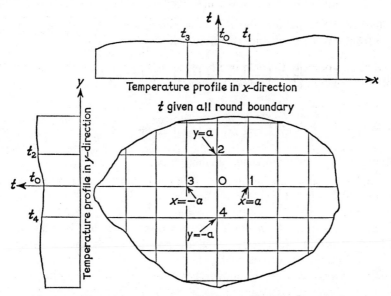

Fig. 21.11   Relaxation method applied to two-dimensional conduction

directions. Taking any mesh point 0 as the origin of the coordinates, it is often possible to express the variation of temperature t in each of the x and y directions by a power series, i.e.

$$t = b_0 + b_1 x + b_2 x^2 + b_3 x^3 + \ldots)_{y=\text{constant}}$$

$$t = c_0 + c_1 y + c_2 y^2 + c_3 y^3 + \ldots)_{x=\text{constant}}$$

This can always be done when t varies continuously in the field. By means of Mac-Laurin's theorem it is possible to express the value of t at any distance x from 0 in terms of differential coefficients of $t = \phi(x)$ at $x = 0$. Thus

$$t_0 = b^0$$

$$\left(\frac{\partial t}{\partial x}\right)_0 = 1b_1$$

$$\left(\frac{\partial^2 t}{\partial x^2}\right)_0 = 1 \times 2b_2$$

$$\left(\frac{\partial^3 t}{\partial x^3}\right)_0 = 1 \times 2 \times 3b_3, \text{ etc.}$$

And the variation of t in the x direction can be written as

$$t = t_0 + \left(\frac{\partial t}{\partial x}\right)_0 \frac{x}{1!} + \left(\frac{\partial^2 t}{\partial x^2}\right)_0 \frac{x^2}{2!} + \left(\frac{\partial^3 t}{\partial x^3}\right)_0 \frac{x^3}{3!} + \cdots$$

The values of the temperature at the mesh points 1 and 3, at distances $x = a$ and $x = -a$ from 0 respectively, are therefore given by

$$t_1 = t_0 + \left(\frac{\partial t}{\partial x}\right)_0 \frac{a}{1!} + \left(\frac{\partial^2 t}{\partial x^2}\right)_0 \frac{a^2}{2!} + \left(\frac{\partial^3 t}{\partial x^3}\right)_0 \frac{a^3}{3!} + \cdots$$

and

$$t_3 = t_0 - \left(\frac{\partial t}{\partial x}\right)_0 \frac{a}{1!} + \left(\frac{\partial^2 t}{\partial x^2}\right)_0 \frac{a^2}{2!} - \left(\frac{\partial^3 t}{\partial x^3}\right)_0 \frac{a^3}{3!} + \cdots$$

Hence, neglecting fourth power and higher terms

$$t_1 + t_3 = 2t_0 + a^2 \left(\frac{\partial^2 t}{\partial x^2}\right)_0 \qquad (21.22)$$

This approximation can be made if the chosen mesh width a is sufficiently small; it will be explained in Example 21.3 how it is possible to tell when a is small enough for the fourth power term to be negligible.

By similar reasoning it can be shown that

$$t_2 + t_4 = 2t_0 + a^2 \left(\frac{\partial^2 t}{\partial y^2}\right)_0 \qquad (21.23)$$

where t_2 and t_4 are the temperatures at the mesh points 2 and 4 respectively. Substituting (21.22) and (21.23) in (21.21), we get

$$\left(\frac{\partial^2 t}{\partial x^2}\right)_0 + \left(\frac{\partial^2 t}{\partial y^2}\right)_0 + \frac{q_g}{k} = \frac{t_1 + t_2 + t_3 + t_4 - 4t_0}{a^2} + \frac{q_g}{k} = 0 \qquad (21.24)$$

The finite difference relation (21.24) can also be deduced in a somewhat easier way. The homogeneous conducting material can be imagined to be replaced by conducting rods connecting the mesh points (Fig. 21.12), and the conductivity of the rods is chosen to be equivalent to the conductivity of the shaded region. Considering *unit depth of material* (i.e. perpendicular to the paper), the flow through the rod from mesh point 1 to 0 is

$$Q_1 = -k(1 \cdot a)\frac{t_0 - t_1}{a} = -k(t_0 - t_1)$$

The total flow toward the mesh point 0 from 1, 2, 3, 4 is therefore

$$Q = Q_1 + Q_2 + Q_3 + Q_4$$
$$= -k(t_0 - t_1) - k(t_0 - t_2) - k(t_0 - t_3) - k(t_0 - t_4)$$
$$= k(t_1 + t_2 + t_3 + t_4 - 4t_0)$$

If the heat generated per unit volume is q_g, the equivalent heat that is generated at each mesh point is $q_g a^2$. With steady flow, for reasons of continuity, it follows that

$$k(t_1 + t_2 + t_3 + t_4 - 4t_0) + q_g a^2 = 0$$

which is equivalent to (21.24).

Fig. 21.12 The conducting rod analogy

An equation of the form (21.24) can be written down for each mesh point in the field. Thus the differential equation, which must be satisfied everywhere in the field, can be replaced by a finite difference relation which must be satisfied at a finite number of points in the field. The accuracy of the numerical solution depends on the density of the mesh chosen.

In principle, to obtain a numerical solution it is necessary to write down a linear equation corresponding to (21.24) for each mesh point, and the solution of the differential equation is thus reduced to a solution of a finite number of linear simultaneous equations. In practice the solution of these equations by conventional methods is not feasible, because the large number of simultaneous equations would result in considerable labour. When several problems of the same type are to be solved, it is probably worth writing a computer programme, making of matrices. For the occasional problem, or for checking a programme, a convenient procedure of successive approximation by the *relaxation method* can be used: this method is best explained by a numerical example.

Fig. 21.13 Relaxation of the 6 cm mesh

Example 21.3 The temperature of one side of a long 18 cm thick wall is 100 °C, and of the other is 400 °C. Metal ribs, 48 cm apart, project 6 cm into the high-temperature side. The conductivity of the ribs is very high compared with the conductivity of the wall, so that their temperature may be assumed to be 400 °C. Calculate the steady rate of heat flow through each 48 cm segment of the wall, and also find the heat flow if the ribs were omitted.

A typical segment covered with a 6 cm square mesh is drawn in Fig. 21.13. To solve the problem the boundary conditions all round the segment must be known. The insert shows the approximate shape of the isothermals, and it is evident that it is reasonable to assume that at sections 1–1 and 3–3 the temperature gradient perpendicular to the wall is constant. The boundary values of temperature around the segment are thus fully determined, and are underlined.

The next step in the solution is to guess the temperatures at all the mesh points. Although the process of successive approximation will finally provide the best answer obtainable with the coarse mesh chosen, a reasonably close estimate will of course shorten the procedure. By a rough process of interpolation it is possible to select a reasonable set of temperatures, and their values are shown in the top left-hand corners of each mesh point. Due to the symmetry of the segment about section 2–2, only half of it need be considered.

The finite difference relation (21.24) must be satisfied at each mesh point. Checking this at point b, for example, we find that

$$t_c + 100 + t_a + t_f - 4t_b = 220 + 100 + 200 + 310 - (4 \times 205) = 10$$

The error, of magnitude 10, is called the *residual* at point b, and it is entered into the top right-hand corner of the mesh point. Similarly all other residuals are found and entered in corresponding positions. The object of the process of relaxation is to modify all the guessed temperature values until the residuals have vanished, i.e. until all the residuals are 'relaxed'.

The alterations to the initial guesses are written above the temperatures, and one alteration is carried out at a time. The procedure is to start with the mesh point which has the numerically largest residual, i.e. point d. It is clear from the finite difference relation (21.24) that if we increase the value of the temperature by 1 °C at a mesh point, we alter the residual by -4 at that point. Also, applying (21.24) to the four neighbouring points, we see that their residuals are altered as a result of this change by $+1$ each. The basic operation of $+1$ °C applied to any mesh point is illustrated in Fig. 21.14. To eliminate the residual at point d, we must change the temperature by $-20/4 =$

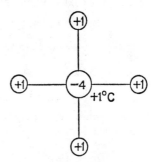

Fig. 21.14 The relaxation operator

-5 °C. The residual 0 is entered at d, and the residuals at c and c' are changed from -5 to -10. As the other two neighbouring points of d are fixed boundary temperatures, i.e. 100 °C and 400 °C, no alteration is made at these points. From the new set of residuals, again the largest is selected and the procedure is repeated. The relaxation is continued in the order shown in the table given below. It should be noted that when t_c is altered by -1 °C, the residual at d is changed by -2 and not by -1; we have to consider point c' at the same time for reasons of symmetry. It will also be noted that to avoid fractional changes in temperature, we do not always reduce the residuals to zero in any one operation. To avoid fractional temperatures in the final solution, the residuals are not completely relaxed, but their algebraic sum for the whole segment should be made as near zero as possible. The changes in temperature are added to the initial guesses at each mesh point, and the algebraic sum is located in the bottom left-hand corner. (There is only one change of temperature at each mesh point in this example, but in other cases more than one change at some of the points may be necessary before the mesh is relaxed.)

Residual at	a	b	c, c'	d	e	f	g
Initial set of residuals	10	10	−5	−20	−10	0	10
Operation:							
(1) −5 °C at d			−10	0			
(2) +3 °C at a	−2	13			−7		
(3) +3 °C at b	1	1	−7			3	
(4) +2 °C at g				−5		5	2
(5) −2 °C at e	−1				1	3	
(6) −1 °C at c		0	−1	−2			1
(7) +1 °C at f		1			2	−1	2
Sum of final set of residuals = 0	−1	1	−1	−2	2	−1	2

To calculate the heat flow through the wall, we may make use of the 'conducting rod' analogy described previously. Thus the heat flow from b to the wall surface (per metre height of wall, perpendicular to the paper) is $-k(100 - t_b)$. Hence the total outflow from the section, taking account of its symmetry, is

$$Q_{out} = k(200 - 100) + 2k(203 - 100) + 2k(208 - 100)$$
$$+ 2k(219 - 100) + k(235 - 100) = 895k$$

To check, we may calculate the inflow from the hot face, which is

$$Q_{in} = k(400 - 300) + 2k(400 - 303) + 2k(400 - 311)$$
$$+ 4k(400 - 332) + k(400 - 235) = 909k$$

The '$4k$' term arises because of the contribution from the metal rib to points g and g'. We may take the average value

$$Q = \frac{Q_{in} + Q_{out}}{2} = \frac{909k + 895k}{2} = 902k$$

as the answer.

In order to check whether the mesh is sufficiently fine, i.e. that a is small enough for the fourth power term to be neglected in the MacLaurin expansion, the process must be repeated with a finer mesh. If the new temperatures differ only slightly from the old values the mesh is sufficiently fine; otherwise we must proceed to an even finer mesh. Labour is saved if the original points are featured in the new mesh, and we may choose one of $6/\sqrt{2}$ cm width.

as shown in Fig. 21.15. The values of temperature at a, b, c, d, e, f, g, are taken from the previous solution, and the values of the new points are found by interpolation. It follows from the derivation of equation (21.21) that the rotation of the coordinates (termed 'transformation') has no effect on this differential equation, and hence the same finite difference relation applies.

After relaxing, we find that t_d undergoes the largest change, i.e. from 235 °C to 226 °C. Recalculating the average heat flow, we find that

$$Q = \frac{Q_{in} + Q_{out}}{2} = \frac{862k + 880k}{2} = 871k$$

Fig. 21.15   Relaxation of the $6/\sqrt{2}$ cm mesh

Using a mesh of 3 cm, t_d becomes 225 °C, and the average heat flow

$$Q = \frac{Q_{in} + Q_{out}}{2} = \frac{875k + 877k}{2} = 876k$$

Thus, by progressively choosing a finer mesh, we can converge as close to an accurate solution as we wish. Without the ribs the heat flow trough a 48 cm segment of the wall would be

$$Q = -k\frac{0·48 \times 1(100-400)}{0·18} = 800k.$$

If it is impossible to fit a square net accurately into the boundary of the region considered, there may be mesh points near the boundary with links shorter than a

491

(Fig. 21.16). It can be shown that the finite difference relation to be satisfied for such a point is

$$\frac{2t_1}{\xi(1+\xi)} + \frac{2t_2}{\eta(1+\eta)} + \frac{2t_3}{(1+\xi)} + \frac{2t_4}{(1+\eta)} - 2t_0\left(\frac{1}{\xi}+\frac{1}{\eta}\right) + \frac{q_g a^2}{k} = 0 \quad (21.25)$$

There are numerous tricks by means of which the process of relaxation may be shortened. For these, and more complicated types of problem, the reader is referred to Ref. (1).

Fig. 21.16 A mesh point near an irregular boundary

At the end of section 21.4 we saw how a simple electrical analogue can be constructed to provide a solution to the two-dimensional form of equation (21.19). Often it is preferable to have, not an analogue providing a 'continuous' solution, but one which is equivalent to the 'lumped' approximation represented by Fig. 21.11 or the conducting rod picture of Fig. 21.12. A mesh of resistances connecting junctions 0, 1, 2, 3, 4, etc., can be used, and voltage and current measurements obtained from this provide an equivalent to the approximate numerical solution. The advantage of a 'lumped' circuit is that 'heat sources' can be easily incorporated, and unsteady flow can be simulated by inserting capacitances at the mesh points [Ref. (29)].

21.6 Unsteady conduction

It is frequently of interest to know the heat flow through a solid when the flow is not steady, e.g. through the wall of a furnace that is being heated or cooled, or through the cylinder wall of an internal-combustion engine. To calculate the heat flow under these conditions it is again first necessary to find the temperature distribution in the solid; in this case, however, the distribution varies continuously with time. We shall consider two simple cases of some practical importance.

(i) *The quenching of a billet*

When a metal billet of uniform temperature t is suddenly plunged into a large liquid bath of temperature t_s, the internal resistance to heat flow is often small compared with the resistance between the billet surface and the fluid. This may be the case in the quenching of small steel components during heat treatment. It is then reasonable to assume that the temperature t, though changing with time τ, is nevertheless uniform throughout the billet at any instant.

Let the billet in Fig. 21.17a have a volume V, surface area A, density ρ, specific heat c_p, and let us assume that the surface heat transfer coefficient between the billet and the liquid is constant and equal to h. Further, we may treat the bath as large enough for its temperature to be substantially constant at t_s. The energy lost by the billet during a time interval $d\tau$ is equal to the heat transferred to the liquid and can be expressed by the equation

$$dQ = \rho c_p V \frac{dt}{d\tau} = -hA(t - t_s)$$

or, if we write $\theta = t - t_s$,

$$\frac{d\theta}{\theta} = -\frac{hA}{\rho c_p V} d\tau$$

This equation can be integrated, and with the boundary condition $\theta = \theta_0 = t_0 - t_s$ at $\tau_0 = 0$, the result is

$$\ln \frac{\theta}{\theta_0} = -\frac{hA\tau}{\rho c_p V} \quad \text{or} \quad \frac{\theta}{\theta_0} = e^{-(hA\tau/\rho c_p V)}$$

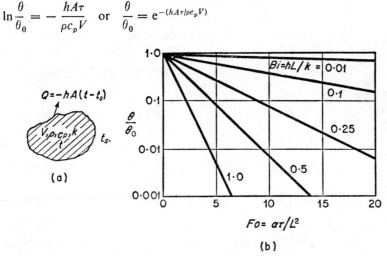

Fig. 21.17 The quenching of a billet

It will be noted that the solution depends on the shape of the billet only in so far as this determines the ratio of volume to surface area, V/A, and this ratio is a linear dimension which we shall denote by L. Introducing L and the thermal conductivity k for the billet in the manner shown below, and writing α for the group $k/\rho c_p$, the index of e can be written as

$$\frac{hA\tau}{\rho c_p V} = \left(\frac{h}{k} \frac{V}{A}\right) \left[\frac{k}{\rho c_p} \left(\frac{A}{V}\right)^2 \tau\right] = \left(\frac{hL}{k}\right) \left(\frac{\alpha\tau}{L^2}\right) = (\text{Bi})(\text{Fo})$$

and hence

$$\frac{\theta}{\theta_0} = e^{-(\text{Bi})(\text{Fo})} \tag{21.26}$$

The solution for this simple case can thus be expressed in terms of two dimensionless groups—the *Biot number* $Bi = hL/k$ and the *Fourier number* $Fo = \alpha\tau/L^2$—which

493

are also important parameters in more complicated situations of unsteady conduction. These groups require further comment.

Bi is a measure of the relative resistance to heat flow of the inside of the solid to that of the adjacent fluid: a low Biot number implies that most of the resistance is in the fluid boundary layer; a high Biot number implies that most of the resistance is within the solid. The solution for the billet given by equation (21.26) only applies to relatively low values of *Bi* because of the initial assumption made about the internal resistance. It is important to note that *Bi* differs from the Nusselt number $Nu = hL/k$, which we shall meet in the study of convection, in that *Nu* contains the *fluid* conductivity.

T/[K]	250	300	400	500	600	800
Aluminium . . .	96·5	93·7	97·0	103	107	128
Copper	114	112	107	103	99·4	94·2
Steel (0·5 % C) . .		15				
Glass		0·35–0·5				
Building brick . .		0·25–0·5				
Concrete . . .		0·5 –0·8				
Mercury liquid . .	3·9	4·3	5·2	6·0	7·3	8·0
Sodium liquid . .	117 (sol.)	114 (sol.)	68	68	66	60
Water (sat.) . .	1·24 (ice)	0·147	0·172	0·165		
Steam (sat.) . .		382	9·39	0·962		
Steam (1 atm) . .			25·2	41·5	62·7	120
Hydrogen (1 atm) .	113	155	257	382	529	892
Air (1 atm) . .	15·7	22·2	37·6	55·6	75·4	119·1
Carbon dioxide (1 atm) .	7·60	11·0	19·5	30·8	44·8	71·5

Fig. 21.18. Thermal diffusivity $\dfrac{k/\rho c_p}{10^{-6}[\text{m}^2/\text{s}]}$

Fo represents a dimensionless time scale. It contains a dimensional group $\alpha = k/\rho c_p$ called the *thermal diffusivity* of the material: the dimensions are $[\text{length}]^2/[\text{time}]$ and typical values in m^2/s are given in Fig. 21.18. As we shall see, α appears in the differential equation of unsteady conduction, and is also an important parameter in forced convection.

The solution (21.26) for the billet can be represented in a very simple graphical way, by plotting θ/θ_0 on a logarithmic scale against a linear scale of *Fo*. Lines of constant *Bi* radiate from $\theta/\theta_0 = 1$ as shown in Fig. 21.17b.

When the internal resistance of the billet is not negligible, the temperature history becomes a function of the spatial coordinates also. To establish the temperature field as a function of time it is necessary to derive and solve the differential equation of unsteady three-dimensional conduction [a derivation of the one-dimensional form

of the equation being given in section 21.6 (ii)]. A solution for the temperature field
then takes the form

$$\frac{\theta}{\theta_0} = f\left(Bi, Fo, \frac{x}{L}, \frac{y}{L}, \frac{z}{L}\right)$$

where L is any suitably chosen linear dimension of the solid (not necessarily V/A
in this case). Solutions have been obtained for a large variety of shapes, for the initial
conditions considered here, namely a uniform initial temperature in the solid which
is then subjected to a step change of ambient temperature. When presented
graphically, such solutions appear very much like Fig. 21.17b, except that each
point (x, y, z) within the solid requires a separate chart. For example, considering
the radial-symmetrical case of a sphere,

$$\frac{\theta}{\theta_0} = f\left(Bi, Fo, \frac{r}{r_0}\right)$$

where r_0 is the outer radius of the sphere; a separate chart is needed for each value of
$0 < r < r_0$ of interest. Typical charts of this kind can be found in Refs. (41) and (42).

(ii) *Numercial solution of unsteady one-dimensional conduction*

In section 21.6 (i) we presented a solution for the case where t is not a function of
position. We shall now proceed with the more general case where t varies with position
and also consider more difficult boundary conditions; but we shall restrict ourselves
to one-dimensional flow. A numerical method will be used to obtain the solution,
first when the boundary condition is given in the form of a prescribed temperature
variation of the solid surface, and then when it is given as a variation of ambient fluid
temperature coupled with a known, but not necessarily constant, surface heat transfer
coefficient.

Before the numerical method can be outlined, it is necessary to modify the differential equation (21.19) derived in section 21.4, to account for the time dependence of
temperature. This can be done by equating dQ in equation (21.18) not to zero but to
the rate of energy stored by the volume element $(dx\, dy\, dz)$ when its temperature rises
at a rate $(dt/d\tau)$; the rate of energy stored is $\rho c_p(dx\, dy\, dz)(dt/d\tau)$. Alternatively, for
the simple case of one-dimensional flow we can work from first principles; this we
shall proceed to do with reference to Fig. 21.19.

The heat inflow into a layer dx during a time interval $d\tau$ through an area A is

$$Q_1 = -kA\frac{\partial t}{\partial x}d\tau$$

The symbol Q is used to denote heat transfer, and *not* heat transfer per unit time, in
the remainder of this section. The corresponding outflow is

$$Q_2 = -kA\frac{\partial}{\partial x}\left\{t + \left(\frac{\partial t}{\partial x}\right)dx\right\}d\tau = -kA\left\{\frac{\partial t}{\partial x} + \left(\frac{\partial^2 t}{\partial x^2}\right)dx\right\}d\tau$$

The difference between the heat inflow Q_1 and the heat outflow Q_2 must be equal to

495

the heat stored* by the conducting layer dx during the time interval $d\tau$. The temperature rise of the layer during this time is $(\partial t/\partial \tau)\,d\tau$, and hence

$$Q_1 - Q_2 = \rho c_p (A\,dx)\left(\frac{\partial t}{\partial \tau}\right) d\tau$$

Thus

$$\rho c_p (A\,dx)\left(\frac{\partial t}{\partial \tau}\right) d\tau = kA\left(\frac{\partial^2 t}{\partial x^2}\right) dx\,d\tau$$

or

$$\frac{\partial t}{\partial \tau} = \alpha \left(\frac{\partial^2 t}{\partial x^2}\right) \tag{21.27}$$

where $\alpha = k/\rho c_p$ is the thermal diffusivity already encountered in section 21.6 (i).

Fig. 21.19   Fig. 21.20

The next step is to develop a finite difference relation which can be used to determine the temperature at a discrete number of sections, such as x_{n-1}, x_n, x_{n+1}, separated by equal space intervals Δx as in Fig. 21.20. Since the temperature at these planes varies with time, it is necessary to find the values at successive intervals of time $\tau_{p-1}, \tau_p, \tau_{p+1}$, which are conveniently chosen to be separated by equal time intervals $\Delta \tau$. Thus each temperature will be denoted by two suffixes: $t_{n,p}$ for example means the temperature in the wall at section x_n at the instant τ_p. By making use of equation (21.22), the differential coefficient $(\partial^2 t/\partial x^2)$ in equation (21.27) can be replaced by

$$\left(\frac{\partial^2 t}{\partial x^2}\right)_{n,p} = \frac{t_{n+1,p} + t_{n-1,p} - 2t_{n,p}}{\Delta x^2} \tag{21.28}$$

---

*It is more correct to say that the net inflow of heat is equal to the increase of internal energy of the layer plus the work done against adjacent layers in expansion or contraction. If the pressure in the wall and outside is constant, the relevant specific heat is the specific heat at constant pressure.

The differential coefficient $(\partial t/\partial \tau)$ can be written*

$$\left(\frac{\partial t}{\partial \tau}\right)_{n,p} = \frac{t_{n,p+1} - t_{n,p}}{\Delta \tau} \tag{21.29}$$

It follows that equation (21.27), applied to section x_n at time τ_p, can be replaced by the finite difference relation

$$t_{n,p+1} - t_{n,p} = \frac{\alpha \Delta \tau}{\Delta x^2}\left(t_{n+1,p} + t_{n-1,p} - 2t_{n,p}\right) \tag{21.30}$$

$\alpha \Delta \tau/\Delta x^2$ will be recognised as a Fourier number similar to that encountered in section 21.6 (i), and equation (21.30) can thus be written as

$$t_{n,p+1} = Fo\left[t_{n+1,p} + t_{n-1,p} + \left(\frac{1}{Fo} - 2\right)t_{n,p}\right] \tag{21.31}$$

Although the Fourier number Fo can be given any desired value by appropriate choice of $\Delta \tau$ and Δx, we shall see later that not all values are acceptable. Taking typical values of Fo used in practice, we obtain

$$\text{For } Fo = \tfrac{1}{2}; \; t_{n,p+1} = \tfrac{1}{2}(t_{n+1,p} + t_{n-1,p}) \tag{21.31a}$$

$$\text{For } Fo = \tfrac{1}{3}; \; t_{n,p+1} = \tfrac{1}{3}(t_{n+1,p} + t_{n-1,p} + t_{n,p}) \tag{21.31b}$$

$$\text{For } Fo = \tfrac{1}{4}; \; t_{n,p+1} = \tfrac{1}{4}(t_{n+1,p} + t_{n-1,p} + 2t_{n,p}) \tag{21.31c}$$

Such equations can be seen to give the 'future' temperature $t_{n,p+1}$ as a weighted average of the 'present' temperatures at x_{n-1}, x_n and x_{n+1}.

In order to find a solution of equation (21.27), whether analytically or via equation (21.30), boundary conditions must be specified. It is essential to know (i) the temperature distribution through the wall at some time τ_0, and (ii) the variation of temperature with time at two boundary planes. A simple example, where the wall surface temperatures are known over the period of time, will show how the computation might be set out for desk calculation.

Example 21.4 A wall is 25 cm thick and initially, at $\tau = 0$, the temperature is uniform throughout at 100 °C. One surface is maintained at 100 °C, while the other increases linearly with time to 700 °C in two hours. Determine the temperature distribution at the end of this period if the thermal diffusivity of the material is 0.52×10^{-6} m²/s.

---

*It may appear more obvious to replace $(\partial t/\partial \tau)_{n,p}$ by a 'time-centred' finite difference ratio over the period τ_{p-1} to τ_{p+1}, viz:

$$\left(\frac{\partial t}{\partial \tau}\right)_{n,p} = \frac{t_{n,p+1} - t_{n,p-1}}{2\Delta \tau}$$

but it is found that solutions involving this approximation become unstable in the sense briefly explained after Example 21.4.

We will choose five space intervals, so that $\Delta x = 0.05$ m. Then if Fo is to be $\frac{1}{4}$ say, the time interval will be

$$\Delta\tau = \frac{Fo\Delta x^2}{\alpha} = \frac{0.05^2}{4 \times 0.52 \times 10^{-6} \times 3600} = 0.334 \text{ h}$$

In this example, with $Fo = 1/4$, six time increments of $\Delta\tau$ fit almost exactly into a 2 h period. If this were not so it would be possible to use a value of Fo not exactly equal to a round figure, but a better way would be to calculate the temperatures just before and just after the required time and interpolate the temperatures obtaining at that time.

Applying equation (21.31c) we get the results shown in the table (the boundary conditions being in bold type): the equation for $t_{2,1\frac{2}{3}}$ for example is

$$t_{2,1\frac{2}{3}} = \frac{1}{4}(101.6 + 207.8 + 2 \times 118.7) = 136.7 \,°\text{C}$$

The final line of figures gives the desired temperature distribution.

Hours	t_0	t_1	t_2	t_3	t_4	t_5
0	**100**	100	100	100	100	**100**
$\frac{1}{3}$	**200**	100	100	100	100	**100**
$\frac{2}{3}$	**300**	125	100	100	100	**100**
1	**400**	162·5	106·2	100	100	**100**
$1\frac{1}{3}$	**500**	207·8	118·7	101·6	100	**100**
$1\frac{2}{3}$	**600**	258·6	136·7	105·5	100·4	**100**
2	**700**	313·4	159·4	112·0	101·6	**100**

The rates of heat flow through planes x_{n-1}, x_n, x_{n+1} are not equal (and vary with time) as a result of the storage of energy in the material between the planes. But once the temperature distributions have been obtained, the amount of heat flowing through some plane x_n during a time interval $\Delta\tau$ can be estimated from the average temperature gradient $(\partial t/\partial x)_n$ at that section during this time interval, i.e.

$$Q_n = -kA \left(\frac{\partial t}{\partial x}\right)_n \Delta\tau \tag{21.32}$$

The average temperature gradient for calculating Q_n during the time interval τ_p to τ_{p+1} would be

$$\left(\frac{\partial t}{\partial x}\right)_{n,p+\frac{1}{2}} = \frac{1}{4\Delta x}[(t_{n+1,p} + t_{n+1,p+1}) - (t_{n-1,p} + t_{n-1,p+1})] \tag{21.33}$$

The question arises whether Fo can be chosen arbitrarily, i.e. whether Δx and $\Delta\tau$ can be chosen independently. If it can, this would imply that the coefficient $[(1/Fo) - 2]$ of the temperature $t_{n,p}$ in equation (21.31) can assume negative values, viz. when $Fo > 1/2$. In fact a choice of $Fo > 1/2$ would imply, from equation (21.31), that the higher the value of $t_{n,p}$, the lower will be the resulting value of $t_{n,p+1}$; it can be shown by the following argument that this may lead to difficulties *vis-à-vis* the law of conservation of energy. Let us assume that at a given instant τ_p the distribution of temperature is as in Fig. 21.21. Referring to equation (21.30), the longer the chosen time interval $\Delta\tau$ and therefore the higher the value of Fo other things being equal, the higher will be the rise $(t_{n,p+1} - t_{n,p})$. Provided t_{n+1} and t_{n-1} do not change during $\Delta\tau$, the maximum permissible rise of $t_{n,p+1}$ is to point A, because then the inflow of heat to section x_n is exactly balanced by the outflow. This can be seen to correspond to a value of $Fo = 1/2$, equation (21.31a). Inspection of the temperature gradients shows that a rise to point B, obtained by making $Fo > 1/2$, would imply that an increase of temperature is obtained with a heat outflow from section x_n which is larger than the heat inflow, and this clearly contradicts the law of conservation of energy. Of course $Fo = 1/2$ does not represent an absolute limiting criterion—because t_{n+1} and t_{n-1} will themselves change during

Fig. 21.21   Criterion for Fo Fig. 21.22

$\Delta\tau$ and, depending upon whether they rise or fall, the criterion could in principle be somewhat larger or smaller than 1/2. Nevertheless, because the method finds $t_{n,p+1}$ in terms of temperatures at instant τ_p, neglecting what happens to t_{n+1} and t_{n-1} during $\Delta\tau$,

$$Fo \le \tfrac{1}{2} \tag{21.34}$$

represents the relevant criterion for self-consistency of the method. Satisfying (21.34) serves a further purpose: it also ensures that the solution is stable (i.e. that the temperature profiles do not diverge from or increasingly oscillate about the exact solution). This is because the criterion for stability is in fact usually less stringent than (21.34).*

We may now turn to an extension of the method to cover the more practical case where it is the temperature variation of a fluid adjacent to the wall which is known, rather than the temperature variation of the wall surface. In such circumstances we also require a knowledge of the film coefficient h between surface and fluid. Referring to Fig. 21.22, we may make use of the condition of continuity of heat flow at the wall surface x_w by writing for unit area

$$-k \left(\frac{\partial t}{\partial x}\right)_{w,p} = -h(t_{w,p} - t_{s,p}) \tag{21.35}$$

*For a discussion of criteria of stability of such solutions the reader may turn to Refs. (22), (25) or (19).

where $t_{s,p}$ is the bulk fluid temperature at time τ_p. This condition is best applied by subdividing the wall as shown, viz. with section x_1 a distance $\Delta x/2$ in from the surface. (The temperature gradients are greater near the surface and a narrower space interval here leads to greater accuracy.) The introduction of an imaginary extension of the wall to plane x_0 in the fluid, also a distance $\Delta x/2$ from the surface, then enables us to satisfy equation (21.35) by writing

$$k \frac{t_{1,p} - t_{0,p}}{\Delta x} = h(t_{w,p} - t_{s,p}) \tag{21.36}$$

Further, putting $t_w = (t_1 + t_0)/2$ and $Bi = h\Delta x/k$, algebraic manipulation yields

$$t_{0,p} = \left(\frac{2Bi}{2 + Bi}\right) t_{s,p} + \left(\frac{2 - Bi}{2 + Bi}\right) t_{1,p} \tag{21.37}$$

$t_{1,p+1}$ can now be found by direct application of equation (21.31) to the sections x_0, x_1 and x_2, viz.

$$t_{1,p+1} = Fo\left[t_{2,p} + t_{0,p} + \left(\frac{1}{Fo} - 2\right) t_{1,p}\right]$$

which, on substituting for $t_{0,p}$ from (21.37), becomes

$$t_{1,p+1} = Fo\left[t_{2,p} + \frac{2Bi}{2 + Bi} t_{s,p} + \left(\frac{1}{Fo} - \frac{2 + 3Bi}{2 + Bi}\right) t_{1,p}\right] \tag{21.38}$$

In this case we avoid a negative coefficient of $t_{1,p}$ by making

$$Fo \le \frac{2 + Bi}{2 + 3Bi} \tag{21.39}$$

This is less stringent than $Fo \le 1/2$ provided that $Bi \le 2$, and the Biot number can always be made less than 2 by a suitable choice of Δx. Note that equation (21.38) is only required for calculating values of t_1: values of t_2, t_3, etc., will be obtained from the appropriate form of equation (21.31) in the manner illustrated in Example 21.4.

When the heat transfer between fluid and wall is required during a period of time, we may make use of the right-hand side of equation (21.35). Thus

$$q = -\sum_{\tau=0}^{\tau_p}\Delta\tau\, h(t_{w,p} - t_{s,p}) \tag{21.40}$$

For this purpose we shall require values of t_w and these can be found by substituting equation (21.37) in the expression $t_{w,p} = (t_{1,p} + t_{0,p})/2$ to yield

$$t_{w,p} = \frac{2t_{1,p} + (Bi)t_{s,p}}{2 + Bi} \tag{21.41}$$

It is necessary to use the average values of t_w and t_s during each period when applying equation (21.40). The following example will make this clear.

Example 21.5 Let the wall in Example 21.4 separate two fluids, one fluid remaining at 100 °C while the other rises to 700 °C in 2 hours. The film coefficient on each side may be assumed constant and equal to 0·32 kW/m² K, and the conductivity of the wall material is $8·7 \times 10^{-4}$ kW/m K. Estimate the heat flow into and out of the wall, and the energy stored in the wall, during the 2 hour period per square metre of wall surface.

If we make the same choice of Δx as before, the Biot number becomes

$$Bi = \frac{h\Delta x}{k} = \frac{0·32 \times 0·05}{8·7 \times 10^{-4}} = 18·4$$

This is greater than 2, but $(2 + Bi)/(2 + 3Bi) = 0·357$ so that a $\Delta\tau$ of 1/3 h, and hence the value of $Fo = \frac{1}{4}$ used before, will satisfy criterion (21.39). The numerical equations required for the computation of the temperature distribution are therefore:
(a) Equation (21.31c) for values of t_2, t_3, t_4.
(b) For t_1 and t_5, from equation (21.38),

$$t_{1,p+1} = \tfrac{1}{4}(t_{2,p} + 1·80 t_{s1,p} + 1·20 t_{1,p})$$

$$t_{5,p+1} = \tfrac{1}{4}(t_{4,p} + 1·80 t_{s2,p} + 1·20 t_{5,p})$$

(The coefficients have been evaluated to two decimal places: when rounding off note that the sum of the coefficients must equal $1/Fo$, i.e. $1 + 1·80 + 1·20 = 4$.)
(c) For t_{w1} and t_{w2}, from equation (21.41),

$$t_{w1,p} = \frac{2t_{1,p} + 18·4 t_{s1,p}}{20·4}$$

$$t_{w2,p} = \frac{2t_{5,p} + 18·4 t_{s2,p}}{20·4}$$

The table gives the results: the columns headed t_{w1} and t_{w2} being completed after the temperatures at the other planes have been computed. Finally, the relevant average values during the time intervals and in the space intervals have been shown in parentheses.

Hours	t_{s1}	t_{w1}	t_1	t_2	t_3	t_4	t_5	t_{w2}	t_{s2}
0	100	100	100	100	100	100	100	100	100
	(150)	(145)							
$\frac{1}{3}$	200	190	100	100	100	100	100	100	100
	(250)	(238)							
$\frac{2}{3}$	300	285	145	100	100	100	100	100	100
	(350)	(333)							
1	400	381	204	111	100	100	100	100	100
	(450)	(429)							
$1\frac{1}{3}$	500	477	269	132	103	100	100	100	100
	(550)	(526)							
$1\frac{2}{3}$	600	574	339	159	110	101	100	100	100
	(650)	(623)							
2	700	672	412	192	120	103	100	100	100
			(542)	(302)	(156)	(112)	(102)	(100)	

The heat inflow to the wall per unit area is, from equation (21.40),

$$q = \tfrac{1}{3} \times 3600 \times 0\cdot32(5 + 12 + 17 + 21 + 24 + 27) = 40\ 700\ \text{kJ/m}^2$$

$(t_{w2} - t_{s2})$ is zero throughout the period, so that there is no heat outflow and the energy stored in the wall must also be 40 700 kJ/m². This can be checked by adding the quantities of energy stored in the three Δx-wide elements and the two $\Delta x/2$-wide elements according to the equation

$$q_{\text{stored}} = \Sigma \rho c_p \Delta x (t_{n,p} - t_{n,0})$$

$$\rho c_p = \frac{k}{\alpha} = \frac{8\cdot7 \times 10^{-4}}{0\cdot52 \times 10^{-6}} = 1673\ \text{kJ/m}^3\ \text{K}$$

$$q_{\text{stored}} = 1673\left[\frac{0\cdot05}{2}(542 - 100) + 0\cdot05\{(302 - 100) + (156 - 100) + (112 - 100)\right.$$

$$\left. + (102 - 100)\} + \frac{0\cdot05}{2}(100 - 100)\right] = 41\ 200\ \text{kJ/m}^2$$

The discrepancy between the results is only about 1 per cent which is an indication that the numerical results are self-consistent and correct. The agreement is not proof that the solution is close to the exact solution; to establish this it is necessary to proceed in stages with smaller space and time intervals until no marked change in the numerical values occurs, taking care that at each stage criteria (21.34) and (21.39) are satisfied. Alternatively, methods are available which enable the errors in the finite difference approximation to be estimated: see Refs. (22) and (25).

It should be clear from the foregoing how the treatment can be modified when the film coefficient is not constant but varies with time. The Biot number will then also carry the suffix p, viz. Bi_p, and the coefficients in equation (21.38) will be numerically different for each time interval. One possible reason why h might be a function of time is that the fluid velocity varies with time: or again, in free convection, it would vary as the temperature difference $(t_w - t_s)$ changed. For extensions of the method to cover other boundary conditions, such as those associated with composite walls, the reader can turn to Refs. (15) and (25).

Finally, it should be noted that the conditions in the examples given here are such that the required restriction on Fo can be met with reasonable values of Δx and $\Delta \tau$. In other circumstances it may be that it can be met only by using too few space and time intervals for accuracy, or such a large number of intervals that the computation is unduly lengthy. Other forms of finite difference relation can be used which overcome such difficulties, and Ref. (25) deals with this point also.

22 Convection

The study of heat transfer by convection is concerned with the calculation of rates of heat exchange between fluids and solid boundaries. It was pointed out in the Introduction that the solution of the differential equations that govern the transfer of heat by convection usually presents considerable mathematical difficulties, and exact solutions can rarely be obtained. This chapter is confined to a description of methods giving approximate solutions to some simple but important cases of steady heat flow.

It is known from experience that the main resistance to heat flow from a wall to a fluid is in a comparatively thin boundary layer adjacent to the wall. The first method of calculating rates of heat transfer presented here makes use of the hydrodynamic concept of a *velocity boundary layer* and the analogous thermal concept of a *temperature boundary layer*. The object of this procedure is to find the value of the temperature gradient $(dt/dy)_0$ in the fluid in the immediate vicinity of the wall. It is assumed that the fluid at the wall adheres to it, and therefore that the heat flow at the wall is by conduction and not by convection, i.e.

$$Q = -kA \left(\frac{dt}{dy}\right)_0$$

The second method presented is like the first in that it studies the behaviour of the boundary layer. It makes use of the similarity, first pointed out by Reynolds, between the mechanism of fluid friction in the boundary layer (i.e. the transfer of fluid momentum to the wall) and the transfer of heat by convection. When this analogy is valid, rates of heat transfer can be predicted from the measurement of the shear stress between a fluid and a wall.

The third method is concerned with model testing used to obtain information in a form which is applicable to similar systems of any scale. It is based on the principle of dynamic similarity, and employs the method of dimensional analysis for the correlation of empirical data.

It was pointed out in section 21.2 [equation (21.8)] that it is customary in engineering to express rates of heat transfer between a fluid and a wall by a relation of the form

$$Q = -hA(t_s - t_w)$$

where h is a *convection* or *film* heat transfer coefficient. (The suffix c on h_c can be dispensed with in this chapter since we shall not be referring here to the equivalent radiation coefficient h_r.) It must be remembered that h is not a physical constant of the fluid. h is a function of all the parameters that affect the heat flow, such as viscosity, velocity, etc., and it may even depend on A and $(t_s - t_w)$ because Q does not necessarily vary linearly with these variables. The results of the three methods of approach just outlined are usually expressed in terms of the coefficient h, and the problem of finding Q is simply transferred to the evaluation of h.

Before the methods of determining h can be discussed, it is necessary to present some facts about the properties of the velocity boundary layer which is formed when a fluid flows past a flat plate, or through a cylindrical tube. No attempt is made here to derive the results quoted, and for their derivation the reader is referred to books on fluid dynamics such as those of Refs. (16), (18) and (8).

22.1 Some results of simple boundary layer theory

The flow of fluids in the vicinity of solid surfaces is discussed briefly in this section. Attention is focused on the velocity distribution near a wall and the shear stress between fluid and wall, both of which are of interest in the study of convection heat transfer. There are two main types of flow. In *laminar* or *streamline* flow, fluid elements move in continuous paths, or streamlines, without mixing with the fluid in adjacent paths. In *turbulent* flow, eddy motion of small fluid elements is superimposed on the main flow; the velocities of individual elements fluctuate in the direction of flow and perpendicular to it, and so result in the mixing of the fluid.

Two cases are described here, viz. (i) two-dimensional flow parallel to a flat plate, and (ii) flow through a cylindrical tube. Historically, the flow in tubes was studied first, and many of the results were applied later to flow past flat plates. A more logical presentation is possible, however, if the flat plate is considered first, and this order is adopted here. The equations presented apply to liquids, or gases flowing with velocities sufficiently low for compressibility effects to be negligible.

(i) *Flow of fluid past a flat plate*

When a fluid flows with a free stream velocity U_s past a flat plate, it is assumed that the fluid velocity adjacent to the surface is zero, i.e. that there is no slip between fluid and surface.* At increasing distances perpendicular to the plate, the stream velocity approaches the free stream velocity U_s asymptotically. In fluid dynamics it is usual to divide the flow region artificially into two parts. One region adjacent to the wall, in which the velocity varies from U_s at some distance δ to zero at the wall, is called the *boundary layer*. The other region, called the *free stream*, lies outside the boundary layer, and there the velocity is assumed to be equal to U_s everywhere.

If the flow of the fluid approaching the leading edge of the plate is laminar, a

---

*This assumption is far from accurate for gases at very low pressures. See Ref. (35) (R. M. Drake and E. D. Kane), p. 117, and Ref. (6), Chap. 10.

laminar boundary layer of thickness δ builds up on the plate (Fig. 22.1). At the edge of the plate δ is zero, and it increases gradually with the distance x. The shear stress τ at any point in the fluid in an x-z plane is proportional to the velocity gradient (dU/dy) at that point. The constant of proportionality is called the *dynamic viscosity* μ of the fluid, so that

$$\tau = \mu \left(\frac{dU}{dy}\right) \qquad (22.1)$$

Or, in terms of the *kinematic viscosity* v,

$$\tau = \rho v \left(\frac{dU}{dy}\right) \qquad (22.1a)$$

Fig. 22.1

Thus the shear stress is zero outside the boundary layer, and increases to a maximum value at the wall where the velocity gradient is greatest. Equation (22.1) is sometimes known as *Newton's law of viscosity*.

The velocity profile in the laminar boundary layer at some distance x from the leading edge can be approximately described by a cubic parabola, i.e.

$$\frac{U}{U_s} = \frac{3}{2}\left(\frac{y}{\delta}\right) - \frac{1}{2}\left(\frac{y}{\delta}\right)^3 \qquad (22.2)$$

It can be seen that this equation satisfies the following conditions: (1) $U = 0$ at the wall, where $y = 0$, (2) $U = U_s$ at the edge of the boundary layer, where $y = \delta$, (3) $(dU/dy)_\delta = 0$ at $y = \delta$, and (4) the curvature of the velocity profile at the wall $(d^2U/dy^2)_0 = 0$. This last condition must be satisfied for the following reason. In the immediate proximity of the wall, velocities and accelerations are small, and the only forces acting on the fluid layers are viscous shear stresses. Hence, for dynamic equilibrium of the layer at the wall (Fig. 22.1), $\tau_1 = \tau_2$ and $(dU/dy)_{y\simeq0}$ is independent of y, i.e. the profile is a straight line in the immediate vicinity of the wall and $(d^2U/dy^2)_0 = 0$.

From the assumed velocity profile, equation (22.2), it can be shown that the thickness of the boundary layer increases with the distance x from the leading edge of the plate according to the relation

$$\frac{\delta}{x} = \frac{4\cdot64}{Re_x^{1/2}} \qquad (22.3)$$

where Re_x is the *Reynolds number* based on the distance x, i.e.

$$Re_x = \frac{\rho U_s x}{\mu} \qquad (22.4)$$

505

The shear stress τ_w at the wall is usually expressed in terms of the dynamic head of the free stream as follows

$$\tau_w = f\frac{\rho U_s^2}{2} \qquad (22.5)$$

where f is a dimensionless *friction factor*, which is a function of the Reynolds number.* τ_w must also be equal to $\mu(dU/dy)_0$, according to equation (22.1), so that an expression for f can be derived from (22.2), (22.3) and (22.5). The local friction factor f_x at x is then found to be

$$f_x = \frac{0\cdot664}{Re_x^{1/2}} \dagger \qquad (22.6)$$

It is essential to distinguish between the local factor, and the average factor f over the length x which can be found by integration. Considering unit width of plate, the total drag force on the length x is

$$\int_0^x f_x \frac{\rho U_s^2}{2}\,dx$$

This is equal to $f(\rho U_s^2/2)x$, by definition of the average factor f, and hence it can be shown that

$$f = \frac{1\cdot328}{Re_x^{1/2}} = 2f_x \qquad (22.7)$$

Beyond a certain critical distance x_{cr} from the leading edge, the flow loses its laminar character, eddies develop, and the flow becomes fully turbulent after a transition region (Fig. 22.2). The critical distance, in terms of the Reynolds number, has been found from experiment to be

$$Re_{cr} = 500\,000 \quad \text{i.e.} \quad x_{cr} = \frac{\mu}{\rho U_s}\,500\,000$$

Although the critical distance x_{cr} may be considerably less than this if the free stream approaching the plate is turbulent, there is always at least a short distance at the nose of the plate over which there is a laminar boundary layer.

The friction factor for a turbulent boundary layer cannot be deduced theoretically, but it has been found experimentally to be given by

$$f_x = \frac{0\cdot0592}{Re_x^{0\cdot2}} \qquad (22.8)$$

---

*Several friction factors, which not only differ in name but are also defined as multiples of f used here, can be found in the literature: caution is necessary when using equations involving such factors.

†The expression actually quoted here has been obtained from the accurate expression for the velocity profile of the boundary layer. If (22.2) is used, the numerator becomes 0.647.

In turbulent flow the eddies act as macroscopic carriers of free stream momentum to the wall; in laminar flow the transfer of momentum is due solely to molecular carriers, i.e. viscous friction. It follows that the velocity gradient near the wall in a turbulent boundary layer is much steeper than in a laminar one. Prandtl has shown that equation (22.8) implies that the velocity profile in the turbulent boundary layer must follow a 7th power law, i.e.

$$\frac{U}{U_s} = \left(\frac{y}{\delta}\right)^{1/7}$$
(22.9)

Fig. 22.2   Transition of the laminar boundary layer into a turbulent boundary layer

The thickness of the turbulent boundary layer can then be shown to be given by

$$\frac{\delta}{x} = \frac{0 \cdot 38}{Re_x^{0 \cdot 2}}$$
(22.10)

The boundary layer condition (4), shown to be satisfied by the laminar velocity profile given by equation (22.2), must also be satisfied by the turbulent profile at the wall. But from equation (22.9) it is clear that $(d^2U/dy^2)_0 \neq 0$ at the wall. This discrepancy need not cause concern because there is always a thin laminar sublayer of thickness δ_b in the immediate proximity of the wall (Fig. 22.2). In this sublayer the velocity increases linearly with y to satisfy boundary condition (4). The velocity at the interface between the turbulent layer and the laminar layer is then found from (22.8), (22.9), (22.10), and the linear relation $(U/U_b) = (y/\delta_b)$, to be given by

$$\frac{U_b}{U_s} = \frac{2 \cdot 11}{Re_x^{0 \cdot 1}}$$
(22.11)

The thickness of the sublayer immediately follows from (22.9) and (22.11) as

$$\frac{\delta_b}{\delta} = \left(\frac{U_b}{U_s}\right)^7 = \frac{186}{Re_x^{0\cdot7}} \tag{22.12}$$

Relations (22.9) to (22.12) are all empirical results in the sense that they are deduced from the empirical law (22.8).

(ii) *Flow of fluid through a cylindrical tube*

When a fluid approaches a tube with a uniform velocity U_s, a boundary layer gradually builds up until it reaches the centre of the tube. The velocity profile is then said to be *fully developed*, and may be assumed not to vary down the tube. Although the velocity

Fig. 22.3 Establishment of a fully developed velocity profile in a tube

profile actually approaches its final shape only asymptotically, it has been found that for practical purposes it is fully developed in laminar flow at a distance l from the entrance of the tube (Fig. 22.3), where

$$\frac{l}{d} = 0\cdot0288 Re_d \tag{22.13}$$

l is called the *entrance* or *approach length*. The Reynolds number is defined by

$$Re_d = \frac{\rho U_m d}{\mu} \tag{22.14}$$

i.e. it is calculated on the basis of the tube diameter d and the weighted mean velocity U_m. The weighted mean velocity at any cross-section is defined by the equation

$$\int_A \rho U \, dA = \rho U_m A \tag{22.15}$$

where $\rho U_m A$ is the total rate of mass flow through A. If the Reynolds number exceeds the critical value for flow through tubes (≈ 2300), the flow becomes turbulent in the approach length, and the velocity profile becomes fully established in a distance less than that predicted by (22.13). This is shown in Fig. 22.3b.

For fully developed laminar flow the velocity profile is parabolic, and is given by

$$\frac{U}{U_r} = 2\left(\frac{y}{r}\right) - \left(\frac{y}{r}\right)^2 \tag{22.16}$$

where r is the radius of the tube, U_r is the velocity at the axis of the tube, and U the velocity at a distance y from the wall. It is often convenient to work in terms of the weighted mean velocity U_m, as defined by (22.15), because this value can be obtained from the mass flow which is often more easily measured than U_r. The value of the velocity for any annulus of width dy at radius $(r - y)$ may be found from (22.16), and if this is substituted in (22.15) we get

$$\int_0^r \rho U_r \left\{2\left(\frac{y}{r}\right) - \left(\frac{y}{r}\right)^2\right\} 2\pi(r - y)\, dy = \rho U_m \pi r^2$$

This reduces to

$$\frac{U_m}{U_r} = 0 \cdot 5 \tag{22.17}$$

If the shear stress τ_w at the wall is expressed in terms of the dynamic head of the fluid as follows

$$\tau_w = f\frac{\rho U_m^2}{2} \tag{22.18}$$

then by making use of equation (22.1), (22.16) and (22.17) we get

$$f = \frac{16}{Re_d} \tag{22.19}$$

For fully developed turbulent flow the shear stress at the wall can also be expressed by (22.18), where

$$f = \frac{0 \cdot 0791}{Re_d^{1/4}} \tag{22.20}$$

Equation (22.20) cannot be deduced theoretically; it is an experimental relation, called the *friction law of Blasius*, and it holds reasonably well up to Reynolds numbers of about 200 000 (Fig. 22.4). Prandtl has shown that, as an implication of (22.20), the velocity profile must follow a 7th power law, i.e.

$$\frac{U}{U_r} = \left(\frac{y}{r}\right)^{1/7} \tag{22.21}$$

This equation is similar to (22.9) for the flat plate, except that the boundary layer thickness δ is replaced by the radius r, and the free stream velocity U_s by the velocity at the axis U_r.

The weighted mean velocity for the turbulent velocity profile can be calculated from (22.15) and (22.21) as

$$\int_0^r \rho U_r \left(\frac{y}{r}\right)^{1/7} 2\pi(r-y)\,\mathrm{d}y = \rho U_m \pi r^2$$

which reduces to

$$\frac{U_m}{U_r} = 0.817 \tag{22.22}$$

Fig. 22.4   Friction factor for a smooth tube: (1) laminar flow equation (22.19), (2) transition range, (3) turbulent flow equation (22.20)

As with the turbulent boundary layer near a flat plate, there is always a thin laminar sublayer at the tube wall. It can be shown that the velocity at the edge of the sublayer is given by

$$\frac{U_b}{U_r} = \frac{1.99}{Re_d^{1/8}} \tag{22.23}$$

and the thickness of the sublayer follows immediately from (22.21) and (22.23) as

$$\frac{\delta_b}{d} = \frac{1}{2}\left(\frac{U_b}{U_r}\right)^7 = \frac{61.8}{Re_d^{7/8}} \tag{22.24}$$

Like relations (22.9) to (22.12), the relations (22.21) to (22.24) are essentially empirical, because they have been deduced from the empirical friction law—the law of Blasius (22.20).

510

(iii) *Shear stress due to eddy motion*

So far we have only considered the magnitude of the shear stress *at the wall*, expressing the result in terms of the friction factor *f*. The analysis has been based on equation (22.1), which was introduced when discussing laminar flow. Since even with turbulent flow there is always a laminar sublayer next to the wall, the question of the general validity of equation (22.1) has not arisen. However, before we can discuss the analogy between the transfer of momentum and the transfer of heat (section 22.4), it is necessary to know something about the shear stress in regions where the flow is turbulent.

The shear stress in the turbulent boundary layer at section 1–1 (Fig. 22.2), a distance y_1 from the wall, is not given by equation (22.1) because there is an exchange of mass across the surface 1–1 and the momentum is transferred by both macroscopic and molecular carriers. Due to eddy motion the fluid velocity fluctuates and the instantaneous velocity U_1 in the x-direction can be resolved into two components. If $\bar{U}_1$ represents the mean velocity of the fluid, and U_1' represents the instantaneous superimposed velocity fluctuation, then

$$U_1 = \bar{U}_1 + U_1' \tag{22.25}$$

Similarly, the instantaneous velocity in the y-direction can be written as

$$V_1 = \bar{V}_1 + V_1' \tag{22.26}$$

For steady flow $\bar{U}_1$ and $\bar{V}_1$ are constants and, although the values of U_1' and V_1' fluctuate, their mean values must be zero over a time p which is long compared with the period of fluctuation.

Now let an infinitesimal time interval dp be considered. The mass flow upward (+ve) through unit area of surface 1–1 during time dp is $\rho V_1\, dp$; the momentum in the x-direction carried upward through this surface (+ve) is $(\rho V_1\, dp)U_1$. The shear stress acting along 1–1 due to eddy motion only, $\tau_{\varepsilon 1}$, must be equal to the *average rate* at which momentum is transported through unit surface area along 1–1, i.e.

$$\tau_{\varepsilon 1} = \frac{\rho}{p}\int_0^p (U_1 V_1)\, dp$$

If this integral is positive, the net flow of momentum is upward, i.e. away from the wall, and if it is negative, the net flow of momentum is toward the wall. On substitution of (22.25) and (22.26) this becomes

$$\tau_{\varepsilon 1} = \frac{\rho}{p}\int_0^p (\bar{U}_1\bar{V}_1 + \bar{U}_1 V_1' + U_1'\bar{V}_1 + U_1' V_1')\, dp \tag{22.27}$$

But because the time averages of the fluctuating components of the velocities, U_1' and V_1', are both zero,

$$\frac{\rho}{p}\int_0^p (\bar{U}_1 V_1')\, dp = \frac{\rho \bar{U}_1}{p}\int_0^p V_1'\, dp = 0$$

$$\frac{\rho}{p}\int_0^p (U_1'\bar{V}_1)\, dp = \frac{\rho \bar{V}_1}{p}\int_0^p U_1'\, dp = 0$$

Therefore

$$\tau_{s1} = \frac{\rho}{p} \int_0^p (\bar{U}_1 \bar{V}_1 + U_1' V_1') \, dp$$

Now let us restrict the analysis to 'one-dimensional' flow, i.e. flow for which $\bar{V}_1 = 0$, although V_1' may still assume finite values. This restriction effectively covers the two cases which are considered here: (i) for flow past a flat plate the thickness of the turbulent boundary layer increases very slowly and therefore $\bar{V}_1$ is negligible, (ii) for fully developed flow through a tube $\bar{V}_1$ must be exactly zero. With this restriction then, the first term in the integral of (22.27) is also zero, i.e.

$$\frac{p}{\rho} \int_0^p (\bar{U}_1 \bar{V}_1) \, dp = 0$$

It follows that (22.27) becomes

$$\tau_{s1} = \frac{\rho}{p} \int_0^p (U_1' V_1') \, dp = \rho (U_1' V_1')_{\text{mean}} \tag{22.28}$$

where $(U_1' V_1')_{\text{mean}}$ is the time average of the product $(U_1' V_1')$.

The question arises whether the time average of $(U_1' V_1')$ differs from zero. Although the time averages of U_1' and V_1' are individually zero, the answer to this question is not immediately obvious. Consider a positive V_1', i.e. an instantaneous flow upward through 1–1. As this flow arrives from a region which, *on the average*, has a velocity lower than $\bar{U}_1$, the fluid element arriving at 1–1 produces negative U_1' at 1–1; hence $(U_1' V_1')$ is, *on the average*, negative for positive V_1'. Now consider a negative V_1', i.e. an instantaneous flow downward through 1–1. As this flow arrives from a region in which, *on the average*, the velocity is greater than $\bar{U}_1$, the fluid element arriving at 1–1 produces a positive U_1' at 1–1; hence $(U_1' V_1')$ is, *on the average*, negative for negative V_1'. Thus, on the whole $(U_1' V_1')_{\text{mean}}$ is negative and not zero. This simplified picture indicates that there is a net flow of momentum toward the wall, resulting in a shear stress as indicated in Fig. 22.2. Thus the shear stress due to eddy motion at 1–1 is of the same sense as that due to viscous forces. To be able to sum these separate effects, we may write, by analogy with equation (22.1a),

$$\tau_{s1} = \rho \varepsilon_1 \left(\frac{d\bar{U}}{dy} \right)_1$$

where ε is called the *eddy diffusivity*. This quantity is defined by equating the two expressions for τ_{s1}, i.e.

$$\tau_{s1} = -\rho (U_1' V_1')_{\text{mean}} = \rho \varepsilon_1 \left(\frac{d\bar{U}}{dy} \right)_1 \tag{22.29}$$

The negative sign must be introduced if ε is to be a positive quantity like ν, since $(d\bar{U}/dy)_1$ is always positive and we have just seen that $(U_1' V_1')_{\text{mean}}$ is always negative.

The total shear stress is the result of the viscous and eddy shear stresses, or in other words the flow of momentum toward the wall is due to a molecular diffusion of momentum *and* a macroscopic diffusion of momentum. Hence

$$\tau_1 = \rho \nu \left(\frac{d\bar{U}}{dy}\right)_1 + \rho \varepsilon_1 \left(\frac{d\bar{U}}{dy}\right)_1 = \rho (\nu + \varepsilon_1) \left(\frac{d\bar{U}}{dy}\right)_1 \qquad (22.30)$$

It must be emphasised that the eddy diffusivity* is a function of the turbulence and it is not a property of the fluid, although the turbulence arises as a result of fluid viscosity. ε changes in value with the distance from the wall. In the sublayer where the flow is laminar and $V' = 0$, ε must be zero, but its value rapidly increases to many times the value of ν in the turbulent layer. The table below shows some experimental values for the distance from the wall of a tube at which $\varepsilon = 10\nu$, for various Reynolds numbers. The distance from the wall is expressed as a fraction of the tube radius.

Re_d	4 000	10 000	100 000
y/r	0·23	0·10	0·01

To end this section summarising boundary layer theory, we must note certain restrictions on the application of the equations quoted. Relations (22.8), (22.11), (22.12) for a flat plate, and (22.20), (22.23), (22.24) for a tube, which apply to turbulent flow, are only valid if the surfaces are *hydraulically smooth*. The criterion of smoothness is that the unevenness of the surface must be less than $0·4\delta_b$. Since δ_b decreases with increase in flow velocity, smoothness is a particularly important factor at high velocities. Relations for rough surfaces will not be given here.

In the derivation of the relations given in this section, it has been assumed that there are no temperature gradients in the boundary layer. Since μ and ρ vary with temperature, temperature gradients in the fluid affect the validity of these relations. For fluid flow where the temperature difference between the *free stream temperature* t_s and the wall temperature t_w is not too great, all the equations can be used with reasonable accuracy if the fluid properties are evaluated at a *film temperature* t_f, where t_f is defined as

$$t_f = \frac{t_s + t_w}{2} \qquad (22.31)$$

For flow through a tube, t_f is usually defined in terms of the *weighted mean temperature* t_m and the wall temperature t_w, i.e.

$$t_f = \frac{t_m + t_w}{2} \qquad (22.32)$$

*For a full discussion of the concept of eddy diffusivity see Ref. (18), Chaps. 18 and 19.

t_m, at any cross-section of the tube, is the temperature that the fluid would assume if it were thoroughly mixed, and is in fact the temperature often measured in experimental work.

Example 22.1 Air at 15 °C and 1 atm flows with a velocity of 8 m/s past a flat plate which is maintained at a uniform temperature of 115 °C. Calculate the thickness of the boundary layer 0·6 m from the leading edge, and estimate the position of the point of transition to turbulent flow. Also calculate the total drag force on one side of the plate, per metre width of plate, over the first 0·6 m.

The film temperature t_f is $(15 + 115)/2 = 65$ °C. At this temperature the density of air at atmospheric pressure is $\rho = 1{\cdot}046$ kg/m^3, and its viscosity is $\mu = 2{\cdot}021 \times 10^{-5}$ kg/m s [see Ref. (14)]. Hence the Reynolds number 0·6 m from the leading edge is

$$Re_x = \frac{\rho U_s x}{\mu} = \frac{1{\cdot}046 \times 8 \times 0{\cdot}6}{2{\cdot}021 \times 10^{-5}} = 248\ 400$$

The thickness of the boundary layer at this point is therefore

$$\delta = \frac{4{\cdot}64x}{Re_x^{1/2}} = \frac{4{\cdot}64 \times 0{\cdot}6}{248\ 400^{1/2}} = 0{\cdot}00559 \text{ m} = 0{\cdot}559 \text{ cm}$$

The point of transition, assuming that the approaching air is initially laminar, will be at

$$x_{cr} = \frac{\mu}{\rho U_s} Re_{cr} = \frac{2{\cdot}021 \times 10^{-5}}{1{\cdot}046 \times 8} \times 500\ 000 = 1{\cdot}2 \text{ m}$$

The average friction factor is

$$f = \frac{1{\cdot}328}{Re_x^{1/2}} = \frac{1{\cdot}328}{248\ 400^{1/2}} = 0{\cdot}00266$$

The total drag force over the first 0·6 m, per metre width of plate, is therefore

$$\tau_w A = f \frac{\rho U_s^2}{2} A = 0{\cdot}00266 \times \frac{1{\cdot}046 \times 8^2}{2} (0{\cdot}6 \times 1) = 0{\cdot}0534 \text{ kg m/s}^2 \text{ or N}$$

It is interesting to note how the velocity profile for laminar flow is distorted from its parabolic shape when the fluid flowing through a tube is cooled or heated. Curve *a* in Fig. 22.5 shows the simple parabolic profile in *isothermal flow*, i.e. for the limiting case of heat transfer by convection with $(t_w - t_m) \to 0$, when there are negligible temperature gradients in the fluid. When the fluid is heated $t_w > t_m$, and when it is cooled, $t_w < t_m$. As the temperature is increased, the viscosity of a liquid decreases and the viscosity of a gas increases. Thus when a liquid is heated the viscosity at the wall is lower than at the axis, and the gradient at the wall is increased compared with the parabolic profile, as is shown by curve *b*. Curve *b* also serves to represent the velocity profile of a gas when it is cooled. Curve *c* shows the distortion of the profile for a liquid that is being cooled or a gas that is being heated.

In the following sections two simple cases are considered: (a) the heating or cooling of a fluid flowing parallel to a plate, and (b) the heating or cooling of a fluid flowing through a cylindrical tube with a fully established velocity profile. It will be assumed

(a) 'Isothermal' flow

(b) Heating a liquid or cooling a gas

(c) Cooling a liquid or heating a gas

Fig. 22.5   Distortion of the parabolic velocity profile in laminar flow through a tube

that the fluid is either a liquid or a gas and that the fluid properties are independent of temperature.

22.2 Forced convection in laminar flow past a flat plate*

When a plate is heated or cooled to a uniform temperature t_w, starting at a distance x_a from the leading edge of the plate, temperature gradients are set up in the fluid. The temperature of the fluid immediately adjacent to the wall is t_w, and far away from it the fluid temperature is equal to the free stream temperature t_s. It is possible

Fig. 22.6   The development of the temperature boundary layer along a flat plate

to divide the fluid temperature field artificially into two regions. The region adjacent to the wall, in which the temperature varies from t_s to t_w, is called the *temperature boundary layer*. The other region lies outside the boundary layer, and there the temperature is assumed to be everywhere equal to the temperature t_s. Thus the concept of the temperature boundary layer is analogous to that of the velocity boundary layer.

At any section $x > x_a$ (Fig. 22.6) the profile of the temperature boundary layer

*The method given in sections (22.2) and (22.3) was first developed and applied by G. Kroujiline, Ref. (27). The presentation in this chapter is based on the work of E. R. G. Eckert, Ref. (6), Chap.7.

may be expressed as a polynomial in y, and it may be assumed that a cubic parabola adequately describes the profile. Hence

$$t = t_w + ay + by^2 + cy^3$$

It is more convenient to work in terms of the temperature difference between fluid and wall. If we write θ for $(t - t_w)$, the expression becomes

$$\theta = ay + by^2 + cy^3 \tag{22.33}$$

a, b, and c can be found by applying the boundary conditions as follows. At the wall, where $y = 0$, the fluid layer is at rest and can have no velocity component perpendicular to the wall. Therefore at the wall the heat transfer is entirely by conduction and not by transport of fluid, i.e.

$$q_w = -k \left(\frac{\mathrm{d}t}{\mathrm{d}y}\right)_0 = -k \left(\frac{\mathrm{d}\theta}{\mathrm{d}y}\right)_0 \tag{22.34}$$

where k is the thermal conductivity of the fluid. Hence from equation (22.34)

$$\left(\frac{\mathrm{d}^2\theta}{\mathrm{d}y^2}\right)_0 = 0 \quad \text{and therefore} \quad b = 0$$

If the thickness of the temperature boundary layer is denoted by δ_t, then the following further two boundary conditions must be satisfied at $y = \delta_t$, namely

$$\theta = \theta_s \quad \text{and} \quad \left(\frac{\mathrm{d}\theta}{\mathrm{d}y}\right)_{\delta_t} = 0$$

On substituting the three boundary conditions in the cubic parabola for the temperature profile (22.33), this becomes

$$\frac{\theta}{\theta_s} = \frac{3}{2}\left(\frac{y}{\delta_t}\right) - \frac{1}{2}\left(\frac{y}{\delta_t}\right)^3 \tag{22.35}$$

The rate of heat flow per unit area at the wall at section x can be calculated by finding the value of $(\mathrm{d}\theta/\mathrm{d}y)_0$ from equation (22.35). Substituting this value in equation (22.34), we have

$$q_w = -k \frac{3}{2\delta_t} \theta_s \tag{22.36}$$

The value of δ_t in (22.36) is still unknown, and must be found before q_w can be calculated. It is convenient to introduce into this equation the ratio of the temperature boundary layer thickness δ_t to the velocity boundary layer thickness δ. Denoting this ratio by ψ we have

$$\psi = \frac{\delta_t}{\delta} \tag{22.37}$$

so that equation (22.36) can be written as

$$q_w = -\frac{3}{2\psi\delta}k\theta_s \qquad\qquad (22.38)$$

The value of δ is known from equation (22.3), i.e.

$$\frac{\delta}{x} = \frac{4\cdot64}{Re_x^{1/2}}$$

and hence

$$q_w = -\frac{3}{2\times4\cdot64\psi x}Re_x^{1/2}k\theta_s \qquad\qquad (22.39)$$

The problem is thus reduced to the determination of the value of ψ. It will be assumed initially that $\delta_t \leqslant \delta$, i.e. $\psi \leqslant 1$; this assumption will be justified later.

(i) *Derivation of the heat flow equation of the boundary layer*

Consider a region near the surface of the plate of unit width perpendicular to the paper in the z direction (Fig. 22.7). Let this region be bounded by two surfaces 1–1 and 2–2, a distance dx apart. The other surfaces limiting this region are the surface

Fig. 22.7

of the plate 0–0, and a surface 1–2 outside the edges of the temperature and velocity boundary layers a distance l away from the wall.

The continuity and energy equations can now be applied to this open system. The streamlines are of the form shown by the dotted lines in Fig. 22.7. For continuity of mass, the difference between the mass entering the region across plane 1–1 and the mass leaving across plane 2–2 is equal to the mass leaving across plane 1–2. Since no external work is done on or by the fluid passing through the region, the steady-flow energy equation states that the net rate of heat flow into the region must equal

$$\sum_{2-2}\delta m(h + \tfrac{1}{2}U^2) + \sum_{1-2}\delta m(h + \tfrac{1}{2}U^2) - \sum_{1-1}\delta m(h + \tfrac{1}{2}U^2)$$

where δm is the mass entering or leaving per unit time with a specific enthalpy h and a specific kinetic energy $U^2/2$. We shall assume, in what follows, that the kinetic energy of the fluid is insignificant compared with the enthalpy; only in high-speed flow need the kinetic energy be taken into account.* Furthermore, since the fluids considered are either gases or liquids (i.e. flow without change of phase), we may substitute $c_p t$ for h because the enthalpy varies but little with pressure.

With these modifications, we have

$$\sum_{1-1} \delta m(h + \tfrac{1}{2}U^2) = \rho c_p \int_0^l tU \, \mathrm{d}y$$

and

$$\sum_{2-2} \delta m(h + \tfrac{1}{2}U^2) = \rho c_p \int_0^l tU \, \mathrm{d}y + \rho c_p \frac{\mathrm{d}}{\mathrm{d}x} \left(\int_0^l tU \, \mathrm{d}y \right) \mathrm{d}x$$

ρ and c_p are assumed not to vary with temperature, and are therefore written outside the integrals. From the continuity equation, the mass of the fluid leaving 1–2 must be equal to the difference between the inflow through 1–1 and the outflow through 2–2, i.e.

$$-\rho \frac{\mathrm{d}}{\mathrm{d}x} \left(\int_0^l U \, \mathrm{d}y \right) \mathrm{d}x$$

The mass outflow through surface 1–2 is at a temperature t_s, since $l > \delta_t$, and hence

$$\sum_{1-2} \delta m(h + \tfrac{1}{2}U^2) = -\rho c_p t_s \frac{\mathrm{d}}{\mathrm{d}x} \left(\int_0^l U \, \mathrm{d}y \right) \mathrm{d}x$$

Because the temperature gradient across 1–2 is zero, there can be no conduction of heat across this surface. Also, since the gradients $(\mathrm{d}\theta/\mathrm{d}x)$ across 1–1 and 2–2 are small compared with $(\mathrm{d}\theta/\mathrm{d}y)$, and since the heat conducted across these surfaces is negligible compared with the flow of energy due to transport of fluid across them, we can neglect any heat conduction across these two surfaces. We may assume therefore that the only heat which crosses the boundary of the region is that which flows across the wall surface 0–0. The flow of heat into the region from the wall 0–0 is given by

$$\mathrm{d}Q_w = -k \, \mathrm{d}x \left(\frac{\mathrm{d}\theta}{\mathrm{d}y} \right)_0$$

If the flow of heat is away from the wall, θ decreases with increase of y and hence the negative sign. For steady heat and fluid flow, the equation expressing the First Law becomes

$$-k \, \mathrm{d}x \left(\frac{\mathrm{d}\theta}{\mathrm{d}y} \right)_0 = \rho c_p \frac{\mathrm{d}}{\mathrm{d}x} \left(\int_0^l tU \, \mathrm{d}y \right) \mathrm{d}x - \rho c_p t_s \frac{\mathrm{d}}{\mathrm{d}x} \left(\int_0^l U \, \mathrm{d}y \right) \mathrm{d}x$$

*See Introduction to Part IV, and Ref. (6), Chap.10.

If α is substituted for $k/\rho c_p$, and θ for $(t - t_w)$, this reduces to

$$\frac{\mathrm{d}}{\mathrm{d}x} \int_0^l (\theta_s - \theta)U \, \mathrm{d}y = \alpha \left(\frac{\mathrm{d}\theta}{\mathrm{d}y}\right)_0 \tag{22.40}$$

α is the *thermal diffusivity* of the fluid, and equation (22.40) is called the *heat flow equation* of the boundary layer.

(ii) *Solution of the heat flow equation*

It is now possible to substitute the equations of the velocity and temperature profiles (22.2) and (22.35) in the heat flow equation. The right-hand side of (22.40) becomes

$$\alpha \left(\frac{\mathrm{d}\theta}{\mathrm{d}y}\right)_0 = \alpha \frac{3}{2\delta_t} \theta_s = \frac{3}{2} \frac{\alpha\theta_s}{\psi\delta} \tag{22.41}$$

The left-hand side of (22.40) can be split into two integrals, i.e.

$$\int_0^l (\theta_s - \theta)U \, \mathrm{d}y = \int_0^{\delta_t} (\theta_s - \theta)U \, \mathrm{d}y + \int_{\delta_t}^l (\theta_s - \theta)U \, \mathrm{d}y$$

The second integral, between limits $\delta_t > y > l$, must be zero, because between these limits $(\theta_s - \theta) = 0$. The integral in the left-hand side of (22.40) can therefore be written

$$\int_0^l (\theta_s - \theta)U \, \mathrm{d}y = \theta_s U_s \int_0^{\delta_t} \left\{ 1 - \frac{3}{2}\left(\frac{y}{\delta_t}\right) + \frac{1}{2}\left(\frac{y}{\delta_t}\right)^3 \right\}\left\{ \frac{3}{2}\left(\frac{y}{\delta}\right) - \frac{1}{2}\left(\frac{y}{\delta}\right)^3 \right\} \mathrm{d}y$$

$$= \theta_s U_s \delta \left(\frac{3}{20}\psi^2 - \frac{3}{280}\psi^4\right) \approx \theta_s U_s \delta \frac{3}{20}\psi^2 \tag{22.42}$$

Note that the ratio $\psi = \delta_t/\delta$ is less than 1, according to the assumption made at the beginning of this section, so that $3\psi^4/280 \ll 3\psi^2/20$ and can therefore be neglected.

Substituting (22.41) and (22.42) in the heat flow equation (22.40), this can now be rewritten as

$$\frac{3}{20} \theta_s U_s \frac{\mathrm{d}}{\mathrm{d}x}(\psi^2\delta) = \frac{3}{2}\frac{\alpha\theta_s}{\psi\delta}$$

or

$$0\cdot1 U_s \psi\delta \frac{\mathrm{d}}{\mathrm{d}x}(\psi^2\delta) = \alpha$$

Hence

$$0\cdot1 U_s \left(\psi^3\delta \frac{\mathrm{d}\delta}{\mathrm{d}x} + 2\psi^2\delta^2 \frac{\mathrm{d}\psi}{\mathrm{d}x}\right) = \alpha$$

The terms containing the velocity boundary thickness δ can be found from equation (22.3). Since

$$\frac{\delta}{x} = \frac{4\cdot64}{Re_x^{1/2}}, \quad \delta^2 = \frac{21\cdot5x\mu}{U_s\rho} \quad \text{and} \quad \delta\frac{d\delta}{dx} = 0\cdot5\frac{21\cdot5\mu}{U_s\rho}$$

The heat flow equation therefore becomes

$$0\cdot1U_s\left(\psi^3 0\cdot5\frac{21\cdot5\mu}{U_s\rho} + 2\psi^2\frac{21\cdot5x\mu}{U_s\rho}\frac{d\psi}{dx}\right) = \alpha$$

or

$$1\cdot075\left(\psi^3 + 4\psi^2 x\frac{d\psi}{dx}\right) = \frac{k}{\rho c_p}\frac{\rho}{\mu} = \frac{k}{c_p\mu}$$

The group $c_p\mu/k$ occurs frequently in heat transfer work and is called the *Prandtl number*; it is a dimensionless group involving three fluid properties, and is thus itself a property of the fluid. Writing Pr for $c_p\mu/k$, the heat flow equation reduces to

$$\psi^3 + \frac{4}{3}x\frac{d(\psi^3)}{dx} = \frac{1}{1\cdot075Pr} \tag{22.43}$$

The differential equation (22.43) can now be solved for ψ. Dividing throughout by $4x^{1/4}/3$, the left-hand side of the equation is transformed into a perfect differential. Thus

$$\frac{3}{4}\psi^3 x^{-1/4} + x^{3/4}\frac{d(\psi^3)}{dx} = \frac{3}{4}\frac{x^{-1/4}}{1\cdot075Pr}$$

and hence

$$\frac{d(x^{3/4}\psi^3)}{dx} = \frac{3}{4}\frac{x^{-1/4}}{1\cdot075Pr}$$

Integrating, we have

$$x^{3/4}\psi^3 = \frac{x^{3/4}}{1\cdot075Pr} + C$$

and therefore

$$\psi^3 = \frac{1}{1\cdot075Pr} + Cx^{-3/4}$$

The value of the constant C in the solution can be found from the boundary condition $\psi = 0$ when $x = x_a$; thus $C = -x_a^{3/4}/1\cdot075Pr$. The solution is therefore

$$\psi = \frac{1}{(1\cdot075Pr)^{1/3}}\left\{1 - \left(\frac{x_a}{x}\right)^{3/4}\right\}^{1/3} \tag{22.44}$$

The solution will now be restricted to the special case of a plate heated all along its length, i.e. when $x_a = 0$. In this case

$$\psi = \frac{1}{(1 \cdot 075 Pr)^{1/3}} \approx \frac{1}{Pr^{1/3}} \qquad (22.45)$$

Thus when a plate is heated from the leading edge the ratio of thicknesses of the temperature and velocity boundary layers is constant along the length of the plate. If the value of ψ is now substituted in equation (22.39), the heat flow at section x is given by

$$q_w = -\frac{3(1 \cdot 075 Pr)^{1/3}}{2 \times 4 \cdot 64 x} Re_x^{1/2} k\theta_s = -0 \cdot 331 Pr^{1/3} Re_x^{1/2} \frac{k\theta_s}{x}$$

An exact solution to this problem has yielded a similar answer, but with a numerical constant of $0 \cdot 332$.* The equation can now be divided by $k\theta_s/x$. Remembering also that the heat transfer or film coefficient h is defined by the equation

$$q_w = -h(t_s - t_w) = -h\theta_s$$

the local heat transfer coefficient h_x at section x can be calculated from the relation

$$\frac{h_x x}{k} = 0 \cdot 332 Pr^{1/3} Re_x^{1/2}$$

The group $h_x x/k$ appears frequently in heat transfer equations and is called the *Nusselt number*. It will be denoted by Nu_x, the suffix x indicating that the value refers to a distance x from the leading edge of the plate. The Nusselt number is a dimensionless group, and is a measure of the rate of heat transfer by convection; it has here been shown that it can be expressed as a function of two dimensionless groups, the Reynolds number which describes the flow, and the Prandtl number which is a property of the fluid. Thus

$$Nu_x = 0 \cdot 332 Pr^{1/3} Re_x^{1/2} \qquad (22.46)$$

When the plate is heated from the leading edge over a length l, the average value of h over the heated length is given by

$$h = \frac{1}{l} \int_0^l h_x \, dx$$

If the value of h_x given by (21.46) is substituted in the integral, this yields

$$h = 2h_l \quad \text{or} \quad Nu = 2Nu_l \qquad (22.47)$$

where h_l and Nu_l are the local heat transfer coefficient and local Nusselt number respectively at $x = l$.

It remains to justify the assumption that $\psi = \delta_t/\delta \leqslant 1$. For this purpose let the most unfavourable case be considered, viz. when the plate is heated over its entire

---

*The agreement is fortuitous: the solution given here is approximate because the velocity and temperature profiles were described by cubic parabolas. In fact these profiles can only be described accurately by infinite series, see Ref. (18), Chap. 14.

length. Evidently the result expressed as equation (22.45) is in accord with the assumption provided that the Prandtl number for the fluid is equal to or greater than 1. This is so for most liquids except molten metals as can be seen from Fig. 22.8. The special case of liquid metals is discussed briefly at the end of this section. The Prandtl number for gases and dry vapours varies only slightly, and lies between 0·65 and 1. A brief consideration of (22.42), which is the first equation affected by the assumption $\psi \leqslant 1$, shows that the error introduced by neglecting the fourth power term is small

$T/[K]$	250	300	400	500	600	800
Hydrogen	0·713	0·706	0·690	0·675	0·661	0·633
Air	0·720	0·707	0·688	0·680	0·680	0·690
Carbon dioxide . .	0·773	0·763	0·737	0·703	0·686	0·694
Steam (low p) . . .		0·93	0·94	0·95	0·93	0·89
Steam (sat.) . . .		0·96	1·01	1·30		
Water (sat.) . . .		5·81	1·34	0·85		
Ammonia (sat. liq.) . .	1·70	1·13				
Freon-12 (sat. liq.) . .	3·5	3·5				
Mercury, liquid . . .	0·0353	0·0261	0·0173	0·0129	0·0095	0·0079
Sodium, liquid . . .			0·0097	0·0069	0·0055	0·0046
Mineral oils . . .		10–1000				

Fig. 22.8 Typical values of Prandtl number $c_v\mu/k$

even if ψ is slightly greater than unity. Thus the assumption can be justified even for gases and vapours. We may also infer from (22.45) and Fig. 22.8 that the thicknesses of the velocity and temperature boundary layers for gases are practically equal, whereas for a very viscous lubricating oil, for which $Pr = 1000$ say, δ_t is about $\delta/10$.

Due to the length of the solution for the flat plate, it is worth retracing the principal steps in the argument:

(a) The heat flow at the wall at a distance x from the leading edge of the plate is given by equation (22.34)

$$q_w = -k \left(\frac{\mathrm{d}\theta}{\mathrm{d}y}\right)_0$$

It is therefore necessary to find the temperature profile at x, in the form $\theta = \phi(y)$.
(b) By introducing the concept of the temperature boundary layer of finite thickness δ_t, and by assuming that the function $\theta = \phi(y)$ is a cubic parabola, the heat flow at the wall is found to be given by equation (22.36)

$$q_w = -k \frac{3}{2\delta_t} \theta_s$$

but δ_t has still to be determined.

522

(c) The formation of the temperature boundary layer must depend upon the flow, and therefore upon the profile of the velocity boundary layer of thickness δ. Introducing the ratio $\psi = \delta_t/\delta$ and the value of δ in terms of Re_x into equation (22.36), we derive equation (22.39)

$$q_w = -\frac{3}{2 \times 4.64\psi x} Re_x^{1/2} k\theta_s$$

This equation can be written in terms of the Nusselt number

$$Nu_x = \frac{3}{2 \times 4.64} \frac{Re_x}{\psi}$$

or in general terms

$$Nu_x = \phi(Re_x, \psi)$$

(d) Using the continuity and energy equations, we arrive at the heat flow equation of the boundary layer which enables the value of ψ to be expressed in terms of physical properties of the fluid and the distance x from the leading edge of the plate, i.e.

$$\psi = \phi\left(\frac{c_p\mu}{k}, \frac{x_a}{x}\right) = \phi\left(Pr, \frac{x_a}{x}\right)$$

When the plate is heated from the leading edge, ψ is independent of x, and

$$\psi = \phi(Pr)$$

Hence the final result for this latter case is given in the form

$$Nu_x = \phi(Re_x, Pr) \tag{22.48}$$

In the foregoing we have assumed that the fluid properties are constant, but in practice the properties vary with temperature. The procedure recommended at the end of section 22.1, of using property values at a 'film temperature', can normally be adopted with reasonable accuracy. For large temperature gradients this may not always be satisfactory, however, and this point will be discussed more fully in section 22.5.

Example 22.2 For the conditions given in Example 22.1, calculate the heat transfer co-efficient 0·6 m from the leading edge of the plate. Hence calculate the rate of heat transfer from one side of the plate to the air, per metre width of plate, over the first 0·6 m.

From Ref. (14) Pr for air at 65 °C is 0·699, and from Example 22.1 Re_x is 248 400. Hence the local Nusselt number 0·6 m from the leading edge is

$$Nu_x = 0.332 Pr^{1/3} Re_x^{1/2} = 0.332 \times 0.699^{1/3} \times 248\,400^{1/2} = 147$$

The conductivity of air at 65 °C is $k = 2 \cdot 913 \times 10^{-5}$ kW/m K. Hence the value of the local heat transfer coefficient is

$$h_x = Nu_x \frac{k}{x} = 147 \times \frac{2 \cdot 913 \times 10^{-5}}{0 \cdot 6} = 0 \cdot 007\ 14\ \text{kJ/m}^2\ \text{s K}$$

The average heat transfer coefficient h over the length x is $2h_x$, and hence the rate of heat transfer over the first 0·6 m, per metre width of plate, is

$$Q = -hA(t_s - t_w)$$
$$= -(2 \times 0 \cdot 007\ 14)(0 \cdot 6 \times 1)(15 - 115) = 0 \cdot 857\ \text{kJ/s or kW}$$

(iii) *Solution applicable to liquid metals*

Finally, we must see how the solution presented here can be modified to be valid for liquid metals. Such substances have become of increasing importance in recent years as heat transfer media in nuclear reactors. The point at which we can expect the analysis to be incorrect for liquid metals is where it is assumed that $\delta_t < \delta$ because although equation (22.45) is no longer strictly applicable, it suggests qualitatively that $\delta_t > \delta$ when $Pr < 1$. Let us re-examine equation (22.40) in conjunction with Fig. 22.7 and the assumption that $\delta_t > \delta$.

The integral on the left-side of (22.40) can be split into three parts, viz.

$$\int_0^l (\theta_s - \theta) U \, \mathrm{d}y = \int_0^\delta (\theta_s - \theta) U \, \mathrm{d}y + \int_\delta^{\delta_t} (\theta_s - \theta) U \, \mathrm{d}y + \int_{\delta_t}^l (\theta_s - \theta) U \, \mathrm{d}y$$

Following the argument at the beginning of section 22.2 (ii), the third integral must be zero. Also, U in the second integral is constant and equal to U_s, because the limits of integration lie outside the velocity boundary layer. Further simplification can be achieved by assuming that for liquid metals, which have very low values of Pr, $\delta \ll \delta_t$. We can then ignore the first integral and write finally

$$\int_0^l (\theta_s - \theta) U \, \mathrm{d}y \approx U_s \int_0^{\delta_t} (\theta_s - \theta) \, \mathrm{d}y \tag{22.49}$$

This approximation is tantamount to assuming that the velocity is uniform and equal to U_s from the wall surface: an idealised type of flow often referred to as 'slug' flow. The heat flow equation now becomes

$$U_s \frac{\mathrm{d}}{\mathrm{d}x} \int_0^{\delta_t} (\theta_s - \theta) \, \mathrm{d}y = \alpha \left(\frac{\mathrm{d}\theta}{\mathrm{d}y} \right)_0$$

and on substituting equation (22.35) for θ it is easy to show that integration yields

$$\frac{\delta_t}{x} = \left(\frac{8}{Re_x \times Pr} \right)^{\frac{1}{2}} \tag{22.50}$$

Substitution of this expression for δ_t in (22.36) then gives

$$Nu_x = 0\cdot530(Re_x \times Pr)^{1/2} \tag{22.51}$$

The product $(Re_x \times Pr)$ is a dimensionless group called the *Peclet number*, *Pe*, and it is an important parameter in liquid metal convection.

An exact solution to the problem, when the plate is heated to a uniform temperature, gives

$$Nu_x = 0\cdot487Pe^{1/2} \quad \text{when} \quad Pr = 0\cdot03$$
$$Nu_x = 0\cdot516Pe^{1/2} \quad \text{when} \quad Pr = 0\cdot01$$
$$Nu_x = 0\cdot526Pe^{1/2} \quad \text{when} \quad Pr = 0\cdot006$$

To check that our assumption of $\delta \ll \delta_t$ is reasonable, we may find an expression for $\psi = \delta_t/\delta$ by combining equations (22.3) and (22.50) to give

$$\psi = \frac{0\cdot609}{Pr^{1/2}} \tag{22.52}$$

This is of the same form as equation (22.45), i.e. $\psi = f(Pr)$, and for a typical liquid metal of $Pr = 0\cdot01$ it yields a value of $\psi = 6$. Thus the approximation of 'slug' flow appears not unreasonable.

The approximate solution for liquid metals is based on the heat flow equation (22.40), which itself may not always be valid for such substances. It will be remembered that when deriving (22.40) the temperature gradient $d\theta/dx$ was assumed small compared with $d\theta/dy$, enabling us to ignore any conduction across planes 1–1 and 2–2 in Fig. 22.7. For liquid metals, conduction in the direction of flow may not always be negligible, particularly at low values of Re_x and therefore of *Pe*. On the other hand, the approximate solution is valid some way into the turbulent flow region, where the $\frac{1}{7}$th power velocity profile is closer to the 'slug' flow model, because eddy mixing has relatively little effect on heat transfer over and above the contribution from the high thermal conductivity of the liquid metal.

22.3 Forced convection in fully developed laminar flow through a tube

The problem of calculating the value of the heat transfer coefficient in fully developed laminar flow in a tube is considerably easier than the problem of the flat plate dealt with in the last section. The reason is that, once the flow is fully established, the fluid can have no velocity components normal to the wall anywhere in the cross-section, otherwise successive velocity profiles would not be identical (Fig. 22.9). There is no divergence of the streamlines away from the wall as in Fig. 22.7, and the heat flow in the radial direction must therefore be entirely by conduction.

It will be assumed, as for the flow past a flat plate, that conduction in an axial direction can be neglected, i.e. that anywhere in the fluid $(\partial t/\partial x) \ll (\partial t/\partial y)$, and therefore in the axial direction the energy transport is entirely by convection. It will also

525

be assumed that the properties of the fluid are independent of temperature, or that the temperature difference between fluid and wall is sufficiently small for such variations to be neglected. This last assumption implies that the velocity profile is parabolic, i.e. that equation (22.16),

$$\frac{U}{U_r} = 2\left(\frac{y}{r}\right) - \left(\frac{y}{r}\right)^2$$

is valid.

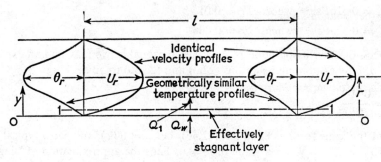

Fig. 22.9 Fully developed flow through a tube

As with the flat plate, the temperature profile may again be approximately expressed by a cubic parabola (Fig. 22.9)

$$t = t_w + ay + by^2 + cy^3$$

or working in terms of the temperature difference $\theta = t - t_w$,

$$\theta = ay + by^2 + cy^3 \tag{22.53}$$

Consider a length l of the cylinder. The rate of radial conduction at any radius $(r - y)$ is given by

$$Q = -k2\pi(r - y)l\left(\frac{d\theta}{dy}\right)$$

In the immediate vicinity of the wall the fluid layers are effectively stagnant, and there is little convection in the axial direction. The only heat flow is by conduction in the radial direction. To satisfy continuity of heat flow, this implies that when $y \simeq 0$, Q is independent of y, i.e. $Q_1 = Q_w$ in Fig. 22.9. Hence near the wall, by differentiating the previous equation with Q treated as constant,

$$\left(\frac{d^2\theta}{dy^2}\right) = -\frac{Q}{2\pi kl}\frac{1}{(r - y)^2} = \frac{1}{(r - y)}\left(\frac{d\theta}{dy}\right)$$

And therefore when $y = 0$, we have the first boundary condition

$$\left(\frac{d^2\theta}{dy^2}\right)_0 = \frac{1}{r}\left(\frac{d\theta}{dy}\right)_0$$

Applying this boundary condition to (22.53), we get

$$\left(\frac{d\theta}{dy}\right)_0 = a = r\left(\frac{d^2\theta}{dy^2}\right)_0 = r2b$$

Two further boundary conditions can be stated; at the axis of the tube where $y = r$,

$$\theta = \theta_r, \quad \text{and from symmetry} \quad \left(\frac{d\theta}{dy}\right)_r = 0$$

Applying these two boundary conditions to (22.53) it follows that

$$\theta_r = ar + br^2 + cr^3$$

and

$$\left(\frac{d\theta}{dy}\right)_r = 0 = a + 2br + 3cr^2$$

The constants a, b, c can now be found from the three boundary conditions, and on substituting their values in (22.53) the equation becomes

$$\frac{\theta}{\theta_r} = \frac{6}{5}\left(\frac{y}{r}\right) + \frac{3}{5}\left(\frac{y}{r}\right)^2 - \frac{4}{5}\left(\frac{y}{r}\right)^3 \qquad (22.54)$$

The rate of heat flow per unit area at the wall is

$$q_w = -k\left(\frac{d\theta}{dy}\right)_0 = -\frac{6}{5}k\frac{\theta_r}{r} = -h_r(t_r - t_w)$$

and therefore the Nusselt number based on the temperature difference between axis and wall $(t_r - t_w) = \theta_r$ is

$$(Nu_d)_r = \frac{h_r d}{k} = 2\cdot 4 \qquad (22.55)$$

The suffix r refers to the fluid temperature used in the definition of h, and suffix d to the characteristic linear dimension. It must be emphasised that this result is the Nusselt number at any section x once the flow has become fully developed, and thus the local and average values are the same in this region.

It is more convenient to work in terms of a temperature difference $\theta_m = (t_m - t_w)$, where t_m is the weighted mean temperature of the fluid, since t_m is usually the temperature measured in practice. By definition, t_m at any cross-section of the tube is the temperature that the fluid would assume if it were thoroughly mixed. By considering a strip dy at radius $(r - y)$, t_m is thus defined by the equation

$$\int_0^r \rho c_p U t 2\pi (r - y)\, dy = \rho c_p U_m t_m \pi r^2$$

The product $\rho U_m \pi r^2 = \rho U_m A$ is given by equation (22.15), and hence

$$\int_0^r \rho c_p U t 2\pi (r - y)\, dy = c_p t_m \int_0^r \rho U 2\pi (r - y)\, dy$$

$U = \phi(y)$ is given by (22.16), and $t = \phi(y)$ by (22.54). Again treating the properties ρ and c_p as constant, and working in terms of θ instead of t, we get

$$\rho c_p 2\pi \int_0^r U_r \left\{ 2\left(\frac{y}{r}\right) - \left(\frac{y}{r}\right)^2 \right\} \theta_r \left\{ \frac{6}{5}\left(\frac{y}{r}\right) + \frac{3}{5}\left(\frac{y}{r}\right)^2 - \frac{4}{5}\left(\frac{y}{r}\right)^3 \right\} (r - y)\, dy$$

$$= \rho c_p 2\pi \theta_m \int_0^r U_r \left\{ 2\left(\frac{y}{r}\right) - \left(\frac{y}{r}\right)^2 \right\} (r - y)\, dy$$

Hence

$$\theta_m = \frac{102}{175}\, \theta_r = 0\cdot583\theta_r$$

The heat transfer coefficient at the wall can now be defined in terms of θ_m, i.e.

$$q_w = -h(t_m - t_w) = -h\theta_m$$

The Nusselt number, defined in terms of h based on θ_m, then becomes

$$Nu_d = \frac{hd}{k} = (Nu_d)_r \frac{\theta_r}{\theta_m} = 2\cdot4 \times \frac{175}{102} = 4\cdot12 \tag{22.56}$$

The method given here is only approximate because it is based on an assumed particular temperature profile, viz. a cubic parabola. An exact solution to the problem of fully developed flow also leads to the result $Nu_d = $ constant, but the value of the constant depends upon a thermal boundary condition along the wall of the tube which did not have to be stipulated in the preceding approximate method. Two alternative boundary conditions of considerable interest are (a) constant wall temperature t_w, and (b) constant heat flux q_w which is equivalent to constant temperature gradient at the wall $(dt/dy)_0$. Boundary condition (a), which may be approached in practice with a tube heated on the outside by a condensing vapour, leads to the result

$$Nu_d = 3\cdot65$$

Boundary condition (b) can be approached with a tube which is heated by an electric current passing through it, and in this case

$$Nu_d = 4\cdot36$$

The approximate constant of 4·12 therefore falls between the two exact results and is in reasonable agreement with both.

The calculation of the heat transfer in the entrance section of a tube is considerably more complicated than that for fully developed flow. In addition to distinguishing between different thermal boundary conditions along the tube wall, it is necessary to consider whether at the beginning of the heated section the velocity profile is already developed or whether it is developing simultaneously with the temperature profile, i.e. whether there is an unheated entrance length or not. Whatever case is considered, the value of the temperature gradient at the wall is infinite at the beginning of the heated section and $Nu_a = \infty$. Along the tube the Nusselt number falls and approaches the appropriate constant value asymptotically. A rough rule is that the local Nu_a effectively reaches the asymptotic value when $l/d = 0.05(Re \times Pr)$. This means that with a fluid of high Pr, like a viscous mineral oil, the entrance length can be very long: and laminar flow conditions are quite likely to be encountered with such fluids.

As has been pointed out in section 22.1, and illustrated in Fig. 22.5, the velocity profile in fully developed flow is not parabolic because of the dependence of viscosity on temperature. Since the temperature profile is a function of the velocity profile, the former will also be changed and even the foregoing accurate solutions are only valid for 'isothermal flow', i.e. for the limiting case of heat transfer by convection as $\theta_m \to 0$. It may also be noted that with a finite temperature difference between fluid and wall, the velocity profile may be further distorted by *free* convection currents superimposed on the flow in the tube: this is only significant in large diameter tubes at low Reynolds numbers.

It will be appreciated that a full treatment of heat transfer in laminar flow in a tube, including effects of an entrance length, different thermal boundary conditions, variable properties, and free convection, is very complex. For further details the reader is referred to Ref. (6).

22.4 Forced convection in turbulent flow—Reynolds analogy

So far it has not been possible to produce a complete analytical solution for heat transfer by forced convection when the flow is turbulent. It was suggested by Reynolds, however, that there exists a similarity between forced convection heat transfer and fluid friction, and that it should be possible to predict rates of heat transfer from hydrodynamic measurements, e.g. from the measurement of viscous drag on a surface.[*]
The reason for this similarity is as follows.

Consider a fluid flowing past a wall as in Fig. 22.10. Along any section such as 1–1 there is a viscous shear stress due to a molecular momentum transfer given by

$$\tau_{v1} = \mu \left(\frac{dU}{dy}\right)_1 = \rho v \left(\frac{dU}{dy}\right)_1$$

[*]Ref. (21). Also see Ref. (10), Chapter 24

There is also heat flow by molecular conduction through 1–1, given by

$$q_{\alpha 1} = -k \left(\frac{\mathrm{d}t}{\mathrm{d}y}\right)_1 = -\rho c_p \alpha \left(\frac{\mathrm{d}t}{\mathrm{d}y}\right)_1$$

The similarity between the foregoing equations is apparent. When considering turbulent flow it is necessary to show that there is also a similarity between the equation expressing the shear stress due to eddy motion, and the equation expressing the heat flow due to eddy motion.

Fig. 22.10 Similarity between heat flow and shear stress in a fluid boundary layer

In section 21.1, we resolved the local instantaneous velocity components U_1, V_1 at 1–1 into mean velocity components $\bar{U}_1$, $\bar{V}_1$, and superimposed fluctuation components U'_1, V'_1. Thus we wrote equation (22.25) and (22.26) respectively as

$$U_1 = \bar{U}_1 + U'_1 \quad \text{and} \quad V_1 = \bar{V}_1 + V'_1$$

By assuming $\bar{V}_1$ to be negligible, it was then shown that, in addition to the viscous shear stress, there must be a shear stress due to eddy motion across 1–1 given by (22.29), viz.

$$\tau_{\varepsilon 1} = -\rho(U'_1 V'_1)_{\text{mean}} = \rho \varepsilon_1 \left(\frac{\mathrm{d}\bar{U}}{\mathrm{d}y}\right)_1$$

The eddy diffusivity ε is a function of the local velocity fluctuations, i.e. it is a quantity depending on the amount of turbulence. The total shear stress is, according to (22.30),

$$\tau_1 = \rho(\nu + \varepsilon_1) \left(\frac{\mathrm{d}\bar{U}}{\mathrm{d}y}\right)_1$$

It is stated in section 22.1, and must be emphasised again here, that for steady flow with $\bar{V}_1$ negligible there is no *net* mass transfer through 1–1, and that the transfer of momentum to the wall is due to fluid elements of various velocities fluctuating in a direction normal to the wall, exchanging momentum with neighbouring elements. Heat is transferred through the fluid by a similar process, and it might be supposed that the rate of transfer via eddy motion can be expressed by an equation analogous to (22.29); such an equation can be deduced as follows.

Let the temperature of the wall be higher than the temperature of the free stream. The instantaneous temperature t_1 at 1–1 fluctuates, due to the occasional arrival of hot eddies from below and cold eddies from above, and we may thus write

$$t_1 = \bar{t}_1 + t_1' \tag{22.57}$$

The mass flow per unit area across surface 1–1, during a time interval dp, is $\rho V_1 \, dp$, and the transport of energy across unit area is thus $(\rho V_1 \, dp)c_p t_1$. Considering a period p, which is long compared with the period of velocity fluctuations, the average rate of heat flow may be found by integration as

$$q_{\varepsilon 1} = \frac{1}{p} \int_0^p \rho c_p V_1 t_1 \, dp$$

By substituting the values of V_1 and t_1 from (22.26) and (22.57) in the integral, we obtain

$$q_{\varepsilon 1} = \frac{\rho c_p}{p} \int_0^p (\bar{V}_1 \bar{t}_1 + \bar{V}_1 t_1' + V_1' \bar{t}_1 + V_1' t_1') \, dp$$

For steady flow the time averages of the velocity and temperature fluctuations are zero, i.e.

$$\int_0^p \bar{V}_1 t_2' \, dp = \bar{V}_1 \int_0^p t_1' \, dp = 0$$

$$\int_0^p V_1' \bar{t}_1 \, dp = \bar{t}_1 \int_0^p V_1' \, dp = 0$$

Further, as has been explained, $\bar{V}_1$ is zero or nearly so for the cases we are considering. Hence

$$\int_0^p \bar{V}_1 \bar{t}_1 \, dp = 0$$

Thus the heat transfer per unit area at 1–1 is

$$q_{\varepsilon 1} = \frac{\rho c_p}{p} \int_0^p (V_1' t_1') \, dp = \rho c_p (V_1' t_1')_{\text{mean}} \tag{22.58}$$

where $(V_1' t_1')_{\text{mean}}$ stands for the time average of the product of the instantaneous velocity fluctuation V_1' and the temperature fluctuation t_1'. This equation is similar to (22.28),

$$\tau_{\varepsilon 1} = \rho (U_1' V_1')_{\text{mean}}$$

which relates the shear stress to the product of the instantaneous velocity fluctuation components.

The energy flow across 1–1 may be written, by analogy with Fourier's law, as

$$q_{\varepsilon 1} = -\rho c_p \varepsilon_{H1} \left(\frac{d\bar{t}}{dy}\right)_1$$

where ε_H is an *eddy diffusivity for heat flow*, defined by

$$q_{\varepsilon 1} = \rho c_p (V_1' t_1')_{\text{mean}} = -\rho c_p \varepsilon_{H1} \left(\frac{d\bar{t}}{dy}\right)_1 \tag{22.59}$$

Two principal questions relating to the value of $(V_1' t_1')_{\text{mean}}$, and therefore to ε_H, remain to be answered.

First, is the time average of the product $(V_1' t_1')$ different from zero? Although the time averages of V_1' and t_1' are individually zero, the answer to this question is not immediately obvious. Consider a positive V_1', i.e. an instantaneous flow upward through 1–1. As this flow arrives from a region which, *on the average*, is hotter than 1–1, $(V_1' t_1')$ is, *on the average*, positive for positive V_1'. Now consider a negative V_1', i.e. an instantaneous flow downward through 1–1. As this flow arrives from a region which, *on the average*, is colder than 1–1, $(V_1' t_1')$ is again, *on the average*, positive. Thus on the whole $(V_1' t_1')_{\text{mean}}$ is greater than zero, except in a region of no temperature gradient where t' must always be zero, or in a region of laminar flow where V' must always be zero. As $(d\bar{t}/dy)$ must be negative according to the adopted sign convention of positive heat flow into the fluid from the wall, the negative sign has been introduced into (22.59) to define ε_H as a positive quantity. The total heat flow through 1–1, which is a result of a molecular energy flow and an energy flow due to eddy motion, is therefore

$$q_1 = -k \left(\frac{d\bar{t}}{dy}\right)_1 - \rho c_p \varepsilon_{H1} \left(\frac{d\bar{t}}{dy}\right)_1$$

or

$$q_1 = -\rho c_p (\alpha + \varepsilon_{H1}) \left(\frac{d\bar{t}}{dy}\right)_1 \tag{22.60}$$

where α is the *thermal diffusivity* of the fluid. (The concept of thermal diffusivity was first introduced in section 21.6.)

The second question of fundamental importance is whether the eddy diffusivity for heat flow, as defined by (22.59), is numerically equal to the eddy diffusivity for momentum transfer, as defined by (22.29). Although an eddy moving across the stream tends to assume the temperature and momentum of the new layer into which it moves, if it returns to its original layer immediately it will not necessarily have assumed the momentum and temperature of its new surroundings completely. The values of ε and ε_H depend on the speed of adaptability of an eddy, and that need not be the same for momentum as it is for temperature. H. Reichardt* investigated the problem in some detail, and concluded that the eddy diffusivities for momentum transfer and heat transfer are not necessarily equal, although they are nearly equal for turbulent flow through tubes and past flat plates. For simplicity we shall assume here that the same value ε can be used in equation (22.29) and (22.59), and these equations thus serve as starting points for our similarity considerations between heat transfer and friction. It is necessary to state here again, however, that the value of ε varies from point to point in a turbulent fluid, although no assumption need be made about the variation of its value with the coordinates (x, y).

(i) *The simple Reynolds analogy*

We now require to find the relation between the heat flow q_w and the shear stress τ_w at the wall, and for this purpose equations (22.30) and (22.60) may be rewritten in the following form

$$dU = \frac{\tau\, dy}{\rho(\nu + \varepsilon)} \tag{22.61}$$

$$dt = -\frac{q\, dy}{\rho c_p(\alpha + \varepsilon)} \tag{22.62}$$

We have dropped the bar across U and t in (22.61) and (22.62) because only mean values are required in what follows and there is no need to retain this distinguishing mark. If it is required to find the relation between shear stress and heat flow in terms of some convenient reference velocities and temperatures, it is first necessary to integrate these equations between limits of y corresponding to the required reference planes. Usually the velocities and temperatures in the free stream and at the wall are the values known. Thus, by integration we obtain

$$U_s - U_w = U_s = \int_0^l \frac{\tau\, dy}{\rho(\nu + \varepsilon)} \tag{22.63}$$

$$t_s - t_w = \theta_s = -\int_0^l \frac{q\, dy}{\rho c_p(\alpha + \varepsilon)} \tag{22.64}$$

The upper limit of the integral $y = l$ is chosen in the free stream, i.e. l is larger than both δ and δ_t.

To evaluate the integrals it is of course necessary to know how τ, q and ε vary with the distance from the wall y. For the purpose of comparing q_w and τ_w, however, this is not necessary. The analysis will be restricted initially to fluids whose kinematic viscosity and thermal diffusivity are equal, i.e. when

$$\nu = \alpha \quad \text{or} \quad \frac{\mu}{\rho} = \frac{k}{\rho c_p}$$

*Ref. (28). Also see Ref. (35) (S. E. Isakoff and Th. B. Drew), p. 405 and (G. I. Taylor), p. 193. For a good discussion of the concept of eddy diffusivity in turbulent heat transfer theory, see Ref. (20).

The Prandtl number of the fluids considered is therefore

$$Pr = \frac{c_p\mu}{k} = 1$$

It follows that whatever may be the variation of ε with y, $(\nu + \varepsilon)$ and $(\alpha + \varepsilon)$ are equal and therefore vary in the same way with y. Further, it is an experimental fact, which can be justified by theoretical considerations [Ref. (28)], that the heat flow normal to the wall and the shear stress τ along surfaces parallel to the wall vary in an approximately similar way with y. This is a particularly reasonable assumption for flow past plates and *turbulent* flow through cylindrical tubes. The assumption implies that in general

$$\frac{q_1}{\tau_1} = \frac{q_w}{\tau_w} \neq \phi(y)$$

Dividing (22.64) by (22.63), and ignoring the negative sign which arises from the sign convention adopted for τ and q, we have

$$\frac{\theta_s}{U_s} = \frac{\displaystyle\int_0^l \frac{q\,dy}{\rho c_p(\alpha + \varepsilon)}}{\displaystyle\int_0^l \frac{\tau\,dy}{\rho(\nu + \varepsilon)}} = \frac{\dfrac{q_w}{\tau_w c_p}\displaystyle\int_0^l \frac{\tau\,dy}{(\alpha + \varepsilon)}}{\displaystyle\int_0^l \frac{\tau\,dy}{(\nu + \varepsilon)}}$$

and hence

$$\frac{q_w}{\tau_w c_p} = \frac{\theta_s}{U_s} \tag{22.65}$$

This relation is called the *Reynolds analogy*. It gives the ratio of the local rate of heat transfer to the local shear stress at the solid boundary.

Let the flow past a flat plate be considered first. Equation (22.65) can be used as it stands, and for smooth plates τ_w can be found from (22.5) and (22.8). It is convenient to put (22.65) into non-dimensional form: first by rewriting it as

$$\frac{q_w}{\theta_s} = \frac{\tau_w}{U_s} c_p$$

then by suitable multiplications and divisions of both sides of the equation we get

$$\frac{q_w x}{\theta_s k} = \frac{\tau_w}{\rho U_s^2}\frac{\rho U_s x}{\mu}\frac{c_p\mu}{k}$$

where x is the distance from the leading edge of the plate. This equation can now be rewritten as

$$Nu_x = \frac{f_x}{2} Re_x Pr$$

where f_x is the local friction factor defined by (22.5). As this relation has been derived on the assumption that the Prandtl number of the fluid is 1, we finally obtain for the local Nusselt number

$$Nu_x = \frac{f_x}{2} Re_x \tag{22.66}$$

Since f_x is merely a function of Reynolds number, equation (22.6), we arrive at the same form of result as that obtained for the laminar flow of fluids having a Prandtl number of unity, i.e. that the Nusselt number is a function of Reynolds number, equation (22.48).

Usually we are interested to find the average Nusselt number, and not its local value. In this case we can calculate, or find from experiment directly, the average shear stress on the plate; this average value can then be used in (22.65) for finding the average rate of heat flow over the plate, and hence the average Nusselt number can be deduced.

To apply the Reynolds analogy to flow through tubes, (22.63) and (22.64) must be considered afresh, although the only modification required is in the upper limit of the integrals. There is no 'free stream', and the appropriate reference section is the tube axis, where $y = r$. Equation (22.65), when applied to flow through a tube, thus becomes

$$\frac{q_w}{\tau_w c_p} = \frac{\theta_r}{U_r}$$

If the upper limit applied to the integrals in equations (22.63) and (22.64) is merely y, where the temperature difference is θ and the velocity is U, the equivalent of (22.65) becomes

$$\frac{q_w}{\tau_w c_p} = \frac{\theta}{U}$$

Since $(q_w/\tau_w c_p)$ is independent of y, this equation implies that the temperature and velocity profiles are geometrically similar. It can be shown that $\theta_m/U_m \approx \theta_r/U_r$, and the ratio of the weighted mean temperature and velocity may be used instead of the ratio of their values at the axis. The Reynolds relation for turbulent flow in tubes then becomes

$$\frac{q_w}{\tau_w c_p} = \frac{\theta_m}{U_m} \tag{22.67}$$

This can again be put into non-dimensional form, as

$$\frac{q_w d}{\theta_m k} = \frac{\tau_w}{\rho U_m^2} \frac{\rho U_m d}{\mu} \frac{c_p \mu}{k}$$

and hence, as the Prandtl number has been assumed to be unity, at any section x

$$Nu_d = \frac{f}{2} Re_d \tag{22.68}$$

535

τ_w (or f) can be found from measurements of the pressure drop, as in the following example, or for smooth tubes it can be found from the friction law of Blasius, equation (22.20).

Example 22.3 At some section of a cooled tube, having an inner diameter of 5 cm, air is at 150 °C and 1 atm and flows with a velocity of 8 m/s. The surface temperature of the tube is maintained at 40 °C. If the static pressure drop per metre length of tube is $1\cdot6 \times 10^{-4}$ bar, calculate approximately the temperature drop of the air over the next metre of tube.

Instead of working out the shear stress and heat transfer per unit surface area of tube, it is more convenient to evaluate them for a surface area corresponding to unit length of tube. The shear force on the wall, per metre length of tube, is therefore

$$\tau_w = \Delta p \, A = 1\cdot6 \times 10^{-4} \left[\frac{\text{bar}}{\text{m}}\right] \times 10^5 \left[\frac{\text{N/m}^2}{\text{bar}}\right] \times 1 \, [\text{m}] \times \pi 0\cdot025^2 \, [\text{m}^2] = 0\cdot0314 \, \text{N}$$

As a first approximation we may assume that for air $Pr = 1$, i.e. that the Reynolds relation (22.67) can be applied. Hence the rate of heat transfer per metre length of tube is

$$q_w = \frac{\tau_w c_p \theta_m}{U_m}$$

$$= \frac{0\cdot0314 \left[\text{N or } \dfrac{\text{kg m}}{\text{s}^2}\right] \times 1\cdot017 \left[\dfrac{\text{kJ}}{\text{kg K}}\right] \times \dfrac{\theta_m}{[\text{K}]} \, [\text{K}]}{8 \left[\dfrac{m}{s}\right]} = 0\cdot003\,99 \frac{\theta_m}{[\text{K}]} \, \text{kJ/s (or kW)}$$

c_p has been taken from Ref. (14) at t_1: when t_2 has been found a more accurate result can be obtained by repeating the calculation with c_p evaluated at $(t_1 + t_2)/2$.

The average value of θ_m over the foot length is

$$\theta_m = \left(\frac{150 + t_2}{2} - 40\right)$$

Fig. 22.11   Examples 22.3 and 22.4

Hence

$$q_w = 0\cdot003\,99 \left(\frac{150 + t_2}{2} - 40\right) = \text{kJ/s}$$

From an energy balance, this must also be equal to the enthalpy drop of the air in cooling from 150 °C to t_2 and hence

$$q_w = \rho c_p U_m \frac{\pi d^2}{4} (150 - t_2)$$

Although the velocity U_m falls along the tube and ρ rises, due to the cooling of the air, for a duct of constant cross-sectional area the product ρU_m remains constant. Thus ρU_m may be evaluated from the known conditions at section 1 and hence

$$q_w = 0\cdot8373 \times 1\cdot017 \times 8 \times \frac{\pi}{4} \times 0\cdot05^2 \times (150 - t_2) = 0\cdot0134(150 - t_2) \, \text{kJ/s}$$

536

Eliminating q_w from this and the previous equation, we get

$$0.003\,99 \left(\frac{150+t_2}{2}-40\right)=0.0314(150-t_2)$$

and hence $t_2 = 121\ °C$. Therefore the temperature drop of the air over the metre length is

$$150 - t_2 = 29\ K$$

(ii) *The Prandtl-Taylor modification of the Reynolds analogy*

We have seen how eddy motion in a boundary layer promotes momentum transfer (resulting in shear stresses) and heat transfer, and we have drawn attention to the similarity between these two processes. Thence, by making certain assumptions which could not be adequately justified here, we derived the relation (22.65) between the shear stress and the heat flow at a solid boundary. This simple relation, called the Reynolds analogy, was derived for fluids of $Pr = 1$, i.e. for cases where $v = \alpha$. Now the assumption essential to the argument is that $(v + \varepsilon) = (\alpha + \varepsilon)$ at any distance from the wall. In a region of high turbulence this equality is still valid, even if $v \neq \alpha$, because normally in such a region $\varepsilon \gg \alpha$ or v. (An important exception is when the fluid is a liquid metal—see end of this section.) But though the flow at some distance from the wall may be highly turbulent, there is a sublayer of thickness δ_b adjacent to the wall in which the flow is laminar and in which ε is zero. Thus when considering a fluid of $Pr \neq 1$, the integration in (22.63) and (22.64) cannot be carried out in one step from the wall to the free stream; the lower limit must be $y > \delta_b$.

Prandtl and Taylor suggested that a good approximation is obtained by choosing the lower limit at $y = \delta_b$. Thus the flow region is divided into two parts; the sublayer in which the flow is laminar and $\varepsilon = 0$, and the region outside the sublayer in which ε is considered large enough for the assumption of $(v + \varepsilon) = (\alpha + \varepsilon)$ to be valid even when $Pr \neq 1$ (Fig. 22.10).

Consider the flow past a flat plate first. By changing the limits of the integrals in (22.63) and (22.64) to $y = \delta_b$ and $y = l$, we obtain

$$U_s - U_b = \int_{\delta_b}^{l} \frac{\tau\,dy}{\rho(v+\varepsilon)} \tag{22.69}$$

$$t_s - t_b = -\int_{\delta_b}^{l} \frac{q\,dy}{\rho c_p(\alpha+\varepsilon)} \tag{22.70}$$

Dividing (22.70) by (22.69), and making the same assumptions as before about the expressions under the integrals, we obtain

$$\frac{q_b}{\tau_b c_p} = \frac{t_s - t_b}{U_s - U_b} = \frac{\theta_s - \theta_b}{U_s - U_b} \tag{22.71}$$

For the laminar sublayer, in which the velocity gradient is constant [see section 22.1 (i)], the shear stress is given by

$$\tau_w = \tau_b = \mu \left(\frac{dU}{dy}\right) = \mu \frac{U_b}{\delta_b} \tag{22.72}$$

The heat flow through the thin sublayer is by conduction only, and thus the temperature gradient through this layer must also be constant. Hence

$$q_w = q_b = -k \left(\frac{dt}{dy}\right) = -k \frac{t_b - t_w}{\delta_b} = -k \frac{\theta_b}{\delta_b} \tag{22.73}$$

Dividing (22.73) by (22.72), and ignoring the negative sign as before, we get

$$\frac{q_w}{\tau_w} = \frac{q_b}{\tau_b} = \frac{k}{\mu} \frac{\theta_b}{U_b} \tag{22.74}$$

Eliminating θ_b from (22.71) and (22.74), the following relation is obtained

$$\frac{q_w}{\tau_w c_p} = \frac{\theta_s}{U_s} \frac{1}{1 + \dfrac{U_b}{U_s}(Pr - 1)} \tag{22.75}$$

This relation is the *Prandtl-Taylor modification of the Reynolds analogy*, and it reduces to (22.65) when $Pr = 1$. The ratio of the velocities U_b/U_s for smooth plates can be taken from (22.11), and (22.75) can be put in non-dimensional form in the same way as (22.65). The relation finally obtained is

$$Nu_x = \frac{f_x}{2} \frac{Re_x Pr}{1 + 2 \cdot 11 Re_x^{-0 \cdot 1}(Pr - 1)} \tag{22.76}$$

When considering turbulent flow through a tube, the integration of (22.69) and (22.70) must be carried out between limits of $y = \delta_b$ and $y = r$. This procedure yields the relation

$$\frac{q_w}{\tau_w c_p} = \frac{\theta_r}{U_r} \frac{1}{1 + \dfrac{U_b}{U_r}(Pr - 1)}$$

The velocity and temperature profiles in the laminar sublayer are linear according to (22.72) and (22.73), and are therefore geometrically similar. When the limits in equations (22.69) and (22.70) are δ_b and y, the equivalent of (22.71) becomes

$$\frac{\theta - \theta_b}{U - U_b} = \frac{q_b}{\tau_b c_p}$$

As $q_b/\tau_b c_p$ is independent of y, this equation implies that the velocity and temperature profiles are also geometrically similar in the turbulent layer. This result was obtained before when considering a fluid of $Pr = 1$. Again $\theta_m/U_m \simeq \theta_r/U_r$, and the ratio of the weighted mean temperature and velocity may be used in the foregoing equation instead of the ratio of their values at the axis. The modified Reynolds relation thus becomes

$$\frac{q_w}{\tau_w c_p} = \frac{\theta_m}{U_m} \frac{1}{1 + \dfrac{U_b}{U_r}(Pr - 1)} \tag{22.77}$$

For fully developed flow in smooth tubes the ratio U_b/U_r is found from (22.23), and putting (22.77) into non-dimensional form, we get finally

$$Nu_d = \frac{f}{2} \frac{Re_d Pr}{1 + 1\cdot99 Re_d^{-1/8}(Pr - 1)} \tag{22.78}$$

Again f can be found measurements of the pressure drop, or for smooth tubes from (22.20) where it is expressed in terms of Re_d. Equations (22.76) and (22.78) show that for forced convection the Nusselt number is a function of the Reynolds and Prandtl numbers.

It is worth noting that the dimensionless combination $Nu/(Re \times Pr)$ is often used as an alternative for Nu when presenting heat transfer data: it is called the *Stanton number St* and is equal to $h/c_p \rho U$. Equation (22.78) can then be written as

$$St = \frac{f}{2} \frac{1}{1 + 1\cdot99 Re_d^{-1/8}(Pr - 1)}$$

It should also be apparent that the simple Reynolds analogy then becomes $St = f/2$. The main reason why the concept of Stanton number has found favour lies in the ease with which it is obtained from experimental data. For example, from the energy balance on a heated or cooled fluid in a tube we can write

$$h \pi dl\, \theta_m = \rho U \frac{\pi d^2}{4} c_p(t_2 - t_1)$$

or

$$\frac{h}{c_p \rho U} = St = \frac{d}{4l} \frac{(t_2 - t_1)}{\theta_m}$$

Thus St can be obtained directly from the tube dimensions and temperature measurements without reference to fluid properties.

Example 22.4   Recalculate Example 22.3 taking the correct Prandtl number for air. [Property values used in the solution are taken from Ref. (14).]

   At the mean film temperature $t_f = (136 + 40)/2 = 88\,°C$, Pr for air is $0\cdot695$. Thus for air the modified relation by Prandtl and Taylor should be used. It is first necessary to calculate the ratio U_b/U_r, and assuming that the tube is smooth and the flow fully developed, this ratio is

$$\frac{U_b}{U_r} = \frac{1\cdot99}{Re_d^{1/8}}$$

The Reynolds number is $Re_d = \rho U_m d/\mu$. Since the product ρU_m remains constant, it may be evaluated for the conditions at section 1 (Fig. 22.11). The viscosity μ at t_f is $2\cdot122 \times 10^{-5}$ kg/m s. Hence

$$Re_d = \frac{0\cdot8373 \times 8 \times 0\cdot05}{2\cdot122 \times 10^{-5}} = 15\ 780$$

and

$$\frac{U_h}{U_r} = \frac{1\cdot99}{(15\ 780)^{1/8}} = 0\cdot594$$

The rate of heat flow per metre length of tube is therefore

$$q_w = \frac{\tau_w c_p}{U_m\left\{1 + \dfrac{U_b}{U_r}(Pr - 1)\right\}}\,\theta_m$$

$$= \frac{0\cdot0314 \times 1\cdot017}{8\{1 + 0\cdot594(0\cdot695 - 1)\}}\left(\frac{150 + t_2}{2} - 40\right) = 0\cdot004\ 88\left(\frac{150 + t_2}{2} - 40\right) \text{ kJ/s}$$

From the energy balance, this must again be equal to

$$q_w = \rho c_p U_m \frac{\pi d^2}{4}(150 - t_2) = 0\cdot0134(150 - t_2) \text{ kJ/s}$$

Eliminating q_w from this and the previous equation, we get

$$0\cdot004\ 88\left(\frac{150 + t_2}{2} - 40\right) = 0\cdot0134(150 - t_2)$$

and hence $t_2 = 116\,°C$. Therefore the temperature drop of the air over the metre length is

$$150 - t_2 = 34\ \text{K}$$

which is about 1/6 greater than the result obtained by assuming $Pr = 1$.

The Prandtl-Taylor modification of the Reynolds analogy permits the calculation of heat transfer data from friction data for fluids having Prandtl numbers from about 0·5 to 2. The limited range of application of this modification is due to the choice of $y = \delta_b$ as the lower limit in the integration of (22.63) and (22.64); the assumption that outside the sublayer $\varepsilon \gg \nu$ or α is not accurate, and the larger the Prandtl number the larger is the error introduced by assuming that $(\alpha + \varepsilon) = (\nu + \varepsilon)$. Von Kármán improved on the Prandtl-Taylor approach by considering three layers, viz. the laminar sublayer, the fully turbulent boundary layer, and a *buffer layer* in between in which ε, ν and α are all of the same order of magnitude. Only in the fully turbulent region, outside the buffer layer, can it be normally assumed that $\varepsilon \gg \nu$ or α.

Extreme conditions exist when the Prandtl number of a fluid is very high, e.g. several thousand as for a very viscous oil, or when it is very low, e.g. of the order of 0·01 as for a liquid metal. Fig. 22.12 illustrates the simple case of $Pr = 1$ side by side with these two extremes, for flow past a flat plate. For the case of $Pr = 1$ the velocity and temperature profiles are identical. We have already seen that this is so for laminar flow, i.e. equation (22.45) with Pr put equal to 1 yields $\delta_t = \delta$ and this, together with the similarity of equations (22.2) and (22.35), means that the boundary layers are identical. It can be shown that this identity is also true for a turbulent boundary layer when $Pr = 1$.

When Pr is very large (i.e. when k is very small), δ_t is much smaller than δ [as in laminar flow, equation (22.45)] and the temperature boundary layer may lie almost

Fig. 22.12   The effect of Prandtl number on the temperature profile

entirely in the laminar sublayer of thickness δ_b. When δ_b is known, q_w can simply be calculated from (22.73) by putting $t_b = t_s$; alternatively it is possible to obtain a fair estimate of δ_b from experimental values of q_w. Thus turbulence, and the value of ε, will have little effect on the temperature profile, but will merely affect the velocity profile.

When Pr is very small, as for liquid metals, the molecular diffusivity α ($= k/\rho c_p$) is relatively large and may predominate over the eddy diffusivity ε even when turbulence is extreme, i.e. even when $\varepsilon \gg \nu$. In other words, the effective conductivity $\rho c_p(\alpha + \varepsilon) \approx \rho c_p \alpha$ is fairly constant throughout the boundary layer, even though the effective viscosity $\rho(\nu + \varepsilon)$ may vary appreciably. Under such conditions the temperature gradient varies only slightly through the boundary layer, and again turbulence and ε play a small part in determining the temperature profile, which in fact extends beyond the edge of the velocity boundary layer. With fully developed flow in a tube, the boundary layer reaches the axis, and it is not difficult to predict the rate of turbulent convection. Since the effect of ε on the heat transfer is small, the rate of turbulent convection should be governed by a relation similar to that for laminar flow, i.e.

Nu_d = constant; the constant will not be quite the same as in equation (22.56) because the velocity profile is very different and therefore the temperature profile will also be altered slightly. A fair estimate of this constant for turbulent flow can be obtained from a 'slug' flow model for which U is uniform across the tube, but in which the radial conductivity is put equal to $\rho c_p \alpha$, i.e. k. In fact it has been found that for liquid metals Nu_d rises only moderately over a wide range of Re_d well into the turbulent regime, and only extreme turbulence raises Nu_d significantly.

The analogy between heat and momentum transfer has been developed to cover the whole range of possible Prandtl numbers, and for the most general theories the reader is referred to Ref. (10), pp. 500–21, and Refs. (31) and (36). The problem of liquid metals is treated specifically in Ref. (38).

The Reynolds analogy can only be applied if it may be assumed that the eddy diffusivity for momentum transfer is about equal to the eddy diffusivity for heat transfer. As already pointed out, this is a valid assumption for the boundary layer in a tube or near a flat plate. It does not apply, for example, in the turbulent wake formed when a fluid flows across a cylinder, and the analogy cannot be applied to find the rate of heat transfer in such cases. In general it may be said that the Reynolds analogy applies when the drag forces on a surface are due to *skin friction* alone, but not when some of the drag is due to flow separation from the surface as occurs in the flow round a cylinder or, to take a more interesting example, near the outlet of a cascade of axial-flow compressor blades.*

22.5 The principles of dynamic similarity and dimensional analysis applied to forced convection

We have seen that analytical solutions to the problem of heat transfer by forced convection can be found for laminar flow, although they are difficult even for simple geometric shapes. The Reynolds analogy provides a semi-analytical approach to turbulent convection problems, but this has also a limited range of application. The direct measurement of heat transfer coefficients is still the main approach to many problems concerning forced convection, and this section describes the method by which this empirical information can be correlated systematically so that a few experiments can produce results suitable for general application.

Initially we shall assume that the physical properties of the fluid do not vary with temperature; later the effect of such variation is discussed in some detail.

Certain generalisations, which are not proved here, can be made about the patterns of fluid and heat flow in forced convection, and these generalisations follow from the study of the complete set of differential equations (not given here) that describe forced convection.† Consider two regions, each within a specified *boundary* enclosing a solid body and a fluid flowing around it. As an example we may think of two cylinders

---

*Ref. (32) (A. G. Smith), p. 247
†Ref. (10), Chapter 23, and Ref. (18), Chap. 14.

and limited regions around them in which fluids flow; these regions are referred to as the *fields* (Fig. 22.13). The two fields can easily be imagined to be similar in the sense that

(a) the bodies and boundaries are geometrically similar,
(b) the velocity distributions around the boundaries are similar (in magnitude, direction and turbulence pattern),
(c) the temperature distributions around the boundaries are similar.

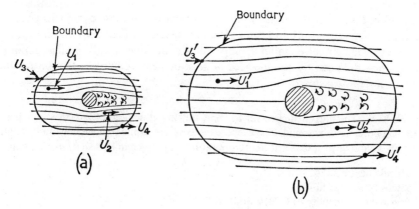

Fig. 22.13   The principle of similarity in forced convection

Once (a), (b) and (c) are satisfied, it is possible to prove that the following statements are valid for both laminar and turbulent flow.

(1) The velocity distributions *within* the fields will be similar when the Reynolds numbers $(\rho Ul/\mu)$ are the same for both fields.
(2) The temperature distributions *within* the fields will be similar when, in addition to condition (1), the Prandtl numbers $(c_p\mu/k)$ are the same for both fluids.
(3) When (1) and (2) are satisfied, then the Nusselt numbers (hl/k) for corresponding surface elements will be the same for both solid bodies, and hence the average Nusselt numbers will be the same for both surfaces.

When the Reynolds numbers are the same and the Prandtl numbers are the same, for both fields, the fields are said to be *dynamically* or *physically similar* from the point of view of forced convection heat transfer. Conditions of similarity (a), (b) and (c) by themselves do not ensure complete similarity.

These conclusions may be summarised by writing

$$Nu = \phi(Re, Pr) \tag{22.79}$$

i.e. the heat transfer expressed in terms of the Nusselt number Nu, for any particular geometrical arrangement, can be expressed as a function of the Reynolds number

Re and the Prandtl number *Pr*. The general deduction (22.79), although not proved so far, is in general agreement with the particular solutions obtained in the preceding sections.

When comparing the Reynolds numbers in the two fields, any pair of reference velocities, such as U_1, U_1', or U_2, U_2' can be chosen (Fig. 22.13), provided that the velocities are in corresponding positions in the two fields. The same remark applies to the choice of the linear dimension *l* in the Reynolds number and the Nusselt number; thus either the cylinder diameter or cylinder radius may be chosen in the example considered.

It follows that empirical data, which have been obtained at certain Reynolds and Prandtl numbers for a certain geometrical arrangement of surfaces and flow, may be extended to geometrically similar arrangements under dynamically similar conditions. For instance, data can be obtained for large scale heat exchangers from experiments on small scale apparatus.

Example 22.5 The film coefficient for toluene, flowing at a speed of 2·6 ft/s through a pipe of 0·5 inch bore, is to be found. The average temperature of the toluene flowing through the pipe is 86°F, and its properties at this temperature, expressed in British (ft lb h R) units, are:

Specific heat c_p: 0·41 Btu/lb R
Viscosity μ: 1·23 lb/ft h
Thermal conductivity k: 0·0859 Btu/ft h R
Density ρ: 55·6 lb/ft³

It is desired to obtain the information from experiments carried out with water flowing through a pipe of 2 cm bore. Specify the conditions of the experiment from which the information is to be obtained, i.e. the temperature and velocity of the water, in SI units.

Also, if the film coefficient for water found from the correct experiment is 3·61 kW/m² K, what is the value for toluene under the above conditions?

The Reynolds number and Prandtl number respectively must be the same for the water experiment as for the toluene. For toluene

$$Pr = \frac{c_p\mu}{k} = \frac{0·41 \times 1·23}{0·0859} = 5·87$$

$$Re_d = \frac{\rho U d}{\mu} = \frac{55·6(2·6 \times 3600)(0·5/12)}{1·23} = 17\,630$$

Previously we have assumed that the fluid properties do not vary with temperature. The conclusions reached so far in this section will be assumed correct, provided the temperature differences in the fluid are not too great, even if the properties are functions of temperature. All non-dimensional groups must then be evaluated at the mean fluid temperature. The Prandtl number is a property of the fluid, and is therefore dependent on temperature. The Reynolds number depends on the fluid properties ρ, μ, and on the fluid velocity U. It is therefore first necessary to specify the correct temperature for the experiment, so that it can be carried out at the right Prandtl number. The Reynolds number can then be adjusted by a suitable selection of the velocity U.

Since the properties of water vary but little with pressure, the values for saturated water in Ref. (14) may be used. The relevant values are:

t	20	25	30	35	°C
Pr	6·95	6·09	5·39	4·80	
c_p	4·183	4·181	4·179	4·178	kJ/kg K
$\mu \times 10^6$	1002	890	797	718	kg/m s
$k \times 10^6$	603	611	618	625	kW/m K
v	0·001	0·001	0·001	0·001	m³/kg

By interpolation, the correct average temperature for the Prandtl numbers to be the same is therefore 26·6 °C. To find U,

$$Re_d = 17\,630 = \frac{1000U \times 0·02}{861 \times 10^{-6}}$$

and hence $U = 0·759$ m/s

The Nusselt number for the water experiment was found to be equal to

$$Nu_d = \frac{hd}{k} = \frac{3·61 \times 0·02}{613 \times 10^{-6}} = 118$$

Hence the film coefficient for toluene is given by

$$\frac{h(0·5/12)}{0·0859} = 118$$

or

$$h = 243 \text{ Btu/ft}^2 \text{ h R}(= 1·38 \text{ kW/m}^2 \text{ K})$$

In order to arrive at the general form of the solution, given by (22.79), it is not necessary to derive the complete set of differential equations governing forced convection, and (22.79) can be obtained by *dimensional analysis* using the method of indices.* The disadvantage of this method is that it is necessary to know from previous experience all the variables on which the heat transfer depends before the analysis can be commenced.

The dimensional analysis which follows will be devoted to finding the form of an expression for the heat transfer coefficient h, defined by

$$h = -\frac{q}{t_s - t_w} = -\frac{q}{\theta_s}$$

*Refs. (4) and (12).

in terms of all the variables that affect the heat transfer. From the previous sections it is clear that the variables that may affect h in forced convection are

viscosity	μ
density	ρ
thermal conductivity . . .	k
specific heat	c_p
temperature (relative to wall) .	θ
fluid velocity	U
characteristic linear dimension .	l

Since the solution to be obtained is to apply to geometrically similar surfaces, it is sufficient to include among the variables a single dimension l, which is characteristic of the field considered. For the example illustrated in Fig. 22.13, the cylinder diameter d is one possible choice. In general, we can write

$$h = \phi(\mu, \rho, k, c_p, \theta, U, l) \tag{22.80}$$

The terms in any physical equation must always be dimensionally homogeneous, and such equations can always be expanded into a sum of a finite or infinite number of terms: e.g. equation (22.80) can be written

$$h = A\mu^{a_1}\rho^{a_2}k^{a_3}c_p^{a_4}\theta^{a_5}U^{a_6}l^{a_7} + B\mu^{b_1}\rho^{b_2}k^{b_3}c_p^{b_4}\theta^{b_5}U^{b_6}l^{b_7} + \ldots \tag{22.81}$$

in which each term has the same dimensions as h.

In the function considered here, there are eight variables, and Buckingham's Π theorem states that if the least number of independent fundamental units is n, then the function can be reduced to one involving $(8 - n)$ non-dimensional groups. Equation (22.79) suggests, however, that the solution can be expressed in terms of three non-dimensional groups. This requires some explanation since it implies the use of five fundamental units instead of the three (mass, length, time) usually accepted.[*]

Thermodynamics suggests that the units of heat are the same as those of mechanical energy, i.e. (mass)(length)²/(time)². When considering phenomena including transformations of energy, it would be impossible to treat heat as a fundamental unit. However, it was pointed out in the Introduction that in many practical heat transfer processes, rates of energy transformation are small compared with rates of energy transfer. In this field of study, therefore, heat may usually be regarded as a fundamental unit. For similar reasons temperature may be regarded as fundamental. Temperature occurs in products such as $h\theta$, $k\theta$ and $c_p\theta$. Such products always have the dimensions of heat, and temperature does not appear explicitly in their units. It is therefore of no consequence what dimensions are given to θ, and for the purpose

---

[*]Dimension is the power of a fundamental unit in a derived unit. For example, the gravitational acceleration g has the units (length)/(time)²; the dimensions of length are 1, those of time −2, and those of mass 0.

of the present analysis it may be regarded as a fundamental unit. The dimensions of
h, k and c_p are then fixed by this choice. We shall discuss later the results of an analysis
in which heat is assigned the dimensions of energy and is not regarded as fundamental.

It is now possible to tabulate the units of the eight variables, each number in the
table indicating the dimension of each fundamental unit.

	mass	length	time	heat	temperature
h	0	-2	-1	1	-1
μ	1	-1	-1	0	0
ρ	1	-3	0	0	0
k	0	-1	-1	1	-1
c_p	-1	0	0	1	-1
θ	0	0	0	0	1
U	0	1	-1	0	0
l	0	1	0	0	0

Five linear simultaneous equations can be written down for each term in (22.81),
which must be satisfied if the function of h is to be dimensionally homogeneous.
Considering the first term,

mass   .   .   . $0 = a_1 + a_2 - a_4$
length .   .   . $-2 = -a_1 - 3a_2 - a_3 + a_6 + a_7$
time   .   .   . $-1 = -a_1 - a_3 - a_6$
heat   .   .   . $1 = a_3 + a_4$
temperature .   . $-1 = -a_3 - a_4 + a_5$

There are seven unknowns in these equations, and a solution can only be obtained
in terms of two of the unknowns which can be arbitrarily chosen. On their choice,
however, will depend the actual non-dimensional groups obtained. It is desirable to
separate the effect of velocity into one group, and to do this it is necessary to solve
in terms of the index a_6 of U. The other variable which it is convenient to keep in
one group is c_p, and therefore the solution should be obtained also in terms of a_4.
Finding the values of a_1, a_2, a_3, a_5, a_7 in terms of a_4 and a_6, the following are obtained.

$a_1 = a_4 - a_6$
$a_2 = a_6$
$a_3 = 1 - a_4$
$a_5 = 0$
$a_7 = a_6 - 1$

Similarly the values of b_1, b_2, b_3, b_5, b_7 can be found in terms of b_4 and b_6, and (22.81)
then becomes

$$h = A\frac{k}{l}\left(\frac{\rho Ul}{\mu}\right)^{a_6}\left(\frac{c_p\mu}{k}\right)^{a_4} + B\frac{k}{l}\left(\frac{\rho Ul}{\mu}\right)^{b_6}\left(\frac{c_p\mu}{k}\right)^{b_4} + \ldots$$

or

$$Nu = ARe^{a_6}Pr^{a_4} + BRe^{b_6}Pr^{b_4} + \ldots \qquad (22.82)$$

The coefficients A, B, . . ., and the indices a_4, a_6, b_4, b_6, . . ., are constants which must be found experimentally for any particular geometric arrangement of the heat transfer surfaces and flow.

If this analysis is carried out in a similar way, but with heat given the dimensions of energy, we arrive at a relation between four non-dimensional groups, i.e. Nu, Re, Pr, and the additional group $U^2/c_p\theta$. It is worth looking into the significance of this group a little more closely. In the analysis it is not essential to take the temperature θ relative to the surface, but the absolute temperature T can be taken instead: the group U^2/c_pT, called the *Eckert number*, is then obtained. When the fluid is a gas this group has a

Fig. 22.14   Forced convection in fully developed flow through a tube

well known significance. By dividing the group by the pure number $(\gamma - 1)$ its character is unchanged, and it then becomes equal to $U^2/\gamma RT$. This will be recognised as the square of the Mach number of a perfect gas, and we know from experiments with high-speed gas flow that the Mach number has a considerable effect on the heat transfer. This is to be expected because large amounts of kinetic energy are dissipated by viscous friction in the boundary layer. At low velocities, when the dissipation of kinetic energy is negligible, the influence of the Eckert number is small and the analysis presented here is valid.

It has been found that for many common configurations the experimental results fall on a single curve if they are plotted in an appropriate way. One procedure is illustrated in Fig. 22.14 for flow through a tube. First of all, log Nu_d is plotted against log Pr for several values of Re_d and it is found that approximately straight lines are obtained with a slope n which is nearly the same for all Reynolds numbers in the turbulent regime. Then by plotting log (Nu_d/Pr^n) against log Re_d, a single curve is obtained. For a limited range of Reynolds number this plot is also found to be a straight line, of slope m say, and the Nusselt number can then be expressed by the relation

$$Nu_d = CRe_d{}^m Pr^n \qquad (22.83)$$

The values of C and m for the appropriate range of Re_d can easily be taken from the logarithmic plot. Equation (22.83) implies that in practice only the first term in the series of equation (22.82) need be used for a limited range of Reynolds number.

The advantages of presenting heat transfer data in terms of non-dimensional groups are several. Mainly, this method of presentation allows empirical data to be applied to a wide range of physical conditions by application of the principle of dynamic similarity, and requires few experiments to cover a wide range of fluid properties. Also, considerable saving can be achieved by experimental work on small-scale plant. Lastly, data presented in non-dimensional form are suitable for use in any consistent system of units, whether metric of British, and the likelihood of errors arising from confusing the units of the large number of variables involved is minimised.

Example 22.6 Calculate the heat transfer coefficient for water flowing through a 2 cm diameter tube with a velocity of 2·5 m/s. The average temperature of the water is 50 °C, and the surface temperature of the tube is slightly below this temperature. For turbulent flow through tubes

$$Nu_d = 0{\cdot}023 Re_d{}^{0{\cdot}8} Pr^{0{\cdot}4}$$

As the surface temperature of the tube differs only slightly from the mean bulk temperature of the water, all properties may be taken at the latter temperature. At 50 °C, $c_p = 4{\cdot}182$ kJ/kg K, $k = 643 \times 10^{-6}$ kW/m K, $\mu = 544 \times 10^{-6}$ kg/m s, $\rho = 1/v = 988$ kg/m^3 [from Ref. (14)]. Hence

$$Re_d = \frac{\rho U d}{\mu} = \frac{988 \times 2{\cdot}5 \times 0{\cdot}02}{544 \times 10^{-6}} = 90\,800$$

and therefore the flow is turbulent. Also

$$Pr = \frac{c_p \mu}{k} = \frac{4{\cdot}182 \times 544}{643} = 3{\cdot}54$$

and hence the Nusselt number is

$$Nu_d = 0{\cdot}023(90\,800)^{0{\cdot}8}(3{\cdot}54)^{0{\cdot}4} = 353$$

The heat transfer coefficient becomes

$$h = \frac{k}{d} Nu_d = \frac{643 \times 10^{-6}}{0{\cdot}02} 353 = 11{\cdot}35 \text{ kW/m}^2 \text{ K}$$

Dimensional analysis has shown that the relative temperature θ does not appear in the solution (22.82). When the fluid properties are not constant, but vary with temperature, it may not always be possible to correlate experimental data by this

549

equation satisfactorily. Such lack of correlation would be apparent by a scatter of experimental points on logarithmic plots such as Fig. 22.14, and the scatter would be particularly noticeable for points obtained from experiments with large temperature differences.

Nusselt has shown* that if the variation of properties with temperature can be written as functions of the absolute temperature in the form

$$\mu = \mu_0 \left(\frac{T}{T_0}\right)^a, \quad \rho = \rho_0 \left(\frac{T}{T_0}\right)^b, \quad k = k_0 \left(\frac{T}{T_0}\right)^c, \quad c_p = c_{p0} \left(\frac{T}{T_0}\right)^d \tag{22.84}$$

where $\mu_0, \rho_0, k_0\ c_{p0}$ are property values at some reference temperature T_0, then (22.82) can be modified to include such variations of properties by simply adding a non-dimensional group (T_w/T_s). Hence

$$Nu = \phi \left(Re,\ Pr,\ \frac{T_w}{T_s}\right)$$

which for the limiting case of 'isothermal flow' reduces to (22.79). Frequently we can write

$$Nu = CRe^mPr^n \left(\frac{T_w}{T_s}\right)^p \tag{22.85}$$

The ratio (T_w/T_s) is truly non-dimensional, because the ratio of the absolute temperatures is the same on any scale of temperature. Nusselt's method gives particularly good results with gases, as their property variations are well represented by equations (22.84) over a wide range of temperature. For liquids (22.84) can usually be applied over a moderate range of temperature only, and an equation of the form (22.85) is required for each range of temperature.

Other methods than that of Nusselt are often used to allow for variations of fluid properties with temperature. A common procedure is to use an equation of the form of (22.83), and to evaluate the properties used in the Nusselt, Reynolds and Prandtl numbers at a film temperature t_f, this being defined as

$$t_f = \frac{t_b + t_w}{2} \tag{22.86}$$

t_b is the temperature of the bulk of the fluid, and it is equal to the free stream temperature t_s for flow past a flat plate, and to the weighted mean temperature t_m for flow through a tube—see (22.31) and (22.32). When the film temperature is not uniform over the whole surface (it may vary from entrance to exit of a tube), the properties are often evaluated at the arithmetic mean of the inlet and outlet film temperatures.

A word of warning: when using any published equation of the form of (22.83) to obtain a value of h, it is necessary to know how the original experimental correlation was obtained, i.e. how Nu, Re, and Pr have been calculated when producing the

*Ref. (23), and Ref. (10), pp. 496–99.

curves from which C, m, n have been deduced. Some experimenters find it best to correlate data with all or some of the fluid properties taken at the bulk temperature rather than at the film temperature. Thus to find the heat transfer coefficient from an expression of the form (22.83), it is essential to find out first what reference temperatures have been used in deducing C, m, n, and then substitute values of properties corresponding to these temperatures. Ref. (13) is a good source of empirical correlations.

Example 22.7 Air at atmospheric pressure and a temperature of 90 °C flows with a velocity of 8 m/s across a 2·5 cm diameter tube. The tube is maintained at a temperature of 650 °C. Calculate the rate of heat transfer per metre length of tube. For air, the rate of heat transfer can be calculated from the relation

$$Nu_d = CRe_d{}^n \left(\frac{T_w}{T_s}\right)^{n/4}$$

For Reynolds numbers between 40 and 4000, $C = 0\cdot600$ and $n = 0\cdot466$. This empirical relation has been formulated with the fluid properties taken at the film temperature.

The film temperature is $t_f = (90 + 650)/2 = 370\,°C$, and from Ref. (14) we find that at this temperature $k = 4\cdot913 \times 10^{-5}\,\text{kW/m K}$, and $\nu = 5\cdot75 \times 10^{-5}\,\text{m}^2/\text{s}$. The Reynolds number is given by

$$Re_d = \frac{Ud}{\nu} = \frac{8 \times 0\cdot025}{5\cdot75 \times 10^{-5}} = 3480$$

and the ratio of the absolute surface temperature to the free stream temperature by

$$\frac{T_w}{T_s} = \frac{923}{363} = 2\cdot54$$

Hence the Nusselt number is

$$Nu_d = 0\cdot600 \times 3480^{0\cdot466} \times 2\cdot54^{0\cdot1165} = 29\cdot9$$

and the film coefficient is

$$h = \frac{k}{d}Nu_d = \frac{4\cdot913 \times 10^{-5}}{0\cdot025} 29\cdot9 = 0\cdot0588\,\text{kW/m}^2\,\text{K}$$

The rate of heat transfer per metre length of tube is

$$Q = -hA(t_s - t_w) = -0\cdot0588(\pi \times 0\cdot025 \times 1)(90 - 650) = 2\cdot59\,\text{kW}$$

22.6 Heat transfer from a vertical wall by free convection

When a temperature difference is established between a wall and a stationary fluid, the fluid adjacent to the wall will move upward if the wall temperature is higher than that of the fluid, and downward if the wall temperature is lower. The cause of the movement is the temperature gradient itself. This sets up density gradients in the fluid resulting in *buoyancy forces* and free convection currents. The rate of heat transfer depends mainly on the fluid motion, and this is very largely governed by a balance between buoyancy and viscous forces; inertia forces play only a small part in free convection.

Consider any fluid element at a temperature θ above the surrounding fluid; the density ρ of this element is smaller than the density ρ_s of the bulk of the fluid, and it is given by $\rho = \rho_s/(1 + \beta\theta)$, where β is the coefficient of cubical expansion. The velocity of this element is a function of the buoyancy force acting on it, which for unit volume is

$$(\rho_s - \rho)g = \beta g \rho \theta \tag{22.87}$$

It follows that in a solution for the rate of heat transfer by free convection the group of variables $\beta g \rho \theta$ may be expected to appear. In this section the concepts of the velocity and temperature boundary layers are applied to find an analytical solution for the rate of heat transfer from a vertical wall to a fluid, *when the flow is laminar.*

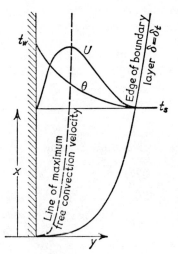

Fig. 22.15 Free laminar convection past a vertical wall

A velocity and a temperature boundary layer will build up at the wall, as shown in Fig. 22.15. The temperature gradually varies from t_w at the wall to t_s at the edge of the temperature boundary layer where $y = \delta_t$. The velocity gradually varies from zero at the wall to some maximum value in the velocity boundary layer, and then to zero at the edge of this boundary layer where $y = \delta$. It has been found from experiment and confirmed by theory that the thickness of the velocity and temperature boundary layers are about equal (except for very viscous fluids). This is to be expected, since the fluid will tend to rise only in the region where the temperature of the fluid is increased. It is beyond the scope of this book to derive the value of δ, and the result* is simply quoted here as

$$\frac{\delta}{x} = 3 \cdot 93 Pr^{-1/2} \left(Pr + \frac{20}{21} \right)^{1/4} Gr_x^{-1/4} \tag{22.88}$$

Here x is the height above the bottom of the heated wall. Gr_x is a dimensionless group called the *Grashof number*, and it is equal to

$$Gr_x = \frac{\beta g \rho^2 x^3 \theta}{\mu^2} \tag{22.89}$$

*The solution, due to H. B. Squire, is given in Ref. (8), p. 641.

The heat flow at the wall at x is

$$q_w = -k \left(\frac{d\theta}{dy}\right)_0 = -h_x \theta_s \tag{22.90}$$

where h_x is the local heat transfer coefficient at x. To find h_x it is first necessary to find the temperature gradient $(d\theta/dy)_0$ at the wall. Following the procedure adopted in section 22.2, the temperature profile is first expressed in terms of a polynomial in y, and then the temperature gradient is deduced from the polynomial. Let the temperature profile be approximated by a quadratic parabola

$$t - t_w = \theta = ay + by^2 \tag{22.91}$$

To find the values of the constants a and b, the following boundary conditions may be used:

$$\theta = \theta_s \quad \text{when} \quad y = \delta$$

and

$$\left(\frac{d\theta}{dy}\right) = 0 \quad \text{when} \quad y = \delta$$

Hence (22.91) becomes

$$\frac{\theta}{\theta_s} = 2\left(\frac{y}{\delta}\right) - \left(\frac{y}{\delta}\right)^2 \tag{22.92}$$

and therefore

$$\left(\frac{d\theta}{dy}\right)_0 = \frac{2\theta_s}{\delta}$$

Substituting the value of the temperature gradient $(d\theta/dy)_0$ in (22.90), we obtain

$$h_x = \frac{2k}{\delta}$$

Since δ clearly increases with increase of x, this equation implies that h_x decreases with increase of x. If the equation is put into non-dimensional form, we have

$$Nu_x = \frac{h_x x}{k} = 2\left(\frac{x}{\delta}\right) \tag{22.93}$$

The value of δ can be taken from (22.88), and hence

$$Nu_x = 0 \cdot 509 Pr^{1/2} \left(Pr + \frac{20}{21}\right)^{-1/4} Gr_x^{1/4} \tag{22.94}$$

In general terms $Nu_x = \phi(Pr, Gr_x)$, suggesting that the Grashof number plays a part in free convection analogous to the part played by the Reynolds number in forced convection.

It is usually of interest to know the average of h over a height l, and this can easily be found by integration. If h is this average value, and h_l the value at height l, then

$$h = \frac{1}{l} \int_0^l h_x \, \mathrm{d}x$$

If h_x is found from (22.94) and substituted in the integral, we find that

$$h = \frac{4}{3} h_l, \quad \text{or} \quad Nu = \frac{4}{3} Nu_l \tag{22.95}$$

For air, as well as for many other gases, $Pr \approx 0\cdot70$. Substituting this value in (22.94), the relation for the local Nusselt number simplifies to

$$Nu_x = 0\cdot376 Gr_x^{1/4} \tag{22.96}$$

It has been shown in Ref. (26) that a horizontal cylinder of diameter d has the same average heat transfer coefficient as a vertical plate of height $l = 2\cdot76d$. Using this result, the expression for the average Nusselt number of the cylinder becomes

$$Nu_d = 0\cdot527 Pr^{1/2} \left(Pr + \frac{20}{21} \right)^{-1/4} Gr_d^{1/4} \tag{22.97}$$

where d is the characteristic dimension used in the Nusselt and Grashof numbers. When $Pr = 0\cdot70$, this reduces to

$$Nu_d = 0\cdot389 Gr_d^{1/4} \tag{22.98}$$

In forced convection the Reynolds number is a criterion of whether the flow is laminar or turbulent. The critical Reynolds number above which the flow is turbulent depends on the geometric configuration; thus for flow through tubes the critical value of Re_d is about 2300, and for flow over a plate the critical value of Re_x is about 500 000 (see section 22.1). In free convection the product $(Pr \times Gr)$ serves as a criterion of turbulence, the exact value depending on the geometric configuration. For a vertical plate the flow becomes turbulent when $(Pr \times Gr_x)$ is about 10^9. $(Pr \times Gr)$ is sometimes called the *Rayleigh number Ra*.

Analytical solutions for free convection problems are usually difficult, and even impossible when the flow is turbulent. As in forced convection, the principle of similarity and the method of dimensional analysis are widely used for the systematic correlation of empirical data. In the following section the method of indices is applied to the dimensional analysis of free convection.

22.7 The principles of dynamic similarity and dimensional analysis applied to free convection

At the beginning of the previous section it was shown that the group of variables $\beta g \rho \theta$ plays an important part in the phenomenon of free convection, and it must therefore be included amongst the variables considered in the dimensional analysis.

As θ and ρ appear already in the list of variables in section 22.5, only the product βg need be added to that list; the units of βg are (length)/(temperature)(time)^2. The velocity U, on the other hand, which is a controlling factor in forced convection, can be omitted from the list for our present purpose. It is true that in free convection the fluid attains some velocity, and this affects the rate of heat transfer, but the free convection velocity is not an *independent* variable. Thus the final list of relevant variables is given by

$$h = \phi(\mu, \rho, k, c_p, \theta, \beta g, l) \tag{22.99}$$

Carrying out a dimensional analysis in the manner described in section 22.5, the functional relation can be reduced to

$$Nu = AGr^{a_1}Pr^{a_2} + BGr^{b_1}Pr^{b_2} + \ldots \tag{22.100}$$

$A, B, \ldots, a_1, a_2, b_1, b_2, \ldots$, are constants which must be found by experiment for any particular configuration of which the characteristic dimension is l. The value of θ in the Grashof number $Gr(= \beta g\rho^2 l^3 \theta/\mu^2)$ must always be measured in a certain reference position for the purpose of consistent correlation. It is usual to take θ as the temperature of the undisturbed fluid relative to the surface, i.e. $\theta = (t_s - t_w)$.

Once again the experimental results may be plotted on logarithmic scales. There are theoretical reasons [see Ref. (23)] why the indices of Pr and Gr in (22.100) should be equal in laminar convection (for any configuration), and in this case we can write that

$$Nu = \phi(Gr \times Pr) = \phi\left(\frac{\beta g\rho^2 l^3 \theta}{\mu^2} \cdot \frac{c_p\mu}{k}\right)$$

When convection is fully turbulent, theory suggests that

$$Nu = \phi(Gr \times Pr^2) = \phi\left(\frac{\beta g\rho^2 l^3 \theta}{\mu^2} \cdot \frac{c_p^2\mu^2}{k^2}\right)$$

Note that μ cancels in the latter case, implying that the heat transfer becomes independent of the viscous stresses in the fluid. Under normal conditions fully developed turbulence exists, if at all, only over part of the heated surface. It is therefore common to correlate experimental results for Nu with $(Gr \times Pr)$, although sometimes a better correlation can be obtained with $(Gr \times Pr^n)$ where $1 < n < 2$. When experimental values are plotted with $\log Nu$ as the ordinate and $\log (Gr \times Pr)$ as the abscissa, the plot for free convection from a wall appears as shown in Fig. 22.16. For a limited range of $(Gr \times Pr)$ it is possible to express the Nusselt number Nu by the relation

$$Nu = C(Gr \times Pr)^m \tag{22.101}$$

and the values of C and m can be taken from the logarithmic plot for the appropriate range.

It is interesting to note that for free convection from a vertical wall the slope of the curve in Fig. 22.16, and therefore the index m, become equal to 1/3 when $(Gr \times Pr)$

is greater than 2×10^9. This is in the range of turbulent convection, as has been explained in section 22.6. Since l appears in the Grashof number to the power of 3, and in the Nusselt number to the power of 1, h_x in the turbulent range is not a function of height. Thus h_x decreases in value as the boundary layer thickens over the lower

Fig. 22.16 Free convection from a vertical wall

section of the wall, where the flow is laminar, and then increases fairly suddenly and assumes a constant value above a certain height where the flow is turbulent.

To allow for the variation of property values with temperature, the procedures recommended at the end of section 22.5 may be used. Most commonly, property values at the film temperature are used in the dimensionless groups. Simplified formulae for free convection for both the laminar and turbulent range can be given for air: those for the more important configurations are summarised in Fig. 22.17.

	average $h/[\text{W/m}^2\,\text{K}]$	
	laminar or transition	turbulent
Vertical plate, or cylinder of large diameter, of height l	$10^4 < Gr < 10^9$ $1\cdot42\{(\theta/l)/[\text{K/m}]\}^{1/4}$	$10^9 < Gr < 10^{12}$ $1\cdot31(\theta/[\text{K}])^{1/3}$
Horizontal cylinder of diameter d	$10^4 < Gr < 10^9$ $1\cdot32\{(\theta/d)/[\text{K/m}]\}^{1/4}$	$10^9 < Gr < 10^{12}$ $1\cdot25(\theta/[\text{K}])^{1/3}$
Square plate $l \times l$: heated plate facing up, or cooled plate facing down	$10^5 < Gr < 2 \times 10^7$ $1\cdot32\{(\theta/l)/[\text{K/m}]\}^{1/4}$	$2 \times 10^7 < Gr < 3 \times 10^{10}$ $1\cdot52(\theta/[\text{K}])^{1/3}$
Square plate $l \times l$: cooled plate facing up, or heated plate facing down	$3 \times 10^5 < Gr < 3 \times 10^{10}$ $0\cdot59\{(\theta/l)/[\text{K/m}]\}^{1/4}$	

Fig. 22.17 Simplified equations for free convection heat transfer coefficients in air at atmospheric pressure

Inspection of the Grashof number *Gr* shows that apart from the properties of the fluid, the linear dimension *l*, and the temperature difference θ, the gravitational acceleration *g* also appears in this group. Full turbulence in free convection can be produced by increasing *g* and thereby the value of *Gr*. An effective increase in *g* always occurs in a vessel rotating at high speed, due to centripetal acceleration. A hollow gas turbine blade internally cooled by a liquid is a practical example [Ref. (39)].

There are instances when heat is transferred by combined free and forced convection, but few experimental data are available. We have already mentioned that free convection may play a significant part in forced convection at low Reynolds numbers. If the dimensional analysis is carried out to allow for both modes of convection, by including the variables *U* and βg simultaneously, the following relation is obtained

$$Nu = \phi(Re, Gr, Pr) \tag{22.102}$$

22.8 Convection with change of phase

Some of the most important industrial processes involving transfers of heat occur when the fluid changes its phase during the process, i.e. when it evaporates or condenses. As these processes usually take place at approximately constant pressure, the temperature of the bulk of the fluid can be assumed to be constant. Condensation and evaporation are now considered in turn.

(i) *Condensation of a vapour on a vertical surface*

In 1916 Nusselt* presented a method for predicting the rate of heat transfer from a vapour condensing in a laminar film and running down a vertical surface, and this method is presented here (Fig. 22.18). One assumption made is that the vapour which flows parallel to the surface, whether up or down, flows so slowly that its drag on the condensed liquid film can be neglected. A further assumption is that the motion of any liquid element in the condensed film is governed solely by a balance of the gravity and viscous forces acting on the element; hydrostatic pressure forces and inertia forces are therefore neglected.

The condensing film has zero thickness at the top of the wall. As the liquid runs down the wall it loses heat to the wall and tends to fall below the saturation temperature. More vapour condenses and prevents the temperature from actually falling at the film surface, so that the film thickens as it moves down the wall. Thus the thickening is due almost entirely to the addition of condensing vapour, and not to a retardation of the fluid in the film; the streamlines in the film can therefore be assumed vertical as indicated in Fig. 22.18a. No heat can flow from the film towards the vapour, because the film surface is essentially at the same temperature as the vapour; all the heat transfer, resulting from the liberation of energy from the vapour condensing on the liquid film, is towards the wall. Also this heat transfer must be entirely by conduction because the streamlines are vertical. The temperature gradient in the film

---

*Ref. (24). For an excellent survey of recent work see A. P. Colburn, Ref. (35), pp. 1–11.

is therefore constant, assuming no variation of the liquid conductivity with temperature within the film. Considering unit width of wall, Fig. 22.18 b, the rate of heat flow towards the wall through a strip of height dx is therefore

$$dQ = -k \, dx \frac{t_w - t_s}{\delta} = -h_x \, dx(t_w - t_s) = -h_x \, dx \, \theta_s \qquad (22.103)$$

Thus

$$h_x = \frac{k}{\delta}$$

Fig. 22.18   Formation of a condensed liquid film on a vertical wall

and putting this in non-dimensional form

$$Nu_x = \frac{h_x x}{k} = \frac{x}{\delta} \qquad (22.104)$$

To find the local heat transfer coefficient h_x, or the corresponding Nusselt number Nu_x, it is first necessary to find the film thickness δ.

Consider the equilibrium of the shaded element of the strip dx; it is assumed that the velocity of such an element is governed by the balance of the shear force and gravity force acting on it. Hence

$$\tau \, dx = \mu \left(\frac{dU}{dy}\right) dx = (\delta - y) \, dx \, \rho g$$

or

$$dU = \frac{\rho g}{\mu} (\delta - y) \, dy$$

Integrating this expression between $y = 0$ and y, we find the velocity at y, and thus the velocity distribution through the film to be

$$U = \frac{\rho g}{\mu} \left(\delta y - \frac{y^2}{2} \right) \tag{22.105}$$

The amount of liquid flowing through a cross-section of the film of unit width and thickness δ, at a distance x from the top of the plate, is

$$m = \int_0^\delta \rho U \, dy = \frac{\rho^2 g}{\mu} \int_0^\delta \left(\delta y - \frac{y^2}{2} \right) dy = \frac{\rho^2 g \delta^3}{3\mu} \tag{22.106}$$

The outflow, through a section $(x + dx)$ from the top of the plate, is $(m + dm)$ and therefore the amount condensing on the film over the height dx is

$$(m + dm) - m = dm = \frac{\rho^2 g \delta^2}{\mu} \, d\delta$$

The heat conducted through the film is equal to the latent heat given up by the condensing mass dm,* and this is equal to

$$dQ = h_{fg} \, dm = h_{fg} \frac{\rho^2 g \delta^2}{\mu} \, d\delta \tag{22.107}$$

From (22.103) and (22.107)

$$dQ = k \, dx \frac{\theta_s}{\delta} = h_{fg} \frac{\rho^2 g \delta^2}{\mu} \, d\delta$$

By integration between the limits $x = 0$, where $\delta = 0$, and x, we find the film thickness δ at section x as

$$\delta^4 = \frac{4 \mu k \theta_s x}{h_{fg} \rho^2 g} \tag{22.108}$$

Substituting (22.108) in (22.104), the Nusselt number at section x is found to be

$$Nu_x = \left(\frac{h_{fg} \rho^2 g x^3}{4 \mu k \theta_s} \right)^{1/4} \tag{22.109}$$

The group of the variables in (22.109) is dimensionless, but it has not yet been given a name. When the plate is inclined at an angle α to the vertical, equation (22.109) can be applied if $g \cos \alpha$ is substituted for g.

It is frequently of interest to know the average value of the heat transfer coefficient

---

*Strictly speaking, the heat conducted through the film is slightly greater than the latent heat: some sensible heat must also be transferred because the average temperature of the film is below the saturation temperature, see Fig. 22.18a.

over the height l. If h is this average value, and h_l the value at a distance l below the top of the wall, then

$$h = \frac{1}{l} \int_0^l h_x \, \mathrm{d}x = \frac{4}{3} h_l, \quad \text{or} \quad Nu = \frac{4}{3} Nu_l \tag{22.110}$$

Nusselt has shown* that a horizontal cylinder of diameter d has the same average heat transfer coefficient as a vertical plate of height $l = 2 \cdot 85d$. Therefore, from (22.109) and (22.110) the expression for the average Nusselt number of a horizontal cylinder becomes

$$Nu_d = 1 \cdot 03 \left(\frac{h_{fg} \rho^2 g d^3}{4 \mu k \theta_s} \right)^{1/4} \tag{22.111}$$

The rate of condensation has in some instances been observed to be considerably higher, and in others lower, than the theory of Nusselt would suggest. There are three principal reasons why it is sometimes higher. Firstly, when the vapour velocity is high, the drag exerted on the film surface will reduce the film thickness and increase its velocity, resulting in an overall increase in the rate of heat transfer, see Ref. (34).

Layer rich in non−condensable gas

Saturation temperature t_s

$t_0' < t_s$

t_w

Liquid film

Fig. 22.19

Secondly, after a certain distance down the wall the rate of condensate flow m may exceed a critical value above which the film becomes rippled and turbulent. This critical value is governed by a film Reynolds number (usually defined as $4m/\mu$), and this in turn is affected by the magnitude of the vapour drag on the film. Thirdly, when the fluid is water vapour it sometimes condenses as droplets on the surface and not as a continuous film. The droplets are initially small, but they grow until they are heavy enough to run down, collecting smaller droplets on the way. This mode of condensation is called *drop-wise condensation*, while the common mode is called *film-wise condensation*; the former mode offers less resistance to heat flow. Clean steam condensing on a clean surface always condenses film-wise; the presence of certain oils and fatty acids, either in the steam or on the surface, promotes drop-wise condensation.

One cause of a lower rate of heat transfer is the presence of non-condensable gases in the vapour; air, for example, is often present in steam condensers. Owing to the condensation of the steam on the water film, the steam-air mixture in the vicinity of the film is relatively richer in air content than the bulk of the mixture. This results in the formation of a non-condensable gas boundary layer of high thermal resistance on the liquid film (Fig. 22.19), and the condensing steam must diffuse through the gas layer. The liquid film surface temperature is below the saturation temperature

*Ref. (10), pp. 667–73.

of the vapour, and the temperature gradient available for conduction through the liquid film is less than it is in the absence of air. This is an additional reason why it is so important to exclude non-condensable gases from condensers (see section 14.4).

Example 22.8   Steam at 0·065 bar condenses on a vertical plate 0·6 m square. If the surface temperature of the plate is maintained at 15 °C, estimate the rate of condensation.

The latent heat h_{fg} of steam at 0·065 bar is 2412 kJ/kg and the saturation temperature is 37·7 °C. The properties of the water film at the mean film temperature of 26·4 °C are [Ref. (14)]: $\rho = 1000$ kg/m³, $\mu = 864 \times 10^{-6}$ kg/m s, $k = 613 \times 10^{-6}$ kW/m K.   Hence the Nusselt number at the bottom of the plate is

$$Nu_l = \left(\frac{h_{fg}\rho^2 g l^3}{4\mu k\theta_s}\right)^{1/4} = \left(\frac{2412 \times 1000^2 \times 9\cdot81 \times 0\cdot6^3}{4 \times 864 \times 613 \times 10^{-12}\,(37\cdot7 - 15)}\right)^{1/4} = 3210$$

The average heat transfer coefficient h, which is equal to 4/3 times the heat transfer coefficient h_l a distance l below the top of the plate, is therefore

$$h = \frac{4}{3}\frac{k}{l}\,Nu_l = \frac{4}{3} \times \frac{613 \times 10^{-6}}{0\cdot6} \times 3210 = 4\cdot37 \text{ kW/m}^2 \text{ K}$$

The total rate of heat transfer is

$$Q = -hA(t_w - t_s) = -4\cdot37 \times 0\cdot6^2(15 - 37\cdot7) = 35\cdot7 \text{ kW}$$

The rate of condensation is

$$m = \frac{Q}{h_{fg}} = \frac{35\cdot7}{2412} = 0\cdot0148 \text{ kg/s}$$

(ii) *Boiling of liquids*

Analysis of heat transfer to boiling liquids is difficult and theories are still incomplete. All that we shall attempt here is a qualitative description of the process to outline the factors governing the rate of heat transfer. The phenomena which occur in *forced-flow boiling* differ from those in *pool boiling* and it will be necessary to discuss these separately.

(a) *Pool boiling:* We speak of pool boiling when a large volume of liquid is heated by a submerged surface and the motion is caused by free convection currents stimulated by agitation of the rising vapour bubbles. Let us imagine an experiment in which the heat flux q from an immersion heater is gradually increased while the corresponding temperature difference θ between surface and bulk of the liquid is recorded. Different regimes may be distinguished.

At very low heat fluxes no boiling occurs, movement of the liquid is by simple *free convection,* and phase change takes place only as evaporation at the free surface of the liquid. In this regime (A–B in Fig. 22.20) the heat flux is approximately proportional to $\theta^{4/3}$: see the turbulent data in Fig. 22.17 from which $h(= q/\theta) \propto \theta^{1/3}$.

Increased heat flux will result in vapour bubble formation at the hot surface even though the bulk of the liquid may still be below saturation temperature. The bubbles which form in the hot boundary layer condense as they rise into the colder liquid, and

Fig. 22.20 Regimes of pool boiling

ultimate phase change is still by evaporation at the free surface. Bubble agitation in this *subcooled boiling* regime is insignificant so that there is little change in the slope of the q-θ curve (B–C in Fig. 22.20).

A further increase in q leads to *nucleate boiling*, where the bulk of the fluid is at saturation temperature. Strictly speaking it is slightly above the saturation temperature as indicated in Fig. 22.21, for a reason which will be apparent later. In this regime,

Fig. 22.21 Temperature distribution in a liquid boiling over a horizontal plate

large numbers of bubbles form on the hot surface and travel through the bulk of the liquid to emerge at the free surface. The resulting increased agitation markedly improves the heat transfer as shown by a steep rise of the q-θ curve C–D.

A point is reached when the bubbles become so large and numerous that liquid has difficulty in flowing back to the surface as the bubbles rise. When this *critical heat flux* is reached (D), the surface suddenly becomes insulated by a continuous vapour blanket and there is a sudden jump in heating surface temperature (E). With water at atmospheric pressure, θ might be about 25 °K at D and more than 1000 °K at E. It will be appreciated that there is a real danger of a heating surface failing or melting when the critical heat flux is reached. The condition of sudden vapour blanketing is referred to as *burn-out* even though with suitable materials physical burn-out need not occur.

When the heating surface can withstand higher temperatures, it is possible to increase q further and the regime is then referred to as *stable film boiling*. Heat transfer is by conduction and radiation across the vapour film with the latter playing an increasingly important role at these large values of θ.

If now the heat flux is reduced below the value at E, the conditions do not jump back to those at D but follow a path to F. At this point the vapour film suddenly collapses and conditions revert to those at point G on the nucleate boiling curve. We may ask whether it is possible to maintain surface temperatures corresponding to values of θ between D and F. It is impossible when the surface is heated in such a way that q is the independent variable, e.g. when the heat flux is being controlled as with an electric heater. It is possible, however, when the surface temperature is controlled as when a condensing fluid at variable pressure is used to provide the necessary heat flux. The dotted curve F–D represents an unstable region where the vapour blanket alternately collapses and reforms.

The heat flux near D may be extremely high (for water at 1 atm it is about 1500 kW/m²) and, because θ is small, the corresponding heat transfer coefficient is very large. It is therefore of interest to be able to operate as close as possible to D without running the risk of burn-out. Various semi-empirical relations have been proposed which enable the critical flux to be predicted—see Refs. (6) and (5).

One of the difficulties in developing a satisfactory theory of boiling is the lack of complete understanding of how vapour bubbles are formed. Also, once a bubble is formed it is difficult to analyse its motion and that of the surrounding liquid: particularly when there is interference between and coalescence of numerous bubbles. The latter problem is one of fluid mechanics which we shall not consider here, but we will look briefly at the mechanism of bubble formation.

Let us consider a spherical vapour bubble of radius r in thermodynamic equilibrium with liquid at pressure p_l, Fig. 22.22a. If σ is the surface tension and p_v the vapour pressure in the bubble, a force balance on an equatorial plane gives

$$(p_v - p_l)\pi r^2 = \sigma 2\pi r \quad \text{or} \quad (p_v - p_l) = \frac{2\sigma}{r} \tag{22.112}$$

It is clear that a finite pressure difference must exist between the vapour and the liquid. Since at equilibrium the liquid temperature t_l equals the vapour temperature t_v, it follows that the liquid must be superheated with respect to the liquid pressure (except for $r = \infty$ when the phase interface is plane). This is the cause of the superheat illustrated in Fig. 22.21.

Equation (22.112) suggests that for an infinitesimal bubble the excess pressure required is infinite, and therefore that bubbles cannot form in the first place. Such extrapolation from the continuum to the molecular level is of doubtful validity. Nevertheless, with pure gas-free liquids on very smooth heating surfaces it is possible to maintain very high degrees of liquid superheat (e.g. 50 K for water at atmospheric pressure). This is an unstable situation leading to sudden explosive bulk boiling or 'bumping', subsequent desuperheating, and cyclic repetition of the process. In practice, dissolved gases, or even vapour trapped in surface roughness cavities, provide nucleation sites from which bubbles can grow with little initial superheat, Fig. 22.22b. Clearly the surface's roughness geometry will be important: and its wettability also because of its effect on the surface tension between liquid and solid surface.

The conditions under which a bubble can exist and grow can be investigated further by applying the Clapeyron equation (7.47). Thus

$$\frac{h_{fg}}{T_{sat}v_{fg}} = \frac{\mathrm{d}p}{\mathrm{d}T_{sat}} \approx \frac{p_v - p_l}{T_v - T_{sat}}$$

where T_{sat} is the saturation temperature at p_l. Combining this with (22.112) we have

$$T_v - T_{sat} = \frac{2\sigma T_{sat}v_{fg}}{h_{fg}} \times \frac{1}{r} \qquad (22.113)$$

Remembering that $T_v = T_l$ for equilibrium, equation (22.113) yields the equilibrium bubble radius for a given liquid superheat $(T_l - T_{sat})$. If not in equilibrium and $(T_l - T_{sat})$ is greater than $(T_v - T_{sat})$ given by (22.113) the bubble will grow by further evaporation whereas if it is less the bubble will collapse. Conditions normally

(a)   (b) Nucleation sites

Fig. 22.22 Bubble formation

exist within the hot liquid boundary layer for a bubble to grow although subsequently it may shrink or collapse as it travels through cooler liquid to the free surface.

The surface tension σ decreases with increase in pressure (or T_{sat}), and a glance at steam tables will show that the other part of the coefficient of $1/r$ in equation (22.113) also decreases with increase in pressure. It follows that, for the formation of bubbles of given r, less liquid superheat is required with increased pressure. Because a given surface provides nucleation sites of a given size, the temperature difference θ required at any value of q can be expected to fall with increase in pressure in the nucleate boiling regime. This tendency is indicated in Fig. 22.20.

Overall heating surface geometry (as distinct from surface roughness) seems to have little effect on q in the nucleate boiling regime: vertical and horizontal planes, and even wires above a certain diameter, give similar results under otherwise identical conditions. In spite of this advantage and much recent research, the art of correlating boiling heat transfer data for different fluids in terms of dimensionless groups is still in its infancy. Such correlations are beyond the scope of this brief summary and for these the reader must turn to Refs. (6), (5) and (7).

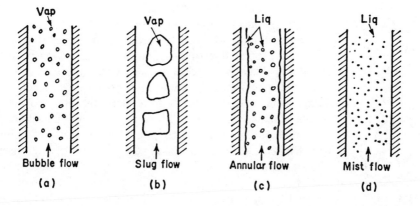

Fig. 22.23 Regimes of two-phase flow

(b) *Forced-flow boiling:* The nature of boiling in forced flow is similar to pool boiling in many respects, the most important difference being in the cause of 'burn-out'. For simplicity we shall concentrate here in describing what happens in a vertical, once-through, boiler tube. Because there is a mixture of liquid and vapour, the flow cannot be symmetrical in a horizontal tube, and the vertical case provides a simpler picture. In a once-through tube, liquid below or at saturation point enters at the bottom, gradually evaporates until a dry or superheated vapour leaves at the top, and no water is recirculated from the exit to the entry: in such a tube the whole range of possible dryness fractions and regimes of boiling can be observed.

Before describing the various regimes it is first necessary to deal briefly with the nature of two-phase flow. This depends mainly on the volume ratio of vapour to liquid, although transition from one type of flow to another at any given ratio will also depend on the velocity of flow. When the quantity of liquid predominates, the vapour flows as dispersed bubbles in the liquid as shown in Fig. 22.23a. With increased vapour content, large vapour slugs tend to form of more or less stable size (b). Even

565

higher vapour content results in annular flow (c), with a thin liquid film on the wall and a vapour core. The liquid film is usually rippled, with liquid droplets blown from the wave crests into the vapour core, so that annular flow is only a rough description. At even higher ratios of vapour to liquid the film may disappear and the liquid flows entirely as a suspended mist (d).

We can now describe what happens in a vertical tube with a constant heat flux along its length, and with a liquid below saturation temperature entering at the bottom. The first section A is a region in which no boiling occurs and heat flow is governed by normal forced convection. In section B the bulk of the liquid is still below saturation, but bubbles form at the wall and collapse as they move inward. Section C covers the regime of simple nucleate boiling, and when the vapour content increases further this changes into the slug flow of section D with nucleation sites in the film on the wall. Further evaporation causes the flow to change to annular flow with suspended mist in section E. At first the film is thick with nucleation sites within the film, but as it thins out evaporation is confined to the free surface of the annular film and the heat is transferred through the film by simple conduction. Eventually, when the film has completely evaporated at the end of section E, there is a sudden jump in wall temperature because of the very poor heat transfer coefficient at the unwetted wall. The temperature rise at 'dry-out' may cause tube failure. (This jump will not, of course, be present when the wall temperature is maintained constant instead of q.) Dry-out differs from burn-out caused by vapour blanketing in pool boiling: the latter type of phenomenon can also occur in forced flow, at section C, if the heat flux is sufficiently high. The dryness fraction at the point of dry-out is not unity because of the suspended mist. Evaporation continues in section F, heat being transferred by radiation from the wall, now at a high temperature, and by convection of the superheated vapour surrounding the mist. Finally the mist disappears and the vapour is superheated in section G.

Forced flow boiling has received considerable attention because of the problems encountered in the design of boiling water nuclear reactors. Not only is it necessary to be able to predict and avoid 'dry-out', but it is also necessary to know the distribution of vapour-liquid ratio along the tube because this affects neutron absorption. The knowledge gained is now being applied to conventional boiler plant, where less critical conditions have provided little incentive to research in this field in the past. Ref. (7) provides a good survey of forced-flow boiling.

Fig. 22.24

	$h/[\text{kW/m}^2\,\text{K}]$
Forced convection	
gases and dry vapours	0·01–1
liquids	0·1–10
liquid metals	5–40
Free convection	
gases and dry vapours	0·0005–1
liquids	0·05–3
Condensation	
film-wise	0·5–30
drop-wise	20–500
Boiling	0·5–20

Fig. 22.25   Typical values of heat transfer coefficient
for different modes of convection

In conclusion, a short table of values of the heat transfer (film) coefficient h (Fig. 22.25) has been compiled to give the reader a picture of the comparative rates of heat transfer which are usually obtained with the various modes of convection discussed in this chapter.

23 Radiation

The laws governing the transfer of heat by radiation are presented in this chapter. First the laws of emission of energy from black and grey surfaces are described. Then methods of calculating energy exchanges between black and grey surfaces at different temperatures are given, assuming that the intervening gas does not absorb any of the radiation. Finally, the radiation from gases and flames is discussed briefly in general terms.

23.1 The laws of black-body radiation

When radiation impinges on matter it may be totally or partially reflected, transmitted through it, or absorbed. The ratio of the reflected to the total incident energy is called the *reflectivity* ρ of the material; the ratio of transmitted to total incident energy is called the *transmissivity* τ; and the ratio of absorbed to total incident energy is called the *absorptivity* α. Therefore

$$\rho + \tau + \alpha = 1 \tag{23.1}$$

Practically all solid materials used in engineering are opaque to thermal radiation (even glass is only transparent to a fairly narrow range of wave-lengths), and thermal radiation is in fact either reflected or absorbed within a very shallow depth of matter. Thus, for solids, it is possible to write

$$\rho + \alpha = 1 \tag{23.2}$$

When radiation is absorbed by a material, it normally increases the random molecular energy and therefore the temperature of the absorbing material. In this way radiation can be said to result in 'heat transfer', although this need not always be the case. Sometimes the energy absorbed is converted into other forms as, for example, when visible radiation falls on a photoelectric cell and some of it is transformed into electrical energy. Such effects are usually of insignificant magnitude and are not considered here.

A material whose absorptivity is 1, i.e. one which absorbs all the radiant energy of whatever frequency incident upon it, is called a *black body*. The laws of radiation for a black body are of fundamental importance, and they are comparatively simple.

This section is devoted to a presentation* of these laws and to a discussion of some of their implications.

It is an experimental fact that unit surface area of a black body emits more energy by radiation than unit area of any other body at the same temperature. (This fact is also a consequence of Kirchhoff's law which is introduced in section 23.2.) The total energy emitted per unit time by unit area of a black surface is proportional to the fourth power of the absolute temperature T. This relation is expressed by the *Stefan-Boltzmann law*

$$q_b = \sigma T^4 \qquad\qquad\qquad (23.3)$$

σ is called the Stefan-Boltzmann constant; it is equal to $56 \cdot 7 \times 10^{-12}\,\text{kW/m}^2\,\text{K}^4$, when the energy emitted per unit area q_b is measured in kW/m^2 and T in K. The suffix b will be used to denote black-body radiation.

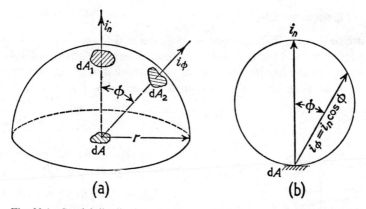

Fig. 23.1   Spatial distribution of energy emitted from a surface element dA

The 'composition' of radiant energy is non-uniform in two respects. Firstly, a surface element does not radiate energy with equal intensity in all directions. Secondly, the radiation emitted consists of electromagnetic waves of various wavelengths and the energy is not distributed uniformly over the whole range of wavelengths. These two aspects of radiation will now be considered in turn.

The spatial distribution of energy emission from an element can be represented in the following way. The total energy dQ_b from an element dA, given by $(q_b\,dA)$, can be imagined to flow through a hemisphere of radius r, Fig. 23.1a. Consider the surface elements dA_1 and dA_2 on this hemisphere, dA_1 lying on the normal to dA, and dA_2 on a line making an angle ϕ with the normal. The solid angle subtended by dA_1 at dA is $d\omega_n = dA_1/r^2$, and by dA_2 at dA is $d\omega_\phi = dA_2/r^2$. The solid angle subtended by the whole hemisphere at dA is $2\pi r^2/r^2 = 2\pi$ steradians (so called by analogy with radians of a circle).

*For further discussion and formal proofs see Ref. (10).

Let the rate of flow of energy through dA_1 be dQ_{bn}, and through dA_2, $dQ_{b\phi}$. Then we shall write

$$dQ_{bn} = i_n \, d\omega_n \, dA \tag{23.4}$$

and

$$dQ_{b\phi} = i_\phi \, d\omega_\phi \, dA \tag{23.5}$$

where i_n is the *intensity of radiation* in the normal direction and i_ϕ is the intensity of radiation in the ϕ-direction. i_n and i_ϕ are defined by (23.4) and (23.5) respectively. The spatial distribution of i_ϕ is expressed by *Lambert's cosine law*

$$i_\phi = i_n \cos \phi \tag{23.6}$$

Lambert's law is illustrated graphically in polar coordinates in Fig. 23.1b. It follows from (23.5) and (23.6) that in general

$$dQ_{b\phi} = i_n \cos \phi \, d\omega \, dA \tag{23.7}$$

for any surface element that subtends an angle $d\omega$ at dA.

To find an expression for i_n in terms of the temperature of the emitting surface, consider a strip on the hemisphere at an angle ϕ to the normal, the strip subtending an angle $d\phi$ at dA (Fig. 23.2). The solid angle subtended by this strip at dA is

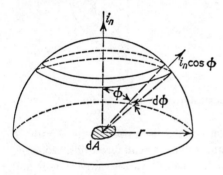

$$d\omega = \frac{2\pi r \sin \phi (r \, d\phi)}{r^2} = 2\pi \sin \phi \, d\phi$$

and thus the radiation passing through this strip is given by (23.7) as

$$dQ_{b\phi} = i_n 2\pi \sin \phi \cos \phi \, d\phi \, dA$$

The total radiation from dA passing through the hemisphere per unit time is

$$Q_b = i_n 2\pi \, dA \int_0^{\pi/2} \sin \phi \cos \phi \, d\phi$$

Fig. 23.2

$$= i_n 2\pi \, dA \int_0^{\pi/2} \tfrac{1}{2} \sin 2\phi \, d\phi = i_n \pi \, dA$$

But Q_b is also given by (23.3) as

$$Q_b = \sigma T^4 \, dA$$

It therefore follows that

$$i_n = \frac{\sigma T^4}{\pi} \tag{23.8}$$

This equation, giving the intensity of normal black radiation, is a consequence of the Stefan-Boltzmann law and Lambert's law.

Energy emitted in the form of radiation (i.e. electromagnetic waves) from a black surface is not emitted at one frequency only, but over a wide and continuous range of frequencies. The wavelength λ of each wave emitted into vacuum, and for all practical purposes into air, is related to the wave frequency ν by the simple expression

$$\nu\lambda = c \tag{23.9}$$

where c is the velocity of propagation of the wave. ν is usually expressed in cycles/s, and if λ is measured in m, c is equal to 3×10^8 m/s.

Consider the energy emitted within a limited range of wavelengths $d\lambda$. The total amount of energy dq_b emitted per unit time and unit area from a black surface within the *waveband* $d\lambda$ can be written as

$$dq_b = q_{b\lambda}\,d\lambda \tag{23.10}$$

where $q_{b\lambda}$, called the *monochromatic emissive power*, is defined by this equation. The spectral distribution of $q_{b\lambda}$, i.e. its variation over all wavelengths, is given by *Planck's law*

$$q_{b\lambda} = \frac{c_1\lambda^{-5}}{e^{c_2/\lambda T} - 1} \tag{23.11}$$

c_1 is a constant equal to $37\cdot4\times10^{-20}$ kW m^2; c_2 is a constant equal to $0\cdot01439$ m K; λ is measured in m, T in K, and $q_{b\lambda}$ in kW/m^3. $q_{b\lambda}$ is evidently a function of the surface temperature T. A typical distribution curve for a given value of T is shown in Fig. 23.3a; the total energy emitted per unit time and unit area of black surface is given by the area under the curve, which must in turn be equal to the energy emission found from the Stefan-Boltzmann law (23.3). Thus

$$q_b = \int_0^\infty q_{b\lambda}\,d\lambda = \sigma T^4 \tag{23.12}$$

The higher the temperature T, the larger is the proportion of energy emitted at the shorter wavelengths, and also the shorter is the wavelength λ_{max} for the peak value of $q_{b\lambda}$. Fig. 23.3b shows $q_{b\lambda}$ distribution curves for various temperatures, clearly showing the shift of λ_{max} to shorter wavelengths with increased temperature. The position of the maximum of $q_{b\lambda}$ in the waveband is expressed by *Wien's displacement law* which states that λ_{max} is inversely proportional to the absolute temperature T, i.e.

$$\lambda_{max}T = 0\cdot00290 \text{ m K} \tag{23.13}$$

(23.13) can be derived by differentiation of (23.11). It is interesting to note that at the temperature of the sun's surface (≈ 6000 K), λ_{max} is in the visible range of wavelengths, i.e. 40×10^{-6} to 80×10^{-6} cm, and about 40 per cent of the energy is emitted in this range. At the average temperature of the tungsten filament of an incandescent lamp (≈ 2800 K) only about 10 per cent of the energy is emitted in the visible range; thus the incandescent lamp is more efficient as a heat source than as a light source.

It must be emphasised that all the foregoing laws are concerned with black-body radiation because their derivation (not given here) is based on the assumption that the body is capable of absorbing completely any radiation incident upon it.

Fig. 23.3 Planck's law of the variation of the monochromatic emissive power with wavelength

Example 23.1 The temperature of a black surface, 0.2 m^2 in area, is $540\,°C$. Calculate (a) the total rate of energy emission, (b) the intensity of normal radiation, and (c) the wavelength of maximum monochromatic emissive power.

(a) From the Stefan-Boltzmann law (23.3), it follows that

$$Q = \sigma T^4 A = 56.7 \left(\frac{813}{1000}\right)^4 0.2 = 4.95 \text{ kW}$$

(b) From (23.8), the intensity of normal black radiation is

$$i_n = \frac{\sigma T^4}{\pi} = \frac{56.7}{\pi} \left(\frac{813}{1000}\right)^4 = 7.88 \text{ kW/m}^2 \text{ steradian}$$

(c) From Wien's law (23.13), it follows that

$$\lambda_{max} = \frac{0.0029}{T} = \frac{0.0029}{813} = 3.57 \times 10^{-6} \text{ m}$$

23.2 Kirchhoff's law and grey-body radiation

No real material emits and absorbs radiation according to the laws of the black body, and only a few, such as lamp-black and platinum-black, approach the black-body condition closely. (It is shown later in this section how an effectively 'black surface' can be obtained for use when checking the laws of black-body radiation experimentally.) Although the laws governing radiation from real bodies are complicated and will not be discussed here, there is another type of 'ideal' body to which many materials approximate in practice and for which the laws of radiation are simple modifications of the black-body laws. Before introducing this second 'ideal' body it is necessary to enunciate an important principle known as Kirchhoff's law.

Kirchhoff's law relates the *absorptivity* and *emissivity* of a body. Before stating the law it is first necessary to extend the notion of absorptivity α given at the beginning of the previous section, and define the term emissivity. The capacity of a body to absorb thermal radiation is not the same for all wavelengths and all angles of incidence: the *monochromatic absorptivity* at any wavelength λ and angle of incidence ϕ is defined as the ratio of the energy absorbed in the waveband $d\lambda$ through the solid angle $d\omega$ to the total incident energy within that waveband and solid angle. Generally it is found that the absorptivity so defined is a function not only of λ and ϕ but also of the surface temperature T, i.e.

$$\alpha_{\lambda\phi T} = f(\lambda, \phi, T) \tag{23.14}$$

The *total hemispherical emissivity* ε, or simply the *emissivity* ε, is defined as the ratio of the total energy q emitted by a surface to the total energy q_b emitted by a black surface at the same temperature. Thus

$$\varepsilon = \left(\frac{q}{q_b}\right)_T \tag{23.15}$$

As with the absorptivity, the emissivity is not the same at all wavelengths and angles of emission, and we may define the *monochromatic emissivity* at any wavelength λ and angle of emission ϕ as the ratio of the energy emitted in the waveband $d\lambda$ through the solid angle $d\omega$ to the energy that would be emitted by a black surface at the same temperature over $d\lambda$ and through $d\omega$. Generally it is found that the emissivity defined in this way is a function of λ, ϕ and the surface temperature T, i.e.

$$\varepsilon_{\lambda\phi T} = f(\lambda, \phi, T) \tag{23.16}$$

A material whose monochromatic emissivity is not constant, but varies with wavelength, angle of incidence or surface temperature, is called a *selective emitter*. A typical distribution curve of monochromatic emissive power q_λ for a body whose monochromatic emissivity varies with wavelength is shown in Fig. 23.4 a. The distribution curve for a black surface at the same temperature is included for comparison.

We are now in a position to state *Kirchhoff's law*, viz. *the monochromatic emissivity of a surface 1 at T_1 is equal to its monochromatic absorptivity for radiation received from a surface 2 at the same temperature $T_2 = T_1$.* This law may be expressed symbolically as

$$\varepsilon_{\lambda\phi} = \alpha_{\lambda\phi}]_{T_1 = T_2} \tag{23.17}$$

573

Kirchhoff's law is true for any wavelength λ and angle ϕ, and it is therefore also true for the average values, i.e. total hemispherical values, of α and ε. For simplicity we shall give a proof only for these average values.

Fig. 23.4   Selective and grey emission

Consider a black body of area A_1 at temperature T_1 in a black enclosure of the same uniform temperature T_1 (Fig. 23.5). The energy emitted by this body is $\sigma T_1^4 A_1$. This must be equal to the energy radiated by the enclosure that impinges on the body A_1, if the system is to maintain thermal equilibrium. Now suppose the black body is replaced by a non-black body of emissivity ε_1, but otherwise of exactly the same shape and temperature as the body it replaces; the radiation falling on A_1 will still be equal to $\sigma T_1^4 A_1$, some of which is absorbed while the remainder is reflected into the black enclosure. The radiation emitted by A_1 is reduced to $\varepsilon_1 \sigma T_1^4 A_1$ and, by definition of α, the energy absorbed is $\alpha_1 \sigma T_1^4 A_1$. For the temperature of the body to remain the same, it must absorb as much radiation as it emits. Thus the absorptivity α_1 of the non-black body A_1 must be equal to its emissivity ε_1, which proves Kirchhoff's law.

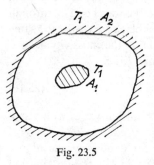

Fig. 23.5

It is important to emphasise that Kirchhoff's law does *not* apply to selective emitters when $T_1 \neq T_2$, the deviation increasing with the difference in temperature between the surfaces exchanging energy. If it did apply, there would be no point in painting the roof of a house white in the tropics; this is an advantage only because white paint has a low value of α for the sun's high-temperature radiation (of mainly short wavelength), while having a high value of ε for its own low-temperature radiation (of mainly very long wavelength). The laws of energy exchange for selective emitters are difficult, and for these the reader is referred to Refs. (10) and (13).

For many purposes it is sufficiently accurate to assume that materials are non-selective emitters. A body whose monochromatic emissivity does not depend on λ,

ϕ and T is called a *grey body*. For such a body the symbols ε and α can always be substituted for $\varepsilon_{\lambda\phi T}$ and $\alpha_{\lambda\phi T}$, e.g. the emissivity at any wavelength $(q_\lambda/q_{b\lambda})$ is independent of λ and therefore equals the average emissivity (q/q_b). A typical radiation distribution curve for a grey body is shown in Fig. 23.4 b, where it may be compared with the corresponding curve for a selective emitter shown in a.

For a grey body the following laws are valid.

(a) Kirchhoff's law can immediately be written in the simpler form

$$\varepsilon = \alpha \tag{23.18}$$

Equation (23.18) is true whatever the composition by wavelength of the incident radiation, and therefore it is independent of the temperature of the surface from which the radiation is received.

(b) The Stefan-Boltzmann law applies with $(\varepsilon\sigma)$ replacing the constant σ, i.e.

$$q = \varepsilon q_b = \varepsilon\sigma T^4 \tag{23.19}$$

[This equation is in a sense also true for a selective emitter, but then ε is a complicated function of T and not a constant.]

(c) Lambert's cosine law applies unchanged, i.e.

$$i_\phi = i_n \cos\phi \tag{23.20}$$

The only difference between a grey body and a black body with respect to this law is that i_n is smaller than i_{bn}. The distribution of i with ϕ is unchanged because for a grey body ε does not vary with ϕ.

(d) It follows from (b) and (c) that the relation for the normal intensity of radiation equivalent to (23.8) is

$$i_n = \frac{\varepsilon\sigma T^4}{\pi} \tag{23.21}$$

In the following sections we shall only consider energy changes between black or grey bodies, but it must be remembered that great care must be taken in applying grey-body rules, particularly when temperature differences are large. We shall also assume that any intervening gas does not absorb or emit radiation. This assumption is permissible if the gas is air, but when the gas contains appreciable amounts of carbon dioxide or water vapour, as it does in furnaces, this assumption is not admissible (see section 23.5).

Some selected values of emissivities are tabulated in Fig. 23.6. For real bodies ε is a function of λ, ϕ and T, but only the normal values ε_n, averaged over all wavelengths, are given in the table. It is possible to make a few generalisations about the behaviour of various materials. (a) Electrical non-conductors (including non-metallic liquids) have emissivities mostly above 0·8 at ordinary temperatures, the value usually falling with increase of temperature. The ratio of their total hemispherical emissivities to their normal emissivities, $\varepsilon/\varepsilon_n$, fall within the range 0·95 — 1, whereas for a grey body $\varepsilon/\varepsilon_n$ would be unity: nevertheless Lambert's cosine law can be applied with negligible error. (b) Clean, polished, metallic solids have very low emissivities, often below 0·10 at room temperature, but their emissivities are substantially proportional to the absolute temperature. They are, on the whole, selective emitters. The ratios

$T/[\text{K}]$		300	500	800
Aluminium	polished	0·04	0·04	0·06
	oxidised	0·09	0·12	0·17
Chromium polished		0·08	0·17	0·27
Copper	polished	0·02	0·02	0·03
	oxidised	0·56	0·61	0·83
Steel	polished	0·07	0·10	0·14
	oxidised	0·79	0·79	0·79
Mercury liquid		0·10		
Silver polished		0·02	0·02	0·03
Tungsten	polished	0·04	0·06	0·08
	filament	0·32	0·32	0·32
White paint		0·96	0·91	0·71
Lampblack paint		0·96	0·97	0·97
Asbestos paper		0·93	0·94	0·94
Quartz glass		0·93	0·89	0·68
Plaster		0·92	0·92	
Building brick		0·93		
Water (and ice)		0·96		

Fig. 23.6 Normal emissivities ϵ_n for various materials

of their total hemispherical emissivities to their normal emissivities, $\varepsilon/\varepsilon_n$, lie between 1·1 and 1·3, and Lambert's cosine law can be applied with moderate accuracy only. (c) Oxidised or greasy metal surfaces have emissivities which are several times higher than those of clean metal surfaces.

Fig. 23.7 Effect of surface finish on the emissivity of a material

Evidently the emissivity ε of a body is a property of surface finish as well as of the material itself. Fig. 23.7 explains why a matt surface is 'blacker' than a smooth or specular surface. Due to surface irregularities, any ray striking a matt surface is not likely to be reflected away from the surface after one incidence only, but is likely to suffer multiple reflections first. The irregularities in the surface act as 'hollow spaces' of high absorptivity. Consequently it follows from Kirchhoff's law that the emissivity of a matt surface is also higher than that of a polished surface made of the same material.

Although there is no perfectly black material, it can now be seen how a small surface which emits 'black radiation' can be obtained. Consider a hollow sphere with a relatively small aperture δA, the inner surface of which is at a uniform temperature (Fig. 23.8). Any ray entering the aperture δA is partly absorbed and partly reflected. The reflected ray, considerably reduced in intensity compared with the incident ray, is again partly absorbed and partly reflected on a second incidence, and so on for successive reflections. Due to the small-ness of the aperture it is improbable that any

Fig. 23.8   The 'perfect' black body

appreciable amount of reflected radiant energy escapes through it. This means that the surface element δA is effectively black for incident radiation, whatever the wave-length. According to Kirchhoff's law, however, the emissivity of δA must be equal to its absorptivity, and therefore δA will emit black radiation corresponding to the inner surface temperature of the sphere.

23.3 Radiation exchange between two black surfaces

When two surfaces can 'see' one another, each absorbs some of the radiation emitted by the other. If the surfaces are at different temperatures, the net result is an energy flow from the hotter to the colder body. Let us consider two surface elements of areas

Fig. 23.9   Radiation exchange between two elemental black areas

dA_1 and dA_2 on two *black* bodies at temperatures T_1 and T_2 respectively (Fig. 23.9.) The two surface elements are a distance x apart, and their normals are inclined at angles ϕ_1 and ϕ_2 respectively to the line x joining them. The energy exchange can be calculated in the following way.

577

dA_2 subtends a solid angle $d\omega_1 = dA_2 \cos \phi_2/x^2$ at dA_1. The intensity of radiation from dA_1 in the x-direction is $i_1 = i_{n1} \cos \phi_1$, and therefore the radiation incident upon and absorbed by dA_2 is

$$dQ_{b1} = i_1 \, d\omega_1 \, dA_1$$

$$= \frac{i_{n1} \cos \phi_1 \cos \phi_2 \, dA_1 \, dA_2}{x^2} \tag{23.22}$$

dA_1 subtends a solid angle $d\omega_2 = dA_1 \cos \phi_1/x^2$ at dA_2. Therefore the radiation emitted by dA_2 and absorbed by dA_1 is

$$dQ_{b2} = i_2 \, d\omega_2 \, dA_2$$

$$= \frac{i_{n2} \cos \phi_1 \cos \phi_2 \, dA_1 \, dA_2}{x^2} \tag{23.23}$$

But the normal intensity of black body radiation is given by the relation (23.8) as $i_n = \sigma T^4/\pi$, and hence the exchange of radiation between the two surface elements is

$$dQ_b = dQ_{b2} - dQ_{b1} = \frac{\sigma \cos \phi_1 \cos \phi_2 \, dA_1 \, dA_2}{\pi x^2} (T_2^4 - T_1^4) \tag{23.24}$$

The solution of any particular problem of heat exchange between black bodies depends upon whether the double integration of equation (23.24) can be carried out. When the surfaces are small compared with their distance apart, fairly flat, and uniform in temperature, an approximate solution can be obtained without integration by substituting areas A_1 and A_2 in (23.24), together with the approximate values of ϕ_1 and ϕ_2. This is illustrated in the following example, together with a method involving a single integration.

Example 23.2 A circular disk of 3 m diameter is exposed to radiation escaping from a furnace through an opening of 0.1 m^2 area. The disk is parallel and coaxial with the opening, and is placed 5 m away from it. The opening may be assumed to radiate as a black body at 1540 °C. Calculate the rate at which energy from the opening falls on the disk.

1st approximate solution: To obtain a very approximate solution, it is assumed that the areas of both opening and disk are small. The energy falling on the disk A_2 (Fig. 23.10) is given by combining (23.22) and (23.8)

$$Q_{b1} = \frac{i_{n1} \cos \phi_1 \cos \phi_2 A_1 A_2}{x^2} = \frac{\sigma \cos \phi_1 \cos \phi_2 A_1 A_2}{\pi x^2} T_1^4$$

A_1 is 0.1 m^2, and for A_2 we write $\pi 3^2/4 = 7.07$ m^2. For $\cos \phi_1$ and $\cos \phi_2$ we may write 1 as an approximation, although some surface elements see one another at considerable inclinations. Thus

$$Q_{b1} = \frac{56.7 \times 1 \times 1 \times 0.1 \times 7.07}{\pi 5^2} \left(\frac{1813}{1000}\right)^4 = 5.51 \text{ kW}$$

2nd approximate solution: In this solution it is still assumed that the opening A_1 is small, but the area of the disk is divided into elemental areas, the energy falling on each area is calculated separately, and the sum total is found by a single integration. Instead of calculating the energy falling on the disk, it is easier to calculate the energy falling on the cap of the sphere seen by the opening within the half-angle $\phi_a = \tan^{-1}(1 \cdot 5/5) = 16 \cdot 7°$ (Fig. 23.10). Consider a strip on the sphere at an angle ϕ_1, subtending an angle $d\phi_1$ at dA_1. The area of the strip is

$$dA_2 = 2\pi r \sin \phi_1 (r \, d\phi_1) = 2\pi r^2 \sin \phi_1 \, d\phi_1$$

and the energy falling on this strip is

$$dQ_{b1} = \frac{\sigma \cos \phi_1 \cos 0 \, A_1 \, 2\pi r^2 \sin \phi_1 \, d\phi_1}{\pi r^2} T_1^{\,4}$$

$$= \sigma A_1 T_1^{\,4} \, 2 \sin \phi_1 \cos \phi_1 \, d\phi_1 = \sigma A_1 T_1^{\,4} \sin 2\phi_1 \, d\phi_1$$

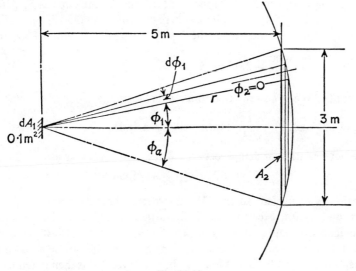

Fig. 23.10

The total energy falling on the cap of the sphere, and therefore on the disk, is

$$Q_{b1} = \int_0^{\phi_a} \sigma A_1 T_1^{\,4} \sin 2\phi_1 \, d\phi_1 = -\sigma A_1 T_1^{\,4} \left[\tfrac{1}{2} \cos 2\phi_1 \right]_0^{\phi_a}$$

$$= -56 \cdot 7 \times 0 \cdot 1 \left(\frac{1813}{1000} \right)^4 \tfrac{1}{2}(0 \cdot 835 - 1) = 5 \cdot 05 \text{ kW}$$

This answer is 8 per cent lower than the one given by the 1st approximation. To obtain an exact answer it is necessary to divide both disk and opening into elemental areas, and find the value of Q_{b1} by a double integration. This, though feasible, would be extremely laborious and the change in the answer would be very small in this particular case.

579

When the foregoing approximate solution is not permissible, the following procedure may be adopted. Equation (23.22) can be rewritten in a slightly different form

$$dQ_{b1} = \sigma T_1^4 \, dA_1 \frac{\cos \phi_1 \cos \phi_2}{\pi x^2} \, dA_2 = \sigma T_1^4 \, dA_1 \, dF_{1-2}$$

F_{1-2} is called the *geometric* or *configuration factor* of surface A_2 with respect to surface
dA_1, and it is found by the single integration of dF_{1-2} over the area, A_2, i.e.

$$F_{1-2} = \int_{A_2} \frac{\cos \phi_1 \cos \phi_2}{\pi x^2} \, dA_2 \qquad (23.25)$$

It is obvious that F_{1-2} is merely a function of the geometric configuration of the two
surfaces dA_1 and A_2. It is a measure of how much of the field of view of dA_1 is occupied
by A_2. If A_2 (or any part of it) occupies the whole field of view of dA_1, F_{1-2} assumes its
maximum value of unity; this can be easily visualised because F_{1-2} for any surface
A_2 which occupies the whole field of view of dA_1 must be the same as F_{1-2} for a hemisphere A_2 surrounding dA_1. For a hemisphere A_2, equation (23.25) can be integrated
(this is similar to the integration for the cap of the sphere, Example 23.2) to give

$$F_{1-2} = \int_0^{\pi/2} \frac{\cos \phi_1 \cos 0}{\pi r^2} 2\pi r^2 \sin \phi_1 \, d\phi_1 = 1$$

Equation (23.23) can be rewritten in a similar form, i.e.

$$dQ_{b2} = \sigma T_2^4 \, dA_1 \frac{\cos \phi_1 \cos \phi_2}{\pi x^2} \, dA_2 = \sigma T_2^4 \, dA_1 \, dF_{1-2}$$

Hence the net radiation exchange between dA_1 and dA_2 follows as

$$dQ_b = dQ_{b2} - dQ_{b1} = \sigma(T_2^4 - T_1^4) \, dA_1 \, dF_{1-2} \qquad (23.26)$$

This equation is equivalent to (23.24). In order to find the total energy exchange
between A_1 and A_2, it is necessary to divide surface A_1 into a number of small, equal,
surface elements dA_1, and find the geometric factor of surface A_2 with respect to each
of these elements. This can sometimes be done analytically, or else a numerical step-
by-step method or a graphical method may be employed.* It is then possible to find an
arithmetic mean value F'_{1-2} of the individual values of F_{1-2} for all the elements on
A_1, and the total energy exchange is finally obtained from

$$Q_b = \sigma(T_2^4 - T_1^4) A_1 F'_{1-2} \qquad (23.27)$$

The procedure for finding Q_b can be reversed by writing (23.24) in the form

$$dQ_b = \sigma(T_2^4 - T_1^4) \, dA_2 \, dF_{2-1}$$

where F_{2-1} is the geometric factor of surface A_1 with respect to dA_2, given by

$$F_{2-1} = \int_{A_1} \frac{\cos \phi_1 \cos \phi_2}{\pi x^2} \, dA_1 \qquad (23.28)$$

*Ref. (6), Chap. 14, and Ref. (37).

An average geometric factor F'_{2-1} for equal elemental areas dA_2 can then be found in the same way as F'_{1-2}, and the total energy exchange can then be found from

$$Q_b = \sigma(T_2{}^4 - T_1{}^4)A_2F'_{2-1} \tag{23.29}$$

The two procedures outlined above merely represent ways of integrating (23.24) and they differ only in the order of integration. Since A_1 and A_2 are independent variables, Q_b must be independent of the order of integration, and it therefore follows from (23.27) and (23.29) that

$$A_1F'_{1-2} = A_2F'_{2-1} \tag{23.30}$$

Sometimes it is easier to follow one order of integration rather than another, and consideration should be given to this choice before a solution is attempted.

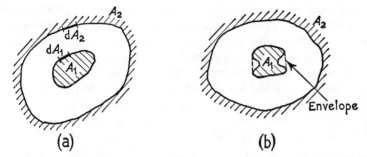

(a)   (b)

Fig. 23.11   Radiation exchange between a black body and a black enclosure

A particularly common case is the one where area A_2 completely encloses area A_1. Let it first be assumed that A_1 consists of convex and flat surfaces only as in Fig. 23.11 a, so that no two surface elements of A_1 can see one another. Now consider an element dA_2 on A_2; the geometric factor of A_1 with respect to dA_2 is less than unity because A_1 does not occupy the whole field of view of dA_2. Now consider any element dA_1 on A_1; the geometric factor of A_2 with respect to dA_1 must be unity, because the whole field of view of dA_1 is occupied by A_2. Hence the average geometric factor F'_{1-2} is equal to 1, and the radiation exchange follows from (23.27) as

$$Q_b = \sigma(T_2{}^4 - T_1{}^4)A_1 \tag{23.31}$$

It also follows from (23.30) that $F'_{2-1} = A_1/A_2$.

Equation (23.31) can also be deduced by a different argument. A_1 emits energy equal to $\sigma T_1{}^4A_1$. A_2 emits energy $\sigma T_2{}^4A_2$, but only $\sigma T_2{}^4A_1$ falls on A_1, the rest being reabsorbed by A_2. If this were not so, the net energy exchange would not be zero when $T_2 = T_1$, as it must be when the system is in equilibrium. This argument assumes that the spatial distribution of energy emission from A_2 is the same at all temperatures, i.e. whether $T_2 = T_1$ or $T_2 \neq T_1$, and the assumption is valid because Lambert's law implies that the spatial distribution of radiation does not vary with temperature. Hence in general

$$Q_b = \sigma T_2{}^4 A_1 - \sigma T_1{}^4 A_1 = \sigma(T_2{}^4 - T_1{}^4)A_1$$

which is in agreement with (23.31).

When the area A_1 has concave elements as in Fig. 23.11 b, so that parts of A_1 can see one another, less than $\sigma T_1{}^4 A_1$ escapes into the enclosure. However, when any concave portion δA_1 can be sealed by a plane surface of area $\delta A_1'$ (e.g. a cavity with a circular perimeter on a sphere or any dent on a plane), it is not difficult to show by considering the thermal equilibrium of this element that what escapes the cavity must be equal to $\sigma T_1{}^4 \delta A_1'$. Thus in such cases the envelope area can be used instead of A_1 in equation (23.31). The calculation is much more difficult when a cavity cannot be sealed by a plane surface, but in many cases equation (23.31) with A_1 taken as the minimum possible envelope area provides a good approximation to the exact answer. An extreme case where this approximation would lead to a completely erroneous result is a sphere with a long rod projecting from it.

Equation (23.31) can easily be extended to the case when A_1 is grey, and although the behaviour of grey bodies in general is left to the next section, we may consider the simple case of a grey body in a black enclosure here. We have seen that when A_1 is black, the radiation emitted by A_1 and absorbed by A_2 is $\sigma T_1{}^4 A_1$, and the radiation emitted by A_2 and absorbed by A_1 is $\sigma T_2{}^4 A_1$. Let A_1 now be replaced by an identical body of temperature T_1, but of emissivity ε_1. The radiation emitted by A_1 is $\varepsilon_1 \sigma T_1{}^4 A_1$, and it is all absorbed by the black enclosure. The radiation emitted by A_2 and falling on A_1 is still equal to $\sigma T_2{}^4 A_1$, but only a fraction ε_1 is absorbed (since $\alpha_1 = \varepsilon_1$). The reflected radiation is then reabsorbed by the black enclosure. Thus the net energy exchange is

$$Q = \varepsilon_1 \sigma T_2{}^4 A_1 - \varepsilon_1 \sigma T_1{}^4 A_1 = \varepsilon_1 \sigma(T_2{}^4 - T_1{}^4)A_1 \tag{23.32}$$

Equation (23.32) can also be used if the enclosure is not black but grey, provided the enclosure A_2 is very large compared with the body A_1. The radiation emitted by A_1 is $\varepsilon_1 \sigma T_1{}^4 A_1$, and all of this is absorbed by the enclosure even though it is not black, because any reflected radiation is finally absorbed by A_2 after multiple reflections and only a negligible fraction falls again on A_1. This would not be so if the body and enclosure were of comparable size. Since the relatively large enclosure is *effectively* black as far as absorption is concerned, Kirchhoff's law implies that it must be black as far as emission is concerned. Equation (23.32) can thus be applied to a grey body in either a black, or relatively large, enclosure.

Except for the simple case of radiation exchange between a grey body and an enclosure, the treatment of grey-body radiation is usually difficult. This is chiefly due to the fact that it is necessary to consider the effect of reflected energy, as will be apparent from the next section.

23.4 Radiation exchange between two grey surfaces

In order to illustrate the method of finding the heat exchange when reflected radiation is significant, it is best to choose an example in which the geometric factor presents no additional difficulty, i.e. when $F_{1-2} = F_{2-1} = 1$, and therefore $A_1 = A_2$ from (23.30). A case that satisfies this condition is that of two very large parallel plates of area A, whose distance apart is small compared with the size of the plates so that the radiation escaping at their edges is negligible. Another case is that of two concentric bodies which are a small distance apart compared with their dimensions, so that their areas are effectively equal (Fig. 23.12).

Consider the radiation emitted by the grey surface 1; it is equal to $\varepsilon_1 \sigma T_1^4 A$. The amount absorbed by the grey surface 2 is $\alpha_2(\varepsilon_1 \sigma T_1^4 A) = \varepsilon_1 \varepsilon_2 \sigma T_1^4 A$ (since $\varepsilon = \alpha$). The amount reflected by surface 2 is $\rho_2(\varepsilon_1 \sigma T_1^4 A)$, of which surface 1 absorbs

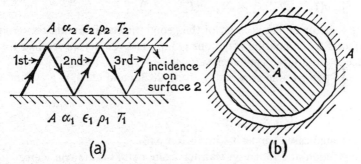

Fig. 23.12 Radiation exchange between two grey surfaces with a geometric factor of 1, i.e. $A_1 = A_2$

$\alpha_1 \rho_2(\varepsilon_1 \sigma T_1^4 A)$ and reflects $\rho_1 \rho_2(\varepsilon_1 \sigma T_1^4 A)$. With the second incidence on surface 2, this surface absorbs $\alpha_2 \rho_1 \rho_2(\varepsilon_1 \sigma T_1^4 A) = \rho_1 \rho_2(\varepsilon_1 \varepsilon_2 \sigma T_1^4 A)$. Similarly it can be shown that on a third incidence, after a further reflection from surface 1, surface 2 absorbs $\rho_1^2 \rho_2^2(\varepsilon_1 \varepsilon_2 \sigma T_1^4 A)$, and so on. The net amount of energy leaving surface 1 and being absorbed by surface 2 is therefore given by the series

$$Q_1 = \varepsilon_1 \varepsilon_2 \sigma T_1^4 A\{1 + (\rho_1 \rho_2) + (\rho_1 \rho_2)^2 + \ldots\}$$

Similarly it can be shown that of the energy $\varepsilon_2 \sigma T_2^4 A$ radiated by surface 2, a net amount

$$Q_2 = \varepsilon_1 \varepsilon_2 \sigma T_2^4 A\{1 + (\rho_1 \rho_2) + (\rho_1 \rho_2)^2 + \ldots\}$$

is absorbed by surface 1. The energy exchange is therefore

$$Q = Q_2 - Q_1 = \varepsilon_1 \varepsilon_2 \sigma(T_2^4 - T_1^4)A\{1 + (\rho_1 \rho_2) + (\rho_1 \rho_2)^2 + \ldots\}$$

The series in brackets is converging, because $\rho_1 \rho_2 < 1$, and the sum of the terms is $1/(1 - \rho_1 \rho_2)$. The net energy exchange is therefore

$$Q = \frac{\varepsilon_1 \varepsilon_2}{1 - \rho_1 \rho_2} \sigma(T_2^4 - T_1^4)A$$

Remembering that $\rho = (1 - \alpha) = (1 - \varepsilon)$, this becomes

$$Q = \frac{\varepsilon_1 \varepsilon_2}{1 - (1 - \varepsilon_1)(1 - \varepsilon_2)} \sigma(T_2{}^4 - T_1{}^4)A$$

$$= \frac{1}{1/\varepsilon_1 + 1/\varepsilon_2 - 1} \sigma(T_2{}^4 - T_1{}^4)A \qquad (23.33)$$

When ε_1 and ε_2 are equal to 1, this reduces to the corresponding expression for black body radiation (23.31).

When the geometric factor is not unity, and the surfaces exchanging energy by radiation are grey, the calculations are usually very difficult. In general it is possible to write the energy exchange between two surfaces as

$$Q = \mathscr{F}_{1-2}\sigma(T_2{}^4 - T_1{}^4)A_1 = \mathscr{F}_{2-1}\sigma(T_2{}^4 - T_1{}^4)A_2 \qquad (23.34)$$

where $\mathscr{F}$ is a factor taking account of the geometric configuration and the greyness of the surfaces. By comparing equations (23.32) and (23.34) we see that for a small grey surface A_1 in a large grey enclosure A_2, $\mathscr{F}_{1-2} = \varepsilon_1$, and $\mathscr{F}_{2-1} = \varepsilon_1 A_1/A_2$. And for the case of two parallel grey plates, from (23.32) and (23.33),

$$\mathscr{F}_{1-2} = \mathscr{F}_{2-1} = \frac{1}{1/\varepsilon_1 + 1/\varepsilon_2 - 1}$$

More complicated cases are dealt with in Ref. (13).

When convection and radiation transfer from a surface are considered simultaneously, as in section 21.2, it is convenient to be able to express the radiation heat transfer in terms of a heat transfer coefficient h_r, which is defined by equation (21.9), i.e.

$$Q = h_r A_1(T_2 - T_1)$$

Equating this to (23.34), we have h_r for the two grey surfaces given by

$$h_r = \mathscr{F}_{1-2}\sigma(T_2 + T_1)(T_2{}^2 + T_1{}^2) \qquad (23.35)$$

For the particularly important case of a grey body A_1 in a black or large enclosure A_2, the energy exchange is given by (23.32), and hence

$$h_r = \varepsilon_1 \sigma(T_2 + T_1)(T_2{}^2 + T_1{}^2) \qquad (23.36)$$

Example 23.3 A hot-water radiator of overall dimensions $2 \times 1 \times 0.2$ m is used to heat a room at 18 °C. The surface temperature of the radiator is 60 °C, and its surface is nearly black. The actual surface area of the radiator is 2·5 times the area of its envelope, and the convection heat transfer coefficient is $h_c = 0.001\,31(\theta/[\mathrm{K}])^{1/3}$ kW/m² K. Calculate the rate of heat loss from the radiator.

As the surface of the radiator is nearly black, the area of the envelope may be assumed to radiate as a black body. The area of the envelope is

$$A_e = 2\{(2 \times 1) + (2 \times 0.2) + (1 \times 0.2)\} = 5.2 \text{ m}^2$$

and the effective radiation heat transfer coefficient is given by putting $\varepsilon_1 = 1$ in (23.36) as

$h_r = 56 \cdot 7 \times 10^{-12}(291 + 333)(291^2 + 333^2) = 0 \cdot 006\ 92\ \mathrm{kW/m^2\ K}$

The convection heat transfer is from the whole radiator surface, which is $2 \cdot 5A_e$, and the convection heat transfer coefficient is

$h_c = 0 \cdot 001\ 31(60 - 18)^{1/3} = 0 \cdot 004\ 55\ \mathrm{kW/m^2\ K}$

The total rate of heat loss is

$Q = h_r A_e \theta + h_c 2 \cdot 5A_e \theta = (h_r + 2 \cdot 5h_c)A_e \theta$

$= \{0 \cdot 006\ 92 + (2 \cdot 5 \times 0 \cdot 004\ 55)\}5 \cdot 2(60 - 18) = 4 \cdot 0\ \mathrm{kW}$

23.5 Radiation from gases and flames

The quantitative calculations of radiation from gases and flames is beyond the scope of this book, and only a brief discussion is given here.

Previous sections have dealt with the calculation of energy exchange between solid surfaces. The assumptions made were that (a) the radiating materials were either black or grey, (b) the materials were opaque, and (c) any gases between the surfaces were transparent. When the transmissivity τ is unity, the absorptivity α must be zero from equation (23.1), and hence from Kirchhoff's law the emissivity ε is zero also. Therefore assumption (c) implies that the gas takes no part in the energy exchange. Air is effectively transparent, as are other diatomic and monatomic gases. Water vapour and carbon dioxide are the main gases met in industrial processes which are imperfect transmitters and radiate to an appreciable extent. However, even these gases only affect radiation exchange significantly if the volume of radiating gas is of considerable thickness. This is due to the fact that gases emit and absorb 'in depth'; the energy is not merely emitted or absorbed by the surface of a gas volume, and successive layers of the gas take part in the process of radiation. With solids only a few molecular layers near the surface are involved. When the intervening gases do play a significant part in the transfer of heat, the treatment is difficult because gases are highly selective emitters and hence cannot be considered grey.*

Flames are hot volumes of gas in which chemical reactions are taking place. The radiation from flames is due to two sources. Some radiation, manifested by a faint blue colour, arises from the reacting gases; such radiation is usually of low intensity. Secondly, when flames are obtained from the combustion of hydrocarbons, solid incandescent carbon particles appear in the combustion region as intermediate products of the combustion process. Ash particles may also be suspended in the flame. Such particles radiate as solid bodies, giving flames their yellow colour, and provide the major contribution to the radiation heat transfer. The radiation from flames can only be calculated if the density and distribution of the suspended solid matter is

*See Refs. (13) and (17).

known. In practice, for a flame with substantial amounts of suspended matter, a reasonable approximation is obtained by assuming the flame to be black.

A yellow flame, rich in suspended solid particles, is called a *luminous flame*, and a blue flame, poor in suspended particles, a *non-luminous flame*. Though a non-luminous flame may be appreciably hotter than a luminous one (the latter indicating incomplete combustion), a luminous flame may be more efficient because it more readily transfers its energy to other bodies by radiation. In gas fires used for space heating, the jets of non-luminous flame are partly surrounded by mantles of refractory material so that these mantles, when heated, make the radiation from the fire more effective.

24 Combined Modes of Heat Transfer

Many problems in which conduction, convection and radiation occur simultaneously can be treated with the aid of an overall heat transfer coefficient in the manner suggested in section 21.2. The overall coefficient can be found once the relevant expressions for the individual heat transfer coefficients have been obtained. In this chapter we shall present examples involving this method of approach.

It was noted in the Introduction to Part IV that although the simple addition of the separate effects is often permissible, it does not always provide an exact solution. Under some conditions there is a significant interaction between the different modes. For example, when a liquid flows over a heated plate, the radiation emitted by the plate is absorbed gradually by successive layers of fluid and this alters the temperature and velocity profiles in the boundary layer, and hence also the convection heat transfer. It is difficult to allow for such interaction and it is normally neglected in practical calculations.

24.1 The heat exchanger

One of the most common examples of heat exchange is the heat transfer from one fluid to another when the fluids are separated by a solid wall. The apparatus in which this type of process occurs is called a *heat exchanger*. In a heat exchanger the temperature of the fluids along the surface of the wall is rarely uniform. To predict the rate of heat transfer, it is necessary to know the overall heat transfer coefficient U for the wall and fluid boundary layers. The calculation is simplified if U can be assumed independent of the distance along the heat exchanger, and if a mean temperature difference Δt_m between the fluids can be found such that the rate of heat transfer can be expressed by the equation

$$Q = UA\Delta t_m \tag{24.1}$$

The assumption that the overall heat transfer coefficient does not vary along the heat exchanger is never strictly true, but it may be a good approximation if at least one of the fluids is a gas. For a gas, the film coefficient is fairly constant over the whole surface because it depends on the properties of the gas (section 22.5) which do not vary appreciably over moderate ranges of temperature. Furthermore, the resistance of the gas film is considerably greater than that of the metal wall or the liquid film,

587

and the value of the gas film resistance effectively determines the value of the overall heat transfer coefficient U.

Only the two simplest forms of heat exchanger are considered here in any detail. Fig. 24.1 shows diagrammatically a *parallel-flow* and a *counter-flow* heat exchanger, together with typical axial temperature distributions for the fluids flowing through the passages. The temperature distribution is plotted against heat exchange area, i.e. against the area of the surface that separates the two fluids. The purpose of the following analysis is to find a suitable mean temperature difference Δt_m, for use when calculating the total rate of heat exchange for a parallel-flow or counter-flow heat exchanger from equation (24.1).

Fig. 24.1   The parallel-flow and counter-flow heat exchanger

The heat transferred through any short section of heat exchanger tube of surface area dA is

$$dQ = U\,dA(t_1 - t_2) = U\,dA\,\Delta t \tag{24.2}$$

Consider the parallel-flow heat exchanger first. While flowing over this section the hot fluid cools, and the cold fluid is heated, in the direction of increasing area. Assuming that no heat is lost to the surroundings, the heat exchange must also be equal to

$$dQ = -m_1 c_{p1}\,dt_1 = +m_2 c_{p2}\,dt_2$$

where m_1 and m_2 are the rates of mass flow of the two fluids, and c_{p1} and c_{p2} are the respective specific heats. Hence

$$d(\Delta t) = d(t_1 - t_2) = dt_1 - dt_2 = -\left(\frac{1}{m_1 c_{p1}} + \frac{1}{m_2 c_{p2}}\right)dQ \tag{24.3a}$$

Considering the counter-flow heat exchanger next, it can be seen that the only difference is that the cold fluid cools in the direction of increasing area. Hence

$$dQ = -m_1 c_{p1}\, dt_1 = -m_2 c_{p2}\, dt_2$$

and thus

$$d(\Delta t) = dt_1 - dt_2 = -\left(\frac{1}{m_1 c_{p1}} - \frac{1}{m_2 c_{p2}}\right) dQ \qquad (24.3b)$$

Integrating equations (24.3a) and (24.3b) between inlet i and outlet o, assuming the specific heats are constant, we get

$$-\left(\frac{1}{m_1 c_{p1}} \pm \frac{1}{m_2 c_{p2}}\right) Q = \Delta t_o - \Delta t_i \qquad (24.4)$$

The positive sign refers to the parallel-flow exchanger, and the negative sign to the counter-flow type. Also, substituting for dQ in equation (24.3a, b) from (24.2), we obtain

$$-\left(\frac{1}{m_1 c_{p1}} \pm \frac{1}{m_2 c_{p2}}\right) U\, dA = \frac{d(\Delta t)}{\Delta t}$$

By integration between inlet i and outlet o, since U is assumed constant, this becomes

$$-\left(\frac{1}{m_1 c_{p1}} \pm \frac{1}{m_2 c_{p2}}\right) UA = \ln \frac{\Delta t_o}{\Delta t_i} \qquad (24.5)$$

Finally, dividing (24.4) by (24.5), we obtain

$$Q = UA \frac{\Delta t_o - \Delta t_i}{\ln (\Delta t_o / \Delta t_i)} \qquad (24.6)$$

Comparing (24.6) with (24.1), it is clear that the required mean temperature difference is

$$\Delta t_m = \frac{\Delta t_o - \Delta t_i}{\ln (\Delta t_o / \Delta t_i)} \qquad (24.7)$$

This is called the *logarithmic mean temperature difference*, and the expression is the same for both types of heat exchanger.

Example 24.1   0·5 kg/s of ethylene glycol flow through a thin-walled copper tube of diameter $d_1 = 1·25$ cm; 0·375 kg/s of water flow in the opposite direction through the annular space formed by this tube and a tube of diameter $d_2 = 2$ cm. The ethylene glycol, which enters at

100 °C, is required to leave at 60 °C, while the water enters at 10 °C. Calculate the length of tube required. Also calculate the length of tube required if the water were flowing in the same direction as the ethylene glycol.

For turbulent flow through a tube

$$Nu_d = 0.023 Re_d^{0.8} Pr^{0.4} \qquad (24.8)$$

and for turbulent flow through an annulus

$$\frac{h}{c_p \rho C} (= St) = 0.023 Re_{d'}^{-0.2} Pr^{-2/3} \qquad (24.9)$$

with all properties taken at bulk temperature. (C in this example denotes velocity.) In (24.9) $Re_{d'}$ is based on the equivalent diameter

$$d' = \frac{4 \text{ (cross-section area)}}{\text{wetted circumference}} = \frac{4\pi(d_2^2 - d_1^2)}{4\pi(d_2 + d_1)} = (d_2 - d_1)$$

The product ρC, which appears in the Reynolds number Re and in the group $(h/c_p \rho C)$ must be constant for each fluid along the tube and equal to m/A. Thus, for ethylene glycol

$$\rho C = \frac{0.5 \times 4}{\pi (0.0125)^2} = 4070 \text{ kg/m}^2 \text{ s}$$

and for water

$$\rho C = \frac{0.375 \times 4}{\pi\{(0.02)^2 - (0.0125)^2\}} = 1960 \text{ kg/m}^2 \text{ s}$$

Calculation of film coefficient for ethylene glycol h_e

The relevant properties and dimensionless groups are set out below, for the bulk temperature at inlet and outlet, and for the mean bulk temperature.

	60 °C	80 °C	100 °C
$c_p/[\text{kJ/kg K}]$ . . .	2·56	2·65	2·74
$\mu/10^{-4}[\text{kg/m s}]$ . . .	51·7	32·1	21·4
$k/10^{-4}[\text{kW/m K}]$ . . .	2·60	2·61	2·63
$Pr = c_p\mu/k$ . . .	50·9	32·6	22·3
$Pr^{0.4}$ . . .	4·82	4·03	3·46
$Re_d = \rho Cd/\mu$ . . .	9840	15 800	23 800
$Re_d^{0.8}$ . . .	1560	2280	3170
Nu_d from (24.8) . . .	173	211	252
$h_e/[\text{kW/m}^2 \text{ K}]$ . . .	3·60	4·41	5·30

Calculation of film coefficient for water h_w

The outlet temperature of the water is found from an energy balance, assuming no heat losses to the surroundings. Thus

$$2.65 \times 0.5(100 - 60) = 4.18 \times 0.375(t - 10) = 53.0 \text{ kW}$$

Hence $t = 43\cdot8\ °C$. The properties and dimensionless groups are set out below, as for ethylene glycol.

				10 °C	27 °C	44 °C
$c_p/[\text{kJ/kg K}]$	.	.	.	4·19	4·18	4·18
$\mu/10^{-4}[\text{kg/m s}]$	.	.	.	13·0	8·53	6·05
$k/10^{-4}[\text{kW/m K}]$	.	.	.	5·87	6·14	6·37
$Pr = c_p\mu/k$	.	.	.	9·29	5·81	3·97
$Pr^{2/3}$	.	.	.	4·42	3·23	2·51
$Re_{d'} = \rho C(d_2 - d_1)/\mu$	.	.	11 300	17 200	24 300
$Re_{d'}{}^{0\cdot2}$	.	.	.	6·47	7·03	7·54
$h_w/[\text{kW/m}^2\text{ K}]$ from (24.9)	.	6·61	8·30	9·96

Calculation of tube length for counter-flow

The resistance of the thin copper tube separating the two fluids can be neglected in comparison with the film resistance. The overall heat transfer coefficient U at any section along the tube is therefore given by

$$\frac{1}{U} = \frac{1}{h_e} + \frac{1}{h_w}$$

At the ethylene glycol inlet (or water outlet) we therefore have

$$\frac{1}{U_i} = \frac{1}{(h_e)_{100}} + \frac{1}{(h_w)_{44}}$$

$$= \frac{1}{5\cdot30} + \frac{1}{9\cdot96}, \quad \text{or} \quad U_i = 3\cdot46\ \text{kW/m}^2\ \text{K}$$

Similarly, at the ethylene glycol outlet (or water inlet),

$$\frac{1}{U_o} = \frac{1}{3\cdot60} + \frac{1}{6\cdot61}, \quad \text{or} \quad U_o = 2\cdot33\ \text{kW/m}^2\text{K}$$

Evidently U is by no means constant along the heat exchanger surface, which is the assumption upon which equations (24.1) and (24.7) are based. These equations will yield an approximate solution, however, if an average value of U is used, i.e. if we take $U = (3\cdot46 + 2\cdot33)/2 = 2\cdot90\ \text{kW/m}^2\ \text{K}$.

The logarithmic mean temperature difference, using the notation of Fig. 24.1, is

$$\Delta t_m = \frac{\Delta t_o - \Delta t_i}{\ln(\Delta t_o/\Delta t_i)} = \frac{(60 - 10) - (100 - 43\cdot8)}{\ln\{(60 - 10)/(100 - 43\cdot8)\}} = 53\cdot1\ \text{K}$$

Hence, from (24.1), the required tube length l is given by

$$Q = U(\pi d_1 l)\Delta t_m$$

$$l = \frac{53\cdot0}{2\cdot90 \times \pi \times 0\cdot0125 \times 53\cdot1} \approx 8\cdot8\ \text{m}$$

Calculation of tube length for parallel-flow

In this case we have, at the ethylene glycol inlet (and water inlet),

$$\frac{1}{U_i} = \frac{1}{5\cdot30} + \frac{1}{6\cdot61}, \quad \text{or} \quad U_i = 2\cdot94 \text{ kW/m}^2\text{K}$$

And at the outlet,

$$\frac{1}{U_o} = \frac{1}{3\cdot60} + \frac{1}{9\cdot96} \quad \text{or} \quad U_o = 2\cdot64 \text{ kW/m}^2\text{K}$$

The average value of U is therefore $2\cdot79$ kW/m^{2}K.

$$\Delta t_m = \frac{(60 - 43\cdot8) - (100 - 10)}{\ln\{(60 - 43\cdot8)/(100 - 10)\}} = 43\cdot0 \text{ K}$$

Fig. 24.2 Heat exchanger with one fluid condensing over the whole heat exchange area

$$l = \frac{53\cdot0}{2\cdot79 \times \pi \times 0\cdot0125 \times 43\cdot0} \approx 11\cdot2 \text{ m}$$

[N.B. For various reasons explained in the following paragraphs, this solution is only approximate. It is more usual to adopt the quicker method of obtaining an average U from h_e and h_w at the mean bulk temperatures of 80 °C and 27 °C respectively. The reader can easily verify that virtually the same lengths are obtained. The method of obtaining U used in this example is preferable, however, because it entails finding the extreme values of the Reynolds number. Hence it is possible to see whether the flow is laminar or turbulent, and thereby verify whether the appropriate equations for h are being used.]

Example 24.1 shows that although the expression for the logarithmic mean temperature difference is the same for both types of heat exchanger, its value is higher for the counter-flow type. Thus for given mass flows and temperature changes, the counter-flow heat exchanger requires less surface area than its parallel-flow equivalent. The difference is most marked if the heat capacity mc_p of the two fluids is the same. When the heat capacity of one fluid is infinite, as occurs in a heat exchanger with one fluid condensing or evaporating, the merits of these two types are equal. This is evident from Fig. 24.2, since the logarithmic mean temperature difference is obviously independent of the direction of flow of fluid 1.

In Example 24.1 it was assumed that the resistance of the metal tube separating the two fluids is negligible compared with the resistances of the two fluid boundary layers. This is usually true when the metal wall is very clean, but with many fluids deposits form gradually on the wall. If such deposits are likely to be formed, it is necessary to allow for their resistance, and this is usually done by means of 'fouling factors'. The allowance made depends upon the nature of the fluid (e.g. whether hard, soft, or distilled water is used), and also on the frequency with which the heat exchanger can be put out of commission for cleaning. A fouling factor is usually quoted as a resistance to be added to the 'clean' overall resistance $1/U$.

In some problems it is not permissible to take an average value of U and assume that it is constant over the whole area of the heat exchanger; this is the case when the temperature range through which a fluid is heated or cooled is very large. A more accurate equation for Q can be derived by assuming that U varies linearly with temperature through the heat exchanger, from U_i at inlet to U_o at outlet. It can then be shown that the rate of heat transfer is given by

$$Q = A \frac{U_i \Delta t_o - U_o \Delta t_i}{\ln (U_i \Delta t_o / U_o \Delta t_i)} \qquad (24.10)$$

In view of the relative inaccuracy of the empirical equations from which the film coefficients are obtained, and the uncertainty about entrance length and fouling effects, the simpler analysis is usually adequate.

Fig. 24.3   Cross-flow heat exchanger, showing typical temperature distributions at inlet and outlet

The *cross-flow* heat exchanger, illustrated in Fig. 24.3, occupies a position intermediate between the parallel-flow and counter-flow types. The analysis for finding the mean temperature difference is more complicated in this case, and the reader is referred to more advanced texts [Refs. (11) and (33)]. Typical temperature distributions at inlet and outlet are shown in the figure. The cross-flow exchanger, although less economic in size than the counter-flow type, is sometimes preferred when it makes the layout of ducts and piping more convenient and accessible.

Many heat exchangers do not fall into any of the simple categories mentioned. Fig. 24.4 shows a typical shell-type heat exchanger with parallel-, cross- and counter-flow paths; such an exchanger is usually referred to as a *mixed-flow* type. Rather than carry out the analysis from first principles, designers usually refer to published curves of mean temperature difference for the principal types of mixed-flow heat exchanger.*

*Ref. (35) (T. Tinker), pp. 89–116, and Ref. (11). Ref. (2) is a good introduction to design calculations.

The heat exchangers discussed so far are sometimes called *recuperators*. In them heat flows steadily and continuously from one fluid to another through a containing wall. An entirely different type of heat exchanger, in which the flow of heat is intermittent, is known as the *regenerator*. In a regenerator a matrix of metal comes into

Fig. 24.4   Shell-type mixed-flow heat exchanger

contact alternately with the hot and the cold fluid, first absorbing heat for a time, and then rejecting it. In the regenerator shown in Fig. 24.5 the matrix is rotating, the hot fluid flowing through the top half, and the cold fluid through the bottom half of the matrix [see Ref. (33)]. The flow passages are suitably sealed from each other. In other types, the fluids are passed alternately through two stationary matrices by opening and closing valves.

Fig. 24.5   A regenerator

24.2 Heat flow through a wall

When a wall separates two fluids of known temperature, it was shown in section 21.2 that the heat flow through the wall can be calculated from equations (21.12) and (21.13). These equations require a knowledge of the surface coefficients, which we have seen depend upon the temperature of the wall surfaces. Since the wall temperatures are unknown and the coefficients are not simple functions of these temperatures, there is in fact no direct easy way of calculating the heat flow. The simplest method is to guess these temperatures, and proceed by a method of successive approximation. The procedure is best illustrated by a numerical example.

Example 24.2 A brick wall of conductivity $k = 6 \cdot 92 \times 10^{-4}$ kW/m K, and of thickness 23 cm, separates two rooms. The mean temperature of the air and other surfaces with which the wall is exchanging heat on the left is 65 °C, and on the right it is 10 °C (Fig. 24.6). Find the heat flow per m² of wall.

Let us assume initially that the surface temperatures of the wall are $t_1 = 55$ °C on the left, and $t_2 = 20$ °C on the right. The free convection heat transfer coefficients can be found from the appropriate equation in Fig. 22.17. To decide whether the laminar or turbulent equation is applicable we must make a rough estimate of Gr. We shall calculate its value midway up a wall of 4 m height, taking $\beta = 1/T \approx 1/310$ K^{-1}, $\theta = 10$ K, and $\nu = 1 \cdot 66 \times 10^{-5}$ m²/s. Thus

$$Gr = \frac{\beta g x^3 \theta}{\nu^2} \approx 9 \cdot 2 \times 10^9$$

The flow is evidently turbulent over most of the wall and the appropriate equation is $h_c = 0 \cdot 001\ 31 (\theta/[\text{K}])^{1/3}$ kW/m² K (Fig. 22.17). For our guess of the surface temperatures,

$$h_{c1} = h_{c2} = 0 \cdot 001\ 31 (65 - 55)^{1/3} = 0 \cdot 001\ 31 (20 - 10)^{1/3} = 0 \cdot 002\ 82 \text{ kW/m}^2 \text{ K}$$

From Fig. 23.8 the normal emissivity ε_n for brick is 0·93. We pointed out in section 23.2 that $\varepsilon/\varepsilon_n$ is $0 \cdot 95 \rightarrow 1$ for this kind of material. Hence the total hemispherical emissivity $\varepsilon \approx 0 \cdot 9$. The radiation heat transfer coefficient to the left of the wall, from (23.36) is

$$h_{r1} = 0 \cdot 9 \times 56 \cdot 7 \times 10^{-12} (338 + 328)(338^2 + 328^2) = 0 \cdot 007\ 54 \text{ kW/m}^2 \text{ K}$$

On the right, the radiation coefficient is

$$h_{r2} = 0 \cdot 9 \times 56 \cdot 7 \times 10^{-12} (293 + 283)(293^2 + 283^2) = 0 \cdot 004\ 88 \text{ kW/m}^2 \text{ K}$$

The overall heat transfer coefficient is then given by equation (21.13),

$$\frac{1}{U} = \frac{1}{h_{c1} + h_{r1}} + \frac{\Delta x}{k} + \frac{1}{h_{c2} + h_{r2}}$$

$$= \frac{1}{0 \cdot 002\ 82 + 0 \cdot 007\ 54} + \frac{0 \cdot 23}{6 \cdot 92 \times 10^{-4}} + \frac{1}{0 \cdot 002\ 82 + 0 \cdot 004\ 88} = 559$$

$$U = 0 \cdot 001\ 79 \text{ kW/m}^2 \text{ K}$$

To check the accuracy of our original guess of t_1 and t_2, we can apply the equation of continuity. For steady flow

$$q = -U(10 - 65) = 0 \cdot 001\ 79 \times 55 = 0 \cdot 0985 \text{ kW/m}^2$$
$$= -(h_{c1} + h_{r1})(t_1 - 65) = -0 \cdot 010\ 36 (t_1 - 65)$$
$$= -(h_{c2} + h_{r2})(10 - t_2) = -0 \cdot 007\ 70 (10 - t_2)$$

Hence we find that

$$t_1 = 55 \cdot 5 \,°C, \quad \text{and} \quad t_2 = 22 \cdot 8 \,°C$$

The procedure can now be repeated with the new values of the surface temperatures. This second approximation gives the following results:

$$h_{c1} = 0 \cdot 002\,77 \text{ kW/m}^2\text{ K}, \quad h_{c2} = 0 \cdot 003\,06 \text{ kW/m}^2\text{ K}$$
$$h_{r1} = 0 \cdot 007\,56 \text{ kW/m}^2\text{ K}, \quad h_{r2} = 0 \cdot 004\,95 \text{ kW/m}^2\text{ K}$$
$$U = 0 \cdot 001\,81 \text{ kW/m}^2\text{ K}, \quad q = 0 \cdot 0996 \text{ kW/m}^2$$
$$t_1 = 55 \cdot 4 \,°C \qquad t_2 = 22 \cdot 4 \,°C$$

Fig. 24.6

It is evident from the last values of q, t_1 and t_2 that a further approximation is unnecessary, and the required heat flow is therefore $0 \cdot 0996$ kW/m^2.

24.3 Heat flow through a cooling fin

When heat has to be transferred from a metal surface to a fluid, the effective surface area is often increased by fins projecting from the surface, as in air-cooled petrol engines. Most of the effective resistance is in the fluid boundary layer adjacent to the

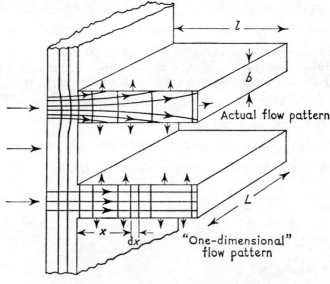

Fig. 24.7   Heat flow through a cooling fin

wall, and this is materially reduced by an increase in surface area; the increase in resistance due to the additional metal is normally relatively small. Here only the simple rectangular fin projecting from a flat surface is considered in any detail. The flow pattern, i.e. the lines of flow and isothermals, is shown in the top fin of Fig. 24.7.

If the fin is slender, it may be assumed that the lines of flow are parallel to the flanks of the fin, and the case is reduced to a 'one-dimensional' problem as shown in the lower fin. We shall assume that the combined convection and radiation heat transfer coefficient h is constant over the fin surface.

It will be remembered from section 21.4 that the initial step in any conduction problem is to find the temperature distribution in the conductor. In order to do this for the fin, let us consider an element dx a distance x from the root of the fin. The heat flow into the element is then

$$Q_x = -kA \left(\frac{dt}{dx}\right) = -kLb \left(\frac{dt}{dx}\right)$$

The heat leaving the element through a section $(x + dx)$ from the root is

$$Q_{(x+dx)} = Q_x + \left(\frac{dQ}{dx}\right) dx = -kLb \left(\frac{dt}{dx}\right) - kLb \left(\frac{d^2t}{dx^2}\right) dx$$

The net outflow to the surroundings at t_s from the element dx, which is equal to

$$dQ = Q_x - Q_{(x+dx)} = kLb \left(\frac{d^2t}{dx^2}\right) dx$$

must also be equal to

$$h\{2(L + b) \, dx\}(t - t_s) \approx h2L(t - t_s) \, dx \qquad \text{(when } b \ll L\text{)}$$

Putting $\theta = t - t_s$, and remembering that $(d^2t/dx^2) = (d^2\theta/dx^2)$, we get

$$\frac{d^2\theta}{dx^2} = \frac{2h}{kb} \theta \tag{24.11}$$

The general solution of this differential equation is

$$\theta = Me^{mx} + Ne^{-mx} \tag{24.12}$$

where

$$m = \left(\frac{2h}{kb}\right)^{1/2} \tag{24.13}$$

The values of the constants of integration M and N can be found from the boundary conditions. At the root, where $x = 0$, the fin temperature is t_0 say, and hence

$$\theta_0 = t_0 - t_s = M + N \tag{24.14}$$

The heat loss from the tip of a slender fin is very small compared with that from the flanks, and can be neglected. This gives us the second boundary condition; i.e. when $x = l$,

$$Q_l = -kLb \left(\frac{d\theta}{dx}\right)_l = 0$$

and hence

$$\left(\frac{d\theta}{dx}\right)_l = 0 \tag{24.15}$$

Differentiating (24.12) therefore, and putting $x = l$,

$$Mme^{ml} - Nme^{-ml} = 0 \tag{24.16}$$

The constants of integration are found from (24.14) and (24.16) as

$$M = \theta_0 \frac{e^{-ml}}{e^{ml} + e^{-ml}} \quad \text{and} \quad N = \theta_0 \frac{e^{ml}}{e^{ml} + e^{-ml}} \tag{24.17}$$

Substituting these values in (24.12), we get the solution for the temperature profile along the fin as

$$\theta = \theta_0 \frac{e^{m(l-x)} + e^{-m(l-x)}}{e^{ml} + e^{-ml}} = \theta_0 \frac{\cosh m(l - x)}{\cosh ml} \tag{24.18}$$

For steady conduction, the heat flow from the fin surface is equal to the heat flow through the base of the fin. Thus the solution for the heat flow is

$$Q_0 = -kLb \left(\frac{d\theta}{dx}\right)_0 = mkLb\theta_0 \left[\frac{\sinh m(l - x)}{\cosh ml}\right]_{x=0}$$

$$= mkLb\theta_0 \tanh ml \tag{24.19}$$

A conduction problem related to the fin is the determination of the temperature in a solid or hollow prismatic rod fixed at one end to a plate of temperature t_0 and projecting into a fluid of temperature t_s. The variation of temperature with distance x from the root of the rod, given in terms of $\theta = t - t_s$, can be shown to be given by equation (24.18), but with m modified to

$$m = \left(\frac{h \times \text{circumference of rod}}{k \times \text{solid cross-sectional area}}\right)^{1/2} \tag{24.20}$$

It is interesting to note that, as in other conduction problems with a boundary condition given in terms of a surface heat transfer coefficient (e.g. section 21.6), the solution for the fin contains implicitly the Biot number Bi. Thus, if we denote by l' the ratio of solid cross-sectional area to circumference, then

$$ml = \left(\frac{hl'}{k}\right)^{1/2} \frac{l}{l'} = (Bi)^{1/2} \frac{l}{l'}$$

A typical application for a hollow rod is provided by a thermometer pocket immersed in a fluid stream as shown in Fig. 24.8. Evidently the thermometer will indicate a temperature t_l somewhere between t_s and t_0. The following numerical example illustrates the order of error that may be involved when using this method of temperature measurement.

Example 24.3 Superheated steam at a mean temperature $t_s = 200\,°C$ flows through a 10 cm diameter pipe. A brass pocket dips radially into the pipe with its closed end on the centre line, and the root of the pocket is at $t_0 = 140\,°C$. The diameter of the pocket is $d = 1\cdot25$ cm and its wall thickness is $s = 1$ mm. The conductivity k for brass is $0\cdot112$ kW/m K and it is estimated that the combined convection and radiation heat transfer coefficient h for the pocket surface is $0\cdot397$ kW/m² K. Predict the thermometer reading t_l. Assume that the thermometer reads the temperature of the bottom plate of the pocket, and that conduction in the axial direction through the thermometer stem and surrounding oil is negligible compared with the conduction along the pocket wall (Fig. 24.8).

Fig. 24.8   The thermometer pocket

The temperature at the end of the pocket is given by putting $x = l$ in (24.18), i.e.

$$t_l - t_s = (t_0 - t_s)\frac{\cosh m(l - l)}{\cosh ml} = -60\frac{1}{\cosh ml}$$

The effective cross-sectional area to be used in (24.20) is approximately equal to πds, and the circumference of the rod is πd. Hence

$$m = \left(\frac{h\pi d}{k\pi ds}\right)^{1/2} = \left(\frac{h}{ks}\right)^{1/2} = \left(\frac{0\cdot397}{0\cdot112 \times 0\cdot001}\right)^{1/2} = 59\cdot5 \text{ m}^{-1}$$

Since

$$\cosh ml = \cosh (59\cdot5 \times 0\cdot05) = 9\cdot87$$

$$t_l - t_s = -\frac{60}{9\cdot87} = -6\cdot1\,\text{K}, \quad \text{or} \quad t_l = 193\cdot9\,°C$$

Clearly the error can be reduced by lagging the pipe to increase t_0, and also by increasing the product (ml). (a) m can be increased by using a thinner tube, or by using a metal of lower conductivity. If the conductivity is reduced excessively, however, the radial resistance to heat flow increases to such an extent that the temperature distribution is no longer 'one-dimensional' and the foregoing treatment is not applicable. (b) l may be increased by inclining the pocket and letting it project beyond the pipe axis as shown by the dotted pocket in Fig. 24.8: the tip should of course remain clear of the boundary layer near the wall.

24.4 Thermocouple in a gas stream

When a thermocouple is used to measure the local temperature T_g of a hot gas flowing through a pipe of lower temperature T_w, the thermocouple indicates a temperature T_p, such that $T_w < T_p < T_g$ (Fig. 24.9). The exact value of T_p is determined by a balance of heat transfers to and from the couple junction. The junction gains heat by convection (assuming that radiation from the gas stream is negligible), and it loses heat by radiation to the wall and by conduction along the wire. The conduction loss can often be neglected if fine-gauge wire is used, particularly if the wire is laid along an

Fig. 24.9   Different types of thermocouple

isothermal for some distance before it is taken out of the pipe radially, as shown in Fig. 24.9. The heat balance between the convection gain and radiation loss can then be written as

$$h_c A(T_g - T_p) = h_r A(T_p - T_w)$$

and hence the true gas temperature T_g can be written in terms of the thermocouple reading T_p and a correction term as

$$T_g = T_p + (T_p - T_w)\frac{h_r}{h_c} \tag{24.21}$$

Example 24.4   Air at atmospheric pressure flows over a thermocouple junction with a velocity of 3 m/s. The temperature of the pipe wall is $T_w = 670$ K, and the couple indicates an air temperature of $T_p = 850$ K. The couple is butt welded, as shown in Fig. 24.9a, and its diameter at the junction is $d = 1 \cdot 2$ mm. The emissivity of the wire is $\varepsilon = 0 \cdot 2$. For the flow across wires, the convection heat transfer can be found from

$$Nu_d = 0 \cdot 5 Re_d^{0 \cdot 5} \tag{24.22}$$

Calculate the true temperature of the air.

As we do not know the exact air temperature, we may use its property values at 850 K, which are [from Ref. (14)]

$$k = 6 \cdot 03 \times 10^{-5} \text{ kW/m}^2 \text{ K} \quad \text{and} \quad \nu = 9 \cdot 06 \times 10^{-5} \text{ m}^2/\text{s}$$

The heat transfer to the wire by convection can be found from

$$Nu_d = 0.5 \left(\frac{3 \times 0.0012}{9.06 \times 10^{-5}}\right)^{0.5} = 3.15$$

and hence

$$h_c = \frac{k}{d} Nu_d = \frac{6.03 \times 10^{-5}}{0.0012} \times 3.15 = 0.158 \text{ kW/m}^2\text{K}$$

The radiation heat transfer coefficient is from (23.36),

$$h_r = 0.2 \times 56.7 \times 10^{-12}(850 + 670)(850^2 + 670^2) = 0.0202 \text{ kW/m}^2\text{K}$$

Hence the true air temperature according to (24.21) is

$$T_g = 850 + (850 - 670)\frac{0.0202}{0.158} = 873 \text{ K}$$

[Since the temperature range between couple and air is only 23 K, it is immaterial whether the properties in (24.22) are evaluated at the bulk, film or couple temperature.]

From (24.21) it can be seen that the accuracy of the thermocouple reading can be improved by the following methods.

(a) By reducing h_r. This can be achieved by shielding the junction with one or more concentric shields, as is shown in Fig. 24.9 b. The shields assume temperatures intermediate between T_w and T_p, and thus reduce the radiation loss.

(b) By increasing h_c. This can be done by using finer wire because, from (24.22), $h_c \propto (1/d)^{0.5}$. It is also possible to increase h_c by increasing the gas velocity over the junction. The thermocouple wire can be surrounded by a tube through which some gas from the main stream is sucked at a high velocity (Fig. 24.9 c). Such an instrument is called a suction pyrometer.

The simple analysis given in this section is only applicable when the velocity of the gas stream is low, i.e. not more than about 50 m/s. When a thermocouple is placed in a gas stream, the gas is brought to rest at the 'stagnation point' of the junction and the loss of kinetic energy appears as a local rise in pressure and temperature. This effect is appreciable with high gas velocities and special designs of thermocouple have to be used [see Ref. (30) for example].

Appendix A   Dimensions and Units

Introduction

Even though the principles of thermodynamics be well understood, difficulties may be encountered when solving numerical problems because of the different systems of units in which data in the literature have been expressed in the past. The purpose of this appendix is to make clear what is involved in the concept of a system of units, and to compare the SI (Système Internationale d'Unités) with older systems. References for further reading on these topics are recommended at the end of this appendix.

First it should be appreciated that the measurement of a physical quantity, such as a length l, is accomplished by comparing it with some standard measure of the quantity, l_u say. Measurements are ratios such as l_1/l_u, l_2/l_u ... and are thus pure numbers. When l_u is elevated to a universal standard of comparison it is called a *unit*. Similar considerations apply to derived quantities such as velocity and pressure which are measured in terms of *derived units* such as m/s and N/m^2 (or Pa) respectively: they are derived from *base units* of mass, length, time, etc. It follows that a physical quantity can be regarded as a product of a pure number and a unit, i.e.

$$\text{physical quantity} = \text{number} \times \text{unit} \tag{A.1}$$

This is a widely accepted convention, and it follows that neither the physical quantity *nor the symbol used to denote it* should imply a particular choice of unit. Thus if we write

$$l_1 = l_2 \tag{A.2}$$

we imply that the two lengths are equal in an absolute sense irrespective of the system of measurement employed. One must, of course, use the same units for similar quantities when applying any such physical equation, i.e. the units used must be *consistent*.

Operations on equations involving physical quantities, units and numerical values, should follow the ordinary rules of algebra. For example, to convert a length measured in feet to one in metres the following simple procedure can be used. The conversion relation $1\,\text{ft} = 0{\cdot}3048\,\text{m}$ can be written

$$1 = \frac{0{\cdot}3048\,\text{m}}{1\,\text{ft}}$$

Then if $l_1 = 7\,\text{ft}$ we may multiply by unity without affecting the equation and so write

$$l_1 = 7\,\text{ft} \times \frac{0{\cdot}3048\,\text{m}}{1\,\text{ft}} = 2{\cdot}1336\,\text{m} \tag{A.3}$$

The convention implied by equation (A.1) leads to another recommended practice which is adopted in this book. The numerical values along the coordinate of a graph, or entered in a table, can be regarded as pure numbers. When so regarded, it follows that the correct coordinate label or column heading should be the ratio of the physical quantity (or its symbol) and the unit; e.g.

for length $\dfrac{l}{[\text{m}]}$ and for pressure $\dfrac{p}{[\text{N/m}^2]}$

If the numerical values have been divided by a power of 10, say by 10^5, the label should be divided similarly. Thus, a point on a pressure coordinate read off as 4·5 and labelled as

$$\frac{p}{10^5[\text{N/m}^2]}$$

means that

$$\frac{p}{10^5[\text{N/m}^2]}=4\cdot5 \quad \text{or} \quad p=4\cdot5\times10^5\,\frac{\text{N}}{\text{m}^2}$$

No ambiguity can arise as to whether 10^5 or 10^{-5} is intended in the last expression when the ordinary rules of algebra are applied in this way.

The requirement of consistency is more far-reaching than implied by equation (A.2). A physical equation will normally involve several terms such as

$$P_1+P_2+P_3+\ldots=0$$

where $P_1, P_2, P_3, \ldots$ may be base or derived quantities. It is not sufficient that consistent units be used, but also that all P's must be of the same *kind*: it is nonsense to write $8\,\text{m/s}+6\,\text{s}=\ldots$ because a velocity cannot be added to a time. To make clear what is meant by the 'same kind of physical quantity' it is necessary to introduce the idea of a *fundamental Unit*, such as mass **M**, length **L**, time **T**, and the idea of *dimension*, which is the power of a fundamental Unit in a *derived Unit*. For example, velocity has the derived Unit $\textbf{LT}^{-1}$, comprising the fundamental Units **L** and **T** of dimensions 1 and -1 respectively. *All meaningful physical equations must be dimensionally homogeneous.* In other words, the dimensions of the fundamental Units must be the same for each term in a physical equation. $8\,\text{m/s}+6\,\text{s}=\ldots$ is not dimensionally homogeneous, and is therefore meaningless, because the dimensions of the fundamental Units [**M, L, T**] for the first term are $[0, 1, -1]$, whereas for the second term they are $[0, 0, 1]$.

It should be appreciated that we have now used the word 'unit' in two different senses: firstly to denote a *measure* of a quantity, and secondly to denote a *kind* of quantity. This usage is confusing, and in this appendix we employ a capital letter when using the word in the second sense.

Summarising the points made so far:

(a) All physical equations are dimensionally homogeneous and should be written in such a way that they are correct whatever consistent system of units is employed (but see a further restriction regarding *coherence* below).

(b) Mathematical manipulation of physical quantities and units must follow the ordinary rules of algebra.

(c) A physical symbol should not be used to represent only the numerical part of a physical quantity (e.g. 'a tube length of l m' is incorrect phraseology because from equation (A.1) it evidently contains a square power of the length dimension). If a need arises to denote the numerical part by a symbol, this can be done using a suitably modified symbol, but the recommended practice is to use explicitly the ratio (physical quantity/unit), e.g. $l/[m]$.

SI and coherence

The units in terms of which all derived units in any system are expressed are called base units. Their choice is arbitrary and dictated by convenience. SI employs seven base units, viz. metre (m), kilogram (kg), second (s), ampere (A), kelvin (K), mole (mol) and candela (cd). Let us consider the derived physical quantity of velocity, defined in terms of distance s and time t as

$$v = k\frac{s}{t}$$

where k is a numerical coefficient. In coherent systems, coefficients such as k are made equal to unity, and it immediately follows that the unit of velocity is m/s. This apparently simple idea is particularly important in the definition of force which stems from Newton's Second Law of Motion

$$f \propto ma \quad \text{or} \quad f = kma \tag{A.4}$$

where f is the force, m the mass and a the acceleration. SI differs from many older systems of units in that, being coherent, it makes k equal to unity and thus (A.4) can be written

$$f = ma \tag{A.5}$$

With a coherent system it follows that when the units of any two of the quantities in (A.5) are chosen the unit of the third is fixed. In SI the kilogram (kg) is chosen as the unit of mass, and 1 m/s^2 as the unit of acceleration. The unit of force, the newton (N), is then defined by equation (A.5) as

$$1[N] = 1[kg] \times 1[m/s^2] \tag{A.6}$$

Similarly, on the cm-g-s system we have the dyne defined by

$$1[dyne] = 1[g] \times 1[cm/s^2] \tag{A.7}$$

The difficulty arises when a non-coherent system of units is employed, i.e. when the units of all three quantities in equation (A.5) are chosen independently of one another. We can then no longer put $k = 1$ in equation (A.4). In the past, European engineers used a unit of force called the kilogram-force (written kgf) which is defined as the gravitational force on a mass of 1 kg, viz. the weight of 1 kg. Since the conventionally accepted gravitational acceleration is

$$g_n = 9 \cdot 806\ 65\ \text{m/s}^2 \approx 9 \cdot 81\ \text{m/s}^2$$

we have from equation (A.4)

$$1[kgf] = k \times 1[kg] \times 9 \cdot 81[m/s^2] \tag{A.8}$$

which yields

$$k = \frac{1}{9 \cdot 81} \left[\frac{s^2 \, kgf}{kg \, m} \right] \quad \text{or} \quad \frac{1}{9 \cdot 81} \left[\frac{kgf}{N} \right] \tag{A.9}$$

Using the non-coherent m-kgf-kg-s system of units, the physical equation equivalent to (A.4), expressing the Second Law of Motion, thus becomes

$$f = \frac{1}{9 \cdot 81} \left[\frac{s^2 \, kgf}{kg \, m} \right] ma \tag{A.10}$$

Rearranging (A.10) we have

$$\frac{f}{[kgf]} = \frac{1}{9 \cdot 81} \frac{m}{[kg]} \frac{a}{[m/s^2]} \tag{A.11}$$

In (A.11) the 9·81 is merely a dimensionless constant relating two quantities of the same kind, viz. the kgf unit of force and the kg m/s^2 (or N) unit of force, just as the 0·3048 is a constant relating the metre to the foot in equation (A.3). If we denote by primed symbols the numerical parts of each physical quantity, equation (A.11) can also be written as

$$f' = \frac{1}{9 \cdot 81} m'a' \tag{A.12}$$

together with the statement that f' is in kgf, m' is in kg and a' is in m/s^2. Both (A.11) and (A.12) can be seen to be different ways of writing a purely *numerical equation*. In the past, equations (A.10), (A.11) and (A.12) have often simply been written as

$$f = \frac{1}{9 \cdot 81} ma$$

followed by the specification that f is in kgf, m in kg and a in m/s^2. This is bad practice because it disregards the convention contained in (A.1) and leaves it to the reader to decide from the context whether the symbols stand for physical quantities or pure numbers.

Although not used in this book, it is worth mentioning another coherent metric system—the m-kgf-s system—in which k in equation (A.4) is again equal to unity and the kgf is retained as the unit of force. The unit of mass, sometimes referred to as the *technical unit of mass*, is then defined by equation (A.5), i.e.

$$1[kgf] = 1[\text{technical unit of mass}] \times 1[m/s^2] \tag{A.13}$$

Comparing this with equation (A.8) when $k = 1$, it follows that

$$1[\text{technical unit of mass}] = 9 \cdot 81[kg] \tag{A.14}$$

Here 9·81 is a dimensionless conversion factor between two mass units, arbitrarily chosen to be *numerically* equal to the conventional value of the gravitational acceleration g_n. Some writers use the symbol g_0 when referring to this number.

As an example of how coherent and non-coherent systems are used in conjunction with a physical equation, we may consider a simple mechanical energy equation expressing the fact that when a body undergoes a change of position and velocity in a field of force, the decrease of potential energy is equal to the increase in kinetic energy. For example, in the vicinity of the earth, where the force can be regarded as independent of position, the physical equation is

$$f(z_1 - z_2) = \tfrac{1}{2}m(C_2{}^2 - C_1{}^2) \tag{A.15}$$

where f is the force of gravity acting on a body of mass m as it moves from one height z_1 to another z_2, and C is the velocity of the body (f is equal to mg in this particular case). The equation is dimensionally homogeneous, each term having the derived Unit of energy $\mathbf{ML^2T^{-2}}$. On the coherent m-kg-s system the equivalent numerical equation is

$$\frac{f}{[N]}\frac{(z_1 - z_2)}{[m]} = \frac{1}{2}\frac{m}{[kg]}\frac{(C_2{}^2 - C_1{}^2)}{[m^2/s^2]} \tag{A.16}$$

or

$$f'(z_1' - z_2') = \tfrac{1}{2}m(C_2'^2 - C_1'^2) \tag{A.17}$$

However, when the non-coherent m-kgf-kg-s system is used the physical equation equivalent to (A.15) becomes

$$f(z_1 - z_2) = \frac{1}{9\cdot81}\left[\frac{s^2\,kgf}{kg\,m}\right]\frac{1}{2}m(C_2{}^2 - C_1{}^2) \tag{A.18}$$

Substituting numerical values of f', z', m' and C' into this equation, yields

$$f'[kgf](z_1' - z_2')[m] = \frac{1}{9\cdot81}\left[\frac{s^2\,kgf}{kg\,m}\right]\frac{1}{2}m'[kg](C_2'^2 - C_1'^2)\left[\frac{m}{s^2}\right] \tag{A.19}$$

This makes it possible to check that the units on both sides of the equation do indeed balance.

Referring back to point (a) at the end of the Introduction, it can now be stated that *all physical equations in this book are written in such a way that they are correct whatever consistent and coherent system of units is employed.*

Minor modifications to the pure and coherent SI have been found convenient, such as the use of the *kilojoule* (kJ) instead of the joule (J) for the unit of energy, and the *bar* for $10^5\,N/m^2$ (or Pa) as the unit of pressure. Although these changes are merely a matter of size of unit which is still decimally related to the coherently derived unit, they can occasionally lead to difficulties. Where dimensional checks are helpful, they have been set out in the manner of equation (A.19): particular attention is drawn to equation (8.5) and Examples 10.1, 10.7, 18.4 and 22.3. Where in the course of a numerical solution there is no obvious difficulty with units, it is unnecessary to be so rigorous. For example, in Example 9.1(a) we have written

$$Q = m(u_2 - u_1) = 0\cdot6014(1457 - 2711) = -754\,kJ$$

It would have been pedantic to write

$$Q = m(u_2 - u_1) - 0\cdot6014[kg](1457[kJ/kg] - 2711[kJ/kg]) = -754\,kJ$$

The old British systems of units are analogous to the metric systems already described, with the coherent ft-lb-s system parallel to the SI, and the non-coherent ft-lbf-lb-s system parallel to the m-kgf-kg-s system. The coherent ft-lbf-s system, with the *slug* as the unit of mass, is parallel to the m-kgf-s system with its technical unit of mass.

Remembering that the conventional value of g is

$$g_n = 9\cdot806\,65\,\frac{m}{s^2}\,\frac{1\,\text{ft}}{0\cdot3048\,m} = 32\cdot1740\,\frac{\text{ft}}{s^2} \approx 32\cdot2\,\frac{\text{ft}}{s^2}$$

we can write down the British equivalents of equations (A.6), (A.8), (A.9), (A.13) and (A.14) as follows.

$$1[\text{poundal(pdl)}] = 1[\text{pound(lb)}] \times 1[\text{ft/s}^2] \tag{A.6b}$$

$$1[\text{pound-force(lbf)}] = k \times 1[\text{pound(lb)}] \times 1[\text{ft/s}^2] \tag{A.8b}$$

$$k = \frac{1}{32\cdot2}\left[\frac{s^2\,\text{lbf}}{\text{lb ft}}\right] \quad \text{or} \quad \frac{1}{32\cdot2}\left[\frac{\text{lbf}}{\text{pdl}}\right] \tag{A.9b}$$

$$1[\text{pound-force(lbf)}] = 1[\text{slug}] \times 1[\text{ft/s}^2] \tag{A.13b}$$

$$1[\text{slug}] = 32\cdot2[\text{lb}] \tag{A.14b}$$

The table concluding this appendix summarises the more important SI units used in this book and their usual metric-technical and British equivalents. It is worth pointing out that although SI is not unique in being coherent, it is unique in that its unit of mechanical power is *identical* with the unit of electric power, viz.

$$1\,\text{W}(= 1\,\text{J/s} = 1\,\text{N m/s}) = 1[\text{ampere (A)}] \times 1[\text{volt (V)}] \tag{A.20}$$

This identity leads to numerous simplifications in many branches of engineering and physics.

The following publications are particularly authoritative and useful. For a detailed description of the SI:

The international system of units 1977 (HMSO)

For an explanation of the relation between, and use of, physical quantities, units and numerical values:

Quantities, units and symbols. A report by the Symbols Committee of the Royal Society, 1975

General principles concerning quantities, units and symbols. ISO 31/0-1974(E) General introduction to ISO 31.

For conversion factors between various systems of units:

Conversion factors and tables. BS 350: Part 1: 1974

Physical quantity	SI	Metric units		ft-lb-s	British units	
		m-kgf-kg-s	m-kgf-s		ft-lbf-lb-s	ft-lbf-s
Length	**m**	**m**	**m**	**ft**	**ft**	**ft**
Mass	**kg**	**kg**	$kg \times 9.806\,65$ [eq. A.14]	**lb**	**lb**	$slug = lb \times 32.1740$ [eq. A.14b]
Time	**s**		**s**	**s**		**s**
Temperature[1] (interval)	**K**		**K**	**R**		**R**
Specific volume	$\dfrac{m^3}{kg}$	$\dfrac{m^3}{kg}$	$\dfrac{m^3}{(kg \times 9.806\,65)}$	$\dfrac{ft^3}{lb}$	$\dfrac{ft^3}{lb}$	$\dfrac{ft^3}{(lb \times 32.1740)}$
Force	N[eq. A.6]		**kgf**	pdl[eq. A.6b]		**lbf**
Pressure[2]	$Pa(= 1\ N/m^2)$ or $bar\ (= 10^5\ N/m^2)$		kgf/cm^2	pdl/ft^2		lbf/ft^2
Energy[3]	$J(= 1\ Nm)$ or $kJ(= 10^3\ J)$		mechanical: kgf m thermal: kcal ($\approx$427 kgf m)	ft pdl	mechanical: ft lbf thermal: Btu($\approx$778 ft lbf)	
Rate of energy flow	$W(= 1\ J/s)$ or $kW(= 10^3\ J/s)$		mechanical: kgf m/s or metric hp ($= 75$ kgf m/s) heat: kcal/s	ft pdl/s	mechanical: ft lbf/s or British hp ($= 550$ ft lbf/s) heat: Btu/s	

The base units of each system are in bold type.

(1) Amongst all units, temperature is unique in that, traditionally, the unit symbol is used to denote not only the unit (viz. the degree interval) but also the position of a temperature relative to a specified datum. While as a pure measure of unit size or interval K ≡ °C (or R ≡ °F), as 'scale datum indicators' the two symbols are different. In this book °C is used only to denote a temperature on the Celsius scale, while K is used (a) for the temperature unit, and (b) to denote positions on the Kelvin scale.

(2) This book sometimes uses the standard atmosphere (atm) defined by

$$1\ atm = 1.013\,25\ bar = 101\,325\ N/m^2$$

The kgf/cm² is also called the technical atmosphere and it is usually denoted by 'at'. It is worth noting that

$$1\ atm \approx 760\ mm\ Hg \qquad 1\ bar \approx 750\ mm\ Hg \qquad 1\ at \approx 736\ mm\ Hg$$

$$1\ atm \approx 760\ mm\ Hg \qquad 1\ bar \approx 750\ mm\ Hg$$

(3) See section 2.3 for definitions of the calorimetric units kcal and Btu.

Appendix B Problems

The following problems are numbered by chapters. Answers have been obtained using the abridged tables of properties of Ref. (17), Part II. All pressures are absolute and, unless otherwise stated, constant specific heats have been used in the solutions.

7.1 Derive the Maxwell relation:

$$\left(\frac{\partial T}{\partial v}\right)_s = -\left(\frac{\partial p}{\partial s}\right)_v$$

Using this and the relation

$$\left(\frac{\partial x}{\partial y}\right)_z \left(\frac{\partial y}{\partial z}\right)_x \left(\frac{\partial z}{\partial x}\right)_y = -1$$

derive the three remaining Maxwell relations.

7.2 Show that

$$\left(\frac{\partial \beta}{\partial p}\right)_T = -\left(\frac{\partial \kappa}{\partial T}\right)_p$$

The compressibility of a substance varies linearly with temperature at constant pressure, according to the law $\kappa = A + BT$. Show that the change in the coefficient of expansion during an isothermal process is given by

$$\beta_2 - \beta_1 = -B(p_2 - p_1)$$

7.3 The specific heat of a particular substance is found to be independent of pressure and to vary with temperature according to the equation $c_p = a + bT + cT^2$, where a, b and c are constants. Find the heat required to raise the temperature of the substance from T_1 to T_2 at constant pressure, and hence find the mean specific heat for the range T_1 to T_2.

Show that the difference between this mean specific heat and the arithmetic mean of c_p at T_1 and T_2 is equal to $-c(T_2 - T_1)^2/6$.

7.4 The Clausius equation of state is $p(v - b) = RT$. For a substance obeying this equation, find expressions for β and κ in terms of p and v. Also, given the relation $c_p = T(\partial s/\partial T)_p$, show that c_p can be a function only of T.

609

7.5 The Dieterici equation of state is $p(v - b) e^{a/vRT} = RT$, where a, b and R are constants. Find an expression for the coefficient of cubical expansion, and show that this approximates to the value of β for a perfect gas when T and v are large.

7.6 Show that the difference between the specific heats of a van der Waals gas is given by

$$c_p - c_v = R\frac{1}{1 - 2a(v - b)^2/RTv^3}$$

The general expression $c_p - c_v = \beta^2 Tv/\kappa$ need not be derived.

7.7 Show that the slope on the *h-s* diagram is equal to
   (i) T, for a reversible constant pressure process,
   (ii) $T - (1/\beta)$, for a reversible isothermal process,
   (iii) $T + \{(c_p - c_v)/c_v\beta\}$, for a reversible constant volume process.

The general relations $c_p - c_v = \beta^2 Tv/\kappa$ and $c_v = T(\partial s/\partial T)_v$ need not be derived.

7.8 Starting with the functions $g = h - Ts$ and $h = u + pv$, show that

(i) $\left(\dfrac{\partial s}{\partial p}\right)_T = -\left(\dfrac{\partial v}{\partial T}\right)_p$;   (ii) $c_p = T\left(\dfrac{\partial s}{\partial T}\right)_p$

Hence show that

$$ds = \frac{c_p\,dT}{T} - \left(\frac{\partial v}{\partial T}\right)_p dp$$

Derive expressions for the change of entropy during an isothermal compression from p_1 to p_2 at temperature T, (a) for a perfect gas, and (b) for a gas obeying the law

$$pv = RT + \left(b - \frac{a}{RT^2}\right)p$$

7.9 Given that

$$ds = \frac{c_p\,dT}{T} - \left(\frac{\partial v}{\partial T}\right)_p dp = \frac{c_v\,dT}{T} + \left(\frac{\partial p}{\partial T}\right)_v dv$$

show that

$$c_p - c_v = T\left(\frac{\partial v}{\partial T}\right)_p\left(\frac{\partial p}{\partial T}\right)_v$$

Hence show that for a perfect gas the difference between c_p and c_v is R, and that for a gas having the characteristic equation

$$pv = RT + \left(b - \frac{a}{RT^2}\right)p$$

the difference is approximately $R + (4ap/RT^3)$.

7.10   Prove the following thermodynamic relations:

$$c_p = T \left(\frac{\partial s}{\partial T}\right)_p \quad \text{and} \quad c_v = T \left(\frac{\partial s}{\partial T}\right)_v$$

Show that on a p-v diagram the ratio of the slope of the isentropic through a point to the slope of an isothermal through the same point is equal to the ratio c_p/c_v.

7.11   Deduce the thermodynamic relations

(i) $\left(\dfrac{\partial h}{\partial p}\right)_T = v - T \left(\dfrac{\partial v}{\partial T}\right)_p = -c_p \left(\dfrac{\partial T}{\partial p}\right)_h$,   (ii) $\left(\dfrac{\partial u}{\partial v}\right)_T = T \left(\dfrac{\partial p}{\partial T}\right)_v - p$

Any Maxwell relation used need not be derived. With the aid of equation (ii), show that

$$\left(\frac{\partial u}{\partial p}\right)_T = -T \left(\frac{\partial v}{\partial T}\right)_p - p \left(\frac{\partial v}{\partial p}\right)_T$$

The quantity $c_p(\partial T/\partial p)_h$ is known as the Joule-Thomson cooling effect. Show that this cooling effect is equal to $(3C/T^2) - b$ for a gas having the equation of state

$$(v - b) = \frac{RT}{p} - \frac{C}{T^2}$$

Finally, show that in general the cooling effect can be accounted for by a combination of a departure from Joule's law [$(\partial u/\partial p)_T = 0$] and a departure from Boyle's law [$(\partial pv/\partial p)_T = 0$], and check this by determining these two partial differential coefficients for the foregoing gas. Hence explain why, under certain conditions, a throttling process can be accompanied by a rise of temperature.

7.12   Show that:

(i) $\left(\dfrac{\partial c_p}{\partial p}\right)_T = -T \left(\dfrac{\partial^2 v}{\partial T^2}\right)_p$,   (ii) $ds = c_p \dfrac{dT}{T} - \left(\dfrac{\partial v}{\partial T}\right)_p dp$

There is no need to derive any Maxwell relation used. Hence, find expressions for c_p, s, h and u, for a gas having the characteristic equation

$$v = \frac{RT}{p} + a - \frac{b}{T^3}$$

where R, a, b are constants. c_p at zero pressure has a constant value of c_{p0}, and s and h are arbitrarily chosen to be zero at the state defined by p_1 and T_1.

7.13   The specific volume of liquid water is $0 \cdot 001\,000\,2$ m³/kg and that of ice is $0 \cdot 001\,091$ m³/kg, at $0 \cdot 01$ °C. The latent heat of fusion of ice at this temperature is $333 \cdot 5$ kJ/kg. Calculate the rate of change of melting point of ice with pressure in K/bar. Prove any formula used in the calculation.
    [$-0 \cdot 007\,44$ K/bar.]

7.14 The relation between saturation vapour pressure and saturation temperature for mercury is given by

$$\log_{10}\frac{p}{[\text{bar}]} = 7\cdot0323 - \frac{3276\cdot6}{T/[\text{K}]} - 0\cdot6520\log_{10}\frac{T}{[\text{K}]}$$

Calculate the specific volume v_g of saturated mercury vapour at 0·1 bar. The specific volume of saturated mercury liquid v_f may be neglected, and the latent heat of vaporisation at 0·1 bar is 294·54 kJ/kg.

[2·140 m³/kg.]

9.1 1 kg of air can be expanded between two states as it does 20 kJ of work and receives 16 kJ of heat. A second kind of expansion can be found between the same initial and final states which requires a heat input of only 9 kJ. What is the change of internal energy in the first expansion, and what is the work done by the air in the second expansion?

[−4 and 13 kJ.]

9.2 A closed system receives 168·7 kJ of heat at constant volume. It then rejects 177 kJ of heat while it has 40 kJ of work done on it at constant pressure. If an adiabatic process can be found which will restore it to its initial state, how much work will be done by the system during that process? If the value of the internal energy in the initial state is arbitrarily put equal to zero find the internal energies at the other two states.

[31·7 kJ, 168·7 and 31·7 kJ.]

9.3 1800 kJ of heat is transferred to 1 m³ of air at 14 bar and 200 °C. Find (a) the final temperature and pressure if the volume remains constant, and (b) the final temperature and volume if the pressure remains constant. After which process has the internal energy of the air the greater value?

When the air is then expanded isothermally to a pressure of 7 bar, what is the final volume in each case?

[(a) 445 °C, 21·3 bar, 3·04 m³. (b) 375 °C, 1·37 m³, 2·74 m³.]

9.4 0·75 kg of air has a pressure of 3 bar and a temperature of 125 °C. After it has received 900 kJ of heat at constant volume find the final temperature and pressure. If the air then expands adiabatically doing 915 kJ of work, (a) by how much is its internal energy changed during the expansion, (b) is the value of this property lower or higher than before the expansion, and (c) is the value lower or higher than before the heating process?

[1796 °C, 15·6 bar, (a) and (b) 915 kJ lower, (c) lower.]

9.5 1 kg of gas expands reversibly and adiabatically, its temperature falling from 240 °C to 115 °C while its volume is doubled. The gas does 90 kJ of work in the process. Find the values of the specific heats c_v and c_p, and the molecular weight of the gas.

[0·72 and 1·01 kJ/kg K, 28·7.]

9.6 State the characteristics of (a) an isothermal and (b) an adiabatic expansion.

A bicycle pump which has a stroke of 20 cm is used to force air into a tyre against a pressure of 4 bar. What length of stroke will be swept through before air begins to enter the tyre when the piston is pushed in (i) slowly and (ii) quickly? Assume that the atmospheric pressure is 1 bar.

[15 and 12·57 cm.]

9.7 A certain mass of gas expanding reversibly at constant pressure does 5 kJ of work. Calculate the quantity of heat which must be transmitted to the gas, and how much of this heat is used to raise the internal energy of the gas. Assume only that the gas is perfect and that $\gamma = 1\cdot66$.

[12·58 kJ, 7·58 kJ.]

9.8 A cylinder of 8 cm internal diameter is fitted with a piston loaded by a coil spring of stiffness 140 N/cm of compression. The cylinder contains $0\cdot0005\,m^3$ of air at 15°C and 3 bar. Find the amount of heat which must be supplied for the piston to move a distance of 4 cm.

[0·417 kJ.]

9.9 Starting from the non-flow energy equation show that the specific heat, c_n, of a perfect gas during a reversible polytropic expansion may be expressed by

$$c_n = \left(\frac{\gamma - n}{1 - n}\right) c_v$$

9.10 1 kg of air at atmospheric pressure and 15 °C is compressed isentropically to 6 atmospheres. Find the final temperature and the work done. If the air is then cooled at the upper pressure to the original temperature of 15 °C, what amount of heat is rejected and what further work of compression is done?

The final state could have been reached by a reversible isothermal compression instead of by the foregoing two processes: what would be the work and heat transfers in this case?

[208 °C, $-138\cdot5$ kJ, 194 kJ, 55·4 kJ, $-148\cdot2$ kJ, $-148\cdot2$ kJ.]

9.11 $0\cdot3\,m^3$ of air at 26 °C and 1 bar is compressed reversibly and polytropically to $0\cdot07\,m^3$. Find the final pressure and temperature, work done, changes in internal energy and enthalpy, and the heat transferred, when the index of compression has the values: 0·9, 1·0, 1·4. 1·5. This should be treated as an exercise in methodical computation.

[For $n = 1\cdot5$: 8·87 bar, 346 °C, $-64\cdot0$ kJ, 80·3 kJ, 112·3 kJ, 16·3 kJ.]

9.12 A quantity of air occupying a volume of 1 m^3 at 4 bar and 150 °C is allowed to expand isentropically to 1 bar. Its enthalpy is then raised by 70 kJ by heating at constant pressure. What is the total work done during this process?

If the process is to be replaced by a reversible polytropic expansion which will result in the same final state being reached, what index of expansion is required? Will the work done be greater or less than in the original process?

[346 kJ, 1·305.]

9.13 A certain mass of carbon dioxide is contained in a vessel of 0.5 m^3 capacity, at an atmospheric temperature of $15 \,°\text{C}$ and a pressure of 20 bar. A valve is opened momentarily and the pressure immediately falls to 10 bar. Some time later the temperature of the remaining gas is again atmospheric and the pressure is found to be 11.78 bar. By making suitable assumptions estimate the value of γ for the gas.

Find also the change in internal energy undergone by the mass of gas remaining in the vessel, (a) during the first part of the process, and (b) during the whole process.

[1.303, -290 kJ, 0.]

9.14 Imagine two curves intersecting at some point on a $p - v$ diagram. The curves are to represent reversible processes undergone by a perfect gas: one an adiabatic process and the other an isothermal process. Show that the ratio of the slope of the adiabatic curve to the slope of the isothermal curve, at the point of intersection, is equal to γ.

A gas having a value of γ equal to 1.66 is expanded reversibly from the same initial state (a) isothermally and (b) adiabatically, such that the pressure ratio is 5 in each case. Calculate the ratio of the work done during (a) to the work done during (b).

[2.25.]

9.15 6 kg of air expands reversibly from a pressure of 13 bar to 1.3 bar, the index of expansion being 1.2. Find from first principles the change in entropy of the air, and sketch the process on the $T - s$ diagram.

[1.652 kJ/K.]

9.16 The initial state of a perfect gas is defined by T_1 and v_1, and the final state by T_3 and v_3. Find the change of entropy per unit mass along any three different reversible paths connecting these states (each consisting of two consecutive processes), and show that the same result is obtained in each case.

9.17 Explain briefly why reversibility is used as a criterion of an ideal process, illustrating your answer by referring to the irreversibility of heat transfer across a finite temperature difference.

1 kg of air undergoes a Joule free expansion from a volume of 1 m^3 to a volume of 8 m^3. What is the change of entropy of (a) the air and (b) the surroundings? Had the process been a reversible isothermal expansion between the same two states what would these changes of entropy have been? Hence compare the net change of entropy of the 'universe' in the two cases and comment on the result.

[(a) 0.597 kJ/K, (b) 0 in first part, (a) 0.597 kJ/K, (b) -0.597 kJ/K in second.]

9.18 The partial evaporation of 1 kg of water at 7 bar is accompanied by the transfer of 100 kJ of work. Find the dryness fraction of the steam formed, and the increase in internal energy of the fluid during the change of phase.

[0.526, 987 kJ.]

9.19 A vertical insulated cylinder of 0.1 m^2 cross-sectional area contains 0.92 kg of water at $15 \,°\text{C}$, and a piston resting on the water exerts a constant pressure of 7 bar. An electric heater forms the bottom of the cylinder and communicates heat to the water at the rate of 0.5 kW. Calculate the time taken by the piston to rise through a distance of 1.3 m.

[52.5 min.]

9.20 0·12 kg of steam initially saturated at 10 bar expands reversibly in a cylinder until the pressure is 1 bar. The volume is then found to be 0·17 m³. Assuming that the process is polytropic, find the index of expansion and the heat transferred during the process.

Find also the change of entropy and sketch the process on the *T–s* diagram.

[1·16, −10·4 kJ, −0·0264 kJ/K.]

9.21 3 kg of steam initially saturated at 60 bar expands reversibly and isothermally to 10 bar in a cylinder. Find the heat transferred and the work done.

If the fluid had been a perfect gas the heat and work transfers would have been equal. Why is this not so for steam, and why would the work and heat transfers have been more nearly equal if the steam had expanded from, say, 6 bar to 1 bar?

[1871 kJ, 1382 kJ.]

9.22 A vessel having a capacity of 0·6 m³ contains steam at 15 bar and 250 °C. Steam is blown off until the pressure drops to 4 bar, after which the valve is closed and the vessel is cooled until the pressure is 3 bar. Assuming that the expansion of the gas remaining in the vessel is isentropic during blow-off, calculate (a) the mass of steam blown off, (b) the dryness of the steam in the vessel after cooling, and (c) the heat transferred during the cooling process.

[2·601 kg, 0·736, −625 kJ.]

9.23 2 kg of steam is initially at a pressure of 3 bar. The initial dryness fraction is such that when the steam has been heated at constant volume to 4 bar it is saturated. After this heating process the steam is compressed isentropically to 10 bar. Find the final temperature, and the heat and work transfers. Sketch the process on the *T–s* and *p–v* diagrams.

[243·7 °C, 960 kJ, −292 kJ.]

10.1 Air enters a centrifugal compressor at 1·05 bar and 15 °C, and leaves at 2 bar and 97 °C. The mass flow is 50 kg/min. What power is actually required to drive the compressor, and what would be the power required for the same pressure ratio if the compression had been frictionless?

What is the rate of increase of enthalpy due to friction?

[68·7 kW, 48·6 kW, 20·1 kW]

10.2 8 kg of air per minute enter a nozzle with negligible velocity and expand from a pressure of 4 bar to 2·2 bar. The temperature falls from 900 °C to 750 °C in the process. Find the velocity of the air as it leaves the nozzle, the velocity which would have been reached had the expansion been frictionless, and the nozzle efficiency.

Indicate briefly, with the aid of a sketch of the *T–s* diagram, two ways in which the rate of change of entropy during the real expansion may be calculated and find the change of entropy per minute by one of these methods.

[549 m/s, 608 m/s, 0·815, 0·268 kJ/K min.]

10.3 Air is flowing through a well lagged duct. At some section A the pressure and temperature are 1·5 bar and 187 °C, and at some distance along the duct at section B the pressure and temperature are 1·3 bar and 160 °C. In which direction is the air flowing?

[From B to A.]

615

10.4 18 kg of air per second enter a turbine at 800 °C with a velocity of 100 m/s. The air expands adiabatically, but not isentropically, as it passes through the turbine, and leaves with a velocity of 150 m/s. It then passes through a diffuser wherein the velocity is reduced to a negligible value and the pressure is increased to 1·01 bar. The air is exhausted to atmosphere at this pressure. If the process in the diffuser can be assumed to be isentropic, and the turbine produces 3600 kW, find the pressure of the air between the turbine and the diffuser.

Say, by referring to a sketch of the processes on the *T–s* diagram, why you think a diffuser is added to the turbine.

[0·966 bar.]

10.5 A centrifugal compressor delivers 9 kg of air per second at a pressure ratio of 4:1. The inlet pressure and temperature of the air are 1·01 bar and 15 °C, and the outlet temperature is 177 °C. If the velocity of the air at inlet to the compressor is 120 m/s and at outlet is 60 m/s, find the power absorbed by the compressor.

The compressed air is heated at constant pressure and allowed to expand in a nozzle to 1·01 bar. Neglecting any change in velocity between compressor outlet and nozzle inlet, find the rate of heat transfer necessary for a velocity of 760 m/s to be produced at the nozzle exit. Assume that the expansion in the nozzle is isentropic.

[1417 kW, 3826 kW.]

10.6 Air at 0·83 bar and 5 °C enters the intake (i.e. diffuser) of a jet engine with a velocity of 270 m/s. The air decelerates reversibly in the intake to a negligible velocity before entering the compressor. Calculate the pressure of the air at inlet to the compressor (a) if the process can be assumed adiabatic and (b) if heat is transferred at such a rate that the temperature remains at 5 °C throughout the process.

[1·27 bar, 1·31 bar.]

10.7 Say briefly why 'enthalpy' is a useful concept when dealing with open systems.

Air at atmospheric pressure and 15 °C enters a duct with a velocity of 270 m/s. The duct at first diverges so that the velocity of the air is reduced to a negligible value, and the pressure and temperature are consequently increased. The air then passes into a length of duct where heat is supplied at constant pressure raising the temperature still further to 1000 °C. At this temperature, and still with negligible velocity, the air enters a nozzle wherein it expands to atmospheric pressure. Assuming that the processes in the diffuser and nozzle are isentropic, find (a) the heat supplied per unit mass flow, and (b) the velocity of the air leaving the nozzle.

[953 kJ/kg, 534 m/s.]

10.8 A diffuser reduces the velocity of an air stream from 300 to 30 m/s. If the inlet pressure and temperature are 1·01 bar and 315 °C and the diffuser efficiency is 90 per cent, determine the outlet pressure. Find also the outlet area required for the diffuser to pass a mass flow of 9 kg/s.

[1·271 bar, 0·428 m².]

10.9 The dryness fraction of steam is to be determined by passing a sample first through a separating calorimeter, which merely separates some of the liquid physically from the vapour, and then through a throttling calorimeter. The following results have been noted:

steam main pressure . . . 10 bar
pressure after throttle . . . 1 bar
temperature after throttle . . 128 °C
mass collected in separator . . 0·86 kg
mass collected after throttling . . 11·6 kg

Find the dryness fraction of the sample.

What is the minimum dryness fraction of steam at 10 bar that could be measured when using the throttling calorimeter alone with an outlet pressure of 1 bar?

[0·91, 0·95.]

10.10 Steam enters a turbine with a negligible velocity at 40 bar and 410 °C. It leaves the turbine with a high velocity through a pipe of 0·16 m² cross-sectional area where the pressure is 0·14 bar and the dryness fraction is 0·93. The turbine output is 2000 kW. Deduce an equation for the outlet velocity and verify that it is approximately 156 m/s. Finally, show that the expansion in the turbine is irreversible.

10.11 9 kg of steam per second, initially at 40 bar and 300 °C, expands adiabatically and irreversibly in a turbine to 4 bar. Determine the power produced if the index of expansion is 1·15, and also the isentropic efficiency of the turbine.

[3150 kW, 0·785.]

10.12 A turbine is normally supplied with steam at 40 bar and 300 °C, but to accommodate a reduction in load the steam has been throttled to 7 bar before it enters the turbine. Determine the percentage reduction in power output, assuming that the turbine exhaust pressure remains unchanged at 0·34 bar and that the turbine efficiency and mass flow remain unchanged.

[32·4 per cent.]

10.13 A lagged vessel of 1·4 m³ capacity is full of steam at a pressure of 14 bar and 0·85 dryness fraction. Water at 15 °C is injected under a constant head of 150 m. How much water will it be necessary to inject to reduce the pressure in the vessel to 7 bar, and what will be the final dryness fraction?

[14·9 kg, 0·196.]

10.14 A compressed air bottle, containing air at 34 bar and 15 °C, is to be used to drive a small turbine as an emergency starter for an engine. The turbine is required to produce an average output of 4 kW for a period of 30 seconds. Assuming that the expansion is isentropic, that the pressure in the bottle is allowed to fall to 3·4 bar during the process, and that the turbine exhausts to atmosphere at 1·01 bar, determine the necessary capacity of the bottle.

[0·029 m³.]

11.1 A steam turbine plant operates on the Rankine cycle with a boiler pressure of 50 bar and a condenser pressure of 0·14 bar. The steam leaves the boiler at 400 °C. Draw up a complete energy balance for the ideal plant, including the feed pump term, and find the cycle efficiency and the work ratio. Sketch the cycle on the T–s diagram. Show how the cycle would appear if the processes in the turbine and feed pump were not reversible, and hence explain why the cycle efficiency would be lower in practice.

[0·352, 0·995.]

11.2 Neglecting the feed pump term, compare the Rankine cycle efficiency of a high-pressure steam plant operating at 80 bar with that of a low-pressure plant at 40 bar. In both cases the maximum temperature is 400 °C and the condenser pressure is 0·07 bar. Explain your results briefly.

Also find the ratio of the capacities of the condensing plant (i.e. ratios of heat rejection), assuming that the turbines are to produce the same power.

[0·391, 0·364; 0·893 (H.P./L.P.).]

11.3 A steam turbine plant is throttle governed and operates with a boiler pressure and temperature of 40 bar and 350 °C, and a condenser pressure of 0·2 bar. The isentropic efficiency of the turbine may be assumed constant at 0·80 for all operating conditions. Plot curves showing the variation of cycle efficiency and specific steam consumption with turbine inlet pressure, as the steam is throttled from 40 bar to 10 bar. The h–s chart should be used, and the feed pump term may be neglected.

$$\begin{bmatrix} p & 40 & 30 & 20 & 10 & \text{bar} \\ \eta & 0·261 & 0·249 & 0·232 & 0·203 & \\ \text{s.s.c.} & 4·85 & 5·08 & 5·45 & 6·23 & \text{kg/kWh.} \end{bmatrix}$$

11.4 A reheat cycle works between pressures of 120 and 0·035 bar. The steam is superheated to 570 °C, and after expansion to the dry saturated state is reheated to 500 °C. Calculate the ideal thermal efficiency of the cycle taking account of the feed pump work.

[0·453.]

11.5 A steam turbine is supplied with steam at 40 bar and 450 °C, and exhausts at 0·035 bar. Neglecting the feed pump term, find the ideal Rankine cycle efficiency of the plant, and compare it with the ideal efficiency which could be obtained if regenerative feed heating were employed. Assume the use of open heaters and (a) one bleeding point at 3 bar, (b) three bleeding points at 15, 3 and 0·5 bar.

[0·391, (a) (0·173 kg) 0·415, (b) (0·114, 0·083, 0·080 kg) 0·429.]

11.6 Calculate the efficiency of a regenerative cycle using a single feed water heater, neglecting the feed pump term. The steam leaves the boiler at 30 bar superheated to 450 °C, and the condenser pressure is 0·035 bar. The heater is of the *closed* type, and the bled steam is flashed into the condenser.

[(Bleeding pressure 2·7 bar, 0·165 kg) 0·401.]

11.7 A steam turbine plant works between the limits of 20 bar and 300 °C, and 0·035 bar. Compare the following three schemes:

(a) A simple Rankine cycle.

(b) A reheat cycle, with the steam reheated to 300 °C at the pressure where it becomes saturated.

(c) A regenerative cycle, with one bleed point (open heater) at the pressure where the steam becomes saturated.

Assume in all cases that the isentropic *stage* efficiency of the turbines is 0·71. For each scheme find the cycle efficiency, the work done, the heat rejected per kg of steam leaving the boiler, and the final dryness fraction. (N.B. To solve the problem it is necessary to draw the line of condition on the *h–s* diagram—see section 19.2 (v). In drawing the line, select about ten stages of about equal work output for the complete expansion; the exact number chosen will not affect the position of the line of condition significantly, provided that the number is not too small.)

$$\begin{bmatrix} \text{(a) } 0\cdot260, \ 756 \text{ kJ/kg, } 2157 \text{ kJ/kg, } 0\cdot885; \\ \text{(b) } 0\cdot268, \ 874 \text{ kJ/kg, } 2392 \text{ kJ/kg, } 0\cdot981; \\ \text{(c) } 0\cdot275, \ 684 \text{ kJ/kg, } 1808 \text{ kJ/kg, } 0\cdot885. \end{bmatrix}$$

11.8 Saturated steam at 30 bar enters a high-pressure turbine and expands isentropically to a pressure at which its dryness fraction is 0·841. At this pressure the moisture is extracted and returned to the boiler via a feed pump. The remainder, assumed to be dry steam, is expanded isentropically to 0·04 bar in a low-pressure turbine, and the condensate is returned to the boiler via a second feed pump. Calculate the ideal cycle efficiency, taking into account the feed pump terms.

Calculate the cycle efficiency when the isentropic efficiency of each turbine and feed pump is 0·80. What is then the dryness fraction at the exit of each turbine? Compare results with those of Example 11.3.

[0·357; 0·291, 0·880 and 0·871.]

11.9 The mercury cycle of a binary vapour plant works between pressures of 10 bar and 0·08 bar and the mercury enters the turbine dry saturated. The feed water, pre-heated to boiling point in an economiser, is evaporated in the mercury condenser, and the temperature difference between mercury and steam must be about 16 K. After evaporation the steam is superheated to 450 °C and expanded to a condenser pressure corresponding to about 33 °C. Calculate the correct mercury to steam flow ratio and the cycle efficiency.

[8·43, 0·504.]

11.10 A Rankine cycle operates with steam between limits of 40 bar and 450 °C, and 0·035 bar. The cycle is topped with a mercury cycle that operates between the limits of 12 bar (saturated at turbine entry) and 0·2 bar. The isentropic turbine efficiencies are 0·75 in both cycles. The heat to the steam economiser and superheater is supplied externally. Calculate the mercury to steam flow ratio and the cycle efficiency.

[6·88, 0·389.]

11.11 A factory requires 2250 kW from its electric motors for driving machinery, and 15 000 kW at a temperature of 139 °C from process steam. Two alternative schemes are considered:

(a) The power is taken from the main electricity grid with an overall thermal efficiency from fuel to motors of 20 per cent. The process steam is raised in a special boiler plant with an overall efficiency of 80 per cent.

(b) The power is generated in the factory's steam power plant; the initial steam conditions are 27 bar and 320 °C, and the condenser temperature is 29 °C. The overall efficiency from turbine rotor output to electric motor output is 70 per cent; the internal isentropic efficiency of the turbine is 80 per cent; the boiler efficiency is 85 per cent. Process steam is obtained by bleeding off the required amount of steam from some intermediate stage of the turbine. (Assume a straight line of condition for the turbine on the *h–s* diagram.)

Estimate the relative merits of the two schemes from the point of view of fuel economy.

[Scheme (b) 19·0 per cent saving over scheme (a).]

11.12 A factory requires power at the rate of 3750 kW and heat from process steam at 107 °C at the rate of 5500 kW. Assuming that the boiler efficiency is 0·9 and the turbine overall isentropic efficiency is 0·75, compare the heat requirements for the following two schemes;

(a) A single boiler generates steam at 20 bar and 320 °C supplying a turbine which exhausts into a condenser at 0·07 bar. The process steam is passed out at the required pressure, and after condensation it is returned to the boiler. (Assume a straight line of condition on the *h–s* diagram for the turbine.)

(b) A boiler supplies a turbine as in (a), and a separate process steam boiler generates steam at 107 °C.

[(a) 19 900 kW, (b) 23 170 kW.]

12.1 In a gas turbine plant, air is compressed from 1·01 bar and 15 °C through a pressure ratio of 4:1. It is then heated to 650 °C in a combustion chamber, and expanded back to atmospheric pressure. Calculate the cycle efficiency and work ratio if a perfect heat exchanger is employed. The isentropic efficiencies of the turbine and compressor are 0·85 and 0·80 respectively.

[0·319, 0·319.]

12.2 In a gas turbine plant, air is compressed through a pressure ratio of 6:1 from 15 °C. It is then heated to the maximum permissible temperature of 750 °C and expanded in two stages each of expansion ratio $\sqrt{6}$, the air being reheated between the stages to 750 °C. A heat exchanger allows the heating of the compressed gases through 75 per cent of the maximum range possible. Calculate the cycle efficiency, the work ratio and the work per kg of air. The isentropic efficiencies of the compressor and turbine are 0·80 and 0·85 respectively.

[0·324, 0·385, 152 kJ/kg.]

12.3 In a gas turbine plant, air at 10 °C and 1·01 bar is compressed through a pressure ratio of 4:1. In a heat exchanger and combustion chamber the air is heated to 700 °C while its pressure drops 0·14 bar. After expansion through the turbine the air passes through a heat exchanger which cools the air through 75 per cent of the maximum range possible, while the pressure drops 0·14 bar, and the air is finally exhausted to atmosphere. The isentropic efficiency of the compressor is 0·80, and that of the turbine 0·85. Calculate the efficiency of the plant.

[0·228.]

12.4 In a gas turbine, air at 15 °C is compressed from 1·01 bar through a pressure ratio of 6:1 after which it is heated to a temperature of 750 °C. It is then expanded in a turbine to atmospheric pressure. Assume the following:

  isentropic efficiency of compressor.   . . 0·80
  isentropic efficiency of turbine   . . 0·85
  c_p in compressor . . . . . 1·005 kJ/kg K
  c_p in combustion chamber and turbine . 1·130 kJ/kg K
  pressure loss in combustion chamber . 0·14 bar
  the gas constant for the products is the same as that for air.

Calculate the cycle efficiency and work ratio.
  [0·204, 0·320.]

12.5 A closed cycle gas turbine plant comprises a compressor, a heat exchanger, a heater, a two-stage turbine with reheater, and a cooler. The maximum and minimum pressures and temperatures in the cycle are 30 bar and 570 °C, 7·5 bar and 15 °C. The pressure in the re-heater is 15 bar. Sketch the layout of the plant and indicate pressures and temperatures between the components if (a) the heat exchanger is used, (b) the heat exchanger is by-passed. Calculate the ideal cycle efficiencies and work ratios in both cases. The gas used in the circuit is helium, which has a molecular weight of 4 and a specific heat $c_v = 1·5R$.
  [(a) 0·478, 0·478; (b) 0·359, 0·478.]

12.6 In an air-standard cycle heat is supplied at constant volume to raise the temperature of the air from T_1 to T_2. The air then expands isentropically until its temperature falls to T_1; and after this it is returned to its original state by a reversible isothermal compression. Show that the cycle efficiency is

$$\eta = 1 - \frac{T_1}{(T_2 - T_1)} \ln \frac{T_2}{T_1}$$

Calculate the efficiency when the pressure rises from 10 to 35 bar during the supply of heat.
  Explain briefly (a) why you would expect the efficiency of this cycle to be less than the Carnot efficiency, and (b) why this cycle would not be used in practice.
  [0·498.]

12.7 An air-standard cycle consists of the following processes: (i) Isentropic compression from 15 °C, 1·01 bar through a compression ratio of 5:1. (ii) Heat addition at constant volume of 2600 kJ/kg. (iii) Isentropic expansion to the initial volume. (iv) Heat rejection at constant volume. Sketch the cycle on p–v and T–s diagrams, and calculate its ideal efficiency, mean effective pressure and peak pressure.
  [0·475, 18·9 bar, 73·0 bar.]

12.8 The cycle in problem 12.7 is modified so that the heat is added (a) at constant pressure, (b) half at constant volume and half at constant pressure. The compression ratio is increased so as to allow the same peak pressure as before. Sketch the two cycles on p–v and T–s diagrams, and calculate their ideal efficiencies and mean effective pressures.
  [(a) $r_v = 21·3$, 0·593, 19·8 bar; (b) $r_v = 8·37$, 0·552, 19·9 bar.]

12.9 In a Carnot cycle 1 kg of air is compressed isothermally from 1·0 to 4·0 bar, the temperature being 10 °C. The maximum temperature reached in the cycle is 400 °C. Find the efficiency and mean effective pressure of the cycle.

[0·579, 1·96 bar.]

12.10 Compare the net work outputs of a Carnot cycle and Ericsson cycle. The following conditions are common to both cycles: air is the working fluid; the maximum and minimum temperatures are 800 °C and 60 °C; and the ratio of the maximum volume to the minimum volume is 25.

Explain briefly why the use of a perfect regenerator can bring the efficiency of the Ericsson cycle up to the Carnot efficiency whereas it cannot do this for the Joule cycle.

[62·8 and 435 kJ/kg.]

12.11 An ideal heat engine cycle consists of the following processes. Air initially at 1·01 bar and 15 °C is compressed to 20 bar according to the law $pv^{1·2}$ = constant; heated at constant pressure until the temperature is 800 °C; expanded isentropically to its initial volume; and finally cooled at constant volume until the pressure is again 1·01 bar. Find the efficiency and work ratio of the cycle.

[0·465, 0·512.]

13.1 A vapour compression refrigerator using Freon-12 works between temperature limits of −5 °C and 40 °C. The refrigerant leaves the compressor dry saturated. Calculate the refrigeration effect and coefficient of performance if (a) the refrigerant leaves the condenser saturated, and (b) the refrigerant is subcooled to 20 °C before entering the throttle valve.

[(a) 106·3 kJ/kg, 4·78; (b) 126·05 kJ/kg, 5·66.]

13.2 In a refrigerator, Freon-12 is compressed isentropically from a saturated state at −5 °C to a pressure of 10·84 bar. The refrigerant is then cooled at constant pressure to 25 °C, and is throttled down to a temperature of −5 °C at which it is evaporated. To calculate the temperature and enthalpy after compression (a) use the columns in the tables, (b) assume that the specific heat of the superheated vapour is $c_p = 0·791$ kJ/kg K. Sketch the cycle on T–s and p–h diagrams and calculate the coefficient of performance.

[4·96, 4·98.]

13.3 The working fluid in a heat pump installation is ammonia. The ammonia, after evaporation to a dry saturated state at 2 °C, is compressed to a pressure of 12·37 bar at which it is cooled and condensed to a saturated liquid state. It then passes through a throttle valve and returns to the evaporator. Sketch the cycle on T–s and p–h diagrams and calculate the coefficient of performance, assuming that the isentropic efficiency of compression is 0·85.

Describe briefly the mode of operation of an absorption refrigerator, and discuss its advantages and disadvantages *vis-à-vis* the compression type.

[7·80.]

13.4 Background heating for a house is provided by a heat pump which, by virtue of its refrigeration capacity, also services a cold room used as a larder. The heating required is

2·5 kW, which is provided by hot water circulating through the heat pump condenser, and the evaporator takes its heat from the cold room. The heat pump/refrigerator uses Refrigerant-12 and operates with a condenser pressure of 20·88 bar and an evaporator pressure of 2·61 bar. The refrigerant enters the compressor, which has an isentropic efficiency of 0·85, with 5 K superheat, and enters the throttle valve as a saturated liquid. Calculate the mass flow of refrigerant and rate of heat extraction from the cold room.

The heat pump is driven by an electric motor of 0·90 efficiency. Calculate the thermal advantage of the plant as a heat pump over that of direct heating by electricity. Finally, if electricity is produced from primary fuel with an efficiency of 0·35, what is the overall thermal advantage over that of heating the house by primary fuel?

[0·0208 kg/s, 1·568 kW; 2·41, 0·845.]

13.5 An ammonia refrigerator works between the temperature limits of $-4\,°C$ and $26\,°C$. After isentropic compression the refrigerant leaves the compressor saturated and is subcooled to $18\,°C$. Calculate the coefficient of performance. Illustrate the cycle on T–s and p–h diagrams. (Use the values given below only.)

t	p	h_f	h_{fg}	h_g	s_f	s_g
[°C]	[bar]	[kJ/kg]	[kJ/kg]	[kJ/kg]	[kJ/kg K]	[kJ/kg K]
−4	3·691	162·8	—	—	0·647	—
26	10·34	303·7	1162·8	—	1·140	—

The specific heat of liquid ammonia is $c_p = 2·68$ kJ/kg K.

[8·48.]

13.6 A heat pump using ammonia as a working fluid operates between temperature limits of $4\,°C$ and $50\,°C$ (saturation temperature in condenser). If the working fluid enters the compressor saturated, and is condensed to a saturated liquid, calculate the coefficient of performance of the ideal cycle.

If the heat pump is driven by a steam turbine, with inlet conditions 40 bar and $300\,°C$, calculate the ideal advantage of the combined turbine–heat pump over a direct heater to supply heat at $50\,°C$. Assume that (a) the turbine condenser works at $29\,°C$, and (b) the turbine condenser works at $50\,°C$ and the rejected heat is used. In either case a heat source at over $300\,°C$ is available.

[5·95, (a) 2·19:1, (b) 2·66:1.]

13.7 Water is drawn from a river at $7\,°C$ and has to be heated to $64\,°C$. Calculate the advantage of using the heat pump plant described below over direct heating of the water. Assume that a heat source above $450\,°C$ is available.

Freon-12 Heat Pump: Compressor driven by steam turbine. Condenser temperature $60\,°C$, and evaporator temperature $-5\,°C$. Freon-12 enters compressor as saturated vapour and enters throttle valve as saturated liquid.

Steam Turbine Plant: Steam pressure and temperature leaving boiler 30 bar and $450\,°C$ respectively. Condenser temperature $76\,°C$.

Take 100 per cent isentropic efficiencies for turbine and compressor, and neglect all heat losses. The water drawn from the river passes first through the Freon condenser and then through the steam condenser.

[1·86 : 1.]

13.8 An ammonia refrigeration cycle works between a condenser saturation temperature of 50 °C and an evaporator temperature of −50 °C. The liquid entering the throttle valve is saturated. The vapour entering the compressor has an entropy of 5·365 kJ/kg K, i.e. it will reach saturation at −2 °C. Calculate the refrigeration effect and coefficient of performance.

The cycle is modified as shown in Fig. 13.5. The temperature between the two stage throttle is −2 °C, and the saturated vapour bled off is mixed with the saturated vapour leaving the first stage compressor. Calculate the correct amount to be bled, the refrigeration effect, and the coefficient of performance, all quantities to be based on 1 kg passing through the condenser.

[774·2 kJ/kg, 1·58; 0·197 kg, 822·4 kJ/kg, 1·86.]

14.1 The analysis of air by volume was found to be N_2 78·09 per cent, O_2 20·95 per cent, Argon (molar mass 40) 0·96 per cent. Find the molar mass and gas constant of air. Also, find the values of c_p and c_v for air at 300 K. The values of c_p for the constituents at this temperature are:

N_2 1·040, O_2 0·918, Ar 0·5203 kJ/kg K
[28·95 kg/kmol, 0·287 kJ/kg K, 1·005 and 0·718 kJ/kg K.]

14.2 A mixture of 6 kg of oxygen and 9 kg of nitrogen has a pressure of 3 bar and a temperature of 20 °C. Determine for the mixture (a) the mole fraction of each component, (b) the molar mass, (c) the specific gas constant, (d) the volume and density, (e) the partial pressures and partial volumes.

[(a) 0·368, 0·632; (b) 29·48 kg/kmol; (c) 0·282 kJ/kg K; (d) 4·13 m³, 3·63 kg/m³; (e) 1·10 and 1·90 bar, 1·52 and 2·61 m³.]

14.3 A mixture of 2 kg of hydrogen and 4 kg of nitrogen is compressed in a cylinder so that the temperature rises from 22 °C to 150 °C. The mean values of c_p over this temperature range for the constituents are 14·45 (H_2) and 1·041 (N_2) kJ/kg K. Assuming that the process is reversible and that the polytropic index of compression is 1·2, find the heat transferred during the process. Find also the change in entropy of each constituent and of the mixture.

 [-3067 kJ; $-7\cdot564$ (H_2), $-1\cdot071$ (N_2), $-8\cdot635$ kJ/K.]

14.4 The products of combustion of a fuel have the following composition by volume:

 N_2 0·77, CO_2 0·13, O_2 0·10

How much heat must be transferred, per kg of mixture, to cool the gas from 820 °C to 100 °C in steady flow at constant velocity? The average molar heats at constant pressure of the constituents, over this temperature range, are

 N_2 31·08, CO_2 50·51, O_2 33·38 kJ/kmol K

 [-799 kJ.]

14.5 A vessel contains one kmol of oxygen and two kmol of nitrogen each at 100°C and 2 bar but separated by a wall. Calculate the total entropy of the system assuming that the entropy of each gas is zero at 0 °C and 1 atm.

 If the separating wall is removed so that the two gases can mix freely, calculate the final temperature and entropy of the mixture. Assume that there is no heat exchange with the surroundings, that the volume of the vessel is unchanged by the removal of the partition, and that $\gamma = 1\cdot40$ for oxygen and nitrogen.

 [10·27 kJ/K, 100 °C, 26·15 kJ/K.]

14.6 The figure illustrates an idealised machine for mixing gases reversibly. A volume V_1 of gas A is contained to the left of a piston which is permeable only to this gas, and a volume V_2 of gas B is contained to the right of a second piston permeable only to gas B. Both gases are at pressure p and temperature T, and initially the pistons are only an infinitesimal distance apart. To accomplish the mixing the pistons are allowed to move apart slowly, and the temperature of the surroundings is not allowed to differ by more than an infinitesimal amount from T. Show that the gases can be mixed reversibly in this manner and that the work done is given by

$$W = pV_2 \ln \frac{V_1 + V_2}{V_2} + pV_1 \ln \frac{V_1 + V_2}{V_1}$$

$$= R_0 T \left(n_2 \ln \frac{n_1 + n_2}{n_2} + n_1 \ln \frac{n_1 + n_2}{n_1} \right)$$

Gas A	Gas B
V_1, p, T	V_2, p, T

Also express the partial pressures of the gases in the mixture in terms of p, V_1 and V_2.

14.7 A closed vessel of 0·3 m³ capacity contains dry air at 1 bar and 31·0 °C. Water is allowed to drip into the vessel until the air is saturated. Find the mass of water added and the resultant pressure in the vessel, assuming that the temperature remains at 31·0 °C.

If the contents of the vessel are then raised to 65 °C and more water is admitted until the air is again saturated, what will be the total pressure in the vessel?

[0·009 62 kg, 1·045 bar, 1·362 bar.]

14.8 A closed vessel of 0·3 m³ contains 0·06 m³ of water, the remaining volume being occupied by a mixture of air and water vapour. The pressure in the vessel is 1 bar and the temperature 29·0 °C. Heat is then supplied to raise the temperature to 138·9 °C. Calculate the mass of water evaporated, the final pressure in the vessel and the amount of heat that must have been transferred. Use accurate values of v_f from p. 10 of the tables. (Note that although some of the water evaporates, the volume of vapour actually decreases because of the thermal expansion of the liquid remaining.)

[0·443 kg, 4·83 bar, 28 560 kJ.]

14.9 Air, just saturated with water vapour at 17·5 °C and at a total pressure of 1 bar, is compressed reversibly and isothermally in a cylinder to a total pressure of 2 bar. Assuming that the volume of any water present can be neglected, determine the change in entropy of the mixture per kg of dry air.

[−0·2561 kJ/K.]

14.10 A cylinder of 0·1 m³ capacity is filled with a mixture of air and water vapour at a temperature of 80 °C and a total pressure of 1·5 bar. The vapour has a dryness fraction of 0·9. The mixture is then compressed so that the condition of the vapour follows an isentropic path until the vapour is just dry saturated. Find the final volume and pressure of the mixture.

[10·0 bar.]

14.11 Wet steam enters a condenser at 0·07 bar and is condensed at the rate of 2200 kg/h. If the air leakage is 12 kg/h, and the temperature of both condensate and air extracted is 35 °C, find the mass of water vapour carried away by the air pump per hour. The partial pressure of the air at entry to the condenser may be neglected.

If the rate of flow and temperature rise of the cooling water are 800 kg/min and 20 K respectively, calculate the dryness fraction of the steam at inlet to the condenser.

[30·6 kg/h, 0·764.]

14.12 The temperature of the windows in a house on a day in winter is 5 °C. When the temperature in the room is 23 °C, and the barometric pressure is 74·88 cm Hg, what would be the maximum relative humidity that could be maintained in the room without condensation on the window panes? Under these conditions find the partial pressures of the water vapour and air, the specific humidity, and the density of the mixture.

[31·05 per cent, 0·008 72 and 0·9896 bar, 0·005 48 kg_s/kg_a, 1·172 kg/m³.]

14.13 20 kg/min of air at 1·012 bar, 30 °C and 80 per cent relative humidity, enter a mixing chamber and mix adiabatically with 8 kg/min of dry air at 1·012 bar and 5 °C. The mixture leaves the chamber at 0·97 bar. Determine the temperature and the specific and relative humidities of the mixture leaving the chamber.

[23·0 °C, 0·0153 kg_s/kg_a, 82·9 per cent.]

14.14 60 m³ of moist air is passed through a heating chamber per minute. It enters at 1 bar and 13 °C, with a relative humidity of 80 per cent, and leaves at 24 °C without change of pressure. Determine the final relative humidity of the air and the rate of heat transfer to the air

[0·402, 807 kJ/min.]

14.15 A fan discharges 300 m³ of moist air per minute at 36 °C, 1·10 bar and 92 per cent relative humidity. Find the specific humidity of the mixture, and the mass of water vapour and dry air passing through the fan per minute.

The moist air is passed over refrigerator coils in order to dehumidify it, and leaves the refrigerator chamber with a pressure and temperature of 0·98 bar and 7 °C. The condensate also leaves the chamber at 7 °C. Determine the final specific humidity of the mixture, the mass of water condensing on the cooling coils per minute, and the rate of heat transfer to the cooling coils.

[0·0325 kg$_s$/kg$_a$, 11·49 kg$_s$/min, 353·6 kg$_a$/min, 0·00642 kg$_s$/kg$_a$, 9·22 kg/min, −33 820 kJ/min.]

14.16 1000 kg/min of turbine condenser cooling water are sprayed into a natural draught cooling tower at a temperature of 26 °C, and the water returns to the condenser at 12 °C. Air is drawn into the tower at 15 °C, with a relative humidity of 55 per cent, and leaves the tower saturated at 24 °C. Calculate the mass flow of moist air into the tower and the rate of loss of cooling water by evaporation. It may be assumed that the total pressure throughout the tower is 1 atm.

[1411 kg/min, 18·31 kg/min.]

14.17 Water from a condensing plant is cooled in a forced-draught cooling tower. The draught is produced by a fan absorbing 4 kW as it induces an airflow of 110 m³/min at atmospheric conditions of 0·98 bar, 17 °C, and relative humidity 60 per cent. The air leaves the tower saturated and at a temperature of 30 °C. If the temperature of the water entering the tower is 50 °C and the rate of flow from the tower is 80 kg/min, find the temperature of the water leaving the tower and the rate of flow of water entering the tower.

[27·2 °C, 82·65 kg/min.]

15.1 A gaseous fuel has the following volumetric composition: CH$_4$ 0·38, H$_2$ 0·26, N$_2$ 0·06, CO 0·26, CO$_2$ 0·04. Find the stoichiometric air/fuel ratio by volume and the corresponding wet and dry volumetric analyses of the products of combustion.

[4·86; CO$_2$ 0·121, H$_2$O 0·182, N$_2$ 0·697; CO$_2$ 0·1485, N$_2$ 0·8515.]

15.2 Ethyl alcohol has the chemical formula C$_2$H$_6$O. Calculate the stoichiometric air/fuel ratio by mass and the corresponding wet volumetric analysis of the products of combustion.

Determine also the wet volumetric analysis of the products when 10 per cent excess air is supplied with the fuel.

[8·96; CO$_2$ 0·123, H$_2$O 0·184, N$_2$ 0·693; CO$_2$ 0·113, H$_2$O 0·169, O$_2$ 0·017, N$_2$ 0·701.]

15.3 Find the stoichiometric air/fuel ratio by mass for benzene (C$_6$H$_6$). In an engine test

627

using benzene the air/fuel ratio was observed to be 12:1. Estimate the wet gravimetric composition of the exhaust gases.

[13·2; CO_2 0·201, CO 0·038, H_2O 0·053, N_2 0·708.]

15.4 The composition by mass of a fuel oil is 0·857 carbon, 0·142 hydrogen and 0·001 incombustibles. Find the stoichiometric air/fuel ratio. If the dry exhaust gas analysis is found to be CO_2 0·1229, N_2 0·8395 and O_2 0·0376 by volume, determine the actual air/fuel ratio used.

[14·68; 17·82 by carbon balance.]

15.5 In an engine test the dry volumetric analysis of the products was CO_2 0·0527, O_2 0·1338 and N_2 0·8135. Assuming that the fuel is a pure hydrocarbon and that it is completely burnt, estimate the ratio of carbon to hydrogen in the fuel by mass, and the air/fuel ratio by mass.

[5·33; 39·6.]

15.6 The dry volumetric analysis of the flue gases during a boiler trial showed CO_2 0·062, O_2 0·139 and N_2 0·799. The ultimate analysis of the coal used was: C 0·80, H 0·055, O 0·065, Ash 0·08. Determine the excess air per kg of coal.

[20·6 kg by carbon balance.]

15.7 Find the net and gross calorific values at constant pressure per kg of mixture at 25 °C, for stoichiometric mixtures of (a) air and benzene vapour (C_6H_6), and (b) air and octane vapour (C_8H_{18}). The enthalpies of combustion at 25 °C are −3 169 500 kJ/kmol of C_6H_6, and −5 116 200 kJ/kmol of C_8H_{18}; both figures are for the case where the water in the products is in the vapour phase.

[(a) 2861 and 2981 kJ/kg; (b) 2796 and 3012 kJ/kg.]

15.8 A mixture, having a volumetric analysis of CO_2 0·10, CO 0·40 and air 0·50, is contained in a vessel of fixed capacity at a temperature of 0 °C. When the mixture is exploded by a spark estimate (a) the gravimetric composition of the products, and (b) the temperature they would reach, assuming that no dissociation takes place, that the process is adiabatic, and that the specific heats of the gases are constant. Take c_v as 1·045 (CO_2), 0·888 (CO), 0·830 (O_2) and 0·870 (N_2) kJ/kg K; and the *enthalpy* of combustion of CO at 25 °C as −10 107 kJ/kg.

[CO_2 0·4543, CO 0·1773, N_2 0·3684; 2069 °C.]

15.9 A cylinder contains a stoichiometric mixture of gaseous ethylene (C_2H_4) and air. The mixture occupies a volume of 14 litres and the pressure and temperature are 3·3 bar and 90 °C. The enthalpy of combustion of gaseous ethylene at 25 °C is −47 250 kJ/kg when all the water in the products is in the vapour phase. Determine the heat transferred when complete combustion takes place at constant pressure and the final temperature is 280 °C. The appropriate mean specific heats may be taken from tables.

Find also the temperature to which the products must be cooled for the water vapour to begin to condense. Assume that the total pressure in the cylinder remains constant at 3·3 bar.

[−123·1 kJ, 77·7 °C.]

15.10 Octane vapour (C_8H_{18}) at 25 °C is supplied to a combustion chamber with air at 25 °C. The combustion takes place adiabatically in steady flow, the velocities are negligible and no work is transferred. Assuming complete combustion, determine the air/fuel ratio by mass if the products temperature is to be 790 °C. Use appropriate mean specific heats taken from tables. The enthalpy of combustion of octane vapour at 25 °C is −5 116 200 kJ/kmol when all the water in the products is in the vapour phase.
[51·9.]

15.11 The number of kilomoles of products of combustion of a hydrocarbon fuel and air, when at 1 atm pressure and a certain temperature, are

$$9·01 \ CO_2 + 2·99 \ CO + 5·72 \ H_2O + 0·28 \ H_2 + 1·635 \ O_2 + 56·4 \ N_2.$$

Determine the equilibrium constants, based on pressures in atmospheres, for the reactions

$$CO + \tfrac{1}{2}O_2 \rightleftharpoons CO_2 \quad \text{and} \quad H_2 + \tfrac{1}{2}O_2 \rightleftharpoons H_2O.$$

Finally, set up the simultaneous equations necessary for determining the numbers of kmol of product when the mixture of products is compressed to 10 atm pressure without change of temperature.
[20·55 and 139·3 $atm^{-1/2}$.]

15.12 A mixture of carbon monoxide and air has 20 per cent more carbon monoxide than would be required for a stoichiometric mixture. It is compressed to a temperature and pressure of 550 K and 10 atm, and combustion then occurs at constant volume. If no heat is transferred to the surroundings, show that the temperature reached will be about 2975 K. The following data may be used:

At 2975 K and with pressures in atmospheres,

$$K = \frac{p_{CO_2}}{p_{CO}(p_{O_2})^{1/2}} = 3·35 \ (\text{atm})^{-1/2}$$

The internal energy of combustion of CO at 550 K is −281 380 kJ/kmol. The internal energies of the relevant gases are

	CO	CO_2	O_2	N_2
550 K	2850	6010	3080	2820 kJ/kmol
2975 K	67 880	126 570	72 360	67 080 kJ/kmol

15.13 Heptane (C_7H_{16}) is burnt with 10 per cent less than the stoichiometric amount of air. Determine the number of kilomoles of each product per kilomole of heptane burnt when the products temperature is 2000 K. At this temperature the water gas dissociation constant $(p_{H_2O})p_{CO})/p_{H_2})(p_{CO_2})$ is 4·259.
[$5·32CO_2 + 7·48H_2O + 1·68CO + 0·52H_2 + 37·24N_2$.]

16.1 Air is to be compressed through a pressure ratio of 10:1 from a pressure of 1 atm in one stage. Find the power required for a free air delivery of 3 m^3/min. Also find the rotational speed if the swept volume is 14·2 litres. The clearance volume is 6 per cent of the swept volume and $n = 1·3$ for compression and expansion.
[15·4 kW, 298 rev/min.]

16.2 Air is compressed in a reciprocating compressor from a pressure of 1 bar and a temperature of 15 °C to a pressure of 9 bar in one stage. The index of polytropic compression and expansion is 1·3 and the clearance volume is 4 per cent of the swept volume. Calculate the work done per kg of air delivered and the heat transferred during compression and expansion per kg of air.

[23·6 kJ, −57·4 kJ, 11·9 kJ.]

16.3 A two-stage compressor compresses air from 1 bar and 15 °C to 15·5 bar. Calculate the work done per kg of air delivered and the heat transferred through the cylinder walls and in the intercooler. Assume that the polytropic index of compression is 1·3, and state all other assumptions made in the calculation.

[−266·5 kJ, −25·6 kJ/kg, −107·5 kJ/kg.]

16.4 A three-stage, single-acting compressor is required to compress 135 m^3 of free air per hour from 1 bar to 64 bar. Prove that with complete intercooling the stage pressure ratio should be equal to the cube root of the overall pressure ratio for the required power to be a minimum. Calculate the power needed in this case, assuming $n = 1·3$.

If the mean piston speed is to be 140 m/min, calculate the piston areas, neglecting clearance. State clearly any assumptions made.

[−18·4 kW, 321, 80·2 and 20·1 cm^2.]

16.5 When a gas is compressed, the work done per unit mass flow may be given by

$$\text{(i)} \quad -(h_2 - h_1) \quad \text{or} \quad \text{(ii)} \quad -\int_1^2 v \, dp$$

State precisely in which cases (i) and in which cases (ii) can be applied.

Dry saturated steam at 1 bar is compressed isentropically to 7 bar. Calculate the work done using each of the above formulae, and explain any discrepancy in the results. (Use tables where necessary.)

[−421, −415 kJ/kg.]

16.6 Derive an expression by which the indicated power of an engine can be found from an indicator card, engine speed and engine dimensions.

In a test on a vertical double-acting steam engine the following information was obtained:

area of indicator cards $\begin{cases} \text{out-stroke (top)} \\ \text{in-stroke (bottom)} \end{cases}$		10·5 cm^2
		10·2 cm^2
length of indicator card	. . . .	7·6 cm
spring constant .	. . . .	0·8 bar per cm
speed	. . . . .	120 rev/min
piston diameter .	. . . .	21·5 cm
piston rod diameter .	. . . .	3·7 cm
stroke	. . . . .	30·5 cm

Calculate the indicated power. If the mechanical efficiency of the engine is 80 per cent at the given load, find the torque developed.

[4·76 kW, 303 N m.]

16.7 A single-acting steam engine has a swept volume of 50 litres and the cut-off occurs at 0·4 of the stroke. The steam in the main is saturated and at 13 bar, and the engine exhausts to a condenser at 0·7 bar. Neglecting the effects of clearance volume, and assuming that the expansion occurs according to $pV = $ constant, determine the percentage reduction in work output per machine cycle when the steam is throttled to 7 bar before it enters the steam chest. Find the isentropic efficiency for both the unthrottled and throttled condition.

[49·7 per cent; 72·3 per cent, 67·5 per cent.]

16.8 A reciprocating expander, supplied with air at 2 bar and 20 °C is to develop 1 kW. Assuming reversible polytropic expansion to atmospheric pressure of 0·98 bar, with $n = 1·2$, calculate the mass flow of air required.

If the expander is designed to allow reversible expansion in the cylinder to 0 °C followed by blow-down to atmosphere, calculate the quantity of air required to develop 1kW. The effect of clearance volume may be neglected. Also estimate the maximum relative humidity of the supply if condensation must not occur before blow-down.

[1·062 kg/min, 1·088 kg/min, 40 per cent.]

16.9 A rotary vane compressor has a free air delivery of 170 litres/min when it compresses air from 0·98 bar and 15 °C to 2·3 bar. Estimate the power required to drive the compressor when (a) the ports are so placed that there is no internal compression, (b) the ports are so placed that there is a 30 per cent reduction of volume before back-flow occurs. Assume adiabatic compression. What is the isentropic efficiency in each instance?

[(a) −0·373 kW, 0·718; (b) −0·285 kW, 0·938.]

17.1 The stoichiometric mixture of compressed air and octane of Example 17.1 is burnt adiabatically at constant volume. Find the maximum temperature and pressure reached. Take the internal energy of combustion at 25 °C as −5 124 800 kJ/kmol of octane. After combustion the products are expanded isentropically to the original volume. Estimate the final temperature and pressure before blow-down. Neglect dissociation, but allow for variable specific heats. Compare the results with those obtained assuming pure air with constant specific heats throughout the problem.

[3056 K, 68·6 bar; 1893 K, 6·07 bar. 4620 K, 98·0 bar; 2121 K, 6·43 bar.]

17.2 The equilibrium constant for the combustion equation

$$H_2 + \tfrac{1}{2}O_2 \leftrightharpoons H_2O \text{ (vap)}$$

is

$$K = \frac{p_{H_2O}}{p_{H_2}(p_{O_2})^{1/2}} = 10^{40·05} \text{ (atm)}^{-1/2}$$

at 25 °C. Calculate the maximum work that can be derived from this reaction in surroundings at 25 °C, if the reactants are initially at a pressure of 10 atm and the products leave at the same pressure.

[231 340 kJ/kmol of H_2.]

17.3 Calculate the maximum work that can be derived from a stoichiometric supply of air and CO, in surroundings at 25 °C. (a) The air and CO are supplied separately each at atmospheric pressure and the products leave as a mixture at atmospheric pressure. (b) The air and CO are supplied mixed at atmospheric pressure and the products leave mixed at atmospheric pressures. (This problem can be solved by visualising an appropriate van't Hoff box, similar to that discussed in sections 15.7 and 17·5 but with the addition of reversible mixing and unmixing processes achieved by the device described in problem 14.6.)

[(a) $(257\,100 + 1577)$ kJ/kmol of CO, (b) $(257\,100 - 3516)$ kJ/kmol of CO.]

18.1 Steam expands isentropically in a nozzle to 1·2 bar. The initial pressure and temperature are 6 bar and 300 °C, and the initial velocity is zero. Plot curves showing the variation with pressure along the nozzle of nozzle area, mass flow per unit area, velocity and specific volume. Work from first principles and do not assume a value for the index of isentropic expansion.

18.2 A boiler generates 200 kg of saturated steam per hour at 20 bar. Calculate the minimum free cross-sectional area required for a safety valve for the boiler.

[1·93 cm².]

18.3 Steam is supplied to a nozzle at 11 bar, 0·95 dryness fraction, and expands isentropically to 1 bar. Assuming the law of expansion to be pv^n = constant, determine the value of n from the initial and final conditions of the steam. Using this value calculate the pressure and velocity of the steam at the throat of the nozzle, and the mass flow per unit throat area.

[1·130, 6·37 bar, 449 m/s, 0·164 kg/s cm².]

18.4 Starting from the energy equation

$$c_p T_1 = c_p T_2 + \tfrac{1}{2} C_2^2$$

deduce equations (18.17) to (18.21) which describe the flow of a perfect gas through a nozzle when the initial velocity is zero.

18.5 Air at 4 bar and 15 °C flows from a reservoir through a convergent nozzle into a region of low pressure. Calculate the maximum possible rate of discharge per cm² of nozzle throat area, and state what the pressure downstream of the nozzle must be for the maximum discharge to be possible.

[0·0953 kg/s cm², 2·113 bar.]

18.6 Helium at 6 bar and 90 °C expands isentropically through a convergent-divergent nozzle to 1 bar. Determine (a) the mass flow through a throat area of 3 cm², (b) the exit area required for complete expansion and the corresponding exit velocity. The molecular weight of helium is 4 and $\gamma = 5/3$.

[9·03 kg/min, 3·99 cm², 1390 m/s.]

18.7 A convergent-divergent nozzle, having an exit area twice the throat area, is designed to accept air with negligible velocity. Calculate the design value of the pressure ratio of the nozzle assuming isentropic flow. Say what happens when the downstream pressure falls below the design value.

[0·0939.]

18.8 Steam initially saturated at 10 bar expands isentropically to 4 bar. If the steam is supersaturated throughout the expansion find (a) the 'undercooling' in K at the end of the expansion, and (b) the reduction in enthalpy drop due to supersaturation.
[50 K, 10 kJ/kg.]

18.9 Saturated steam at 13 bar enters a convergent-divergent nozzle with negligible velocity. Determine the throat area required to pass a flow of 60 kg/min assuming that the steam expands isentropically in stable equilibrium. What will be the mass flow passed by this nozzle when the steam expands isentropically in a supersaturated condition?
[5·37 cm², 63·2 kg/min.]

18.10 One method of determining the efficiency of a nozzle is to suspend it in such a way that the reactive force in the line of flow can be measured. In an experiment with a convergent-divergent steam nozzle the reactive force is found to be 360 N when initially dry saturated steam at 7 bar is expanded to 1 bar. The cross-sectional area of the throat is 4·5 cm² and the exit area is such that complete expansion is achieved under these conditions. Determine the efficiency of the nozzle, assuming that all the losses occur after the throat and that $n = 1·135$ for an isentropic expansion.

What will be the reactive force when the external pressure is reduced to 0·3 bar?
[0·92, 423 N.]

18.11 Air enters a duct of constant area with a pressure of 3·9 bar, a temperature of 120 °C, and a velocity of 100 m/s. Heat is transferred to the air at such a rate that at a second measuring station along the duct the pressure is 3·4 bar. If the flow were frictionless, what would be the temperature at this measuring section and the rate of heat transfer per unit mass flow?
[565 °C, 482 kJ s/kg.]

19.1 In a single stage impulse turbine the nozzles discharge the fluid on to the blades at an angle of 25° to the plane of rotation, and the fluid leaves the blades with an absolute velocity of 300 m/s at an angle of 120° to the direction of motion of the blades. If the blades have equal inlet and outlet angles and there is no axial thrust, estimate the blade angle, power produced per kg/s of fluid, and diagram efficiency.
[36·3°, 144 kW, 0·762.]

19.2 Steam at 7 bar and 300 °C expands to 3 bar in an impulse stage. The nozzle angle is 20°, the rotor blades have equal inlet and outlet angles, and the stage operates with the optimum blade speed ratio. Assuming that the isentropic efficiency of the nozzles is 0·9, and that the velocity at entry to the stage is negligible, deduce the blade angle used and the mass flow required for this stage to produce 75 kW.
[36°, 0·472 kg/s.]

19.3 The fixed and moving blades in a two-row, velocity-compounded, impulse stage have equal inlet and outlet angles. Show that when the relative velocity is equal at inlet and outlet of each row, the optimum blade speed ratio is equal to $0·25 \cos \alpha_1$ where α_1 is the nozzle angle. Show also that the diagram efficiency is $\cos^2 \alpha_1$ when the optimum blade speed ratio is used.

19.4 Power for auxiliaries on a rocket motor is to be provided by a two-row, velocity-compounded, impulse stage. The pressure ratio across the turbine is 20:1 and the gases, having $\gamma = 1 \cdot 25$ and $c_p = 2 \cdot 38$ kJ/kg K, enter the turbine at 800 °C. The inlet and outlet angles of the rotor blades are 30° in each row. Assuming isentropic flow, negligible inlet velocity to the turbine, and an axial leaving velocity, find the blade speed, power produced per kg/s of gas, and diagram efficiency.

[400 m/s, 1122 kW, 0·976.]

19.5 Steam is supplied to a turbine at 60 bar and 450 °C, and expands to 0·065 bar. The first one-fifth of the overall isentropic drop is to be absorbed in a two-row, velocity-compounded, impulse stage, the blade speed ratio being 0·22. The nozzle and guide blade outlet angles are 20°, and the blade angles at entry and exit of the first rotor row are equal.

The outlet blade angle of the second row is determined by the requirement that the axial thrust for the whole stage is to be zero. If the velocity coefficient of the nozzles is 0·98, and there is a 10 per cent drop in relative velocity in the fixed and moving blades, find the blade speed, and the blade angles for both fixed and moving blades.

[150·8 m/s, $\beta_1 = \beta_2 = 25 \cdot 4°$, $\alpha_2 = 35 \cdot 7°$, $\beta_3 = 35 \cdot 7°$, $\beta_4 = 51 \cdot 7°$.]

19.6 When a reaction turbine runs at 3600 rev/min the steam consumption is 36 000 kg/h. The steam at a certain stage is at 0·34 bar with a dryness fraction of 0·95, and the stage develops 950 kW. The axial velocity of flow is constant and equal to 0·72 of the blade velocity, and the rotor and stator blades are of the same section having an outlet angle of 20°. Estimate the mean diameter of the annulus, and the blade height, at this point in the turbine.

[0·951 m, 11·5 cm.]

19.7 Ten successive reaction stages of a gas turbine run at 30 rev/s and have the same mean diameter of 1·5 m. All blade sections have inlet and outlet angles of 90° and 20° respectively, and the axial velocity is constant. Estimate the power output for a mass flow of 55 kg/s.

At inlet to the turbine the pressure and temperature are 7 bar and 700 °C, and at outlet the pressure is 2·5 bar. Find the isentropic efficiency of the turbine, and the blade heights of the first and last rows. Take $c_p = 1 \cdot 15$ kJ/kg K and $\gamma = 1 \cdot 333$ for the gas.

[11 000 kW, 0·789, 9·05 cm, 20·8 cm.]

19.8 An impulse steam turbine operates with inlet conditions of 20 bar and 300 °C, and an outlet pressure of 2·6 bar. Suggest suitable interstage pressures if the pressure ratios of the stages are to be equal and simple convergent nozzles are to be used. Assuming that the isentropic efficiency of each stage is 0·7, draw the line of condition on the h–s diagram and find the overall isentropic efficiency and the reheat factor.

[0·721, 1·03.]

19.9 A reaction turbine is supplied with steam at 15 bar and 250 °C and exhausts at 0·2 bar. Assuming that the stage efficiency is 0·75 and that the reheat factor is 1·04, find the specific steam consumption.

If the turbine develops 19 000 kW at 1200 rev/min, calculate the blade height and rotor drum diameter at a point in the turbine where the pressure is 1 bar. The stator and rotor blades are of the same section with an outlet angle of 20°, the blade speed ratio is 0·7, and the blade height is one-twelfth of the drum diameter. Assume that the line of condition is straight on the *h–s* diagram.

[6·46 kg/kW h, 15·0 cm, 1·80 m.]

19.10 An axial-flow compressor stage has equal inlet and outlet velocities and a constant axial velocity through the stage. Using the nomenclature of Fig. 19.21, show that the degree of reaction is given by

$$\Lambda = \frac{C_a}{2U}(\tan \beta_0 + \tan \beta_1)$$

Hence show that for 50 per cent reaction $\beta_0 = \alpha_1$, and $\beta_1 = \alpha_0 = \alpha_2$.

19.11 An axial-flow air compressor of 50 per cent reaction design has blades with inlet and outlet angles of 45° and 10° respectively. The compressor is to produce a pressure ratio of 6:1 with an overall isentropic efficiency of 0·85 when the inlet temperature is 40 °C. The blade speed and axial velocity are constant throughout the compressor. Assuming a value of 200 m/s for the blade speed, find the number of stages required when the work done factor is (a) unity, and (b) 0·89 for all stages.

[9, 10.]

19.12 A ten-stage axial flow compressor provides an overall pressure ratio of 5:1 with an overall isentropic efficiency of 0·87, when the temperature of the air at inlet is 15 °C. The work is divided equally between the stages. A 50 per cent reaction design is used, with a blade speed of 220 m/s and a constant axial velocity through the compressor of 170 m/s. Estimate the blade angles.

[42·2° and 21·2°.]

19.13 A centrifugal compressor produces a pressure ratio of 4:1 with an isentropic efficiency of 0·80, when running at 16 000 rev/min and inducing air at 15 °C in an axial direction. Assuming that the whirl velocity of the air leaving the impeller is 0·9 times the impeller tip speed, find the impeller tip diameter.

[52·8 cm.]

21.1 The inner surface of a 23 cm brick wall of a furnace is kept at 820 °C, and it is found that the outer surface temperature is 170 °C. Calculate the heat loss per m² of wall area, given that the conductivity is 0·865 W/m K.

[2450 W/m².]

21.2 The conductivity of a wall of thickness x_{12} is given by the equation $k = k_0 + 2bt + 3ct^2$, where k_0 is the conductivity at temperature $t = 0$. Show that the steady heat flow through the wall, with surface temperatures t_1 and t_2, is given by

$$Q = -\{k_0 + b(t_2 + t_1) + c(t_2^2 + t_2t_1 + t_1^2)\}A\,\frac{(t_2 - t_1)}{x_{12}}$$

21.3 An insulating brick wall 23 cm thick and of conductivity 0.26 W/m K is added to the outside of the furnace wall of problem 21.1. Calculate the reduction in heat loss, and the brick interface and outer surface temperatures. Assume that the inner surface temperature is unchanged, that the surroundings are at $20\,°C$, and that the combined convection and radiation heat transfer coefficient for the outer surface is 11.9 W/m² K.

[73·6 per cent, 648 °C, 74 °C.]

21.4 Water at 65 °C flows through a 5 cm bore steel pipe 1 cm thick. The surroundings are at 20 °C. The pipe conductivity is 52 W/m K, and the surface heat transfer coefficients inside and outside are respectively 1136 W/m² K and 13 W/m² K. Calculate the heat loss per metre length of pipe and the surface temperatures.

An insulating layer 3 cm thick and of conductivity 0.17 W/m K is added to the outside of the pipe. Assuming the same inside heat transfer coefficient and an outside coefficient of 9.7 W/m² K, calculate the heat loss per metre length and the surface temperatures.

[126·3 W/m, 64·29 °C, 64·18 °C; 53·7 W/m, 64·70 °C, 64·64 °C, 33·55 °C.]

21.5 A pipe of outside diameter 9 cm is covered with two layers of insulating material each 3 cm thick, the conductivity of one material being four times that of the other. Show that the combined conductivity of the two layers is 28 per cent more when the better insulating material is put on the outside than when it is put on the inside.

21.6 A thin metal tube of radius r_1 is kept at a constant temperature T_1 and is lagged with a poor insulating material of conductivity k, external radius r. The temperature of the surroundings is T_2. Assuming a constant surface heat transfer coefficient h, find the radius r for maximum heat loss from the pipe. Explain the physical significance of the solutions you obtain.

$[r = k/h.]$

21.7 Heat is transferred from the inside of a hollow sphere having internal and external radii of r_1 and r_2 respectively. Show that the steady rate of heat conduction in terms of the surface temperatures is given by

$$Q = -\frac{4\pi k r_2 r_1}{r_2 - r_1}(t_2 - t_1)$$

21.8 A hollow cylindrical copper bar, having internal and external diameters of 1.3 and 5 cm respectively, carries a current density of 5000 A/cm². Show that the radial temperature distribution in the bar is governed by the differential equation

$$\frac{d^2 t}{dr^2} + \frac{1}{r}\frac{dt}{dr} = -\frac{q_g}{k}$$

provided that the electrical resistivity ($\rho = 2 \times 10^{-6}$ ohm cm) and the thermal conductivity ($k = 381$ W/m K) may be assumed independent of temperature. Hence show that

$$t = -\frac{q_g}{4k}r^2 + A \ln r + B$$

where A and B are constants of integration to be determined from the boundary conditions.

When the outer surface is maintained at 40 °C and no heat is removed through the central hole, find the position and value of the maximum temperature.

If the inner surface is cooled to 26 °C, with the outer still at 40 °C, find the position and value of the maximum temperature. Also calculate the heat removed internally and externally.

[55·4 °C at $r = 0.65$ cm; 41·9 °C at $r = 1.94$ cm; per m of tube: 52 200 W internally and 39 300 W externally.]

21.9 A square chimney has internal and external dimensions of 0·6 m and 1·2 m respectively. The internal surface is at 260 °C and the external surface at 36 °C. Using the relaxation method, plot isothermal surfaces in the chimney wall, and calculate the heat loss per m height. The thermal conductivity of the brick is 0·88 W/m K.

[1900 W.]

21.10 A rectangular aluminium bar of cross-section 6 cm × 8 cm carries a current density of 3354 A/cm². Find the temperature distribution throughout the cross-section, and hence calculate the heat lost to the surroundings per metre length of bar. The surface temperature of the bar is 40 °C.

Find the temperature distribution if the bar is hollow, 1 cm thick, and internally cooled to 50 °C. Assume that the total current carried and the external surface temperature are unchanged. Also calculate the total heat loss externally and internally per metre length of bar. (Electrical resistivity = 4×10^{-6} ohm cm, $k = 225$ W/m K.)

[(a) Maximum temperature 103 °C, 216 000 W/m; (b) Maximum temperature 56 °C, 432 000 W/m.]

21.11 Derive equation (21.25).

21.12 Air at a temperature of 95 °C flows from a pipe. A thermocouple made of 0.9 mm diameter manganin-constantan wire, initially at room temperature of 15 °C, is moved suddenly into the hot air stream. Estimate the time interval after which the thermocouple indicates (a) 55 °C, (b) 94 °C. Neglect the effect of heat conduction away from the thermocouple junction and the effect of radiation, and assume the temperature to be uniform across the wire. The convection heat transfer coefficient between stream and wire can be taken as 0·11 kW/m² K. The following are average properties for manganin and constantan, which are nearly the same: $k = 0.026$ kW/m K, $c_p = 0.42$ kJ/kg K, $\rho = 8640$ kg/m³.

[(a) 5·2 s, (b) 32·7 s.]

21.13 In Example 21.5 the fluid on the hot side of the wall is maintained at 700 °C, after the heating-up period, for a further period of two hours. Plot a curve of rate of heat transfer to the wall against time from which the heating requirements for the whole four hours can be estimated.

[79 400 kJ/m².]

21.14 A 12 mm thick steel plate is in good thermal contact with a thick aluminium plate, both plates being initially at a uniform temperature of 100 °C. The outer surface 1 of the steel plate is suddenly raised to 600 °C and maintained at this temperature. How long will it take for the contact temperature t_c to reach 180 °C, and what will the temperature be at section 3 at that time?

Assume that at any internal section n the 'future' temperature $t_{n,p+1}$ is given by equation (21.31). Use the space subdivisions shown in the figure, a Fourier number of $\frac{1}{2}$ in *one* of the materials, and first prove in any way you wish that for the contact plane

$$t_{c,p+1} = \frac{Fo_S Fo_A}{RFo_S + Fo_A}\left[2t_{2,p} + 2Rt_{3,p} + \left(\frac{1}{Fo_S} + \frac{R}{Fo_A} - 2 - 2R\right)t_{c,p}\right]$$

where $R = k_A \Delta x_S / k_S \Delta x_A$. For aluminium the conductivity $k_A = 235$ W/m K and the diffusivity $\alpha_A = 90 \times 10^{-6}$ m²/s; the values for steel are one-fifth and one-sixth of those for aluminium respectively.

[3·1 s, 130 °C.]

22.1 Assuming a linear velocity distribution $U/U_s = y/\delta$ for the laminar boundary layer of a fluid flowing past a flat plate, it can be shown that the boundary layer thickness is given by

$$\delta/x = (12/Re_x)^{1/2}$$

Using this result and the assumption that the temperature distribution is linear across a boundary layer of thickness δ_t, show that, for a plate of uniform temperature, $\delta_t/\delta = Pr^{-1/3}$. Also show that the local Nusselt number at a distance x along the plate is given by

$$Nu_x = 12^{-1/2}Re_x^{1/2}Pr^{1/3}$$

22.2 Assuming a fourth power series for the velocity profile of a laminar boundary layer of a fluid flowing past a flat plate, it can be shown that

$$\frac{U}{U_s} = 2\left(\frac{y}{\delta}\right) - 2\left(\frac{y}{\delta}\right)^3 + \left(\frac{y}{\delta}\right)^4 \quad \text{and} \quad \frac{\delta}{x} = \frac{5·84}{Re_x^{1/2}}$$

Using this result and the assumption of a fourth power series for the temperature profile across a boundary layer of thickness δ_t, show that for a plate of uniform temperature $\delta_t/\delta \approx Pr^{-1/3}$. Also show that the local Nusselt number Nu_x at a distance x along the plate is given by

$$Nu_x = 0·343 Re_x^{1/2}Pr^{1/3}$$

22.3 Given equation (22.46), derive equation (22.47).

22.4   Derive the heat flow equation of the boundary layer

$$\frac{d}{dx} \int_0^l (\theta_s - \theta) U \, dy = \alpha \left(\frac{d\theta}{dy}\right)_0$$

and apply this equation to 'slug' flow of a liquid metal along a plate of uniform temperature to find the thickness of the temperature boundary layer. θ would be the liquid metal temperature relative to the plate temperature. Assume that the temperature profile in the boundary layer can be described by an equation of the form

$$\theta = a \sin \left[b \left(\frac{y}{\delta_t}\right) + c \right]$$

where a, b and c are constants to be determined from the boundary conditions.

Hence prove that the local Nusselt number Nu_x is given by

$$Nu_x = \sqrt{\left(\frac{\pi - 2}{4}\right)} \sqrt{(Re_x Pr)} = 0.534 \sqrt{(Re_x Pr)}$$

It can be shown that if the velocity profile can be approximated by an equation of the form

$$U = d \sin \left[e \left(\frac{y}{\delta}\right) + f \right]$$

the velocity boundary layer thickness is then given by

$$\frac{\delta}{x} = \sqrt{\left[\frac{2\pi^2}{(4 - \pi) Re_x}\right]}$$

Show that for a liquid metal of $Pr = 0.01$ the temperature boundary layer thickness is approximately equal to 6δ.

22.5   Show that in fully developed flow in a tube the pressure drop is related to the friction factor by

$$\frac{\Delta p}{l} = \frac{4}{d} f \frac{1}{2} \rho U_m{}^2$$

Explain why you would expect in the entrance region (a) the value of f to be larger than the value corresponding to fully developed flow, viz. that given by equation (22.19) or (22.20), and (b) $\Delta p/l$ to be larger than that predicted from the above equation, even if the correct value of f were used.

22.6   Derive an expression for the ratio of heat transfer to power required to maintain the flow, in terms of the mean fluid velocity U_m and the mean temperature difference θ_m, for a tubular heat exchanger. Assume that the simple Reynolds analogy is valid. What deductions can you make from this expression?
   $[Q/W = c_p \theta_m / U_m{}^2.]$

22.7   It is proposed to use the modified Reynolds analogy (22.75) to predict the heat transfer from a cylinder rotating in still air with a turbulent boundary layer (critical $Re_d = \rho U_s d/\mu \approx 300$). Since U_s is the free stream velocity relative to the surface, U_s should be taken as the surface velocity of the cylinder. The shear stress for a rotating cylinder has

639

been found to be

$$\tau_w = f\frac{\rho U_s^2}{2}; \quad \text{where} \quad \frac{1}{\sqrt{f}} = -0 \cdot 6 + 4 \cdot 07 \log_{10}\left(\frac{\sqrt{f}}{2} Re_d\right)$$

It has also been shown that for fluids with Pr not too different from unity, $U_b/U_s \approx 11 \cdot 6\sqrt{(f/2)}$ in any turbulent boundary layer. Using these data, show that

$$Nu_d = \frac{f}{2} \frac{Re_d Pr}{1 + 11 \cdot 6\sqrt{(f/2)}(Pr - 1)}$$

A 5 cm diameter cylinder rotates with a speed of 960 rev/min in air 15 °C. Its surface temperature is 50 °C. Calculate the heat loss per metre length of cylinder.
[106·9 W.]

22.8 Show by dimensional analysis that in forced convection $Nu = \phi(Re, Pr, U^2/c_p T)$, when frictional heating in the fluid cannot be neglected.

22.9 Air at 260 °C flows at 12 m/s through a thin metal tube of 2·5 cm diameter in surroundings at 15 °C. At what rate will the tube lose heat? What would be the percentage reduction in heat transfer if 2·5 cm of lagging of thermal conductivity 0·173 W/m K were put on the tube? The film coefficient inside the tube can be calculated from equation (24.8), and the combined convection and radiation coefficient from the metal and lagging can be taken as 18·7 and 11·4 W/m^2 K respectively.
[246·1 W per metre length of tube, 41·4 per cent.]

22.10 0·3 m^3/min of air at 15 °C and 1 atm are to be heated to 260 °C while flowing through a 2 cm diameter tube which is maintained at a temperature of 480 °C. Calculate the length of tube required. Find the film coefficient from equation (24.8), and take as the mean tube-air temperature difference the arithmetic mean of the difference at inlet and outlet.
[0·872 m.]

22.11 Given equation (22.94), derive equation (22.95).

22.12 An electric convection heater consists of a vertical cylinder 7·5 cm diameter, 1·2 m long, enclosing an element of 100 W rating. Estimate the equilibrium surface temperature of the cylinder when placed in a room at 20 °C. The cylinder may be treated as a vertical flat wall, see equation (22.94). (Neglect radiation.)
[108 °C.]

22.13 A tube of 2·5 cm diameter and surface temperature of 50 °C is losing heat by free convection to air at 15 °C. In order to estimate the heat loss, model tests with a wire heated electrically to 270 °C are to be carried out in compressed air at 15 °C. The wire is to be 0·25 cm in diameter. Calculate the pressure that would be required, assuming that $Nu = \phi(Gr \times Pr)$ with properties evaluated at the film temperature.
[24·0 bar.]

22.14 A 2000 W immersion heater is placed in a 5 cm diameter sealed horizontal tube, and is used to heat water to 60 °C. If the surface temperature of the tube is not to exceed 95 °C, calculate the length of tube required. For free convection $Nu_d = 0.53(Gr_d \times Pr)^{1/4}$, with all properties taken at film temperature. The coefficient of cubical expansion of water may be taken as $6.21 \times 10^{-4}\,K^{-1}$.

[0·355 m.]

22.15 A flue insulated on the outside passes gases which, to all intents and purposes, have properties identical to those of air. The density of the gases in the flue is ρ and of the air outside is ρ_s, so that $(\rho_s - \rho) \approx \rho\beta\theta$. Assuming that the shear stress on the inside surface is given by equations (22.18) and (22.20), show that

$$Gr_d = 2 \times 0.0791(Re_d)^{7/4}$$

Calculate the velocity in a flue 6 cm in diameter when the inside gas temperature is 80 °C and the outside air temperature is 20 °C. Also, assuming the simple Reynolds analogy in the form $St = \frac{1}{2}f$, estimate the inside convection heat transfer coefficient.

[2·53 m/s, 10 W/m² K.]

22.16 Given equation (22.109), derive equation (22.110).

23.1 A gas fire consumes 1 m³ of gas an hour, of calorific value 17 600 kJ/m³. Its radiant surface, measuring 22 cm × 30 cm, has an effective 'black-body' temperature of 590 °C in surroundings at 20 °C. Calculate the radiant efficiency of the fire.

A plate of 900 cm² area is placed 3 m away from the fire:
(a) along a normal to the fire and parallel to the fire, (b) along a direction at 50° to the normal and parallel to the fire, and (c) as (b) but perpendicular to the line joining the fire and the plate. For each position calculate the radiant energy falling on the plate.

[41·9 per cent; (a) 6·62 W, (b) 2·74 W, (c) 4·26 W.]

23.2 An outside wall of a house consists of two brick walls separated by an air gap. Calculate the heat exchange by radiation across the gap, when the inner surfaces are at 13 °C and 7 °C. The emissivity of both surfaces is 0·9.

A thin aluminium sheet of emissivity 0·1 is placed in the air gap. With the inner brick surface temperature unchanged, estimate the radiation heat transfer across the gap.

[25·2, 1·525 W/m².]

24.1 680 kg/h of oil flows through a thin-walled, 1 cm diameter tube of an oil cooler; 550 kg/h of water flows in the opposite direction through the annular space formed by this tube and a larger tube. The oil, which enters the cooler at 150 °C, is required to leave at 70 °C when the cooling water enters at 10 °C. Calculate the length of tube required. Also calculate the length of tube required if the water were flowing in the same direction as the oil. (The film coefficient between oil and tube is 2·3 kW/m² K, and between water and tube 5·7 kW/m² K. The specific heat of the oil is 2·18 kJ/kg K.)

[8·73 m, 13·67 m.]

24.2 Steam condenses on the outside of a thin 2·5 cm diameter copper tube, 5·5 m long, the steam pressure being 0·07 bar. Water flows through the tube at the rate of 1·2 m/s entering at 12 °C and leaving at 24 °C. Calculate the heat transfer coefficient between steam and water. Assuming that the heat transfer coefficient between steam and tube is 17 kW/m^2 K, estimate the coefficient between water and tube.

[3·35, 4·17 kW/m^2 K.]

24.3 A heat exchanger consists of ten 2 m length pipes of 2·5 cm bore and 2·9 cm outer diameter, each running inside concentric pipes. Hot oil at the rate of 900 kg/h and initially at 250 °C passes counter-current to cold oil at the rate of 1350 kg/h and an initial temperature of 15 °C. Assuming an average overall heat transfer coefficient of 170 W/m^2 K and a constant specific heat of 1·68 kJ/kg K, calculate the outlet temperature of the oils.

[147·6 °C 83·3 °C.]

24.4 Derive equation (24.10).

24.5 Recalculate problems 21.1 and 21.3 ignoring the given colder surface temperature in the former and surface heat transfer coefficient in the latter. Take the free convection coefficient as $h_c = 1\cdot31(\theta/[K])^{1/3}$W/m^2 K, the radiation coefficient from equation (23.36) with $\varepsilon = 0\cdot8$, and the temperature of the surroundings as 20 °C in both cases.

[21.1—2445 W/m^2, 170 °C; 21.3—645 W/m^2 or 73·6 per cent saving, 648 °C, 78 °C.]

24.6 Recalculate problem 21.4, ignoring the given outside surface heat transfer coefficient. Take the free convection coefficient from Fig. 22.17, and the radiation coefficient from equation (23.36) with $\varepsilon = 0\cdot8$.

[Uninsulated: $Gr_d \approx 1\cdot6 \times 10^6$, 119·7 W/m, 64·33 °C, 64·21 °C. Insulated: $Gr_d \approx 3\cdot9 \times 10^6$, 52·7 W/m, 64·70 °C, 64·65 °C 34·10 °C.]

24.7 A large volume of air at 49 °C is separated from another large volume of air at 15 °C by a vertical 0·25 cm thick sheet of (a) aluminium of emissivity 0·1 and conductivity 173 W/m K, and (b) cardboard of emissivity 0·9 and conductivity 0·173 W/m K. The free convection heat transfer coefficient is given by $h_c = 1\cdot31(\theta/[K])^{1/3}$W/m^2 K. Compare the overall heat transfer coefficients with the two materials.

[(a) 2·006 (b) 4·270 W/m^2 K.]

24.8 The inner surface of a 1 cm thick metal wall is maintained at a uniform temperature of 120 °C. The combined convection and radiation heat transfer coefficient from the outer surface to surroundings at 15 °C is 57 W/m^2 K (irrespective of whether there are or are not fins on the surface). Estimate the heat flow through the wall for a strip of 1 m long by 0·018 m wide (1) if no fins are attached to the surface, (2) if rectangular 0·5 cm × 5 cm fins are spaced at 1·8 cm pitch on the outer surface. Take the metal to be (a) aluminium with $k = 225$ W/m K, (b) steel with $k = 54$ W/m K. Also estimate the fin root and tip temperatures.

(N.B. This is an 'open-ended' question. It would seem that a two-dimensional solution, e.g. by relaxation, needs to be obtained between the inner surface and the root of the fin. The labour involved would not be justified, and the following semi-intuitive approach could be adopted. Assume initially that the fin root temperature is 120 °C and calculate, as a first approximation, the heat flow Q_0 into the fin. Using this value and a pseudo-one-dimensional approach, estimate the temperature drop across the wall. The final answer obtained will depend on the sophistication adopted and the number of successive approximations used.)

[(a) (1) 107·5 W, (2) 616 W, $t_0 = 118$ °C, $t_i = 106$ °C; (b) (1) 106·6 W, (2) 490 W, $t_0 = 112·5$ °C, $t_i = 77$ °C.]

24.9 When deriving equation (24.19) for heat flow through the root of a fin it was assumed that the heat flow from the tip of the fin was negligible, i.e. $(d\theta/dx)_i = 0$. If the fin is short this assumption is invalid, and the correct boundary condition for the tip of the fin is $-kA(d\theta/dx)_i = -hA\theta_i$. With this boundary condition it can be shown that

$$Q_0 = mkA\theta_0 \frac{(h/mk) + \tanh ml}{1 + (h/mk) \tanh ml}$$

By differentiation show that fins will not increase the heat flow from the surface unless $1/h > b/2k$.

Appendix C   References

The books and papers mentioned in the text are arranged separately for the four parts. *Books:* the dates given refer to the latest known editions (not printings) at the time of going to press. *Periodicals:* the abbreviations used are taken from the *World List of Scientific Periodicals* (Butterworths, 1963–5). Bold figures refer to volumes, and ordinary figures to pages.

Part I

(1) BRAITHWAITE, R. B., *Scientific Explanation* (Cambridge U.P., 1953).
(2) BRIDGMAN, P. W., *The Nature of Thermodynamics* (Harvard U.P., 1941).
(3) BRIDGMAN, P. W., 'Condensed collection of thermodynamic formulas', *Min. Metall.*, *N.Y.*, **7** (1926) 149.
(4) CARNOT, S. N. L., *Réflexions sur la puissance motrice du feu at sur les machines propres à développer cette puissance* (Bachelier, 1824), transl. Thurston, R. H. (*Am. Soc. mech. Engrs*, 1943).
(5) GLASSTONE, S., *Thermodynamics for Chemists* (Van Nostrand, 1947).
(6) KEENAN, J. H., *Thermodynamics* (Wiley, 1941).
(7) PLANCK, M., *Treatise on Thermodynamics*, transl. from 7th German ed. (Longmans, 1927).
(8) ZEMANSKY, M. W., *Heat and Thermodynamics* (McGraw-Hill, 1968).
(9) GILLESPIE, L. J., and COE, J. R., 'The heat expansion of a gas of varying mass', *J. chem. Phys.* **1** (1933) 103–13.
(10 COOK, S. S., 'Thermodynamics in the making', *Trans. NE. Cst Instn Engrs Shipbldrs* **65** (1948–9) 65–84.
(11) KEENAN, J. H., and HATSOPOULOS, G. N., *Principles of General Thermodynamics* (Wiley, 1965).
(12) HORLOCK, J. H., 'The rational efficiency of power plants with external combustion', *Thermodynamics and Fluid Mechanics Convention, Cambridge April 1964, Proc. Instn mech. Engrs* **178** Pt. 3 I(v) (1963–4) 43–52.
(13) BRUGES, E. A., *Available Energy and the Second Law Analysis* (Academic Press, 1959).

Part II

(1) *NEL Steam Tables* 1964 (HMSO, 1964).
(2) CARNOT, S. N. L., *Réflexions sur la puissance motrice du feu at sur les machines propres à développer cette puissance* (Bachelier, 1824), transl. Thurston, R. H. (*Am. Soc. mech. Engrs*, 1943).

(3) COHEN, H., ROGERS, G. F. C., and SARAVANAMUTTOO, H. I. H., *Gas Turbine Theory* (Longmans, 1972).

(4) BETT, K. E., ROWLINSON, J. S., and SAVILLE, G., *Thermodynamics for Chemical Engineers* (Athlone Press, 1975).

(5) FABER, O., and KELL, J. R., *Heating and Air Conditioning of Buildings* (The Architectural Press, 1971).

(6) The CIBS psychrometric chart (Chartered Instn of Building Services, 1975).

(7) GEYER, E. M., and BRUGES, E. A., *Tables of Properties of Gases* (Longmans, 1948).

(8) GOODGER, E. M., *Petroleum and Performance* (Butterworths, 1953).

(9) HILSENRATH, J., *et al.*, *Tables of Thermal Properties of Gases* (Circ. U.S. natn. Bur. Stand. 564, 1955).

(10) HODGE, J., *Cycles and Performance Estimation* (Butterworths, 1955).

(11) KEENAN, J. H., *Thermodynamics* (Wiley, 1941).

(12) KEENAN, J. H., and KAYE, J., *Gas Tables* (Wiley, 1948).

(13) SCOTT, R. B., *Cryogenic Engineering* (Van Nostrand, 1959).

(14) KAUZMANN, W., *Kinetic Theory of Gases* (Benjamin, 1966).

(15) LYDERSEN, A. L., GREENKORN, R. A., and HOUGEN, O. A., 'Generalized thermodynamic properties of fluids', University of Wisconsin, Engineering Experiment Station, Report No. 4 (1955).

(16) MCADAMS, W. H., *Heat Transmission* (McGraw-Hill, 1954).

(17) ROGERS, G. F. C., and MAYHEW, Y. R., *Thermodynamic and Transport Properties of Fluids—SI units* (Basil Blackwell, 1980).

(18) *Modern Power Station Practice* 3 (Central Electricity Generating Board, 1963).

(19) SPIERS, H. M., *Technical Data on Fuel* (The British National Committee, World Power Conference, 1962).

(20) ZEMANSKY, M. W., *Heat and Thermodynamics* (McGraw-Hill, 1968).

(21) JAKOB, M., 'Steam research in Europe and in America', *Engineering, Lond.*, **132** (1931) 143–6, 518–21, 550–1, 651–3, 684–6, 707–9, 744–6, 800–4.

(22) *Temperature, its Measurement and Control in Science and Industry*, American Institute of Physics, **1**, **2** and **3** (Reinhold, 1941, 1955 and 1962), **4** (Instrument Society of America, 1973).

(23) *Code for Temperature Measurement*, B.S. 1041–1943 (Part 1); Parts 2.1, 2.3, 2.4, 2.5 and 2.7 (1964–72).

(24) WALKER, G., *Stirling-cycle Machines* (Oxford U.P., 1973).

(25) *The International Practical Temperature Scale of 1968*, National Physical Laboratory (HMSO, 1976).

(26) WITHERS, J. G., 'Gas turbine combustion efficiency calculations', *Aircr. Engng*, **22** (1950) 218–22.

(27) *Refrigerant Tables and Charts* (ASHRAE Handbook and Product Directory: 1977, Fundamentals Sect. III, Ch. 16 (ASHRAE, 1977).

(28) *Proceedings of the United Nations Conference on New Sources of Energy*, Rome August 1961, **4–6**, Solar energy (United Nations, 1964).

(29) *CIBS Guide*, Pt C 1-2 (Chartered Instn of Building Services, 1975).

(30) EGERTON, A., SAUNDERS, O. A., and SPALDING, D. B., 'The chemistry and physics of combustion', *Proceedings of the Joint Conference on Combustion*, London, October 1955, Institution of Mechanical Engineers, 14–43.

(31) *Measurement of Humidity*, National Physical Laboratory, Notes appl. Sci. No. 4 (HMSO, 1970).

(32) *Calibration of Temperature Measuring Instruments*, National Physical Laboratory, Notes Appl. Sci. No. 12 (HMSO, 1964).

(33) *Handbook of Chemistry and Physics* (Chemical Rubber Publishing Co., Cleveland, Ohio, revised frequently).

(34) WEIR, C. D., 'Optimisation of heater enthalpy rises in feed-heating trains', *Proc. Instn mech. Engrs*, **174** (1960) 769–96.

(35) WOOTTON, W. R., *Steam Cycles for Nuclear Power Plant* (Temple Press, 1958).

(36) ROSSINI, F. D., *et al.*, *Selected Values of Chemical Thermodynamic Properties* (Circ. U.S. natn. Bur. Stand., 500, 1952);
WAGMAN, D. D., *Selected Values of Chemical Thermodynamic Properties* (U.S. natn. Bur. Stand., Tech. Note 270, 1965).

(37) STULL, D. R., and PROPHET, H. (Eds), *Janaf Thermochemical Tables* (U.S.G.P.O., 1971, Supplements 1974, 1975, 1978).

(38) 'District heating combined with electricity generation in the United Kingdom', *Department of Environment Energy Paper* No. 20 (1977).

(39) WILSON, S. S., and RADWAN, M. S., 'Appropriate thermodynamics for heat engine analysis and design', *Int. J. Mech. Engng Ed.*, **5** (1977) 68–80.

Part III

(1) CHARYK, J. V., and SUMMERFIELD, M. (Eds), *High-speed Aerodynamics and Jet Propulsion* **1–12** (Oxford U.P., 1955–64).

(2) KEARTON, W. J., *Steam Turbine Theory and Practice* (Pitman, 1958).

(3) HOTTEL, H. C., WILLIAMS, G. L., and SATTERFIELD, C. N., *Thermodynamic Charts for Combustion Processes* (Wiley, 1949).

(4) RICARDO, H. R., and HEMPSON, J. G. G., *The High-speed Internal-combustion Engine* (Blackie, 1968).

(5) SHAPIRO, A. H., *The Dynamics and Thermodynamics of Compressible Fluid Flow* **1** and **2** (Ronald Press, 1953 and 1954).

(6) SHEPHERD, D. G., *Principles of Turbomachinery* (Macmillan, 1956).

(7) GYARMATHY, G., and MEYER, H., 'Spontaneous Condensation Phenomena' (in German), *ForschHft Ver. dt. Ing.* No. 508(1965) 1–48.

(8) BINNIE, A. M., and WOODS, M. W., 'The pressure distribution in a convergent-divergent steam nozzle', *Proc. Instn mech. Engrs*, **138** (1938) 229–66.

(9) LYSHOLM, A. J. R., 'A new rotary compressor', *Proc. Instn mech. Engrs*, **150** (1943) 11–16.

(10) MILLER, C. D., 'Relation between spark-ignition engine knock, detonation waves, and autoignition as shown by high-speed photography', *NACA Rep.* No. 855 (1946).

(11) LIEPMANN, H. W., and ROSHKO, A., *Elements of Gas Dynamics* (Wiley, 1957).

(12) SWAINE, J., 'Some design aspects of poppet-valve cylinder heads for spark-ignition liquid-cooled engines', *Proc. Auto. Div. Instn mech. Engrs* (1947-8) 51–9.

(13) AINLEY, D. G., 'An approximate method for the estimation of the design point efficiency of axial flow turbines', *Curr. Pap. aeronaut. Res. Coun.*, 30 (HMSO, 1950).

(14) RICARDO, H. R., 'Combustion in Diesel engines', *Proc. Instn mech. Engrs*, **162** (1950) 145–8.

(15) Symposium on Superchargers and Supercharging, *Proc. Auto. Div. Instn mech. Engrs* (1956–7) 217–85.

(16) DOWNS, D., and WHEELER, R. W., 'Recent developments in "knock" research', *Proc. Auto. Div. Instn mech. Engrs* (1951–2) 89–99.

(17) KESTIN, J., and OWCZAREK, J. A., 'The expression for work in a Root's blower', *Proc. Instn mech. Engrs*, **1B** (1952–3) 91–4.

(18) MAYHEW, Y. R., and ROGERS, G. F. C., 'One-dimensional irreversible gas flow in nozzles', *Engineering, Lond.*, **175** (1953) 355–8.

(19) BROEZE, J. J., *Combustion in Piston Engines* (De Technische Uitgeverij H. Stam N. V., Haarlem 1963).

(20) Symposium on Wet Steam, *Thermodynamics and Fluid Mechanics Convention, Liverpool April* 1966, *Proc. Instn mech. Engrs*, **180** Pt. 3J (1965–6).

(21) SCHMIDT, F. A. F., *The Internal Combustion Engine* (Chapman and Hall, 1965).

(22) ANSDALE, R. F., 'N.S.U.-Wankel engine', *Auto. Engr*, **50** (1960) 166–76.

(23) HESSE, W. J., and MUMFORD, N. V. S., *Jet Propulsion for Aerospace Applications* (Pitman, 1964).

(24) ANGRIST, S. W., *Direct Energy Conversion* (Allyn and Bacon, 1971).

(25) KAYE, J., and WELSH, J. A., *Direct Conversion of Heat to Electricity* (Wiley, 1960).

(26) MITCHELL, W., *Fuel Cells* (Academic Press, 1963).

(27) BENSON, R. S., and WHITEHOUSE, N. D., *Internal Combustion Engines* **1** and **2** (Pergamon, 1979).

Part IV

(1) ALLEN, D. N. DE G., *Relaxation Methods* (McGraw-Hill, 1954).

(2) BUTTERWORTH, D., *Introduction to Heat Transfer*, Engineering Design Guides No. 18 (Oxford U.P., 1977).

(3) CARSLAW, H. S., and JAEGER, J. C., *Conduction of Heat in Solids* (Oxford, Clarendon Press, 1959).

(4) DUNCAN, W. J., *Physical Similarity and Dimensional Analysis* (Edward Arnold, 1953).

(5) ROHENOW, W. M., and HARTNETT, J. P., (Eds) *Handbook of Heat Transfer* (McGraw-Hill, 1973).

(6) ECKERT, E. R. G., and DRAKE, R. M., *Heat and Mass Transfer* (McGraw-Hill, 1959).

(7) COLLIER, J. G., *Convective Boiling and Condensation* (McGraw-Hill, 1972).

(8) GOLDSTEIN, S., *Modern Developments in Fluid Dynamics* **1** and **2** (Oxford, Clarendon Press, 1938).

(9) INGERSOLL, L. R., ZOBEL, O. J., and INGERSOLL, A. C., *Heat Conduction* (University of Wisconsin Press, 1954).

(10) JAKOB, M., *Heat Transfer* **1** (Wiley, 1949).

(11) KERN, D. Q., *Process Heat Transfer* (McGraw-Hill, 1950).

(12) LANGHAAR, H. L., *Dimensional Analysis and Theory of Models* (Krieger, 1979).

(13) MCADAMS, W. H., *Heat Transmission* (McGraw-Hill, 1954).

(14) ROGERS, G. F. C., and MAYHEW, Y. R., *Thermodynamic and Transport Properties of Fluids—SI units* (Basil Blackwell, 1980).

(15) MAYHEW, Y. R., 'Numerical method for unsteady one-dimensional conduction' *Bull. Mech. Engg. Ed.*, **3** (1964) 63–74.

(16) PRANDTL, L., and TIETJENS, O. G., *Applied Hydro- and Aeromechanics* (McGraw-Hill, 1934).

(17) WIEBELT, J. A., *Engineering Radiation Heat Transfer* (Holt, Rinehart and Winston, 1966).

(18) SCHLICHTING, H., *Boundary Layer Theory* (McGraw-Hill, 1968).

(19) SCHNEIDER, P. J., *Conduction Heat Transfer* (Addison-Wesley, 1955).

(20) KESTIN, J., and RICHARDSON, P. D., 'Heat transfer across turbulent incompressible boundary layers', *Int. J. Heat Mass Transfer*, **6** (1963) 147–89.

(21) REYNOLDS, O., 'On the dynamical theory of incompressible viscous fluids and the determination of the criterion', *Phil. Trans.* A186 (1895) 123–64.

(22) FOX, L., *Numerical Solution of Ordinary and Partial Differential Equations* (Pergamon Press, 1962).

(23) NUSSELT, W., 'The fundamental law of heat transfer' (in German), *Gesundheitsingenieur* **38** (1915) 477–82, 490–7.

(24) NUSSELT, W., 'The surface condensation of steam' (in German), *Z. Ver. dt. Ing.*, **60** (1916) 541–6, 569–75.

(25) *Modern Computing Methods*, National Physical Laboratory, Notes appl. Sci. No. 16 (HMSO, 1961).

(26) HERMANN, R., 'Heat transfer by free convection from a horizontal cylinder in diatomic gases' (in German), *ForschHft Ver. dt. Ing.* No. 379 (1936) 1–24.

(27) KROUJILINE, G., 'Investigation of the thermal boundary layer' and 'The theory of heat transmission by a circular cylinder in a transverse fluid flow' (in French), *Tech. Phys. U.S.S.R.*, **3** (1936) 183–94, 311–20.

(28) REICHARDT, H., 'Heat transfer through turbulent friction layers' (in German), *Z. angew. Math. Mech.*, **20** (1940) 297–328, transl. Tech. Memo. natn. advis. Comm. Aeronaut., Wash., No. 1047.

(29) GEBHART, B., *Heat Transfer* (McGraw-Hill, 1971).

(30) PROBERT, R. P., and SINGHAM, J. R., 'The measurement of gas temperatures in turbine engines', *J. scient. Instrum.*, **23** (1946) 72–7.

(31) MARTINELLI, R. C., 'Heat transfer to molten metals', *Trans. Am. Soc. mech. Engrs*, **69** (1947) 947–59.

(32) SMITH, A. G., 'Heat flow in the gas turbine', *Proc. Instn mech. Engrs*, **159** (1948) 245–54.

(33) KAYS, W. M., and LONDON, A. L., *Compact Heat Exchangers* (McGraw-Hill, 1964).

(34) MAYHEW, Y. R., GRIFFITHS, D. J., and PHILLIPS, J. W., 'Effect of vapour drag on laminar film condensation on a vertical surface', *Thermodynamics and Fluid Mechanics Convention, Liverpool April 1966, Proc. Instn mech. Engrs*, **180** Pt. 3J (1965–6) 280–7, 342–3.

(35) *Proceedings of the General Discussion on Heat Transfer, London September 1951, Instn mech. Engrs.*

(36) SEBAN, R. A., and SHIMAZAKI, T. T., 'Heat transfer to a fluid flowing turbulently in a smooth pipe with walls at constant temperature', *Trans. Am. Soc. mech. Engrs*, **73** (1951) 803–9.

(37) HAMILTON, D. C., and MORGAN, W. R., 'Radiant-interchange configuration factors', *Tech. Notes natn. advis. Comm. Aeronaut., Wash.*, No. 2836 (1952).

(38) *Liquid Metals Handbook*, The Atomic Energy Commission and The Bureau of Ships, Dept. of the Navy, 2nd and 3rd ed. (U.S. Government Printing Office, 1952 and 1955).

(39) COHEN, H., and BAYLEY, F. J., 'Heat transfer problems of liquid-cooled gas-turbine blades', *Proc. Instn mech. Engrs*, **169** (1955) 1063–74.

(40) KARPLUS, W. J., *Analogue Simulation* (McGraw-Hill, 1958).
(41) HEISLER, M. P., 'Temperature charts for induction and constant-temperature heating', *Trans. Am. Soc. mech. Engrs*, **69** (1947) 227–36.
(42) SCHNEIDER, P. J., *Temperature Response Charts* (Wiley, 1963).

Appendix D Nomenclature

The notation used in this book follows the general principles laid down in the *British Standard* 1991. All letters denoting mathematical functions and operations (sin, log, d, etc.) and all abbreviations of units (kJ, m, kg, etc.) are printed in roman type, and they follow the *British Standard* closely. Symbols representing physical quantities (pressure p, volume V, etc.) are printed in *italic* type, but in a few instances it was necessary to depart from the recommended letters to avoid confusion. The most frequently used symbols representing physical quantities are listed below, arranged separately for the four parts of the book.

Part I

A	area; non-flow availability function
B	steady-flow availability function
C	velocity
c_p	specific heat at constant pressure
c_v	specific heat at constant volume
E, e	internal energy, specific internal energy (in general)
f	specific Helmholtz function
f	some function
G	Gibbs function
g	specific Gibbs function; gravitational acceleration
H, h	enthalpy, specific enthalpy
$\mathcal{J}$	Joule's equivalent
l	length
m	mass; rate of mass flow
n	polytropic index
p	absolute pressure
Q	heat; rate of heat transfer
R	gas constant
S, s	entropy, specific entropy
T	absolute thermodynamic temperature
U, u	internal energy, specific internal energy
V, v	volume, specific volume
W	work; rate of work transfer
z	height above datum

Greek Symbols

β	coefficient of cubical expansion
η	efficiency

650

θ	temperature on any arbitrary scale
κ	compressibility
μ	Joule-Thomson coefficient
ϕ	some function

Part II

A	area
C	velocity; molar heat
c	specific heat
ΔG	Gibbs function of reaction
g	gravitational acceleration
H, h	enthalpy, specific enthalpy
ΔH	enthalpy of reaction
K	equilibrium constant
l	length
M	molar mass
m	mass; rate of mass flow
n	number of mols; polytropic index
p, p_m	absolute pressure, mean effective pressure
Q	heat; rate of heat transfer
R, R_0	specific gas constant, universal gas constant
r_c	cut-off ratio
r_p	pressure ratio
r_v	compression ratio
r_w	work ratio
S, s	entropy, specific entropy
T	absolute thermodynamic temperature (Kelvin)
t	thermodynamic temperature (Celsius)
U, u	internal energy, specific internal energy
ΔU	internal energy of reaction
V, v	volume, specific volume
v_0	molar volume
W	work; rate of work transfer
x	dryness fraction; mol fraction
Z	compressibility factor

Greek Symbols

γ	ratio of specific heats c_p/c_v
η	efficiency
θ	temperature on any arbitrary scale
ϕ	relative humidity; some function
ω	specific humidity (moisture content)

Suffixes

c	a property in the critical state
f	a property of the saturated liquid; a property of formation
g	a property of the saturated vapour

fg	a change of phase at constant pressure
i	a constituent in a mixture
P	the products of a chemical reaction
R	the reactants of a chemical reaction; a property in the reduced state

Part III

A	area
a	area; velocity of sound
C	absolute velocity
c	specific heat
d	diameter
F	force
g	gravitational acceleration
h	specific enthalpy
ΔH	enthalpy of reaction
I	specific impulse
K	equilibrium constant
k	indicator spring constant
k_c	nozzle velocity coefficient
k_d	coefficient of discharge
L	piston stroke
l	length
M	molar mass
m	rate of mass flow; mass
N	machine cycles per unit time
n	polytropic index
P	potential energy
p, p_m	absolute pressure, mean effective pressure
Q	rate of heat transfer; heat
R	gas constant; force
R_0	universal gas constant
$\mathscr{R}$	reheat factor
r	radius
r_p	pressure ratio
r_v	compression ratio
s	specific entropy
T	absolute thermodynamic temperature
t	time
U	blade velocity
u	specific internal energy
V	volume; velocity relative to blade or duct
v	specific volume
W	rate of work transfer; work
x	dryness fraction
x, y	coordinates

Greek Symbols

α	angle of absolute velocity
β	angle of relative velocity
γ	ratio of specific heats c_p/c_v
η	efficiency
Λ	degree of reaction
τ	shear stress
ϕ	angle
ω	angular velocity

Suffixes

a	an axial component
c	a mass in a clearance volume; a critical value
f	a mass flowing through a machine
t	a property at a throat
w	a whirl component

Part IV

A	area
c_p	specific heat at constant pressure
d	diameter
e	base of natural logarithms
F	geometric or configuration factor
f	friction factor ($= 2\tau/\rho U^2$)
f	some function
g	gravitational acceleration
h	heat transfer or film coefficient; specific enthalpy
h_{fg}	latent heat
i	intensity of radiation
k	thermal conductivity
l	characteristic linear dimension; length
m	rate of mass flow; mass
n	coordinate normal to a surface element
p	time
Q	rate of heat transfer; heat
q	rate of heat transfer per unit area
q_v	rate of internal heat generation per unit volume
r	radius
T	absolute thermodynamic temperature
t	temperature
U	velocity in x-direction; overall heat transfer coefficient
V	velocity in y-direction
x, y, z	coordinates

Dimensionless Groups

Bi	Biot number ($= hl/k$)
Fo	Fourier number ($= \alpha\tau/l^2$)
Gr	Grashof number ($= \beta g \rho^2 l^3 \theta / \mu^2$)
Nu	Nusselt number ($= hl/k$)
Pe	Peclet number ($= Re \times Pr = \rho U c_p l / k$)
Pr	Prandtl number ($= c_p \mu / k$)
Ra	Rayleigh number ($= Gr \times Pr = \beta g \rho^2 l^3 \theta c_p / \mu k$)
Re	Reynolds number ($= \rho U l / \mu$)
St	Stanton number ($= Nu/(Re \times Pr) = h/\rho U c_p$)

Greek Symbols

α	thermal diffusivity ($= k/\rho c_p$); absorptivity
β	coefficient of cubical expansion
δ	thickness of boundary layer
ε	eddy diffusivity; emissivity
θ	temperature difference (relative to wall)
λ	wavelength
μ	dynamic viscosity
ν	kinematic viscosity ($= \mu/\rho$)
ρ	density; reflectivity
σ	Stefan-Boltzmann constant; surface tension
τ	time; shear stress; transmissivity
ϕ	angle; function of
ψ	the ratio δ_t/δ
ω	solid angle

Suffixes

b	the edge of a sublayer; the bulk of a fluid; black-body radiation
c	convection
d	diameter
f	film
m	a mean value
p	at time p
r	the axis of a tube; radiation
s	the free stream
t	the temperature boundary layer
w	the wall
λ	the monochromatic value at a wavelength λ
ϕ	radiation at an angle ϕ

Index

Index